ESSENTIALS OF MEDICINAL CHEMISTRY

ANDREJUS KOROLKOVAS

Associate Professor of Pharmaceutical Chemistry
Faculty of Pharmaceutical Sciences
University of São Paulo

JOSEPH H. BURCKHALTER

Professor of Medicinal Chemistry
College of Pharmacy
The University of Michigan

A WILEY-INTERSCIENCE PUBLICATION

JOHN WILEY & SONS, New York · London · Sydney · Toronto

Library of Congress Cataloging in Publication Data:

Korolkovas, Andrejus, 1923-
  Essentials of medicinal chemistry.

  "A Wiley-Interscience publication."
  Includes bibliographical references and index
  1. Chemistry, Medical and pharmaceutical.
  I. Burckhalter, Joseph Harold, 1912-   joint author.
  II. Title. [DNLM: 1. Chemistry, Pharmaceutical.
  QV744 K84e]

RE403.K67   615′.1   75-37801
ISBN 0-471-12325-0

PREFACE

The nature of a textbook should be determined by the demands of a curriculum which, in turn, should fill the needs of the student. During the early part of the century, subject matter which now is under the heading of medicinal chemistry was included in a course in materia medica to fulfill the needs of the practicing pharmacist who prepared many of the medicaments in his shop. Consideration of the pharmacist as a chemist in control of his domain—the repository of drugs—was reflected in the nature of textbooks emerging in the 1940's and bearing the words *pharmaceutical* and *medicinal chemistry* in their titles. These texts, employing a chemical classification of drugs, were essentially classical textbooks in organic chemistry with inclusion of brief descriptions of pharmacological actions and clinical uses. However, in view of the preempting of certain functions of the practicing pharmacist by the pharmaceutical houses which provided virtually all drugs in final dosage form, these texts were slow to admit changing needs and only gradually and partially did they adopt the pharmacological classification of drugs. The late Professor F. F. Blicke of the University of Michigan recognized as early as 1926 the need for a change when he organized his lectures to pharmacy students using a pharmacological classification. The trend has continued until today. In some institutions, courses in medicinal chemistry have been combined with pharmacology with less and less chemistry remaining in the courses.

With the usurping of a principal function of the pharmacist by the manufacturers of drugs, the modern pharmacist with his excellent educational curriculum needed greater professional fulfillment. While for years students were told that they should become consultants to physicians, they were reluctant except in isolated instances to assume that role. Now courses in clinical pharmacy constitute a large part of the pharmaceutical curriculum.

New curricula in general are directing students toward the patient and away from medicaments as chemicals. Similarly, new curricula of the physician involving abbreviated courses in basic medical science are removing the physician still further away from drugs as chemicals. If the trend continues away from emphasis on drugs as chemicals and toward emphasis on the patient, who will fill the void as chemical and pharmacological authority? Thus, despite the trend away from chemistry because of a new concern for the patient, we believe that the pharmacist should not relinquish his historical role as custodian and master of the repository of chemicals.

The aim of this text is to conduct the student of pharmacy from basic chemistry over a bridge of medicinal chemistry to pharmacology. Medicinal chemistry, together with basic chemistry and biological science, is essential to an understanding of bioavailability—the requirement that a drug reach the active site in adequate concentration to be effective. Pharmaceutical subjects such as biopharmaceutics and pharmacokinetics are covered in other texts. Succinctly stated facts and theories are presented in this text in a form which hopefully will be palatable and yet provide information useful to the student in his future mission of helpfulness to the physician and patient, whether his location will be in the community, hospital, or laboratory.

The absence of preparative chemical methods and historical sketches constitutes concessions to limitations of space. The average contemporary student of pharmacy will not regret these omissions. To the student who wishes to know more of chemical methods relating to drugs, *Merck Index* is recommended as an example of an excellent source. Detailed historical and other aspects of the medicinal chemistry of drugs may be found under References in this text.

Monographs of drugs listed in The United States Pharmacopeia XIX (1975) and The National Formulary XIV (1975), as well as the First Supplements to these compendia, were taken into consideration in the preparation of this text.

We trust that this book also will prove to be helpful to chemists and biologists who are interested in facts and theories relating to drugs which are in actual use.

Appreciation is expressed to Dr. Ahmed El Masry for constructive review of the manuscript.

Mrs. Laura E. Bakken is thanked for assistance in preparation of the manuscript.

ANDREJUS KOROLKOVAS
JOSEPH H. BURCKHALTER

São Paulo, Brazil
Ann Arbor, Michigan
May 1976

CONTENTS

PART 1

INTRODUCTION

CHAPTER 1

BASIC CONSIDERATIONS

I. MEDICINAL CHEMISTRY

A. Definition

Medicinal chemistry has been aptly defined by Dr. Glenn Ullyot as a field which applies the principles of chemistry and biology to the creation of knowledge leading to the introduction of new therapeutic agents. Thus, the medicinal chemist must not only be a competent organic chemist but he must have a basic background in the biological sciences, particularly biochemistry and pharmacology.

The relationship of medicinal chemistry to other disciplines is indicated by the following diagram:

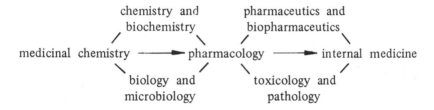

B. Historical Evolution

The beginning of effective therapy by chemicals is lost in antiquity, since it preceded recorded medical history. Early successes in the quest for chemicals effective against disease were predominantly found among antiinfective diseases rather than against those usually accompanying the aging process, such as hypertension or cancer. Also, because chemical processes were largely undeveloped, nature initially provided the more sophisticated and promising agents. Emperor Sheng Nung in 3000 B.C. listed the crude drug ch'ang shan for malaria. Pliny in A.D. 50 introduced aspidium for helminthiasis.

The last several centuries are replete with ironies and paradoxes. Galen is revered today, but his worthless procedures of using small quantities of natural

3

products to treat diseases, introduced during the second century, prevailed for about 1500 years. On the other hand, Theophrastus Paracelsus, originally named Bombastus von Hohenheim, was ironically frustrated by unsuccessful attempts during his lifetime to introduce the useful laudanum (morphine tincture) for relief of pain and tartar emetic, still a useful antimonial, for the dread schistosomiasis. Paracelsus was the first physician or scientist to state the paradox that medicines can be both helpful and harmful. It is ironic that the most visible tribute to Paracelsus in Salzburg, Austria, where he died in 1493, is a painting of him below the apartment where he lived and above the Paracelsus Droggerie, a shop which dispenses only nonprescription drugs.

Another medical irony is that Lord Lister, responsible for the historic use of phenol in aseptic surgery, is remembered chiefly through a proprietary product which bears his name but which has questionable efficacy.

During the seventeenth century, ipecacuanha for amebiasis and cinchona for malaria were introduced. Later simple synthetics, such as ether, chloral, and acetanilid, became available. Just before the turn of the century, Emil Fischer's lock-and-key theory of drug action and, after just the turn of the century, Ehrlich's arsphenamine introduced the rational approach to the search for therapeutic agents. The rational approach has already provided a relatively large number of valuable drugs, as seen in Chapters 2 and 3. However, methods which were largely empirical produced the sulfonamides by Domagk in 1935 and the antibiotics beginning with Fleming in 1929 and revived by Chain and Florey in 1939.

Recent years have seen the creation of new therapeutic agents by medicinal chemists working usually as part of interdisciplinary teams. All their names and contributions are too numerous to mention here. However, two are selected for special recognition. The late Oliver Kamm, Scientific Director of Parke, Davis and Company, left the University of Illinois, where he had been a young instructor in organic chemistry. While there he wrote the first textbook in qualitative organic analysis and helped initiate the valued series entitled *Organic Syntheses*. Kamm, forced to become a pharmacologist as well as chemist, initiated or encouraged the initiation of a remarkable number of successful medicinal agents.

As an illustration of one from the universities who provided leadership in medicinal chemistry, the late Professor F. F. Blicke, of the University of Michigan, founded the first Ph.D. program in the field in the United States, which emphasized a specific curriculum in synthesis. Seventy-eight students received the Ph.D. under his direct supervision. He became the first person in whose name the Division of Medicinal Chemistry of the American Chemical Society held a symposium.

II. CLASSIFICATION OF DRUGS

Drugs can be classified according to various criteria. Usually they are classified according to (*a*) chemical structure or (*b*) pharmacological action.

The first classification was extensively used some years ago and it is still used by some authors. Following this classification, drugs fall in one or more of these categories: acetals, acids, alcohols, amides, amidines, amines, amino acids, amino alcohols, amino ethers, amino ketones, ammonium compounds, azo compounds, enols, esters, ethers, glycosides, guanidines, halogenated compounds, hydrocarbons, ketones, lactams, lactones, mustards, nitro compounds, nitroso compounds, organominerals, phenols, quinones, semicarbazides, semicarbazones, stilbenes, sulfonamides, sulfones, thiols, thioamides, thioureas, ureas, ureides, urethans.

Presently, the second classification is the one preferred by most medicinal chemists. Following this classification, drugs can be divided into these main groups: (*a*) pharmacodynamic agents, used in noninfectious diseases to correct abnormal functions; (*b*) chemotherapeutic agents, used to cure infectious diseases; (*c*) vitamins; (*d*) hormones; and (*e*) miscellaneous agents. Each group can be subdivided. For instance, among pharmacodynamic agents we have (1) drugs acting on the central nervous system; (2) drugs stimulating or blocking the peripheral nervous system; (3) drugs acting primarily on the cardiovascular and renal systems.

In this textbook, drugs are classified mainly according to their pharmacological action. Each one of these many classes is then subdivided according to the chemical structure of the several drugs which comprise it.

III. NOMENCLATURE OF DRUGS

A. Names of Drugs

Drugs have three or more names. These names are the following: code number or code designation; chemical name; proprietary, trivial, brand or trade name, or trademark; nonproprietary, generic, official, or common name; synonym and other names.

The code number usually is formed with the initials of the laboratory or the chemist or the research team that prepared or tested the drug the first time, followed by a number. It does not identify the chemical nature or structure of the drug. It is discontinued as soon as an adequate name is chosen.

The chemical name should describe unambiguously the chemical structure of a drug. It is given according to rules of nomenclature of chemical compounds.

Since the chemical name is sometimes very elaborate, it is not suitable for routine usage.

The proprietary name is the individual name selected and used by manufacturers. If a drug is manufactured by more than one company, as frequently happens, each firm assigns its own proprietary name. *It should be written with capitalization of the first letter of each word of the name.*

The nonproprietary name refers to the common, established name by which a drug is known as an isolated substance, irrespective of its manufacturer. It should be simple, concise, and meaningful but often it is not. The first letter is not capitalized. It is chosen by official agencies, such as the U.S. Adopted Names (USAN) Council, sponsored by the American Medical Association (AMA), the American Pharmaceutical Association (APhA), the U.S. Pharmacopeial (USP) Convention, and the U.S. Food and Drug Administration (FDA). In the *Journal of American Medical Association,* volume 213, page 608, July 27, 1970, appeared guiding principles for coining U.S. adopted names. On a worldwide scale, however, the World Health Organization is the official agency to select, approve, and disseminate the generic names of drugs.

Synonyms are names given by different manufacturers to the same drug or old nonproprietary names. Some drugs may have several names. For instance, diphenhydramine has 13 different names; chlorpromazine, 16; phenobarbital, 22; pethidine, 26; prednisone, 29; chloramphenicol, 45; sulfanilamide, 60; meprobamate, 64; vitamin B_{12}, 82.

B. Generic and Brand Names Not Always Equivalent

It is usually assumed that a drug bearing a generic name is equivalent to the same drug with a brand name. However, this is not always true. Although chemically equivalent, drugs having the same nonproprietary name but different proprietary names, because they are manufactured by different laboratories, can markedly differ in their pharmacological action. Several factors, such as those pointed out by Sadove, account for this difference: size of crystal or particle, its forms, and isomers; form of the agent—solution vs. salt and type of salt; vehicle (primary and secondary), excipient, or binder; coatings, number, and types; degree of hydration of crystal or addition of dehydrating substances to package, or hydration of diluents, vehicles, and the like; diluent; purity—type and number of impurities; viscosity; pH; sustained-release forms; enteric coating; solubility; vehicle, base, or suspending agents; container—stopper, type of glass, whether or not glass is preheated or impervious; package, dating, type, literature enclosed, dehydration of cotton in package—amount of cotton; quality of active ingredient: relative and absolute; contaminants; allergenic substances (primary and secondary) in product; irritation; melting point; ionization of ingredients; surface tension— surface-active agents; storage factors: time, heat, light, vibration; flavoring and coloring agents; dose or quantity of drug, its distribution, and size of tablet or surface-to-tablet ratio; type and characteristics of gelatin capsules; antioxidant included in preparation; dissolution and disintegration rate; buffer type and amount; air, mold, or bacterial contamination of product; antibacterial preservative; metallic contamination in process of manufacture or in packaging.

The following among certainly many other drugs have been found to differ in pharmacological action when supplied by different manufacturers: chloramphenicol, prednisone, p-aminosalicylic acid, phenylbutazone, acetylsalicylic acid, sulfisoxazole, diphenylhydantoin, meprobamate, secobarbital, digitoxin, penicillin G, dextroamphetamine, chloral hydrate, heparin, dicumarol, diethylstilbestrol, digoxin, oxytetracycline, tetracycline, acetaminophen, ampicillin, chlordiazepoxide, erythromycin, nitrofurantoin, riboflavin, sulfadiazine, warfarin.

IV. METABOLISM OF DRUGS

Drugs and other foreign compounds which enter a living organism are stored in the body or removed from it after a period of time. While inside the body, they may remain intact or undergo chemical transformation, giving compounds (a) less active, or (b) more active, or (c) having similar or different activity. This process of chemical alteration of drugs inside the living organism is called *drug metabolism.*

Drugs such as analogues of biogenic amines, steroids, purines, pyrimidines, and amino acids resemble closely substances normally present in animals, including man. For this reason, they may undergo the same specific interactions with enzymes, carrier proteins, and transport systems as their endogenous counterparts.

Most drugs, however, have no relationship whatsoever with normal body substrates. Hence, their metabolism involves nonspecific enzymes, and their movement across membranes and barriers is carried out by either passive diffusion or nonspecific transport systems.

A. Factors Influencing Metabolism

Dearborn has pointed out several factors which affect the metabolism of drugs:

1. Internal environmental factors: sex, age, weight, nutritional state, activity, body temperature, flora and fauna of intestinal tract, pregnancy, emotional state, other chemicals present, hydration, genetic composition, enzyme activity states.

2. Drug administration factors: route of administration, site of administration, rate of administration, volume administered, vehicle composition,

number of doses — duration of therapy, frequency of dosing, physicochemical state of drug.

3. External environmental factors: temperature, humidity, barometric pressure, ambient atmospheric composition, light, other radiation, sound, season, time of day, presence of other animals, habitat, chemicals, handling.

B. Site of Metabolism

Most drugs are metabolized in the liver through the action of microsomal enzymes. The intestines, brain, kidneys, and lungs are other sites of drug metabolism.

C. Phases of Metabolism

According to Williams, drugs which are foreign to the body are generally transformed to metabolites of ever increasing polarity, until they can be easily excreted by the kidneys. Figure 1.1 represents this process, which is

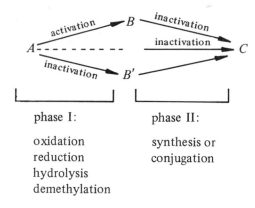

Figure 1.1 Routes of drug metabolism, according to R. T. Williams, *Detoxication Mechanisms,* 2nd ed., Wiley, New York, 1959, p. 735.

usually biphasic. In the first phase, nonpolar drugs are either inactivated, in general, or activated, in some cases, by the introduction of polar groups through oxidation, reduction, and hydrolysis or by the removal of nonpolar (alkyl) groups in order to uncover potential polar groups. In the second phase, the polar compounds are inactivated by a process of synthesis or conjugation, such as methylation, acylation, thiocyanate formation, mercapturic acid formation, glucuronic acid conjugation, amino acid conjugation, and sulfate conjugation. Therefore, the drug administered may be excreted (*a*) unchanged, (*b*) oxidized, reduced, or hydrolyzed, (*c*) conjugated. Some examples of drug metabolism are given in Table 1.1.

Table 1.1 Metabolism of Some Drugs

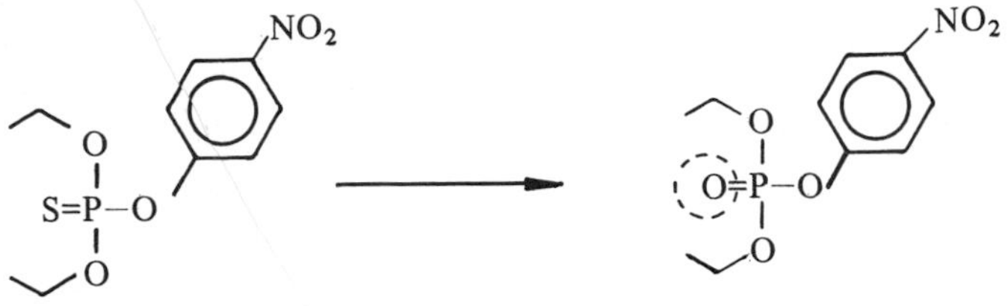

parathion
(inactive)

paraoxon
(active)

chloral hydrate
(active)

trichloroethanol
(more active)

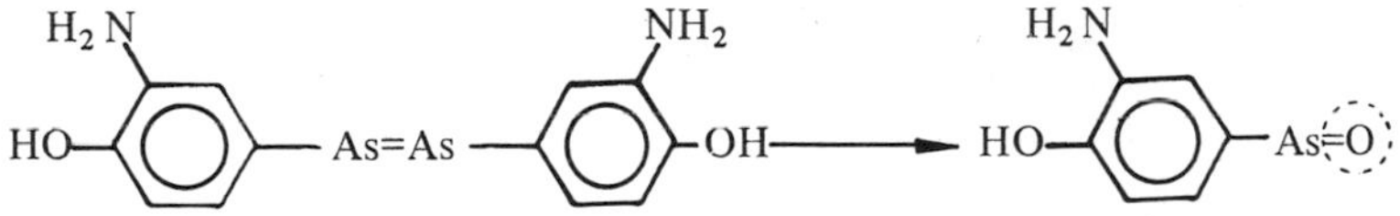

arsphenamine
(active)

oxophenarsine
(more active)

amphetamine
(active)

hydroxyamphetamine
(less active)

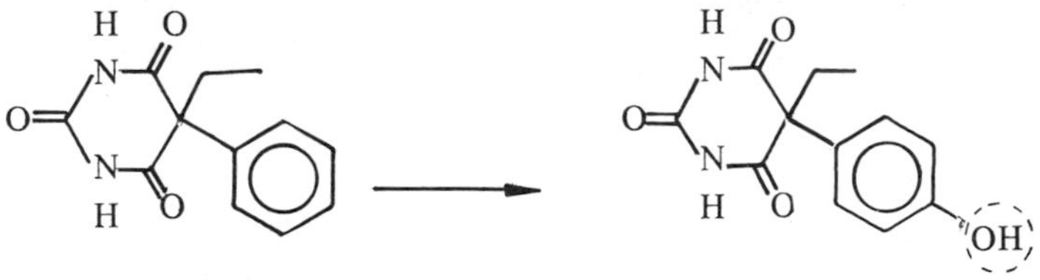

phenobarbital
(active)

p-hydroxyphenobarbital
(inactive)

D. Stimulation or Inhibition of Metabolism

Compounds are known which either stimulate or inhibit drug metabolism. Stimulants shorten the duration of action of a drug by inducing liver microsomal enzymes. More than 200 different compounds have this property; examples are barbiturates, polycyclic hydrocarbons, imipramine, glutethimide, phenylbutazone, nikethamide, glucocorticoids, DDT, chlordane, anabolic steroids, tolbutamide, spironolactone. Inhibitors impair drug metabolism by prolonging the duration of drug action; examples are SKF 525-A (diethylaminoethyl diphenylpropylacetate), Lilly 18947 (2,4-dichloro-6-phenylphenoxyethyl diethylamine), monoamine oxidase inhibitors, thyroxine, alloxan, morphine, DPEA (2,4-dichloro-6-phenylphenoxyethylamine), SKF 26754-A (aminoethyl diphenylpropylacetate), carbon tetrachloride.

REFERENCES

Medicinal Chemistry

L. Mez-Mangold, *A History of Drugs,* Hoffmann-LaRoche, Basle, 1971.

A. Berman, Ed., *Pharmaceutical Historiography,* Proceedings of a Colloquium, American Institute of the History of Pharmacy, Madison, Wis., 1967.

E. Bäumler, *In Search of the Magic Bullet: Great Adventures in Modern Drug Research,* Thames and Hudson, London, 1966.

G. Sonnedecker, revisor, *Kremers and Urdang's History of Pharmacy,* 3rd ed., Lippincott, Philadelphia, 1963.

S. Ross, *Doctor Paracelsus,* Little, Brown, Boston, 1959.

T. S. Work and E. Work, *The Basis of Chemotherapy,* Interscience, New York, 1948.

H. Scott, *A History of Tropical Medicine,* 2nd ed., 2 vols., Williams and Wilkins, Baltimore, 1937-1938.

Classification of Drugs

L. S. Goodman and A. Gilman, Eds., *The Pharmacological Basis of Therapeutics,* 5th ed., Macmillan, New York, 1975.

W. O. Foye, Ed., *Principles of Medicinal Chemistry,* Lea and Febiger, Philadelphia, 1974.

AMA Department of Drugs, *AMA Drug Evaluations,* 2nd ed., Publishing Sciences Group, Acton, Mass., 1973.

D. R. Laurence, *Clinical Pharmacology,* 4th ed., Churchill Livingstone, Edinburgh, 1973.

A. J. Lewis, Ed., *Modern Drug Encyclopedia and Therapeutic Index,* 12th ed., Yorke Medical Group, Dun-Donnelley Publishing Corporation, New York, 1973.

J. R. diPalma, Ed., *Drill's Pharmacology in Medicine,* 4th ed., McGraw-Hill, New York, 1971.

A. Burger, Ed., *Medicinal Chemistry,* 3rd ed., Wiley-Interscience, New York, 1970.

Nomenclature of Drugs

D. J. Chodos and A. R. DiSanto, *Basics of Bioavailability,* Upjohn Company, Kalamazoo, 1973.

E. W. Martin, *Hazards of Medication,* Lippincott, Philadelphia, 1971.

International Nonproprietary Names for Pharmaceutical Substances, Cumulative List No. 3, World Health Organization, Geneva, 1971.

Brands, Generics, Prices and Quality -- The Prescribing Debate After a Decade, Pharmaceutical Manufacturers Association, Washington, 1971.

E. G. Feldmann, *J. Amer. Pharm. Ass.,* NS9, 8 (1969).

M. S. Sadove, *The New Physician,* 15, 257 (1966).

Metabolism of Drugs

J. E. Tomaszewski et al., *Annu. Rep. Med. Chem.,* 9, 290 (1974).

Y. Kobayashi and D. V. Maudsley, *Progr. Med. Chem.,* 9, 133 (1972).

P. Lechat, *Farmaco, Ed. Sci.,* 27, 240 (1972).

B. N. LaDu, H. G. Mandel, and E. L. Way, Eds., *Fundamentals of Drug Metabolism and Drug Disposition,* Williams and Wilkins, Baltimore, Mass., 1971.

E. S. Vesell, Ed., "Drug Metabolism in Man," *Ann. N.Y. Acad. Sci.,* 179, 9-773 (1971).

E. A. Hartshorn, *Handbook of Drug Interactions,* Donald E. Francke, Cincinnati, Ohio, 1970.

J. Hirtz, Ed., *The Fate of Drugs in the Organism,* 2 vols., Masson, Paris, 1970.

D. M. Greenberg, Ed., *Metabolic Pathways,* 3rd ed., Academic, New York, 1968-1970.

A. H. Beckett, *Pure Appl. Chem.,* 19, 231 (1969).

D. Shugar, Ed., *Biochemical Aspects of Antimetabolites and of Drug Hydroxylation,* Academic, New York, 1969.

R. T. Williams, *Pure Appl. Chem.,* 18, 129 (1969).

E. H. Dearborn, *Federation Proc.,* 26, 1075 (1967).

R. T. Williams, *Detoxication Mechanisms,* 2nd ed., Wiley, New York, 1959.

CHAPTER 2

DEVELOPMENT OF DRUGS

I. SOURCE OF DRUGS

Drugs used presently come from the following sources: (1) chemical synthesis, 50%; (2) microorganisms, 12%; (3) minerals, 7%; (4) higher flowering plants, 25%; and (5) animals, 6%. The first three groups, which comprise 69% of drugs, are generally studied under medicinal chemistry. The last two groups, 31% of the total, are generally stressed in pharmacognosy.

II. COST AND PLACE OF DEVELOPMENT OF DRUGS

The introduction of new drugs into therapeutics in the United States is now very expensive. The average cost for one novel drug to emerge on the market is estimated at $25 to $28 million.

There is a strong reason for this high cost. It is more and more difficult to develop new drugs. In 1958, from 14,600 substances synthesized and tested, only 44 found clinical applicability. In 1964, the proportion was even smaller: 150,000 to 17.

During the last 20 years, 90% of the new drugs were developed in industrial firms, 9% in universities and other academic institutions, and 1% in government research laboratories. This is in contrast to earlier decades, when universities contributed about one-half.

The expenditures for research and development of new drugs are steadily growing, going from $12 million in 1940 to $728 million in 1972. Unfortunately, the returns in numbers of new drugs are meager: in 1963, 16; in 1964, 17; in 1965, 23; in 1966, 12; in 1967, 25; in 1968, 11; in 1969, only 9; in 1970, 16; in 1971, 14; in 1972, 14.

Between 1940 and 1966, the number of new chemical compounds introduced as drugs in the United States was 505 (out of a total of 823 in the whole world). The contributions from other countries were as follows: Switzerland, 54; Germany, 39; United Kingdom, 36; France, 22; Denmark, 11; Mexico, 9; Netherlands, 9; Sweden, 8; Belgium, 6; Japan, 6; Austria, 3; Canada, 3; Hungary, 2; Argentina, Australia, Czechoslovakia, India, and Italy, 1 each.

During the period 1958-1970, however, 437 new drug substances were introduced on the world market, with the following percentages according to nations: U. S., 47%; Switzerland, 12%; Great Britain, 12%; West Germany, 8%; France, 5%; Italy, 4%; Sweden, 2%; Belgium, 2%; Holland, 2%; Denmark, 2%; Mexico, 2%; Austria, 1%; Japan, 1%.

III. SEARCH FOR NEW DRUGS

With the aim of discovering new useful therapeutic agents, many substances are being continually synthesized and tested. It is estimated that more than 15,000 sulfonamides, 40,000 potential tuberculostatics, 50,000 potential antimalarials, 50,000 organophosphorous compounds as potential insecticides, 180,000 potential schistosomicides, and 250,000 potential antineoplastics have been screened.

The therapeutic armamentarium is now relatively well supplied with several types of drugs, such as antihistaminics, antispasmodics, ganglionic blocking agents, muscle relaxants, and barbiturates. For this reason, new drugs of these types now attract little interest.

On the other hand, owing to the present status of health of the American people, great effort is being made to introduce new antiinfectious agents, antineoplastic agents, cardiovascular agents, drugs for endocrine systems, and drugs acting on the central nervous system.

IV. GENESIS OF DRUGS

Drugs find their way into therapeutics mainly by one of the following routes: serendipity, random screening, extraction of active principles from natural sources, molecular modification of known drugs, planned synthesis of new chemical compounds on a rational basis.

A. Serendipity

Some drugs or novel uses of known drugs were discovered by accident in the laboratory or clinic by pharmacists, chemists, physicians, and other investigators. It was alert observations that led, for instance, to the introduction in medical practice of acetanilid and phenylbutazone as antipyretics, penicillin as an antibacterial, disulfiram for the treatment of chronic alcoholism, piperazine as an anthelmintic, imipramine and monoamine oxidase inhibitors such as iproniazid as antidepressants, chlorothiazide as a diuretic, mecamylamine as the first hypotensive drug of a novel group, sulfonylureas as oral hypoglycemics, benzodiazepines such as chlordiazepoxide as antianxiety agents.

B. Random Screening

In this approach to discover new drugs, all available chemical substances are submitted to a variety of biological tests in the hope that some may show useful activity. As an essentially empirical method, this way of finding new drugs is not very rewarding. It is estimated that for a new anticonvulsant to emerge through this process, it might be necessary to screen 500,000 chemical compounds.

A variation of this approach is *rationally directed* random screening. It was used during the Second World War to discover new antimalarials: from over 14,000 compounds and chemical products prepared and tested by several institutions in 11 countries, just a few were selected for clinical trials. Another example is the search for new antibiotics from microorganisms. Since 1940, when the scientific community learned of the antibacterial action of penicillin, large-scale systematic random screening resulted in the discovery of more than 3000 antibiotics, six or seven dozen of which are used in medicine and agriculture.

A third example of rationally directed random screening is the isolation and identification of products of drug metabolism. It is seen that several drugs are themselves inactive, but they owe their action to their metabolites. Such is the case, for instance, with acetanilid and acetophenetidin: these two long-known drugs are metabolized to acetaminophen, which exerts the main analgetic action. For this reason, acetaminophen itself was introduced into therapeutics:

acetanilid   acetaminophen   acetophenetidin

Several other drugs are products of drug metabolism, such as oxophenarsine from arsphenamine, oxyphenbutazone from phenylbutazone, desipramine from imipramine, hycanthone from lucanthone, and cycloguanil from chlorguanide.

C. Extraction from Natural Sources

For centuries, mankind has used extracts from plant sources or animal organs for the treatment of various diseases. Owing to the good effects produced by these and similar remedies, folk medicine around the world has been extensively explored. Several drugs used today, especially antibiotics, vitamins, and hormones, resulted from purification of such extracts and isolation and identification of their active principles. One hundred and seventy drugs from plants which are, or once were, official in USP or NF were used by North American Indians. In 1960, 47% of drugs prescribed by physicians in the United States were from natural sources, mostly antibiotics. In 1967, 25%, or 824 out of 3,354, of the trade-name or generic-name products which appeared in the 1.05 billion prescriptions filled in the United States contained one or more ingredients derived from higher plants. In general, folk medicine is not a reliable guide in the search for new drugs. For example, natives of certain Pacific islands claim that about 200 different local plants have contraceptive properties. Scientifically conducted research on extracts from 80 of these plants, however, yielded completely negative results.

D. Molecular Modification

This method of obtaining new drugs, also called molecular manipulation, method of variation, mechanistic method, selective process or approach, is the most used and, as of the present, it is the most productive of new drugs. Molecular modification is a natural development of organic chemistry. Basically, it consists of taking a well-established chemical substance of known biological activity as a lead or prototype and then synthesizing and testing its structural congeners, homologues, or analogues. Examples of results from application of this method are seen in Figures 2.1 to 2.5.

Schueler pointed out several advantages of this method:

1. Greater probability of congeners, homologues, and analogues having pharmacological properties similar to those of the prototype than the compounds selected or synthesized at random.
2. Possibility of obtaining pharmacologically superior products.
3. Likelihood of the production of the new drugs being more economical.
4. Synthesis similar to that of the prototype, with saving of time and money.
5. Data gathered may help to elucidate structure-activity relationship.
6. Use of the same methods of biological assay used for the prototype.

The objectives of this method are twofold: (*a*) to discover the essential *pharmacophoric* moiety, that is, that feature of the molecule which imparts pharmacological action to the drug; (*b*) to obtain drugs having more desirable properties than the prototype in potency, toxicity, specificity, duration of action, ease of application or administration or handling, stability, cost of production.

1. General Processes

Two *general* processes may be used in the method of variation: (*a*) disjunction,

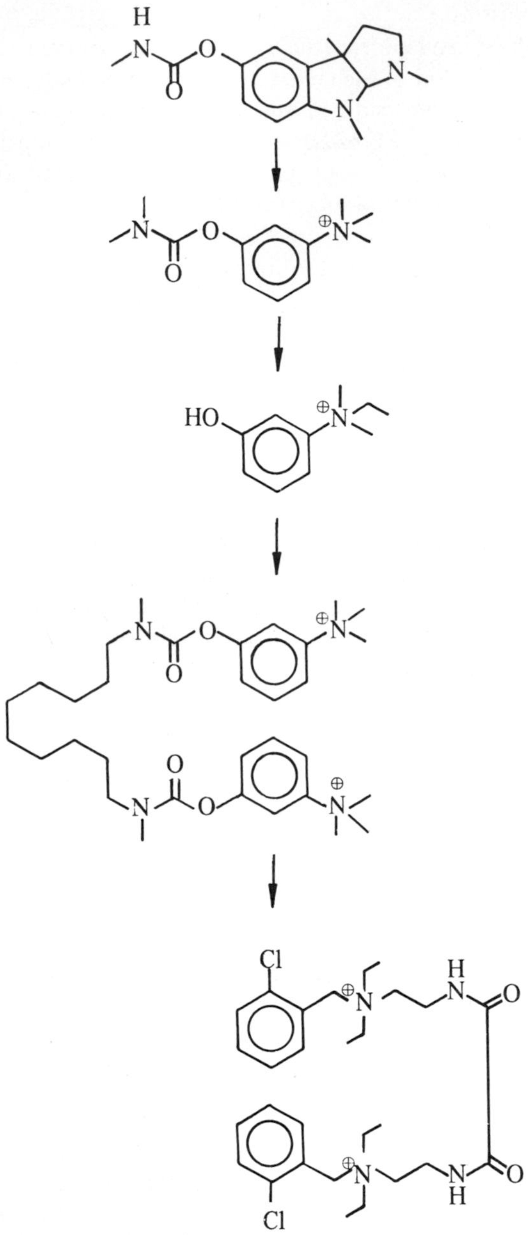

physostigmine (1925)

neostigmine (1931)

edrophonium (1952)

demecarium (1956)

ambenonium (1956)

Figure 2.1 Structural variation performed on the molecule of physostigmine. It was isolated from a plant as a pure alkaloid in 1864 by Jobst and Hesse. Its chemical structure was elucidated in 1925 by Stedman and Barger. Dissociation and association of its pharmacophoric groups resulted in several new and better anticholinesterase agents, some of which are shown above.

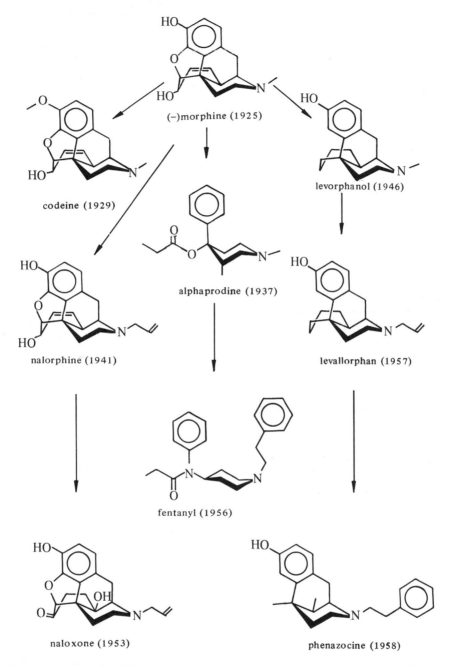

Figure 2.2 Genesis of new hypnoanalgetics and antagonists through molecular modification of (–)-morphine.

dissection, or molecular simplification or dissociation; (b) conjunction or molecular association.

The process of molecular *dissociation* or *disjunction* is the systematic synthesis and evaluation of simpler and simpler analogues of the lead compound. These analogues are partial or virtual replicas of the prototype drug, which is usually a natural product of very intricate chemical structure. Examples of the process of disjunction are the structural variations shown in Figures 2.1 and 2.2.

The process of molecular *association* is the synthesis and evaluation of more and more complex analogues of the prototype. These analogues incorporate certain or all features of the lead compound. Three main types of association can be distinguished:

1. Molecular addition — association of different moieties through weak forces, such as electrostatic attraction and hydrogen bonding
2. Molecular replication — association of identical moieties through covalent bond formation

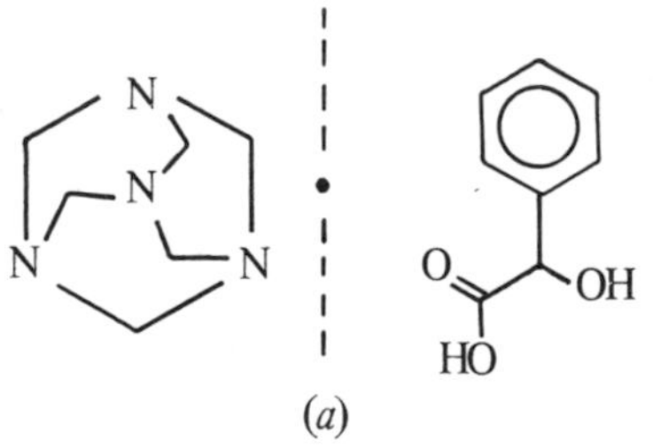

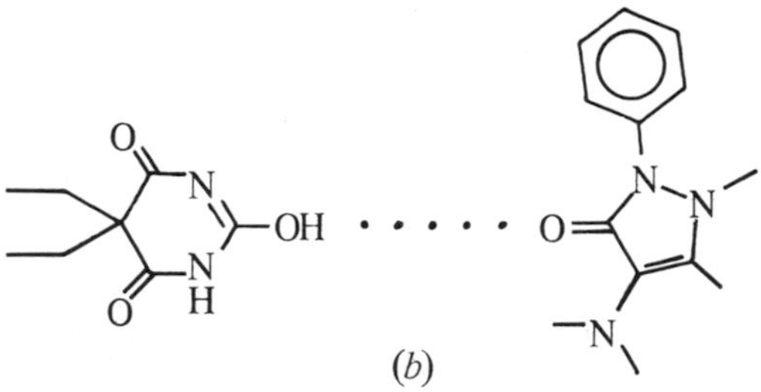

(a)

(b)

Figure 2.3 Examples of molecular addition: (a) methenamine mandelate (methenamine + mandelic acid); (b) Veramon (barbital + aminophenazone). In (a) the salt is formed through electrostatic attraction; in (b) complexation involves hydrogen bonding.

3. Molecular hybridization — association of different or mixed moieties through covalent bond formation

Some of the many compounds of molecular addition used in therapeutics are shown in Figure 2.3.

Examples of molecular replication are shown in Figure 2.4. Another example

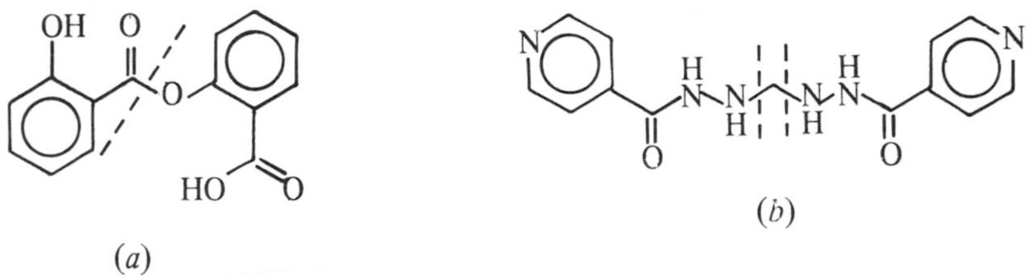

(a)   (b)

Figure 2.4 Examples of molecular replication. Drugs designed by association of identical moities: (a) salicylsalicyclic acid (two molecules of salicyclic acid forming an ester by application of the salol principle); (b) metazide (two molecules of isoniazid joined by a methylene bridge).

is demecarium, product of association of two molecules of neostigmine through a decamethylene bridge (Figure 2.1). Dapsone is also an example of this process, because it can be considered a product of condensation of two molecules of sulfanilamide with loss of sulfamide:

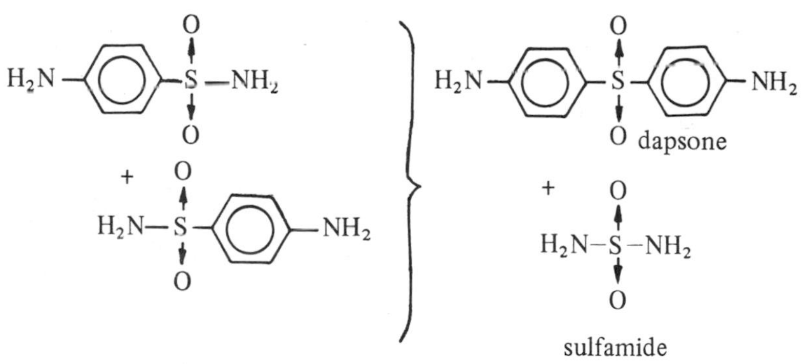

dapsone

sulfamide

There are also several examples of molecular hybridization, some of which are shown in Figure 2.5.

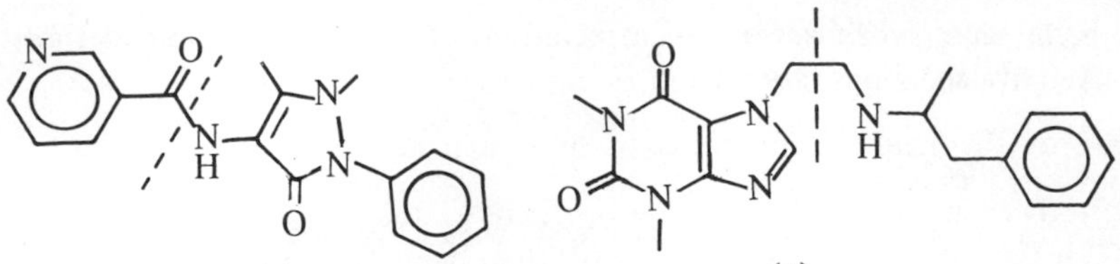

Figure 2.5 Examples of molecular hybridization. Drugs designed by association of mixed moieties: (a) nifenazone (nicotinic acid + aminophenazone); (b) fenetylline (caffeine + methamphetamine).

2. Special Processes

Besides the two general processes, the method of molecular modification uses several special processes, which Schueler grouped in two classes: (a) alterations which increase or decrease the dimensions and flexibility of a molecule; (b) alterations of physical and chemical properties through the introduction of new groups or the replacement of certain moieties by different ones.

The first class comprises processes such as ring closure or opening; formation of lower or higher homologues; introduction of double bonds; introduction of optically active centers; and introduction, removal, or replacement of bulky groups.

The second class includes isosteric substitution; change of position or orientation of certain groups; introduction of alkylating moieties; and modifications toward inhibition or promotion of various electronic states.

Ring Closure or Opening

Several examples are found of new drugs designed by either closure of a chain or opening of a ring. Some are shown in Figures 2.1 (physostigmine and neostigmine).

Formation of Lower or Higher Homologues

Alkane, polymethylene, cyclopolymethylene, and hybrid series of homologues can be easily formed. Unfortunately, it is not possible to establish rigid rules for the pharmacological properties of homologous compounds. However, in the alkane and polymethylene series Ing found the following general types of change:

1. Activity increases regularly until a maximum is reached, higher members being almost or entirely inactive. This can be observed particularly in structurally nonspecific drugs, such as hypnotics, general anesthetics, volatile insecticides, and disinfectants. The same phenomenon also occurs, although

seldom, in structurally specific compounds, as in the case of local anesthetics.

2. Activity increases irregularly, reaches a maximal value, and then decreases, again irregularly. A typical example is found among benzylic esters with atropinic properties.

3. Activity increases (or decreases), reaches a relatively high (or low) value, and then remains more or less constant for a few or many higher members. One example is seen in homologous muscarinic agents of the general formula $R\overset{+}{N}Me_3$, the pharmacological properties of which are greatest when the number of atoms in the chain is five. In ganglionic blocking agents of the general formula $R_3\overset{+}{N}(CH_2)_n\overset{+}{N}R_3$, activity is greatest in those in which n is 4, 5, or 6. In the polymethylene series, corresponding to the general formula $R_3\overset{+}{N}C_6H_4-(CH_2)_n\overset{+}{N}R_3$, activity is maximal in compounds in which n is 2.

4. Activity alternates, members with an odd number of carbon atoms being consistently more active than neighboring members with an even number of carbon atoms, or vice versa. This is observed with antimalarials derived from 6-methoxy-8-aminoquinoline: activity is greater in compounds in which n is an even number (range $n = 2$ to 7).

5. Activity changes, lower members having one type, and higher members a different type, of predominant action. Often higher members are antagonists of the pharmacological effect of lower members, and vice versa. Such is the case in the series of N-alkyl derivatives of norepinephrine, in which alkylation reduces the hypertensive activity of the molecule in the sequence $-NH_2$, $-NHMe$, $-NHEt$, $-NHPr-n$, hypotensive effects occurring when the terminal group is $-NHPr-i$ or $-NHBu$. This anomaly is considered to result from complexation of these compounds with different receptors: in the first case with α receptors as well as with β; in the second mainly with β.

Introduction of Double Bonds

It may cause two main effects:

1. By changing the stereochemistry of the drug, it can give rise to a compound whose activity is different from that of the saturated compound. This happens in cis-cinnamic acid: unlike its dihydro derivative (β-phenyl-propionic acid), it exerts a regulatory activity on plant growth.

2. By altering physicochemical properties, it can modify biological activity. This is observed in hypnotics: ethylenic hydrocarbons are slightly more active than saturated ones.

As to vinyl group and polyvinyl systems, they are present in several drugs which were often introduced in order to get new drugs by application of the principle of vinylogy. For instance, some derivatives of procaine with similar local anesthetic activity were prepared by application of this principle:

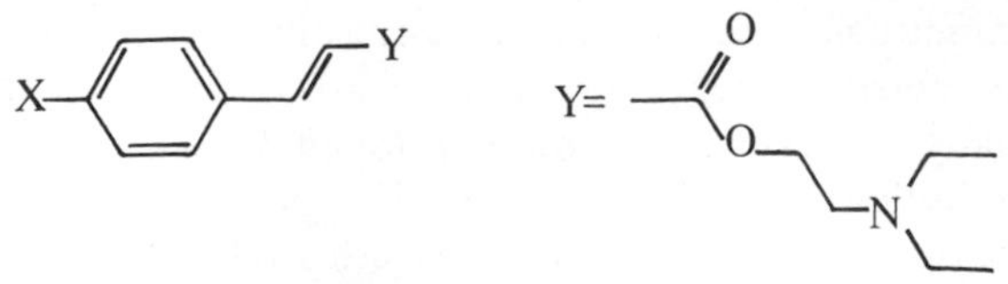

Vinylogy has been applied also in the inverse direction. Thus, removal of the benzene ring from dulcine, which is an edulcorant, resulted in ethoxyurea, a substance likewise sweet:

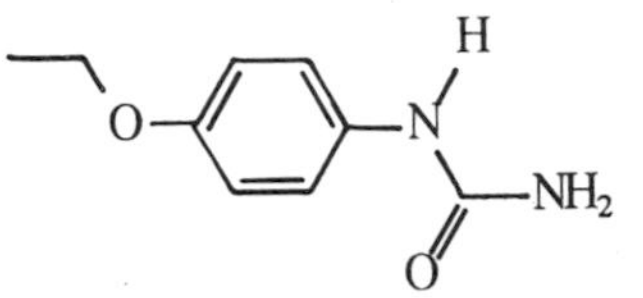

dulcine 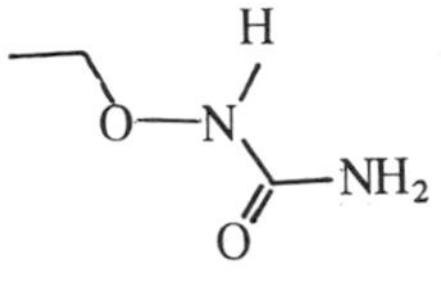   ethoxyurea

Introduction of Optically Active Centers

By changing the stereochemistry of the drug molecule, this modification can alter, sometimes drastically, its pharmacological activity. For instance, of the four isomers of chloramphenicol, only the D-(-)-threo form is active; D-(-)-isoprenaline is 50 to 800 times more active as a bronchodilator than L-(+)-isoprenaline; (-)-norepinephrine is 70 times more active as a bronchodilator than (+)-norepinephrine; L-(+)-acetyl-β-methylcholine is about 200 times more active in gut than D-(-)-acetyl-β-methylcholine; (-)-hyoscyamine is 15 to 20 times more active as a mydriatic than (+)-hyoscyamine; (+)-muscarine has 700 times the muscarinic activity of (-)-muscarine; L-(-)-ascorbic acid has antiscorbutic properties, whereas (+)-ascorbic acid does not; (-)-amino acids are either tasteless or bitter, but (+)-amino acids are sweet; (+)-cortisone is active, but (-)-cortisone is inactive.

Introduction, Removal, or Replacement of Bulky Groups

This special process is used mainly to convert agonists into antagonists, and vice versa. By inspection of Figure 2.6 it can be seen that the difference between agonists and antagonists is the presence of nonpolar bulky groups in antagonists.

Another interesting example is found in β-lactamase-resistant penicillins. It is known that penicillins lose activity when the β-lactam ring is broken. This ring rupture can occur by the catalytic action of β-lactamase, formerly called penicillinase. Bulky groups introduced near this ring prevent by steric hindrance the approach of the enzyme, making the penicillins thus formed resistant to it (see Chapter 33).

Isosteric Substitution

Isosteric and bioisosteric groups are extensively applied in drug design, not only in molecular modification of known drugs but also in rational design of antimetabolites.

In 1919 Langmuir defined as *isosteres* those compounds or groups of atoms that have the same number and arrangement of electrons; for example, N_2 and CO, N_2O and CO_2, N_3^- and NCO^-. Isosteres are characterized by similar physical properties.

Later on Erlenmeyer redefined isosteres as "atoms, ions or molecules in which the peripheral layers of electrons can be considered to be identical."

Considering the vast application of the concept of isosterism in molecular pharmacology, Friedman introduced the term *bioisosteres* as meaning "compounds which fit the broadest definition for isosteres, and have the same type of biological activity," even antagonistic.

In its broadest sense, therefore, the term *isosteres* may be applied to groups that merely bear resemblance in their external electronic shells or more restrictedly to groups with similar localization of regions of high or low electron density in molecules of similar size and shape. According to this criterion there are at least two types of isosteres:

1. *Classical isosteres:* those encompassed by Erlenmeyer's definition; that is, those represented in the hydride-displacement law, the elements of every group of the periodic table, and annular equivalents such as $-CH=CH-$ and $-S-$ (Table 2.1).

Table 2.1 Classical Isosteres

Monovalents	Bivalents	Trivalents	Tetra-Substituted Atoms	Annular Equivalents
F, OH, NH_2, CH_3	$-O-$	$-N=$	$=C=$ $=Si=$	$-CH=CH-$
Cl, SH, PH_2	$-S-$	$-P=$	$=N^+=$	$-S-$
Br	$-Se-$	$-As=$	$=P^+=$	$-O-$
I	$-Te-$	$-Sb=$	$=As^+=$	$-NH-$
		$-CH=$	$=Sb^+=$	

2. *Nonclassical isosteres:* those that, substituted in a certain molecule, give origin to a compound whose steric arrangement and electronic configuration are similar to those of the parent compound; examples of pairs of these isosteres are H and F, $-CO-$ and $-SO_2-$, $-SO_2NH_2$ and $-PO(OH)NH_2$ (Table 2.2).

AGONISTS | ANTAGONISTS

Cholinergics

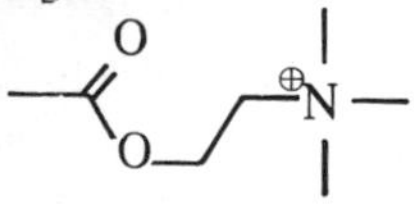

acetylcholine

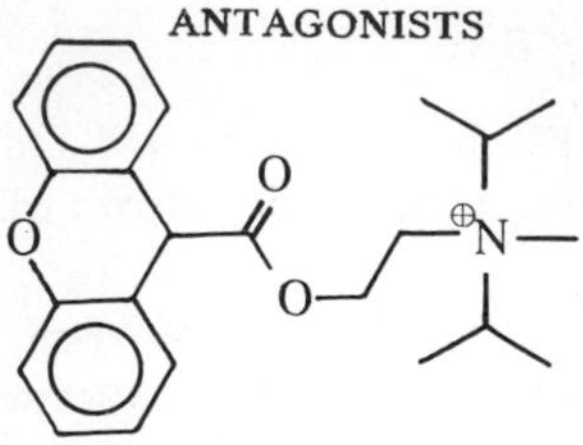

propantheline

α-Adrenergics

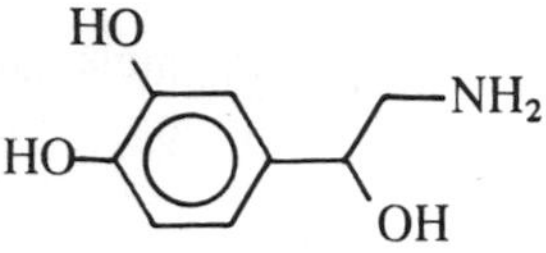

norepinephrine

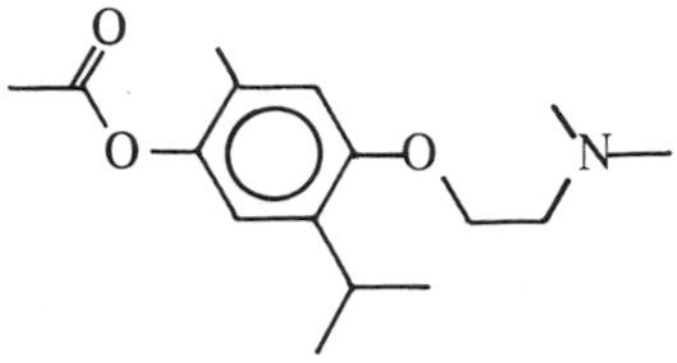

moxisylyte

β-Adrenergics

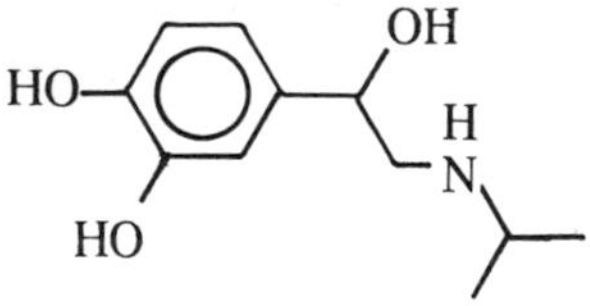

isoprenaline

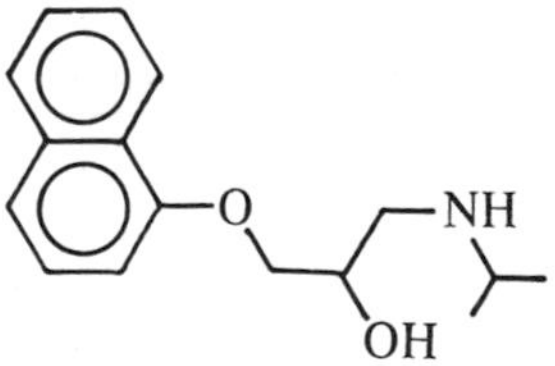

propranolol

H_1-Histaminergics

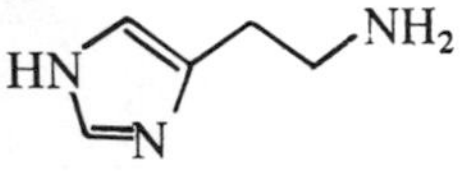

histamine

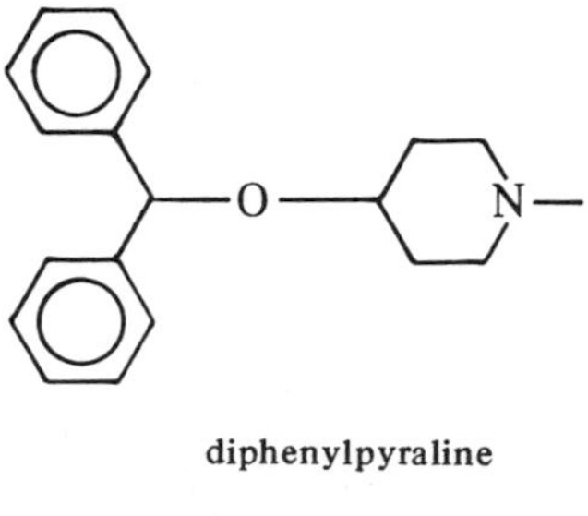

diphenylpyraline

Figure 2.6 Introduction of bulky groups converts agonists into antagonists.

H₂-Histaminergics

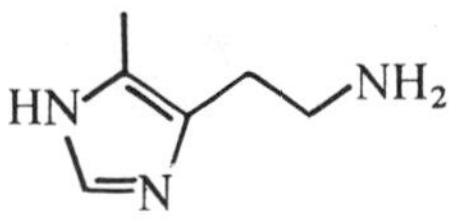

4-methylhistamine	burimamide

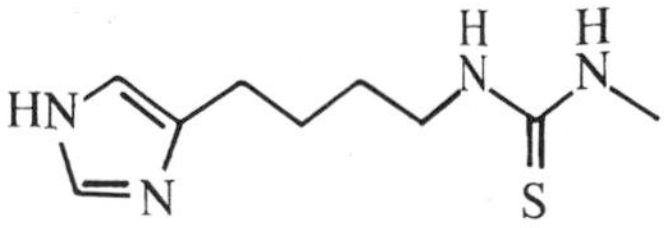

Serotoninics

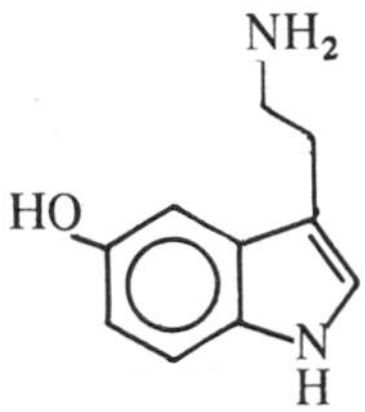

serotonin	methysergide

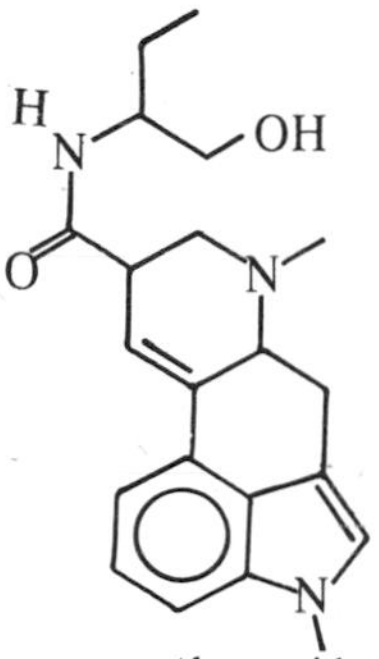

Figure 2.6   (continued)

Table 2.2   Nonclassical Isosteres

$-CO-$	$-COOH$	$-SO_2NH_2$	H	Annular structures	$-O-\overset{\overset{O}{\|\|}}{C}-$	$-OH$
$-SO_2-$	$-SO_3H$	$-PO(OH)NH_2$	F	Open structures	$-\overset{\overset{O}{\|\|}}{C}-O-$	$-NH_2$

Although it is not possible to achieve pure isosterism, the principles of isosterism and bioisosterism are extensively used to modify the structure of biologically active compounds. This substitution yields not only products whose action is identical to that of compounds taken as models but antagonists as well. Several examples may be cited of natural product equivalents, parametabolites, paravitamins, parahormones, and mimetics, as well as their specific antagonists, antimetabolites, antivitamins, antihormones, and lytics developed by applying the concept of isosterism:

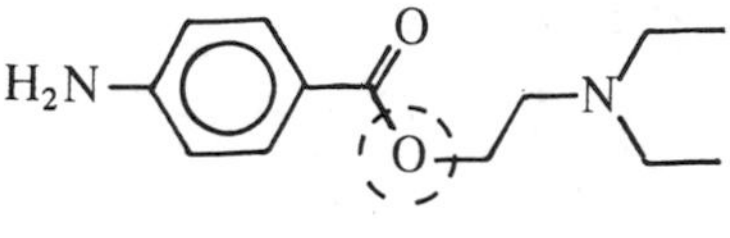

procaine	procainamide

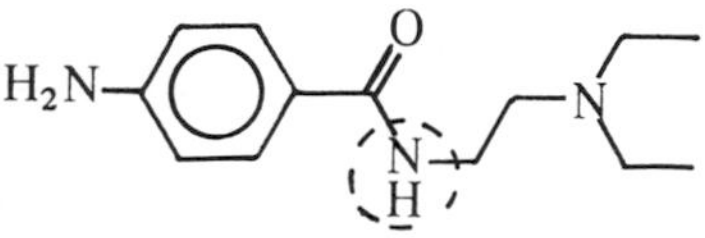

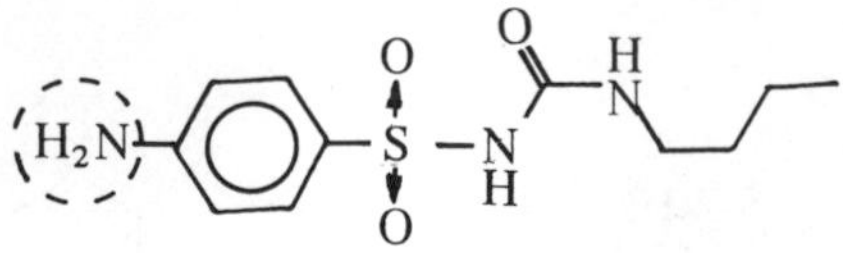

carbutamide 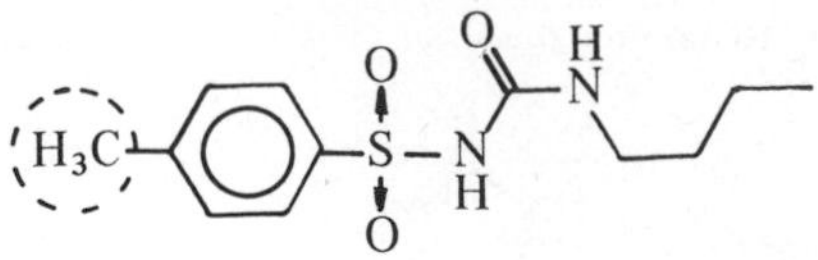   tolbutamide

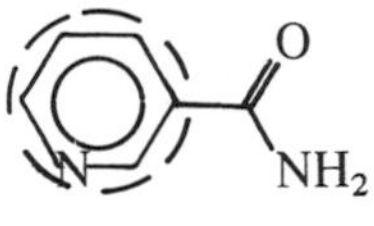

nicotinamide 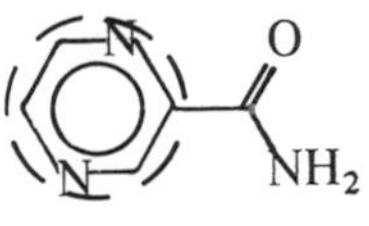   pyrazinamide

In recent years, the interchange of Si for C in certain drugs is being studied. The results have been interesting in several cases, such as in derivatives of choline, barbiturates, penicillin, chloramphenicol, meprobamate, and insecticides.

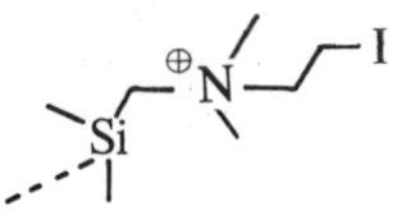

trimethylsilyl
derivative of
choline
(hypotensor)

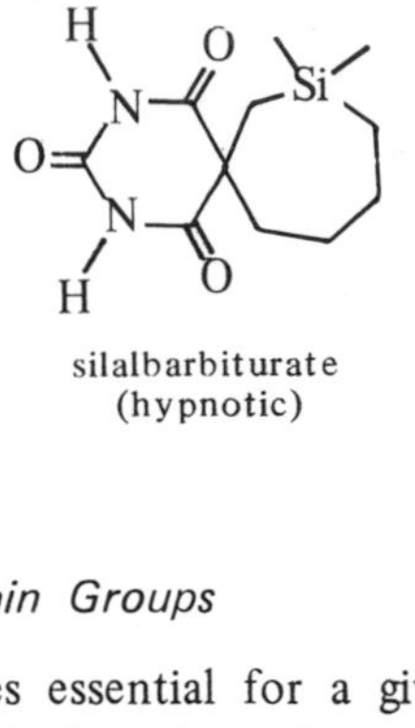

silalbarbiturate
(hypnotic)

Change of Position or Orientation of Certain Groups

The position of certain groups is sometimes essential for a given biological activity. For instance, of the 3 isomers of hydroxybenzoic acid only the *o*-hydroxy is active, because it can form an intramolecular hydrogen bond and, in this way, it can act as chelating agent:

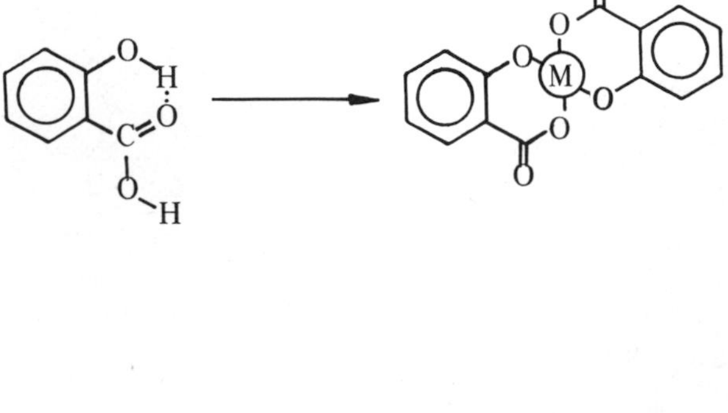

Modifications toward Inhibition or Promotion of Various Electronic States

Certain chemical groups produce two important electronic effects: inductive and conjugative. These effects can alter deeply physical and chemical properties and, therefore, biological activity.

Inductive (or electrostatic) effects result from electronic shifts along simple bonds, owing to the attraction exerted by certain groups because of their electronegativity. Thus, groups that attract electrons more strongly than hydrogen display negative inductive effects ($-I$), whereas those that attract electrons less strongly than hydrogen manifest positive inductive effects ($+I$). Groups that manifest a $-I$ effect are called electron acceptors; those that show a $+I$ effect, electron donors.

The following groups have a $-I$ effect: $-\overset{+}{N}H_3$, $-\overset{+}{N}R_3$, $-NO_2$, $-CN$, $-COOH$, $-COOR$, $-CHO$, $-COR$, $-F$, $-Cl$, $-Br$, $-OH$, $-OR$, $-SH$, $-SR$, $-CH=CH_2$, $-CR=CR_2$, $-C\equiv CH$. The groups having a $+I$ effect are $-CH_3$, $-CH_2R$, $-CHR_2$, $-CR_3$, and $-COO^-$.

In accordance with the intensity of inductive effects, it is possible to place certain groups in decreasing order of the $-I$ effect or in increasing order of the $+I$ effect:

$$F > Cl > Br > I > OCH_3 > C_6H_5 \quad \Big| \quad -I \text{ effect}$$

$$CH_3 < C_2H_5 < CH(CH_3)_2 < n-C_3H_7 < C(CH_3)_3 \quad \Big| \quad +I \text{ effect}$$

Conjugative (or resonance) effects result from the delocalization and high mobility of π electrons and are manifested in compounds with conjugated double bonds. Groups that *increase* the *electronic density* in conjugated systems have $+R$ character; those that decrease such density, $-R$ character.

The following groups have a $-R$ (and simultaneously a $-I$) effect: $-NO_2$, $-CN$, $-CHO$, $-COR$, $-COOH$, $-COOR$, $-CONH_2$, $-SO_2R$, $-CF_3$. Groups that manifest a $+R$ (and, at the same time, a $+I$) effect are $-O^-$, $-S^-$, $-CH_3$, $-CR_3$. The following groups have a $+R$ and a $-I$ effect: $-F$, $-Cl$, $-Br$, $-I$, $-OH$, $-OR$, $-OCOR$, $-SH$, $-SR$, $-NH_2$, $-NR_2$, $-NHCOR$.

Halogens are inserted in several drugs with the purpose of obtaining structurally analogous compounds with modified biological activity. They exert three main effects: steric, electronic, and obstructive. An example of obstructive effect is halogenation in the para position of aromatic rings of some drugs (phenobarbital, for example) in order to prevent hydroxylation, because it is known that in the process of detoxication aromatic rings are hydroxylated in this position and afterward conjugated with glucuronic acid.

3. *Exploration of Pharmacological Effects*

A common practice in search for new drugs is to explore the side effects of known drugs through proper molecular modification. Several examples indicate that this is a fruitful approach.

Exploration of the antidepressant action of isoniazid, used as a tuberculostatic agent, led to the hydrazine and nonhydrazine type of MAO inhibitors. Molecular modification of atropine and its oxide, scopolamine, in exploration of their side effects, resulted in a number of new pharmacodynamic agents: mydriatics, antispasmodics, antidiarrheal, antiulcer, antiparkinsonism, and drugs acting on the central nervous system.

Observation that the antihistaminic promethazine produces sedative central effects prompted molecular modification of this drug in order to enhance this property. This gave origin to chlorpromazine and other phenothiazine tranquilizers.

As a consequence of structural variation of steroids with the purpose of exploring their side effects, new drugs were introduced having antiinflammatory, contraceptive, anabolic, estrogenic, and progestational activities.

The classic example of this approach is the case of sulfonamides. By modifying the structure of sulfas that manifested activities other than antibacterial, many new medicinal agents were born: antimalarials (sulfa drugs), antileprotics (sulfones), diuretics (thiazides), antidiabetics (sulfonylureas), antithyroids (methimazol), and uricosurics (probenecid) (Figure 2.7).

4. Evaluation of Intermediate Products

Owing to empirical screening and to their chemical resemblance to the final products of a planned synthesis of new potential drugs, chemical intermediates have become new drugs.

For example, in the synthesis of thiosemicarbazones as potential tuberculostatic agents it was seen that an intermediate product was more active than the final product. This led to its introduction in medicinal practice under the name of isoniazid, now extensively used as a tuberculostatic agent:

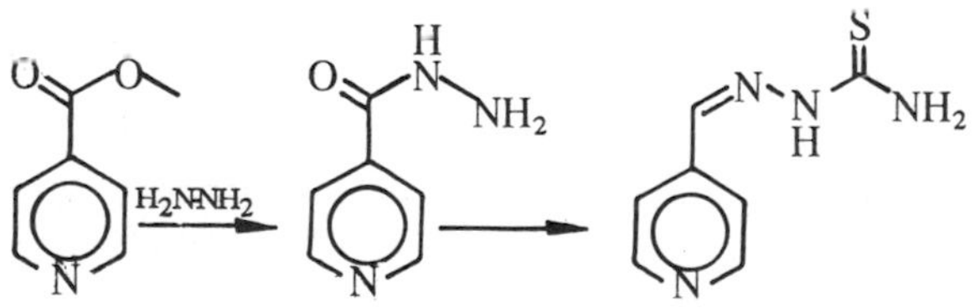

methyl   isonicotinyl-hydrazine   isonicotinaldehyde
isonicotinate   (isoniazid)   thiosemicarbazone

5. Drug Latentiation (Pro-drugs)

Drug latentiation consists essentially in converting, through chemical modification, a biologically active compound to an inactive carrier form that, after

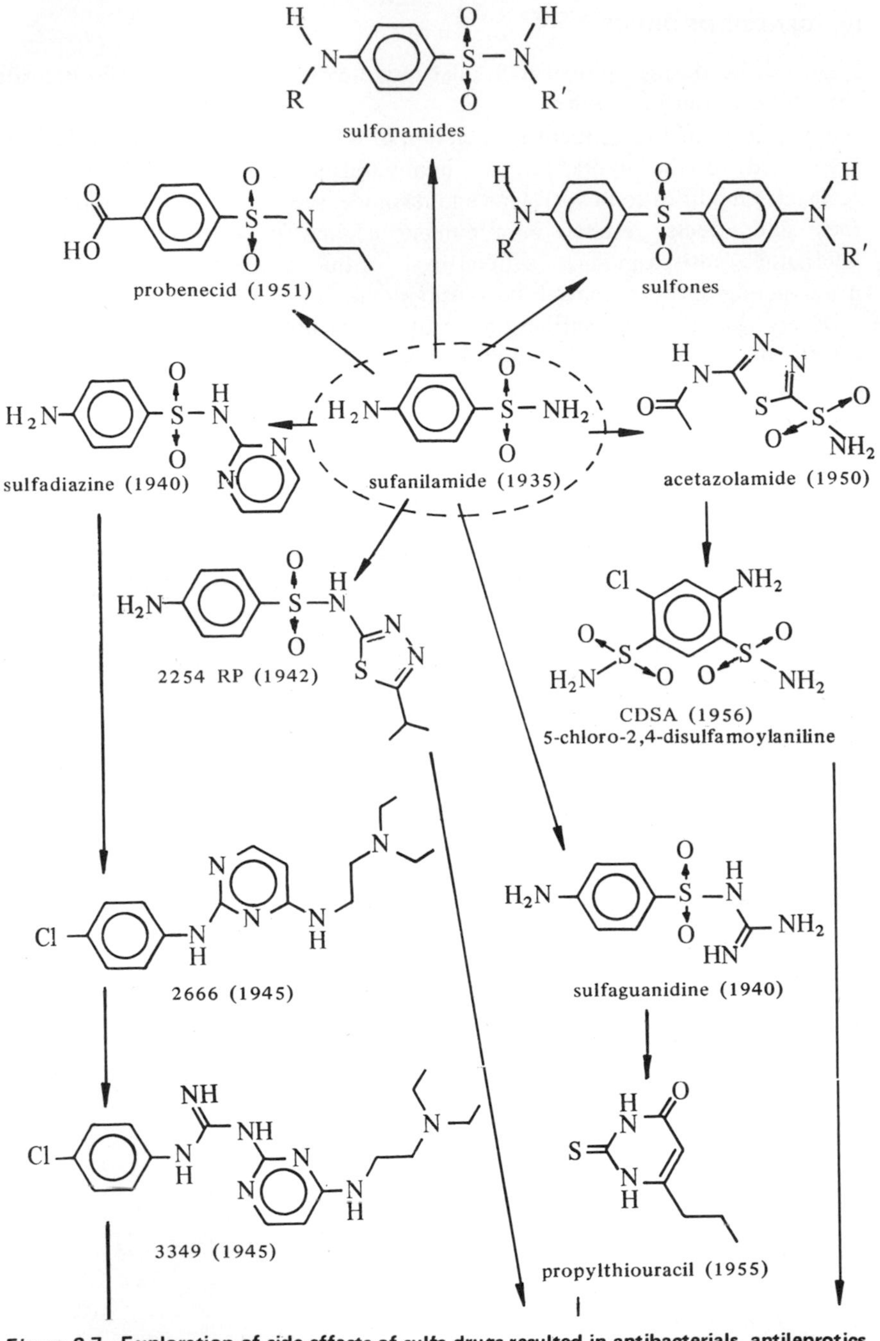

Figure 2.7 Exploration of side effects of sulfa drugs resulted in antibacterials, antileprotics, antimalarials, diuretics, antithyroid drugs, hypoglycemic agents, and drugs for treatment of gout.

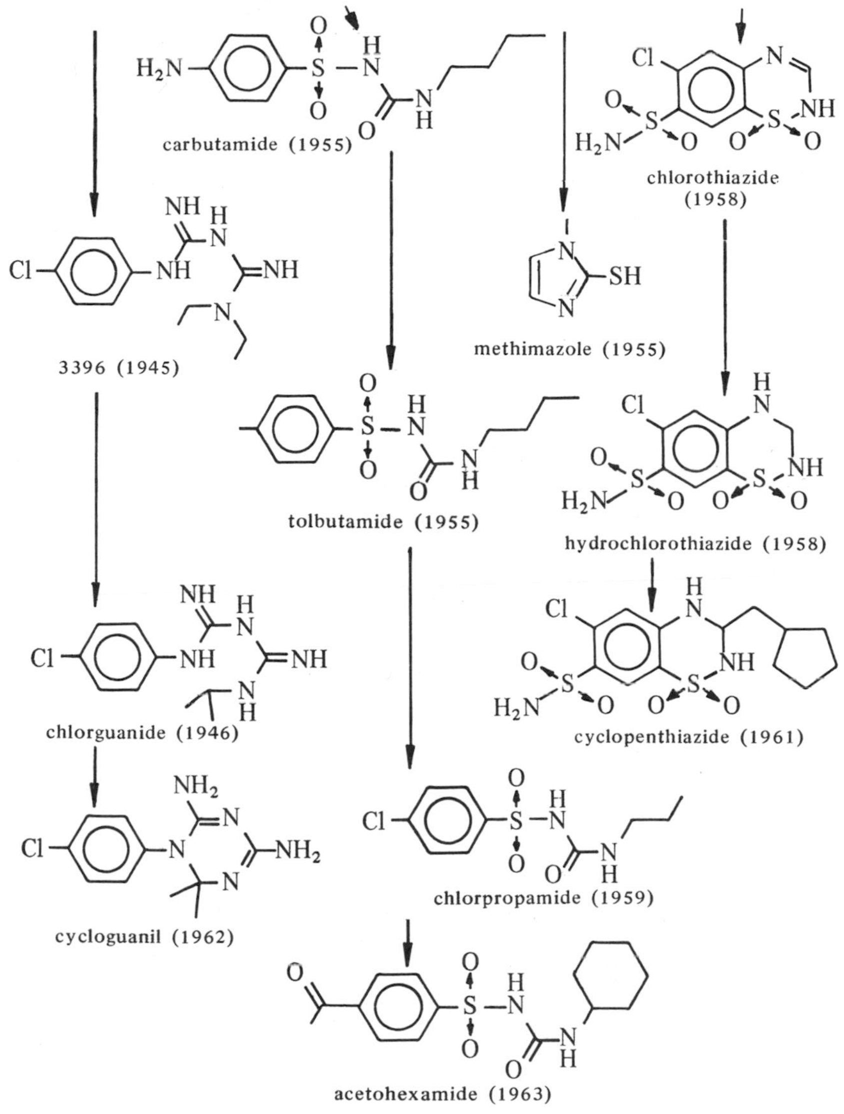

Figure 2.7 (continued)

enzymic or non-enzymic attack, releases the active drug (Table 2.3). The inactive form is often called a pro-drug.

Several objectives are sought in drug latentiation: (*a*) prolongation of action, (*b*) shortening of action, (*c*) drug localization, (*d*) transport regulation, (*e*) adjunct to pharmaceutical formulation, (*f*) lessening of toxicity and side effects.

Table 2.3 Schematic Representation of Some Active Drugs and Their Possible Forms of Transport

Active Drug	Inactive Form of Transport[a]
R—OH	R—O—$\overset{\overset{\text{O}}{\|\|}}{\text{C}}$—X
R—SH	R—S—$\overset{\overset{\text{O}}{\|\|}}{\text{C}}$—X
R—COOH	R—$\overset{\overset{\text{O}}{\|\|}}{\text{C}}$—O—X
R—$\underset{\underset{\text{R}'}{\|}}{\text{N}}$—H	R—$\underset{\underset{\text{R}'}{\|}}{\text{N}}$—$\overset{\overset{\text{O}}{\|\|}}{\text{C}}$—X
R—OH	R—phosphate
R—SH	R—phosphate

Source. Reproduced from N. J. Harper, *Progr. Drug Res.*, **4**, 221 (1962).
[a] X stands for the greater portion of the carrier molecule.

Prolongation of Action

This is accomplished by the following means:

1. Esterification or amidification; example: by acetylation of dapsone, acedapsone, a drug with prolonged action, was obtained.

2. Formation of complex, such as protamine-zinc-insulin, which releases insulin gradually. A recent example is Dapolar, a complex of two known antimalarials: acedapsone and cycloguanil pamoate:

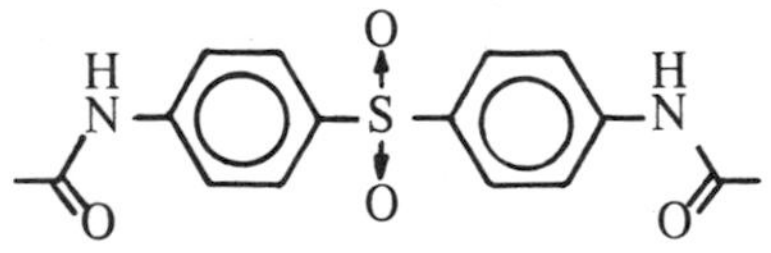

acedapsone

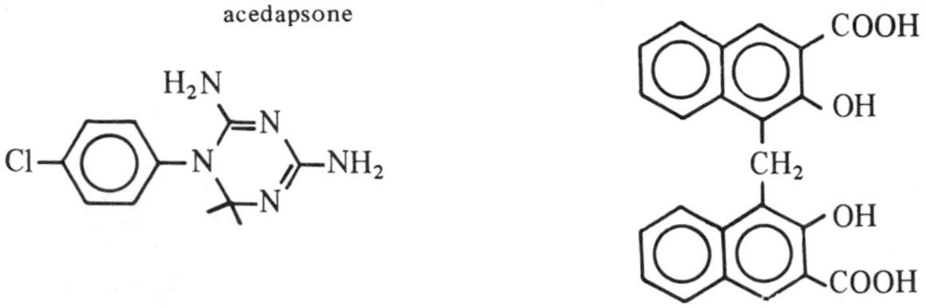

cycloguanyl pamoate

3. Formation of salt; example: intramuscular injection of cycloguanil pamoate showed prophylactic action for up 6 months.

4. Conversion of saturated to unsaturated compounds, such as prednisone and prednisolone, used instead of cortisone and cortisol:

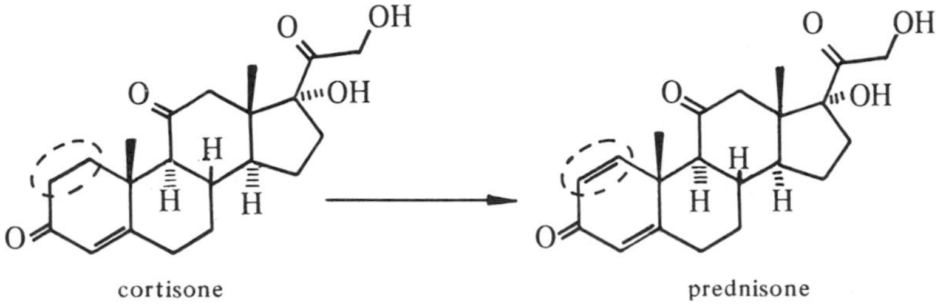

cortisone → prednisone

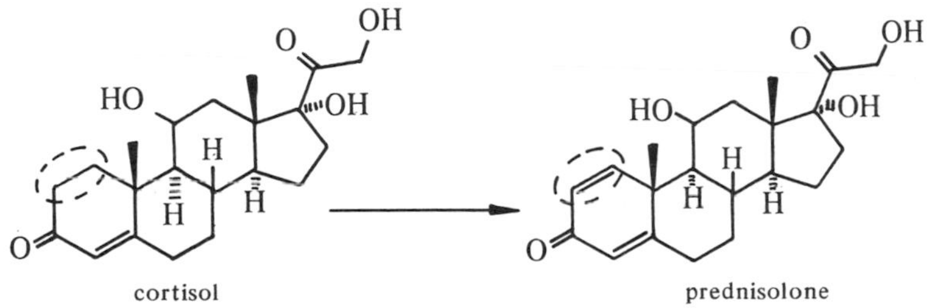

cortisol → prednisolone

Shortening of Action

The action of a drug can be shortened by replacement of a stable moiety by a labile one. Thus, substitution of Cl of chlorpropamide by CH_3, besides the replacement, at the end of the side chain, of C_3H_7 by C_4H_9, results in tolbutamide. Since CH_3 is labile, this group is soon oxidized to COOH, giving an inactive product. As a consequence tolbutamide's half-life is only 5.7 hours, whereas chlorpropamide's is 33 hours:

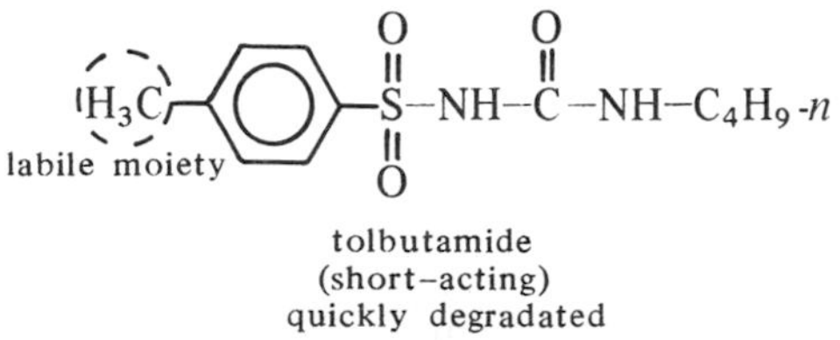

tolbutamide
(short–acting)
quickly degraded

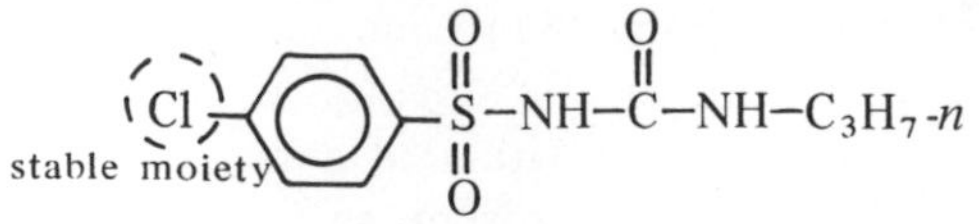

$$Cl-\underset{\text{stable moiety}}{\boxed{\bigcirc}}-\underset{\underset{O}{\overset{O}{\overset{\|}{\underset{\|}{S}}}}}{\overset{O}{\overset{\|}{S}}}-NH-\overset{O}{\overset{\|}{C}}-NH-C_3H_7\text{-}n$$

chlorpropamide
(long–acting)
resistant against oxidative
degradation in the ring

Drug Localization

In the case of compounds with high systemic toxicity but beneficial therapeutic effect in diseased cells, the problem is to add to these drugs a carrier that will transport them to those cells and as a result of enzymic action release them near the receptor site where they will exert their effect *in situ*. Several examples of compounds with latent activity may be cited:

1. Cytostatic or anticancer agents, such as uramustine, and melphalan, in which the carriers are, respectively, uracil and phenylalanine:

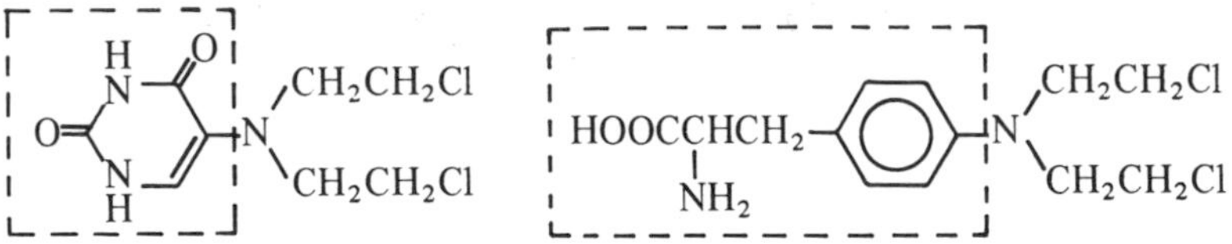

uramustine   melphalan

2. Chelating agents, such as an 8-hydroxyquinoline derivative, in which the carrier is glucuronic acid:

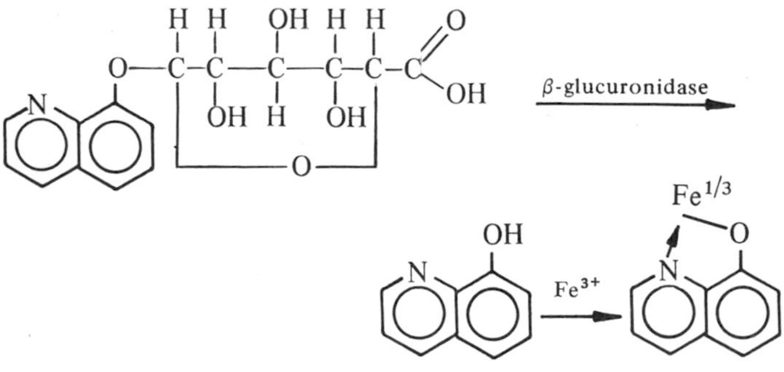

β-glucuronidase

$Fe^{1/3}$

Fe^{3+}

oxine

Transport Regulation

There are a number of recent examples of drug latentiation in which increased efficiency of the active moiety has been brought about through regulation of transport and penetration of the drug within the body. The means used are increase or decrease of volume, alteration of hydrophilicity or liposolubility, introduction or removal of cationic or anionic moieties, change of pH, incorporation of hydrocarbon chain and other suitable stable or labile moieties.

For example, the attachment of strongly hydrophilic moieties to sulfa drugs prevents their transport into the systemic circulation; as a consequence, they act almost exclusively inside the intestine (Table 2.4).

Table 2.4   Sulfonamides Used in Intestinal Infections

Formula[a]	Compound
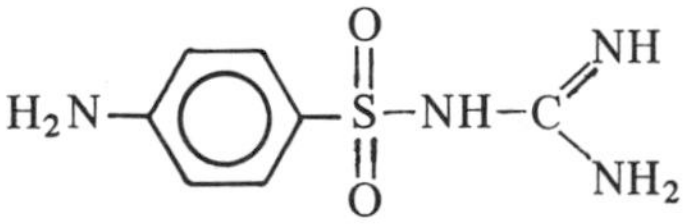 H_2N—⬡—S(O)(O)—NH—C(=NH)(NH$_2$)	sulfaguanidine
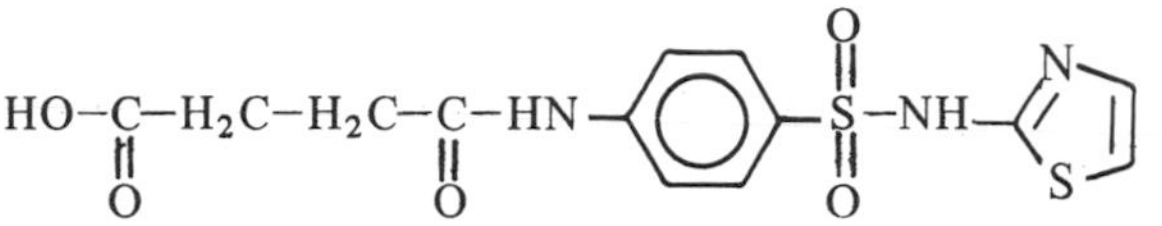 HO—C(O)—H$_2$C—H$_2$C—C(O)—HN—⬡—S(O)(O)—NH—(thiazole)	succinylsulfathiazole
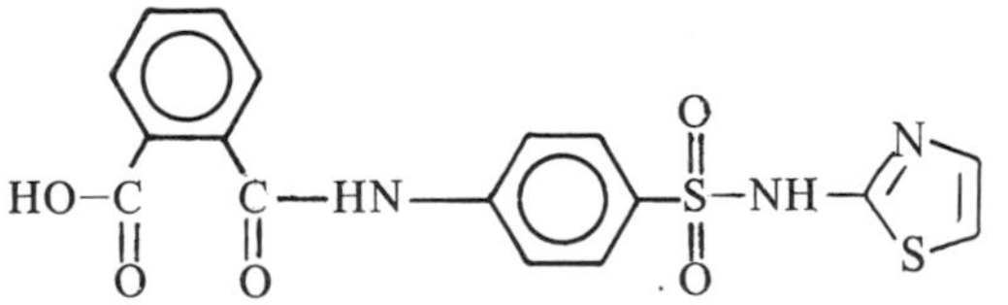 (benzene) HO—C(O) C(O)—HN—⬡—S(O)(O)—NH—(thiazole)	phthalylsulfathiazole
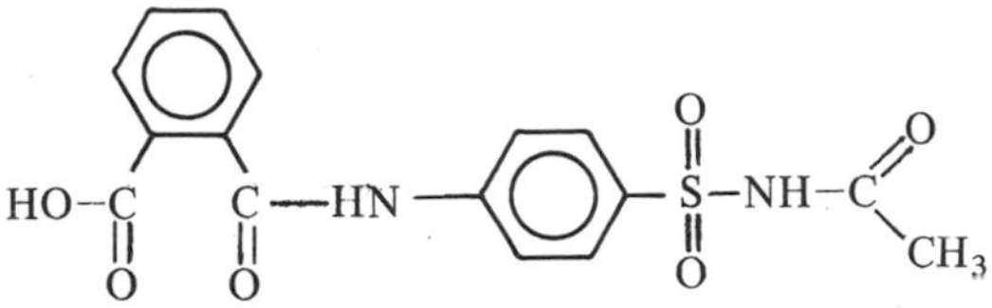 (benzene) HO—C(O) C(O)—HN—⬡—S(O)(O)—NH—C(=O)(CH$_3$)	phthalylsulfacetamide
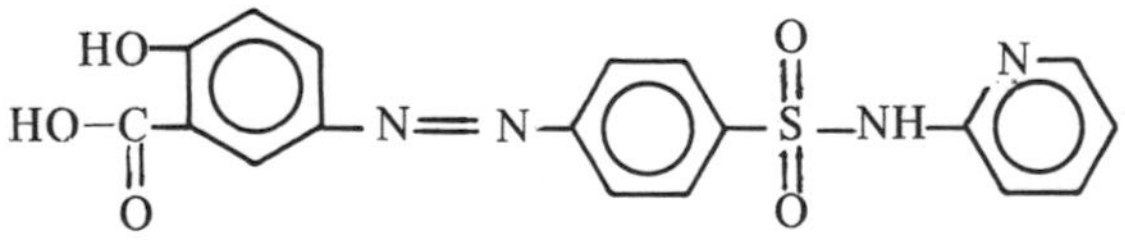 HO—/ HO—C(O)—⬡—N=N—⬡—S(O)(O)—NH—(pyridine)	salazosulfapyridine

Source. After E. J. Ariëns, Progr. Drug Res., 10, 429 (1966).
[a]Strongly hydrophilic transport forms.

Adjunct to Pharmaceutical Formulation

In order to mask the bitter taste of chloramphenicol and make it palatable to children, it was converted to the form of palmitate, which is tasteless. Etisul, a tuberculostatic agent, is a latent form of ethylmercaptan, the active but inflammable and foul-smelling moiety:

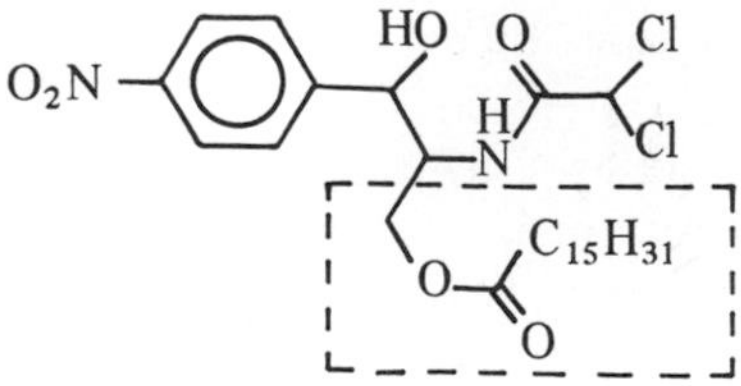

chloramphenicol palmitate

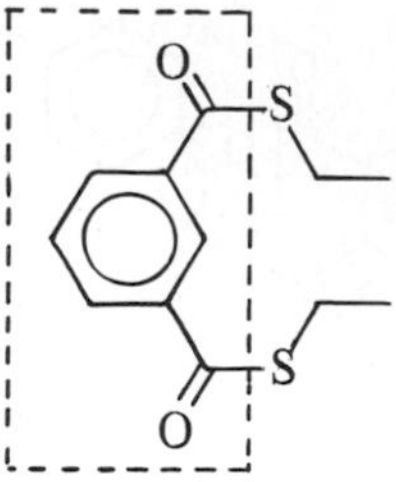

etisul

Lessening of Toxicity and Side Effects

Heavy metals, such as As, Sb, Hg, and Bi, have antiparasitic activity. In inorganic form, however, they are too toxic. Attached to a suitable carrier, as was done in melarsoprol and stibophen, they can be used as chemotherapeutic agents:

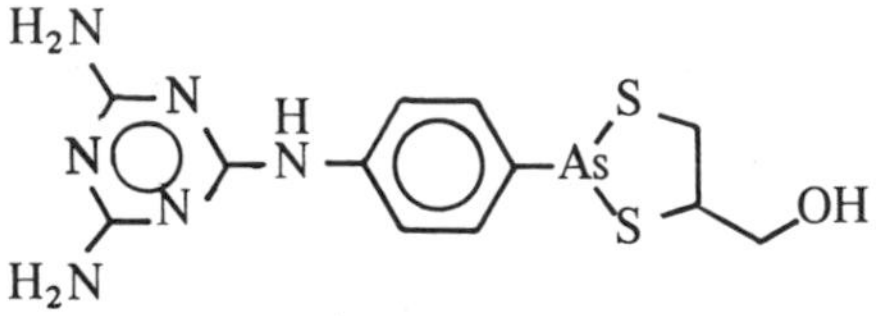

melarsoprol

Certain phenols and carboxylic acids are too toxic to be used as such in medicine. One way to lessen their toxicity is to convert them to esters. *In vivo* these esters undergo hydrolysis regenerating the phenols and carboxylic acids. This process of drug design by molecular association of molecular modification was first applied in 1886 by Necki in the preparation of salol. For this reason the process is called the *salol principle*. When both components of the resulting esters are biologically active compounds, the product of this molecular association is referred to as *true salol* or *full salol*. When only one is active, the molecular modification receives the name of *partial salol*. Examples of drug latentiation involving the salol principle are given in Table 2.5.

E. Planned Rational Synthesis

The great dream of medicinal chemists and pharmacologists has been to design tailor-made drugs, that is, drugs which may have very specific pharmacological action. Several means have been tried to reach this goal, but the probabilities of success are very remote: usually it is necessary to synthesize and test thousands of new substances before one can reach the level of being

Table 2.5   Examples of Application of the Salol Principle

Full Salol	Partial Salol	

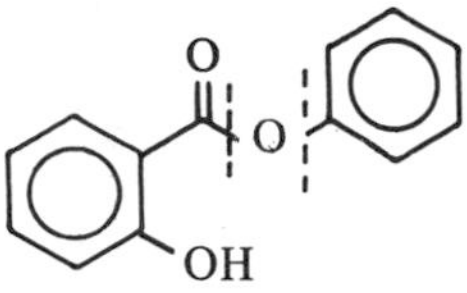
salol
(salicylic acid + phenol)
		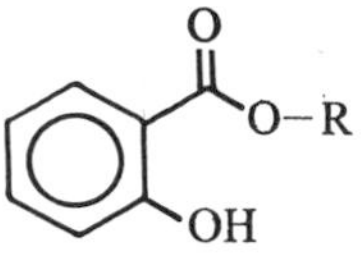 R = CH_3 methyl
		R = C_2H_5 ethyl
		R = CH_2OCH_3
		    methoxymethyl
		    (Mesotan)
		R = CH_2CH_2OH
		    glycol
		    (Spirosal)

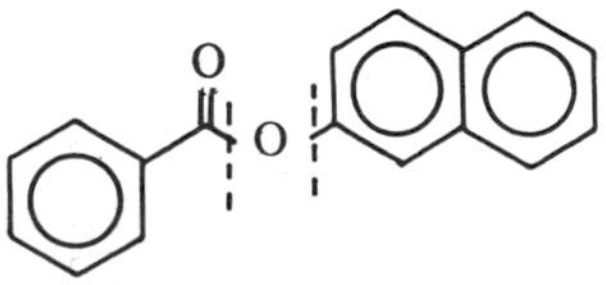 β-naphthol benzoate (benzoic acid + β-naphthol)	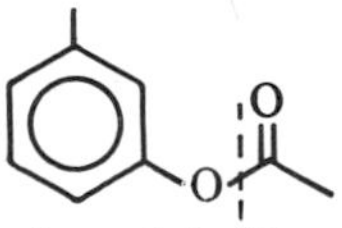 *m*-cresol acetyl (Cresatin)	

introduced into therapeutics. The best probability attained is 3000 to 1.

The rational design of new drugs is, therefore, still in its infancy. Nevertheless, prospects are now brighter than some decades ago. They became greater after the drug designers began to resort to up-to-date knowledge of chemistry, biochemistry, and allied sciences, such as (*a*) organic and bioorganic reaction mechanisms, (*b*) enzymatic mechanisms of action of drugs at molecular and submolecular levels, and (*c*) physicochemical parameters related to drug activity. As a consequence of this truly rational and scientific approach to drug development, in recent years several new drugs have been added to the therapeutic armamentarium.

1. *Enzyme Inhibitors*

In the search for pharmacodynamic agents, the main effort has been put into the synthesis of potential inhibitors of enzymes which catalyze the biochemical reactions leading to the substance responsible for a given physiological role. In this approach it is imperative to know the various steps involved and to try to inhibit preferentially the rate-limiting step.

Some enzyme inhibitors have been introduced by this means, especially through isosteric replacements in the molecules of enzyme substrates:

1. Methyldopa, inhibitor of dopa decarboxylase and used in the treatment of hypertension because it prevents the biosynthesis of precursors of epinephrine and norepinephrine. It also leads to some α-methyldopamine which may act as a false transmitter:

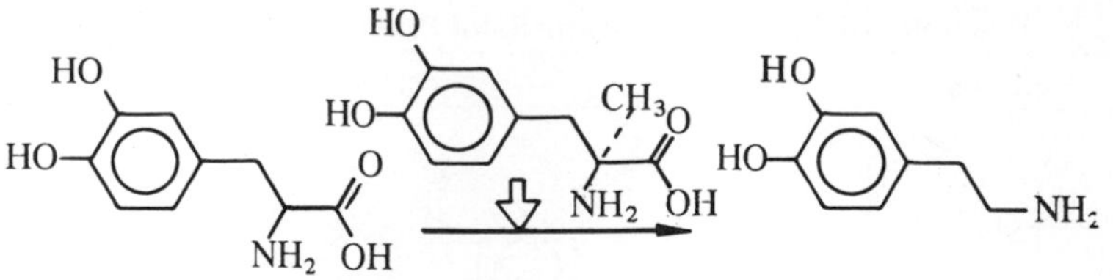

dopa   methyldopa   dopamine

2. Brocresine, inhibitor of histidine decarboxylase and, therefore, of bio-synthesis of histamine:

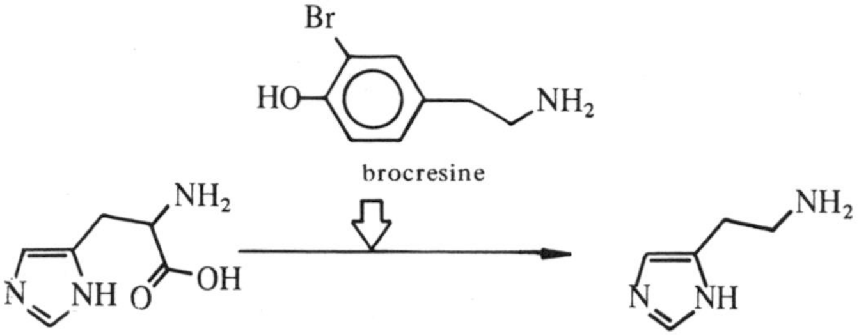

brocresine

histidine   histamine

3. Allopurinol, inhibitor of xanthine oxidase and, by this way, of the uric acid, responsible for gout:

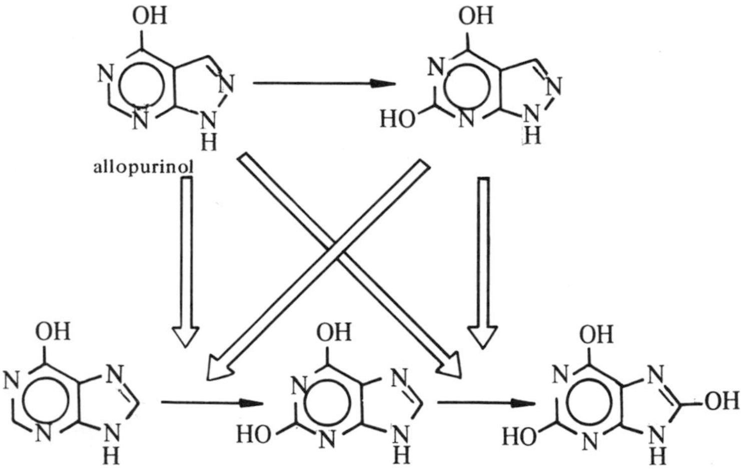

allopurinol

hypoxanthine   xanthine   uric acid

4. Tranylcypromine, inhibitor of monoamine oxidase and used in the treatment of depression:

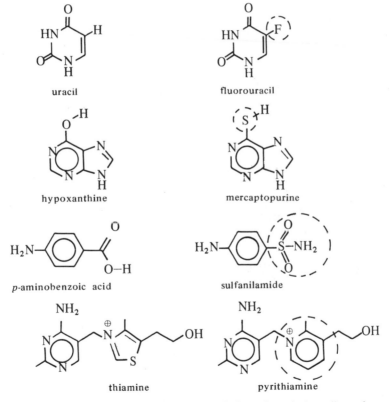

tranylcypromine

morepinephrine → [MAO] → 3,4-dihydroxymandelic acid

2. Antimetabolites

Antimetabolites are drugs that, owing to their structural resemblance to normal cellular metabolites, can replace them in biological processes but cannot carry out their normal role. They are designed usually by isosteric replacement

uracil

fluorouracil

hypoxanthine

mercaptopurine

p-aminobenzoic acid

sulfanilamide

thiamine

pyrithiamine

Figure 2.8 Conversion of metabolites in antimetabolites through isosteric replacement using deceptor groups.

of certain atoms or chemical groups of essential metabolites. Those designed this way are called *classical antimetabolites*. Examples are allopurinol, fluorouracil, mercaptopurine, sulfanilamide, pyrithiamine (Figure 2.8). Incorporation of these antimetabolites in the biological process of a cell ends up with the death of that cell, hence the name *lethal synthesis* given to this process. The isosteric groups used to convert a metabolite in an antimetabolite are called deceptor groups.

After the allosteric phenomenon was described not long ago, a new class of antimetabolites was added, the *nonclassical antimetabolites,* thus called because any resemblance to metabolites is remote. Examples of these metabolites are found in antimalarial agents pyrimethamine and cycloguanil; although they inhibit dihydrofolate reductase, their structural resemblance to folic acid, which is the substrate for this enzyme, is not sharp (compare their structures with those of classical antimetabolites which act on the same enzyme: methotrexate and aminopterin):

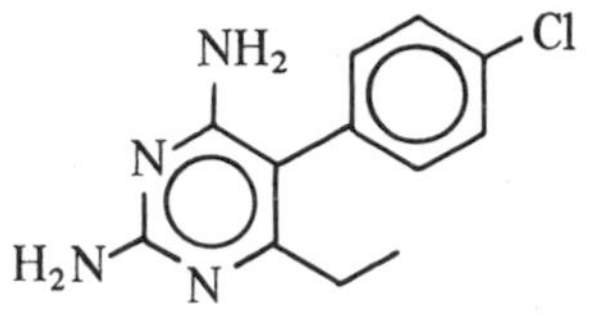

pyrimethamine 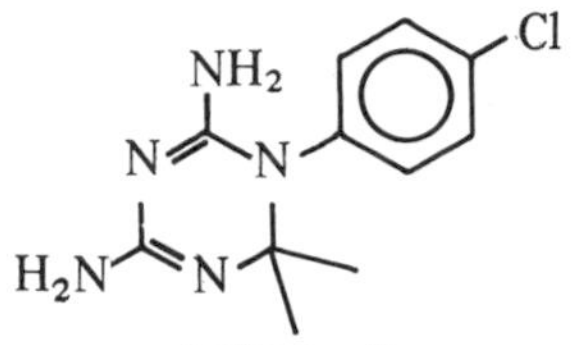   cycloguanil

3. Alkylating Agents

These drugs, whose action was discovered accidentally, are used mostly as antineoplastic agents. Many such compounds have been designed in the hope of alkylating selectively certain groups present in macromolecules of cancer cells, taking into consideration the possibility of forming *in vivo* very reactive intermediates. Unfortunately, they lack desired selectivity and, for this reason, are generally toxic.

Alkylating moieties also explain the prolonged action of β-haloalkylamines used as adrenergic blocking agents:

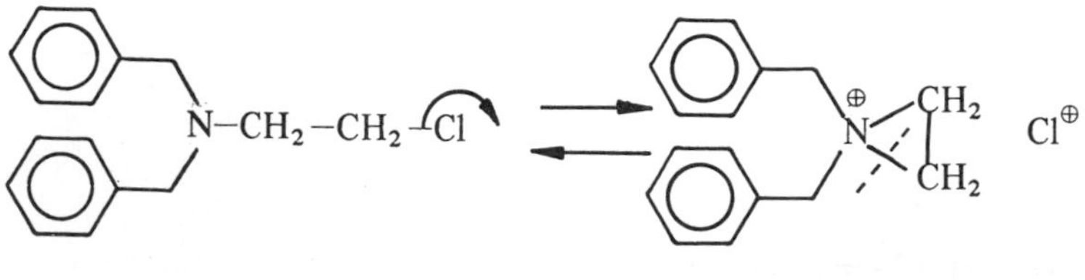

Recently, by using alkylating moieties, Rosen et al. obtained long-acting

local anesthetics:

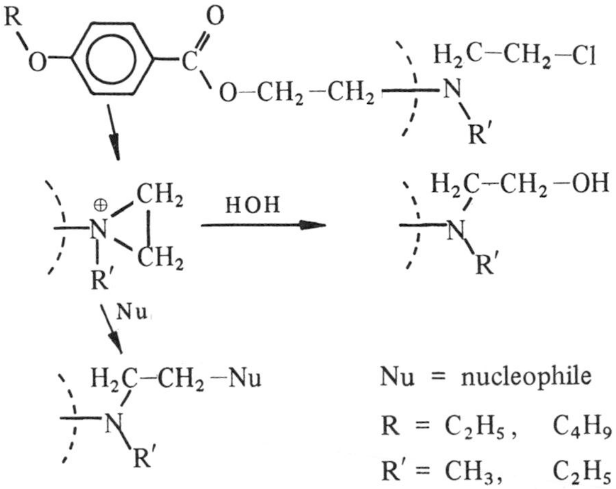

Nu = nucleophile

R = C_2H_5, C_4H_9

R' = CH_3, C_2H_5

4. Antidotes

Certain drugs used as antidotes have resulted from the rational design of chemical compounds. Thus, in order to neutralize the effect of the toxic warfare agent lewisite, α,β-dimercaptopropanol, called British Anti-Lewisite (BAL, for short), and generically dimercaprol, was prepared on the assumption, which proved to be correct, that it would react in the following way:

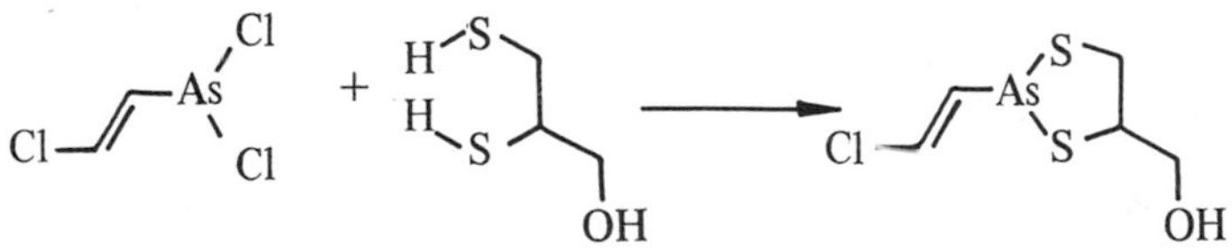

lewisite   dimercaprol

Another example is pralidoxime, designed to be a reactivator of acetylcholinesterase inactivated by organophosphorus compounds through the mechanism shown below:

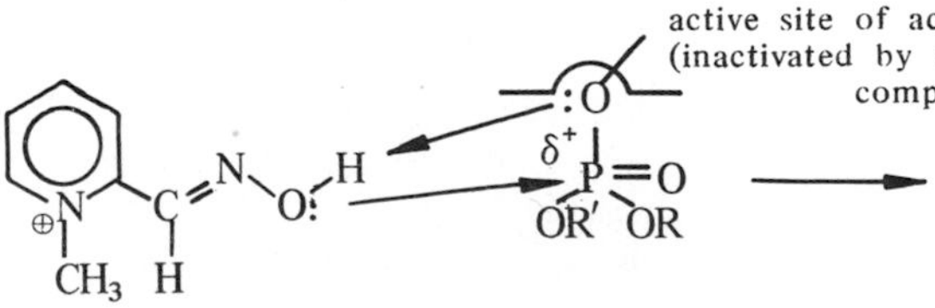

active site of acetylcholinesterase
(inactivated by organophosphorus
compounds)

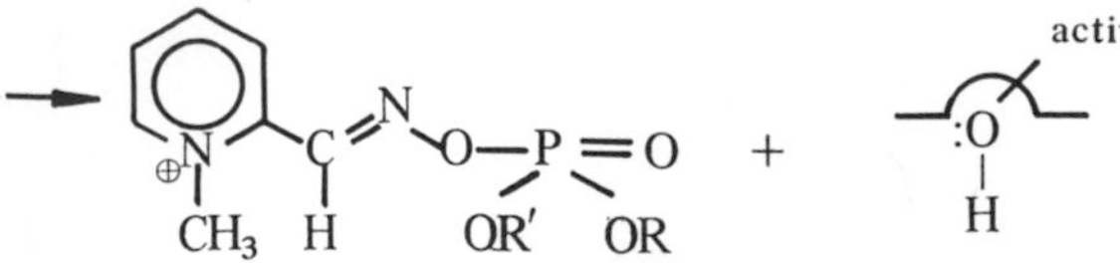

active site of acetylcholinesterase
(reactivated)

REFERENCES

Source of Drugs

W. F. Bynum, *Bull. Hist. Med.*, **44**, 518 (1970).
N. R. Farnsworth, *Tile Till*, **55**(2), 32 (1969).
M. L. Moore, *Ind. Eng. Chem.*, **43**, 577 (1951).

Cost and Place of Development of Drugs

A. Korolkovas, *Rev. Paul Med.*, **83**, 209 (1974).
Prescription Drug Industry Factbook, Pharmaceutical Manufacturers Association, Washington, D.C., 1973.
O. Wintersteiner, *Progr. Drug Res.*, **15**, 204 (1971).
E. Jucker, *Pure Appl. Chem.*, **19**, 249 (1969).
C. J. Cavallito, *Progr. Drug Res.*, **12**, 11 (1968).
R. G. Denkewalter and M. Tishler, *Progr. Drug Res.*, **10**, 11 (1969).
M. Gordon, *Pharm. Ind.*, **24**, 461 (1962).

Search for New Drugs

A. Burger, "Behind the Decline in New Drugs," *Chem. Eng. News,* **53**, (38) 37 (1975).
F. Gross, *Acta Pharm. Suecica,* **10**, 401 (1973).
F. Gross, *Clin. Pharmacol. Ther.,* **14**, 1 (1973).
M. Tishler, *Clin. Pharmacol Ther.,* **14**, 479 (1973).
A. J. Gordon and S. G. Gilgore, *Progr. Drug Res.,* **16**, 194 (1972).
A. A. Rubin, Ed., *Search for New Drugs,* Dekker, New York, 1972.
J. Y. Bogue, *J. Chem. Educ.,* **46**, 468 (1969).

Genesis of Drugs

T. Higuchi and V. Stella, Eds., *Pro-drugs as Novel Drug Delivery Systems,* American Chemical Society, Washington, 1975.
E. J. Ariëns and A. M. Simonis, *Top. Curr. Chem.,* **52**, 1 (1974).
M. S. Amer and G. R. McKinney, *Annu. Rep. Med. Chem.,* **9**, 205 (1974).
E. J. Ariëns, Ed., *Drug Design,* 6 vols., Academic, New York, 1970-1975.
F. H. Clarke, Ed., *How Modern Medicines Are Discovered,* Futura, Mount Kisco, N. Y., 1973.

A. Korolkovas, *Rev. Paul. Med.,* **81**, 105 (1973).

W. P. Purcell, G. E. Bass, and J. M. Clayton, *Strategy of Drug Design: A Molecular Guide to Biological Activity,* Wiley-Interscience, New York, 1973.

Ciba Foundation Symposium, *Carbon-Fluorine Compounds,* Elsevier, Amsterdam, 1972.

G. M. Rosen and S. Ehrenpreis, *Trans. N.Y. Acad. Sci.,* **34**, 255 (1972).

American Chemical Society, "Drug Discovery: Science and Development in a Changing Society," *Advan. Chem. Ser.,* **108** (1971).

A. Korolkovas, *Essentials of Molecular Pharmacology: Background for Drug Design,* Wiley-Interscience, New York, 1970.

B. R. Baker, *Design of Active-Site Directed Irreversible Enzyme Inhibitors,* Wiley-Interscience, New York, 1967.

American Chemical Society, "Molecular Modification in Drug Design," *Advan. Chem. Ser.,* **45** (1964).

H. R. Ing, *Progr. Drug Res.,* **7**, 305 (1964).

R. A. Peters, *Biochemical Lesions and Lethal Synthesis,* Pergamon, Oxford, 1963.

N. J. Harper, *Progr. Drug Res.,* **4**, 221 (1962).

F. W. Schueler, *Chemobiodynamics and Drug Design,* McGraw-Hill, New York, 1960.

R. Hazard et jal., *C. R. Acad. Sci.,* **244**, 2197 (1957).

H. L. Friedman, "Influence of Isosteric Replacements upon Biological Activity," in *First Symposium on Chemical-Biological Correlation* (May 26-27, 1950), National Academy of Sciences – National Research Council, Publication No. 206, Washington, D.C., 1951, pp. 295-358.

H. Erlenmeyer, *Bull. Soc. Chim. Biol.,* **30**, 792 (1948).

F. Y. Wiselogle, Ed., *Survey of Antimalarial Drugs,* 2 vols., Edwards, Ann Arbor, Mich., 1946.

R. C. Fuson, *Chem. Rev.,* **16**, 1 (1935).

C. D. Leake and M. Y. Chen, *Proc. Soc. Exp. Biol. Med.,* **28**, 151 (1930).

I. Langmuir, *J. Amer. Chem. Soc.,* **41**, 868, 1543 (1919).

CHAPTER 3

THEORETICAL ASPECTS OF DRUG ACTION

I. TYPES OF DRUG ACTION

A. The Ferguson Principle

By observing in a homologous series that certain physical properties, such as solubility in water, vapor pressure, capillary activity, and distribution between immiscible phases, change according to geometric progression, Ferguson concluded that "molar toxic concentrations. . . are largely determined by a distribution equilibrium between heterogeneous phases—the external circum-ambient phase where the concentration is measured and a biophase which is the primary seat of toxic action."

According to Ferguson, it is not necessary to define the nature of biophase nor to measure the concentration of the drug at this site. If equilibrium conditions exist between the drug in the molecular biophase and in exobiophase, that is, in extracellular fluids, the tendency for the drug to escape from each phase is the same, even though the concentrations in the two phases are different. This tendency is given the name of *thermodynamic activity.* It is equivalent approximately to the degree of saturation of each phase. Therefore thermodynamic activity in the external phase (exobiophase) corresponds to thermodynamic activity in the molecular biophase; in practice it is the first that is measured, since it is not possible to measure the latter.

In the case of volatile drugs their thermodynamic activity is calculated from the expression p_t/p_s, where p_t is the partial pressure of the substance in solution and p_s the saturated vapor pressure of the substance at the experimental temperature. When the drug is nonvolatile, its thermodynamic activity is calculated by employing the ratio S_t/S_o, where S_t is the drug's molar concentration and S_o its corresponding solubility.

B. Structure and Activity

Based on the mode of pharmacological action, drugs may be divided into two main classes: structurally nonspecific and structurally specific.

44

1. Structurally Nonspecific Drugs

Structurally nonspecific drugs are those in which pharmacological action is not directly subordinated to chemical structure, except to the extent that structure affects physicochemical properties. Among these properties may be cited adsorption, solubility, pK_a, and oxidation-reduction potential, which influence permeability, depolarization of the membrane, protein coagulation, and complex formation. It is assumed that structurally nonspecific drugs act by physicochemical processes for the following reasons:

1. Their biological action is directly related to thermodynamic activity, which is usually high, with values from 1 to 0.01; this means that such drugs act in relatively large doses.
2. Although they vary in chemical structure, they cause similar biological responses.
3. Slight modifications in their chemical structure do not result in pronounced changes in biological action.

2. Structurally Specific Drugs

Structurally specific drugs are those whose biological action results essentially from their chemical structure, which should adapt itself to the three-dimensional structure of receptors in the organism by forming a complex with them. Hence in these drugs chemical reactivity, shape, size, stereochemical arrangement of the molecule, and distribution of functional groups, as well as resonance, inductive effects, electronic distribution, and possible binding with receptors, besides other factors, play decisive roles.

Several considerations suggest that the pharmacological effect produced by these drugs results from complexation with a chemically reactive minuscule area of certain body cells whose topography and functional groups are usually complementary to those of the drugs:

1. Their biological action does not depend solely on thermodynamic activity, which is usually low (less than 0.001); this means that structurally specific drugs are effective in lesser concentration than the structurally nonspecific ones.
2. They have some structural characteristics in common, and the fundamental structure present in all of them, by orienting functional groups into a similar spatial arrangement, is responsible for the analogous biological response they cause.
3. Slight modifications in their chemical structure may result in substantial changes in pharmacological activity, so that substances thus obtained may have actions ranging from antagonistic to similar to that of the parent compound.

3. Distinction between Types of Drug Action

To distinguish structurally nonspecific drugs from those that are structurally specific it is not enough to consider only one or two of several points of differentiation, some of which have been mentioned. *All* of them should be considered. Frequently certain drugs that have no structural similarity nevertheless exhibit similar pharmacological effects, which are not appreciably altered by slight structural variations within each chemical category. For instance, diuretics present a broad variety of chemical structures: methylxanthine, pyrimidine, triazine, sulfamide, organomercurial, benzothiadiazine, thiazide, spirolactonic, ptheridine, acylphenoxyacetic, pyrazine, etc.—and their diuretic action is not much affected by slight structural modifications of the molecule of each group. Yet, contrary to what at first seems to be the case, diuretics are structurally *specific*. Actually they produce *analogous* pharmacological response but by interfering with *different* biochemical processes.

II. PHYSICOCHEMICAL PARAMETERS AND PHARMACOLOGICAL ACTIVITY

A. Parameters Used

The idea that chemical structures of drugs can be correlated mathematically with the biological response that they produce is very old. Until the last 15 years, however, no attempt was made to establish quantitatively structure-activity relationships, since this area was considered to be too complicated. But lately these studies have been intensified, mainly with the aim of designing more specifically active and more potent drugs. Several mathematical equations have been devised which correlate chemical structure with pharmacological activity. In these equations certain parameters are considered which represent the physicochemical properties of drugs. More than 40 parameters have been used. They can be classified in four categories: solubility, empirical electronic, nonempirical electronic, and steric.

1. Solubility Parameters

Also called *lipophilic* or hydrophobic parameters, solubility parameters measure the degree of attraction of drugs by lipidic and hydrophobic regions of macromolecules. They are related, on one side, to the transport of drugs from the exobiophase to the receptor compartment and, on the other side, with the possibility of attraction and interaction between hydrophobic regions of drugs and receptors.

The main parameters of this class are partition coefficient, hydrophobicity constant, solubility, molar mass or number of carbons, chromathographic parameters, substituent constant, molar attraction constant, parachor, surface tension, electronic polarizability, electrophilic polarizability, and electronic charge density.

Solubility

The term *solubility* refers to solubilities in different media and varies between two extremes: polar solvents, such as water, and nonpolar solvents, such as lipids. The name *hydrophilia,* or *lipophobia,* refers to solubility in water, and *lipophilia,* or *hydrophobia,* to that in lipids.

Solubility is particularly relevant in homologous series. For instance, the antibacterial activities of certain normal primary alcohols, cresols, and alkyl phenols; the estrogenic activity of alkyl 4,4'-stilbenediols; and the anesthetic activity of esters of *p*-aminobenzoic acid are directly related to their lipo-solubility.

Partition Coefficients

The biological activity of several groups of compounds can be correlated with their partition coefficients in polar and nonpolar solvents. Overton and Meyer were the pioneers in these studies. They resorted to partition coefficients first to explain the activity of certain narcotics and, later, of general anesthetics. According to these authors, such compounds, because of their greater affinity for lipids (as shown from distribution coefficients in water-oil mixtures), fix primarily to nervous-system cells rich in lipids, and thus they owe their biological action to this phenomenon.

A better correlation was found with the oil/gas partition coefficient. By measuring the minimal alveolar concentration of several anesthetics necessary to produce a standard analgetic effect, Eger and coworkers concluded that anesthetics of high liposolubility are efficient in low alveolar concentrations. According to those authors, anesthesia is induced when anesthetics reach relative saturation in some lipid structure situated in the brain. In recent studies, Eger and collaborators found good correlation between minimal alveolar concentrations and the solubilities of some general anesthetics in olive oil. The results obtained by Eger and coworkers are in agreement with the Ferguson principle, according to which the potency of structurally nonspecific drugs depends on the relative saturation of the biophase, that is, of some cellular compartment.

Surface Activity

Some chemical groups are characterized by the property of conferring water solubility on the molecules to which they belong. Among such groups, called *hydrophilic, lipophobic,* or *polar,* can be cited, in order of decreasing efficiency, the following: $-OSO_2ONa$, $-COONa$, $-SO_2Na$, $-OSO_2H$, and $-SO_2H$. The following groups are less efficient: $-OH$, $-SH$, $-O-$, $=CO$, $-CHO$, $-NO_2$, $-NH_2$, $-NHR$, $-NR_2$, $-CN$, $-CNS$, $-COOH$, $-COOR$, $-OPO_3H_2$, $-OS_2O_2H$, $-Cl$, $-Br$, and $-I$. Furthermore the presence of unsaturated bonds, such as exist in $-CH=CH-$ and $-C\equiv C-$, helps to promote hydrophilicity.

Other groups, called *lipophilic, hydrophobic,* or *nonpolar,* increase the liposolubility of the compounds of which they are part. Examples of these groups are aliphatic hydrocarbon chains, aryl alkyl groups, and polycyclic hydrocarbon groups.

Compounds carrying hydrophilic and lipophilic groups, and provided there is a proper equilibrium between both these groups, have the property of modifying the characteristics of the interface, that is, the boundary surface, between two liquids, or a liquid and a solid, or a liquid and a gas. Some types of molecules decrease the surface tension by concentrating and orienting themselves in a definite arrangement in the interface or on the surface of a solution, and they owe their biological action to this property. Such compounds, called *surfactants,* are used mainly as detergent, wetting, dispersing, foaming, and emulsifying agents.

Surfactant agents present two distinct regions of lipophilic and hydrophilic character. For this reason they are called *amphiphilic,* or *amphiphils* (from the Greek $\alpha\mu\phi\iota$ = both, and $\phi\iota\lambda o\varsigma$ = friend). Their greater or lesser hydrophilicity and lipophilicity depend on the degree of polarity of the groups present. Taking this factor into consideration, they can be classified into four categories:

1. *Nonionic.* They are not ionizable and contain weakly hydrophilic and lipophilic groups, which render them soluble or water dispersible. The hydrophilic group is usually polyoxyethylene ether or polyol. Examples are polysorbate 80, glyceryl monostearate:

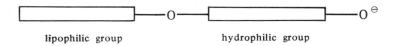

lipophilic group   hydrophilic group

2. *Cationic.* The hydrophilic group bears a positive charge, and it can be quaternary ammonium, sulfonium, phosphonium, iodonium. Examples are benzalkonium chloride, cetylpyridinium chloride, cetramonium bromide, benzethonium chloride:

lipophilic group   hydrophilic group

3. *Anionic.* The hydrophilic group bears a negative charge and it can be carboxyl, sulfate, sulfonate, phosphate. Examples are sodium stearate, sodium

tetradecyl sulfate, sodium xilene sulfonate:

$$
\begin{array}{c}
O \\
\parallel \\
\text{[lipophilic group]} \longrightarrow S \longrightarrow O^{\ominus} \quad Na^{\oplus} \\
\parallel \\
O
\end{array}
$$

lipophilic group   hydrophilic group

4. *Amphoteric.* Also called *ampholytic,* or *ampholytes,* they contain two hydrophilic groups: cationic (amine salt, quaternary nitrogen) and anionic (carboxyl, sulfate). An example is N-lauryl β-propionate:

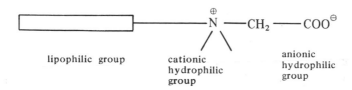

lipophilic group   cationic   anionic
                          hydrophilic   hydrophilic
                          group   group

Because they disorganize cellular membranes and cause hemolysis, besides being easily adsorbed by proteins, surfactant agents are not, in general, applied internally but only topically, as skin disinfectants or sterilizers of instruments. This is the case for cationic surfactants. Nonionic surfactants are largely employed in pharmaceutical preparations for oral (sometimes even parenteral) use as solubilizing agents of water-insoluble or slightly soluble drugs.

2. Empirical Electronic Parameters

They are the constants introduced by organic chemists. They measure the electronic effect of one substituent on the change of energy of a chemical reaction or of a drug-receptor interaction. The most used are the following: Hammett's substituent constant, derivatives of Hammett's substituent constant, pK_a, and Swain and Lupton's constants.

pK_a

The partially lipidic nature of cellular membranes, such as the ones that enwrap the stomach, small intestine, mucosa, nervous tissue, facilitates the passage of drugs with high liposolubility across them. This liposolubility is affected by the pH of the environmental medium and by the degree of dissociation pK_a. Usually drugs are weak acids or weak bases. The degree of dissociation pK_a is calculated

from the following Henderson-Hasselbalch equations:
In the case of acids: $RCOOH \rightarrow RCOO^{\ominus} + H^{\oplus}$

$$pK_a = pH + \log \frac{[\text{undissociated acid}]}{[\text{ionized acid}]}$$

$$= pH + \log \frac{[RCOOH]}{[RCOO^{\ominus}]}$$

In the case of bases: $R\overset{\oplus}{N}H_3 \rightarrow RNH_2 + H^{\oplus}$

$$pK_a = pH + \log \frac{[\text{ionized base}]}{[\text{undissociated base}]}$$

$$= pH + \log \frac{[R\overset{\oplus}{N}H_3]}{[RNH_2]}$$

Weak acids have a high pK_a; weak bases, a low pK_a. The biological activity of certain acids and bases is directly related to their degree of ionization. Whereas some (e.g., phenols, carboxylic acids) act in the molecular form, others (e.g., quaternary ammonium salts) act in an ionized form. In these cases the pH plays an important role: acids are more active at lower pH; bases are more active at higher pH.

Ionization

When the biological activity of a drug results from ions, the activity increases with increase in the degree of ionization. However, if activity results from undissociated molecules, increase in the degree of ionization of active compounds causes a decrease in activity.

Ionization affects other physicochemical properties. Increase in ionization increases a drug's water solubility and decreases its liposolubility and, consequently, its adsorption and passage across lipid barriers and membranes and its concentration in tissues rich in lipids.

In general drugs cross cellular membranes in undissociated forms as intact molecules and act in dissociated forms, as ions. This happens because the passage of ions across the cellular membrane is prevented by two factors:

1. The cellular membrane is made up of layers of electrically charged macromolecules (lipids, proteins, and mucopolysaccharides) which attract or repel ions.

2. Hydration of ions increases their volumes, rendering difficult their diffusion through pores.

It is known that the biological action of amino acridines increases with the degree of ionization. Very likely drugs of this type act at the external part of the cell, since they cannot cross cellular membranes.

3. Nonempirical Electronic Parameters

They are related to π orbitals, electron energies, and certain other electronic indices, because π electrons, in being delocalized, are responsible for the greater part of physicochemical properties of molecules. The main parameters of this class can be assigned to one of the following three factors:

1. Energetic parameters that represent the ability of donating or of accepting π electrons: electronic affinity, lowest empty molecular orbital (LEMO), ionization potential, highest occupied molecular orbital (HOMO), charge-transfer energy, delocalization or resonance energy, miscellaneous experimental parameters
2. Energetic parameters that represent the reactivity of one particular part of the molecule: electrophilic localization energy, nucleophilic localization energy, radicalar localization energy
3. Structural parameters: net electronic charge, electrophilic superdelocalizability, nucleophilic superdelocalizability, dipole moments

4. Steric Parameters

Steric parameters or factors represent the form and the size of the substituent introduced in a parent molecule; that is, they measure the intramolecular steric effect. Examples of these parameters are Taft's substituent constants, Hancock's constant, van der Waals radii, steric constant.

B. Methods of Studying Structure-Activity Relationships

Based on the foregoing four types of parameters, three methods are presently used to study the relationship between chemical structure and pharmacological activity: mathematical models, polarization models, and quantum-chemistry models. All of them rely heavily on multiple-regression analysis to deduce equations that may correlate biological data with physical constants, for the purpose of discovering how molecular properties influence given biological effects and to relate concepts thus obtained to a mathematical, physical, or chemical model.

The biological activity of chemical substances is due not to just one but to all physicochemical properties of the molecule. Therefore, in the study of structure-activity relationship by quantitative methods, it is not possible to obtain perfect correlation of biological action with only one parameter or a reduced number of physicochemical parameters of the molecules considered. The good correlations obtained with only one parameter show simply that this parameter plays a preponderant role.

1. *Mathematical Models*

The method of correlating biological activity with chemical structure by mathematical models consists in expressing the biological activity as a function of parameters attributed to each substituent group or parent part of the molecule. This empirical model is based on an additive mathematical model in which it is assumed that a certain substituent in a specific position contributes additively and constantly to the biological activity of a molecule in a series of chemically related compounds. Studies with this model started with Free and Wilson. They now are being pursued by Purcell, Craig, and other researchers.

More perfect mathematical models are those based on physicochemical parameters. These parameters are more adequately correlated with biological activity, because biological processes are of a physicochemical nature. These models, called *linear free-energy* models, take into account mainly the electronic, steric, and hydrophobic effect of the substituent groups introduced into the parent molecule upon formation of a drug-receptor complex. One of the most active investigators in this field is Hansch. With the purpose of correlating the chemical structure with the physical properties and biological activity of drugs, Hansch has been studying two complex processes:

1. Movement of the drug from the point of application in the biological system to the sites of action

2. Occurrence of a rate-limiting chemical or physical reaction in receptor sites

Both these processes are often distant (in time and space) from the observed biological response, because the drug, before producing an effect, must cross a series of compartments made up essentially of aqueous or organic phases. Owing to the success of Overton and Meyer and their disciples in correlating biological activity with partition coefficients, Hansch has employed a model analogous to the one used by those authors.

Hansch starts from a chemical substance of known biological action and compares its activity with that of compounds of analogous structure, differing from it only in substituent groups. He determines the distribution coefficients of the parent compound and its derivatives between water, a polar solvent, and normal octanol, a nonpolar solvent. The difference between the respective logarithms of the distribution coefficients is called π:

$$\pi_{COOH} = \log P_{COOH} - \log P_H$$

In the foregoing equation π, the hydrophobicity constant, is the measure of contribution of the substituent to solubility in a series of partitions; P_{COOH}

is the partition coefficient of the carboxylic derivative; P_H is the partition coefficient of the parent compound. If π has a positive value, it means that the substituent group increases the solubility of the compound in nonpolar solvents. If it has a negative value, the substituent group will increase the solubility of the compound in polar solvents.

Furthermore in his studies Hansch takes into consideration Hammett's equation, which relates chemical structure both to equilibrium constants and to rate constants. Two parameters enter in this equation: σ, characteristic only of the substituent—it represents the ability of the group to attract or repel electrons through a combination of inductive and resonance effects; ρ, characteristic of the reaction considered—it measures the sensitivity of this type of reaction to substitution in the parent compound.

Hammett's equation is expressed by the formula

$$\log \left(\frac{k}{k_0}\right) = \rho\sigma$$

where k and k_0 refer to the rate constants for reaction of the derivative and the parent compound, respectively; such constants can be substituted by the equilibrium constants K and K_0; in this case the equation takes the form

$$\log \left(\frac{K}{K_0}\right) = \rho\sigma$$

The value of σ can be obtained almost directly by measuring the effect of the substituent in the ionization constant of the parent compound, because

$$\sigma = \log \left(\frac{K_{COOH}}{K_H}\right)$$

in which K_{COOH} and K_H are the ionization constants of the substituted and nonsubstituted compounds, respectively. If a substituent presents a positive σ value, it means that it attracts electrons; if it has a negative σ value, it means that it donates electrons.

By using the parameters π, σ, and ρ and not taking into account steric factors, which were assumed to be constant, Hansch and his coworkers derived the following equation:

$$\log \left(\frac{1}{C}\right) = -k\pi^2 + k'\pi + \sigma\rho + k''$$

where C is the drug concentration necessary to produce the biological effects, and k, k', k'' are constants for the system being studied, determined through regression analysis of the equations corresponding to the derivative biologically tested in the series.

This equation, with slight adaptations in some cases, was applied to various groups of drugs—chloramphenicol derivatives, penicillins, cephalosporins, fungicides, lincomycin and related antibiotics, sulfamidiç diuretics, hemolytic agents, narcotic agents, acetylcholine and histamine antagonists, anticonvulsants, spasmolytics, barbiturates, sulfonamides, a series of monoamine oxidase inhibitors, antihistaminics, β-haloalkylamines, analgetics of the imidazoline series, and several antibacterial agents. Correlations obtained between the observed and calculated activities were good, about 0.8 to 0.9 in most cases. In some cases, a better correlation was found by certain authors (e.g., Cammarata) who used an equation that considered the volume of the substituent and separated its electronic influence into inductive and resonance effects.

Another use of Hansch's equation and its variants is seen in proposals of mechanisms of action of several types of drugs, for example, the antibacterial tetracyclines and the antimalarial chloroquine.

2. *Polarization Models*

Polarization models are based on the theory of intermolecular forces and are used in cases in which the mathematical model of linear free energy, whatever might be the parameters used, does not lead to an appropriate correlation between biological data and chemical structure. They are seldom used.

3. *Quantum-Chemistry Models*

Quantum chemistry models used to correlate chemical structure with biological activity are based on solutions of the Schroedinger equation and application of these solutions to pharmacological systems. The calculations involved are usually performed by computers, since a number of parameters are considered. Those calculations are based mainly on the following theories: Hückel approximate molecular orbital (HMO = Hückel molecular orbital), extended Hückel theory (EHT), omega technique (ω), complete neglect of differential overlap (CNDO), and perturbative configuration interaction using localized orbitals (PCILO).

Molecular orbital calculations, by one or more of the mentioned methods, have been used especially for the following purposes:

1. To determine interatomic distances and electronic density in molecules of biological interest

2. To study the stereochemistry of macromolecules and the preferred conformation of several biologically active compounds

3. To advance a rational explanation for the activities of certain substances and to present hypotheses for the mechanism of action at molecular and electronic levels of several groups of drugs

4. To propose a topography for the hypothetical receptors of several classes of drugs and, in this way, to deduce indirectly how the drug-receptor interaction occurs at molecular and electronic levels

5. To design new drugs, on a rational basis, that may be more specific and more potent

Molecular orbital calculations have been used to calculate a number of indices of chemical and pharmacological interest: HOMO energy, LEMO energy, transition energy, π electron energy, resonance or delocalization energy, localization energy, electronic density, net charge, frontier electron density, superdelocalizability, bond order, free valence.

Two of frequent use in medicinal chemistry are HOMO and LEMO, that measure the electron donor and electron acceptor abilities, respectively. The larger the HOMO energy, the greater the electron donor properties, because the propensity of the molecule to donate electrons is stronger; conversely, the smaller the LEMO energy, the less the resistance to accept electrons.

Since the energies of molecular orbitals are represented in β units, and β is a negative energetic term (varying from -13 to -112 kcal/mole), the more energetic levels of HOMO are indicated by smaller values in β units, and the less energetic levels, by larger values in β units (smaller, therefore, in absolute value). Thus, if the HOMO energy is between 0 and $+0.5\beta$ in a charge transfer interaction, the molecule can function as donor. In an analogous way, if the LEMO energy is between 0 and -0.5β, the molecule can be acceptor. The values

Table 3.1 Energy Levels of HOMO and LEMO of Some Drugs and Related Substances

Substance	E_{HOMO} (β)	E_{LEMO} (β)
9-aminoacridine	+0.456	−0.371
amodiaquine	+0.566	−0.561
amodiaquine, protonated	+0.679	−0.354
chloroquine	+0.648	−0.549
chloroquine, protonated	+0.803	−0.332
guanine	+0.481	−0.754
menadione	+0.972	−0.228
phenazone	+0.248	−0.956
quinacrine	+0.405	−0.421
quinacrine, protonated	+0.507	−0.201
quinine	+0.647	−0.539
serotonin	+0.461	−0.870

of HOMO and LEMO of some substances are in Table 3.1.

III. DRUG RECEPTORS

A. Nature of Drug Receptors

Some drugs, as discussed in Section I, exhibit biological activity in minute concentrations. For this reason they are described as *structurally specific*. The effect produced by them is attributed to interaction with a specific cellular component known as a *receptor*. As a result of this interaction the drug forms a complex with the receptor. (A chemist refers to the receptor in terms of chemical structural components, but the biologist prefers to treat it in micro-anatomical terms.)

The hypothesis of existence of receptors was advanced because of three remarkable characteristics of drug action:

1. High potency. Drugs are known which act at very low concentrations: 10^{-9} M and even 10^{-11} M.
2. Chemical specificity. Evidence for this is the differences in effects produced by optical isomers. Thus, only one of the four isomers of chloramphenicol is active.
3. Biological specificity. Exemplified by epinephrine, which has marked effect on heart muscle but very weak action on striated muscle.

Experimental evidence seems to indicate that receptors are localized in macromolecules most of which have proteinlike properties and exhibit the specific ability to interact at least with natural substrates at their active sites. Their nature is probably similar to that of the active site or allosteric site of enzymes, and they approximate in size the drug molecule that is able to form a complex with them.

Complexation of a drug with special chemical groups on the receptor results in a sequence of chemical or conformational changes that either cause or inhibit biological reactions. Nowadays it is acknowledged that such changes in biopolymers actually occur as an effect of the action of small molecules. A drug's ability to adapt itself to a receptor depends on the structural, configurational, and conformational characteristics of both drug and receptor.

B. Isolation of Drug Receptors

Several attempts have been made to isolate drug receptors, but success as yet has been equivocal. Difficulties in separating the receptor from tissue proteins are great, because during the process of extraction the forces that unite both entities—drug and receptor—are broken. Concomitantly, owing to changes in the structural shape of the macromolecule of which the receptor is an integral part, the functionality of this macromolecule can be destroyed. Furthermore in the isolation process the receptor undergoes changes in its natural spatial arrangement and charge distribution, and both of these factors are essential to its interaction with the drug. Two basic methods—direct and indirect—have been used in the isolation of receptors.

1. Direct Method

In the direct method, attempts are made to label the functional groups of a receptor with substances able to bind irreversibly, that is, by covalent bonding, and then to isolate the resultant drug-receptor complex. Among the chemical reagents used to form covalent bonds are those that can react with the serine hydroxyl group: phosphorylating agents, sulfonyl fluorides, carbamylating agents, alkylating agents, and N-alkylmaleimides, whose general structures are given, respectively, as follows:

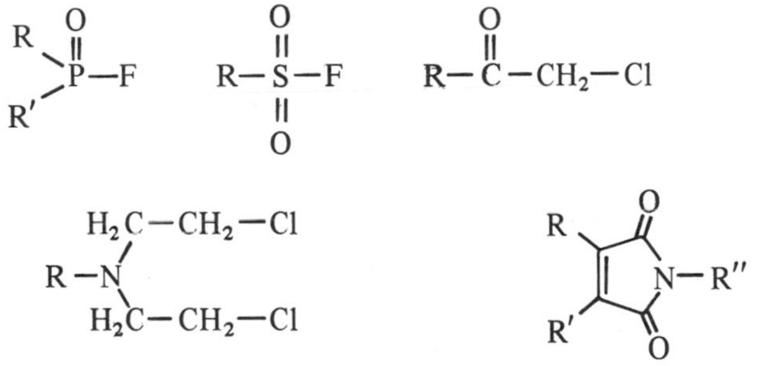

Applied to tissues, the direct method has the inconvenience of being non-specific. Recent attempts with promising results to isolate cholinergic receptors were made by Changeux and colleagues, and Miledi and coworkers.

2. Indirect Method

In the indirect method one tries to identify the macromolecule that contains the receptor through the use of substances able to complex with it reversibly, that is, through weak bonds, and then to isolate the macromolecule and characterize it.

The first attempt to identify the cholinergic receptor by this method was made by Chagas in 1958. Recently, O'Brien and colleagues, Changeux and coworkers, Meunier and collaborators, and De Robertis have tried to reach the same goal, with better success. In 1967, by X-ray diffraction methods, Fridborg and colleagues determined the three-dimensional structure of the complex formed between human carbonic anhydrase C and acetoxymercurisulfonamide, which is a modified inhibitor of this enzyme. Their work gave factual evidence of drug-receptor complexation.

C. Modification of Drug Receptors

Studies have been conducted also with the aim of modifying receptors *in situ,* through physical and chemical means in order to characterize them. Among the former are changes in temperature. Among the latter are alterations in pH, chelating agents, lipid solvents, enzymes, protein denaturating agents, and thiolic reagents.

D. Localization of Drug Receptors

Notwithstanding the great efforts made, accurate and complete topography of receptors is not yet known. This has not prevented, however, the formulation of hypotheses about their structure and stereochemistry. The hypothetical maps of receptors serve very useful purposes, especially the rational explanation of how drugs act and the design of new potential drugs.

Localization of some drug receptors or acceptors has been determined. Most of them are either the active sites or allosteric sites of enzymes or parts of DNA or RNA (Table 3.2). In the case of certain drugs, it is believed that they

Table 3.2 Receptors, Acceptors, or Sites of Action of Some Drugs

Drug	Receptor or Site of Action
acridines	DNA
actinomycins	DNA
alkylating agents	DNA
aminopterin	dihydrofolate reductase
p-aminosalicyclic acid	dihydropteroate synthase
amodiaquine	DNA
anticholinesterases	acetylcholinesterase
antifolate antineoplastics	dihydrofolate reductase
anthracycline	DNA
carcinogenics	DNA
cephalosporins	transpeptidase
chlorguanide	dihydrofolate reductase
chloramphenicol	nucleic acids
chloroquine	DNA
cycloguanil	dihydrofolate reductase
cyclohexemide	nucleic acids
cycloserine	alanine racemase and D-alanyl-D-alanine synthetase
dapsone	dihydropteroate synthase
daunomycin	DNA
epinephrine	adenyl cyclase
erythromycin	nucleic acids

eserine	acetylcholinesterase
ethidium bromide	DNA
fluorouracil	thymidylate synthetase
fluxuridine	thymidylate synthetase
fusidic acid	nucleic acids
hycanthone	DNA
idoxuridine	DNA
kanamycin	nucleic acids
lincomycin	nucleic acids
lucanthone	DNA
MAO inhibitors	monoamine oxidase
mercurial diuretics	carbonic anhydrase
methotrexate	dihydrofolate reductase
mitomycin	DNA
nalidixic acid	DNA
nogalomycin	DNA
organophosphorus insecticides	acetylcholinesterase
penicillins	transpeptidase
porphiromycin	DNA
proflavine	DNA
puromycin	nucleic acids
pyrimethamine	dihydrofolate reductase
quinine	DNA
quinoline antimalarials	DNA
rifampin	RNA polymerase
rifamycins	RNA polymerase
schistosomicidal antimonials	phosphofructokinase
streptomycin	nucleic acids
sulfonamides	dihydropteroate synthase
sulfones	dihydropteroate synthase
sympathomimetics	adenyl cyclase
tetracyclines	nucleic acids
trimethoprim	dihydrofolate reductase
xanthines	phosphodiesterase

act either by intercalation between DNA base pairs, as in the case of chloroquine, or by alkylating and cross-linking of DNA strands, as in the case of mitomycin (Fig. 3.1).

E. Structure

In spite of the little that is known about the subject, it is generally accepted that a receptor is an elastic three-dimensional entity, consisting perhaps in most cases of protein-constituent amino acids, whose stereochemical structure is often complementary to that of the drug and which, sometimes after under-

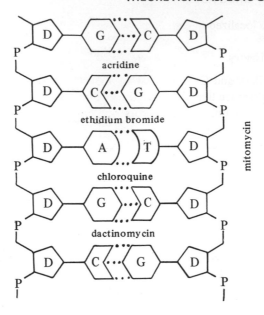

Figure 3.1 Examples of drugs that act by intercalation between base pairs of DNA or by apposition, that is, by alkylating and cross linking of DNA strands:

D = deoxyribose   P = phosphodiester group
G = guanine   A = adenine
C = cytosine   T = thymine

going conformational change, is able to interact with it, usually in its preferred conformation, in order to form a complex held together by various binding forces. As result of this drug-receptor complexation a stimulus is generated and, in turn, causes a biological action or effect.

IV. THEORIES OF DRUG ACTION

A. Nature of Pharmacological Action

The action of drugs results either from their physicochemical properties, as in the case of structurally nonspecific drugs, or from their chemical structure, as in structurally specific drugs. The former, exemplified by general anesthetics and certain hypnotics (*e.g.,* aliphatic alcohols), act in relatively large doses, presumably by forming a monomolecular layer over the whole area of certain cells of the organism. The structurally specific drugs, however, act in very small doses, and it is deduced that their activity results from complexation with specific receptors localized in certain molecules of the organism.

B. Occupancy Theory

Formulated by Clark and Gaddum, the occupancy theory (also called template theory) states in essence that the intensity of pharmacological effect is directly proportional to the number of receptors occupied by the drug. According to this theory, drug-receptor interactions, which comply with the law of mass action, may be represented by the equation where R is a receptor, D a molecule

$$R + D \underset{k_2}{\overset{k_1}{\rightleftharpoons}} RD \longrightarrow E$$

of the drug, RD the drug-receptor complex, E the pharmacological effect, and k_1 and k_2 the rate constants of adsorption and desorption, respectively.

The number of occupied receptors depends on the concentration of the drug in the compartment of the receptor and on the total number of receptors in the unity of area or of volume. The drug effect becomes much more intense as the number of occupied receptors increases; hence maximal action corresponds to the occupation of all receptors.

1. Affinity and Intrinsic Activity

In contradiction to the occupancy theory, not all agonists of a given class of drugs—one example is the alkyltrimethylammonium series of acetylcholine congeners—elicit the same maximal response. Furthermore, this theory does not explain why some drugs act as agonists and other drugs with similar structures act as antagonists.

With the purpose of offering an explanation for this and other incongruences, Ariëns and Stephenson proposed some modifications to the theory of occupancy. According to these authors, drug-receptor interactions comprise two stages: (a) complexation of the drug with its receptor and (b) production of effect. For a chemical compound to manifest biological activity it is necessary not only that it have *affinity* for the receptor, owing to complementary structural characteristics, but also another property called *intrinsic activity* by Ariëns and *efficacy* by Stephenson. This latter property, intrinsic activity or efficacy, would be a measure of the ability of the drug-receptor complex to produce the biological effect.

According to the Ariëns-Stephenson theory, agonists as well as antagonists have strong affinity for the receptor, and this enables them to form the drug-receptor complex. However, only agonists have the ability of giving origin to the stimulus—that is, have intrinsic activity, or efficacy, while antagonists are drugs that bind strongly to the receptor—that is, they have great affinity for it but are devoid of activity. It is possible to transform an agonist into an antagonist through appropriate structural modifications, such as the addition or removal of certain chemical groups. Antagonists, in comparison with agonists, usually have more nonpolar, bulky groups that help to establish a stronger interaction with receptors, as was already seen in Chapter 2 (Figure 2.6).

Rang and Ritter have shown that agonists may change the molecular structure of the receptor and that this change increases the affinity of the receptor toward some antagonists. To this phenomenon they gave the name *metaphilic effect*.

In spite of its appeal, the occupancy theory, even with the additions introduced by Ariëns and Stephenson, cannot explain satisfactorily why drugs vary in their type of action, that is, why one acts as an agonist and another as an antagonist, although both can occupy the same receptor, as the theory assumes. Its failure to elucidate the mechanism of drug action at the molecular level in terms of chemical structure is the chief deficiency of the theory of occupancy. Furthermore, mathematical analysis has shown that drug action cannot be explained by simple receptor-occupation models.

2. *Charnière Theory*

In order to explain why an agonist, although not being able to remove an antagonist from the receptor site, can compete with it according to the law of mass action, Rocha e Silva proposed the so-called charnière theory. It is based on Ariëns and Simonis's hypothesis that in the pharmacological receptor there are two sites: (*a*) specific or critical site, which interacts with the pharmacophoric groups of the agonist; (*b*) nonspecific or noncritical site, which complexes with the nonpolar groups of the antagonist.

According to the charnière theory, both agonist and antagonist fix to the specific site through weak reversible bonds, but the antagonist binds also, and strongly, through hydrophobic and van der Waals interactions, as well as charge transfer, to the nonspecific site. Competition between agonist and antagonist occurs at the specific site of the receptor. Since antagonist is complexed steadily with the nonspecific site of the receptor, even an excess of agonist is unable to dislodge it from there. After removal of the excess agonist, blockade caused by the antagonist returns to the same level as before the addition of the agonist. To his charnière theory Rocha e Silva gave a thermodynamic approach. The phenomenon he described is general. It occurs, for instance, in the competition between diphenhydramine and histamine, atropine and histamine, (+)-tubocurarine and neostigmine, and methantheline and acetylcholine (Figure 3.2).

C. Rate Theory

On the basis of the postulate of Croxatto and Huidobro that a drug is efficient

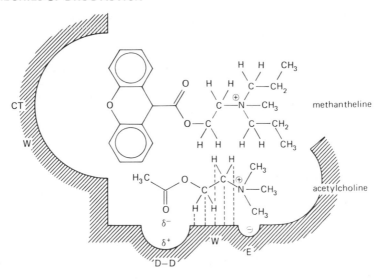

Figure 3.2 Competition between methantheline and acetylcholine for specific site of muscarinic receptor. Bonds involved in the interaction are W, van der Waals and hydrophobic; CT, charge transfer; D-D, dipole-dipole; E, electrostatic. Although firmly bound by nonpolar part of the nonspecific site, antagonist can compete with agonist for specific site of receptor. Adapted from M. Rocha e Silva, *Eur. J. Pharmacol.,* 6, 296 (1969).

only at the moment of encounter with its receptor, Paton and co-workers have advanced a rate theory.

According to Paton, activation of receptors is proportional not to the number of occupied receptors but to the total number of encounters of the drug with its receptor per unit time. Unlike previous theories, the rate theory does not require formation of a stable Michaelis-Menten complex for activation of the receptor by a drug. In accordance with the theory, pharmacological activity is a function only of the *rate* of association and dissociation between molecules of drug and receptor and not of formation of a stable drug-receptor complex. Each association constitutes a quantum of stimulus for the biological interaction.

In the case of agonists, the rates both of association and dissociation are fast (the latter faster than the former) and produce several impulses in unity of time. When we consider antagonists, however, the rate of association is fast but that of dissociation is slow, which explains their pharmacological action. This has some experimental basis, because it has been shown that, before causing blockade, antagonists produce a short stimulating effect. In short, agonists are characterized by high (and variable) dissociation rates, partial agonists by intermediate dissociation rates, and antagonists by low dissociation rates as a consequence of stronger adherence to the receptor and of greater difficulty in being withdrawn from it because they are larger in size when compared with agonists and partial agonists.

As in the case of the occupancy theory, the rate theory has been widely criticized because it presents several inconsistencies, and it does not allow interpretations of various experimentally observed facts. For instance, contrary to what Paton assumes, the agonist has characteristics that favor the formation of a complex that does not rapidly dissociate. Further, the rate theory, as well as the occupancy theory, cannot explain, through interpretation of phenomena that take place at the molecular level, why one drug acts as an agonist while another structurally similar one acts as an antagonist. Both theories, as well as many others that have been proposed, lack a plausible physicochemical basis for the interpretation of phenomena involving receptors at the molecular level.

In an effort to refute criticisms of the rate theory, Paton and Rang proposed as an alternative the *dissociation theory*. In this new theory the dissociation-rate constant is a function not of the intensity of the binding forces but of the extent to which the drug molecule disturbs the secondary protein structure. Relating stimulus to rate of dissociation, and this rate being proportional to the occupation of receptors, the dissociation theory is not formally different from the occupancy theory.

D. Induced-Fit Theory

As it is applied to drug-receptor interaction, the induced-fit theory is based on the hypothesis, for which recent evidence is being accumulated, of induced conformational changes in enzymes. This hypothesis has been advanced by a number of authors. For example, Koshland suggested that the active site of an isolated crystalline enzyme does not necessarily need to have a morphology that is complementary to that of the substrate as a kind of negative to it, but that it acquires such morphology only after interacting with the substrate, which induces such a conformational change. He assumed that the active site of the enzyme is flexible, or, better, plastic or elastic, and not rigid; that is, not only can it be deformed or altered but it also has the ability to return to the original form after being deformed (Figure 3.3).

According to the induced-fit theory the biological effect produced by drugs results from activation or deactivation of enzymes, or even of noncatalytic proteins, through a reversible perturbation or change in tertiary structure of enzymes or proteins. Conformational change is not restricted, however, to proteins. Drugs which have flexible structure can also undergo conformational change as they approach the site of action or the receptor site. Drug-receptor interaction, therefore, can be viewed as a dynamic, and in most cases reversible, topographical and electronic compromise or accomodation between drug and receptor that triggers the stimulus which leads to the biological effect (Figure 3.4).

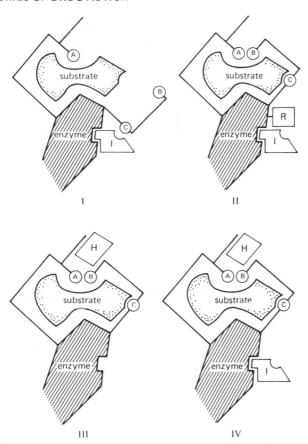

Figure 3.3 The effect of inhibitors and activating agents on the elastic active site of the enzyme. In diagram I the inhibitor I attracts binding group C and prevents proper alignment of a catalytic group B, causing inhibition, which may be competitive, if the B chain is involved in complexation or noncompetitive if it is not involved. In diagram II reagent R prevents juxtaposition of group C with inhibitor I, nulifying its effect without changing its affinity for the active site. In diagram III the hormone H stabilizes the active conformation by attracting chains containing A and B. In diagram IV the hormone H overcomes the effect of the inhibitor I by attracting chains containing catalytic groups A and B. [After D. E. Koshland, *Federation Proc.*, 23, 719 (1964)].

Recently Koshland and coworkers have formulated a variant of the induced-fit theory in order to explain cooperative effects: the phenomenon that binding of one ligand molecule somehow accelerates binding of subsequent ones. Thus, binding of the first ligand to a polymeric protein, which may be an enzyme containing a receptor, induces a conformational change in one of its subunits. A change in the shape of this subunit affects the stability of the remaining

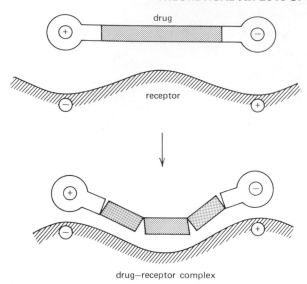

Figure 3.4 Schematic representation of the morphological and charge-induced fit in drug-receptor interaction.

subunits. The resulting energy of stabilization makes possible stronger binding of the next molecules. Analogous theories have been advanced by Changeux and colleagues, Wyman, and Noble.

E. Macromolecular Perturbation Theory

Very similar to the induced-fit theory, the macromolecular perturbation theory was proposed by Belleau in 1964. It may be considered an application of that theory to some classes of drugs.

Taking into account the conformational adaptability of enzymes (and the receptors would be enzymes of a particular sort), Belleau reasoned that in the interaction of a drug with the protein component two general types of perturbations can occur in the complex:

1. Specific conformational perturbation, or specific ordering, which makes possible the adsorption of certain molecules related to the substrate; this is the case of an agonist.

2. Nonspecific conformational perturbation, or nonspecific disordering, which may serve to accommodate other classes of extraneous molecules; this is the case of an antagonist.

If the drug has both characteristics—that is, if it contributes to specific as well as to nonspecific macromolecular perturbation—a mixture of two complexes will result. This explains the partial stimulating action of the drug or the case of the partial agonist or antagonist.

Belleau's hypothesis does not need to assume the existence of affinity and intrinsic activity, and it is in complete agreement with several subsequent experimental data and results, because it offers a plausible physicochemical basis for the explanation of phenomena that involve a receptor at the molecular level.

V. MECHANISM OF DRUG ACTION

Many attempts have been made to propose a general theory of mechanism of drug action. This desideratum, however, becomes more and more remote as new knowledge accumulates. But, although there are a number of drugs whose mechanism of action cannot be included in the following classification, most drugs act at the molecular level by one of the following mechanisms: activation or inhibition of enzymes; suppression of gene function; metabolic antagonism; chelation; alteration of biological membrane permeability; and nonspecific action.

Several drugs act by various mechanisms. There are also drugs whose mechanism of action can be classified in two or three of the foregoing classes; a typical example is the sulfonamides or antifolics: they can be considered either as inhibitors of enzymes or as metabolic inhibitors.

A. Action of Drugs on Enzymes

Drugs acting on enzymes can either activate or inhibit them.

1. Activation of Enzymes

Drugs that can supply inorganic ions act by activation of enzymatic systems. This process can occur in two ways: the ion can interact with an enzyme inhibitor preventing enzyme deactivation, or it can interact with an enzyme directly altering its conformation and charge.

Other types of drugs increase enzymic activity through an induced adaptation mechanism. This phenomenon acquires special relevance in microbial systems. A classical example is the activation of penicillinase induced by penicillin itself.

2. Inhibition of Enzymes

In biochemistry the effect produced by an inhibitor is called a *biochemical lesion*. It refers to any disturbance of metabolism by agents acting directly on metabolic systems.

The inhibition caused by drugs can be *reversible* or *irreversible*. It is reversible when it is characterized by an equilibrium between the enzyme and the inhibitory drug. It is irreversible when it increases with time, provided that the inhibitory drug is present in excess.

There are two main types of inhibition: competitive and noncompetitive.

Competitive Inhibition

In competitive inhibition the drug competes with the substrate for the same site of the enzyme and combines with it reversibly. In this process, therefore, the relative concentrations of the substrate and of the drug are of fundamental importance, because they determine the degree of inhibition. Actually in the presence of excess substrate the drug is displaced from the receptor, which is then occupied by the substrate.

Noncompetitive Inhibition

In noncompetitive inhibition the drug combines with the enzyme or with an enzyme-substrate complex with equal ease but at a site different from that to which the substrate is attracted. This indicates that the inhibitor binds itself to different sites on the enzyme, and not to the catalytic center, that is, the active site. Such inhibition, which is usually reversible and not affected by substrate concentration, depends solely on drug concentration and on the dissociation constant K_1 of this inhibitor; a substrate never displaces the inhibitor even at high concentrations. Though purely noncompetitive inhibition is very unusual, the effect of isoflurophate on cholinesterase action is an example. Presently noncompetitive inhibition is considered as being related to allosteric phenomena.

3. Allosteric Inhibition

The classical concept of enzymic inhibition by an antimetabolite can be represented by the following scheme:

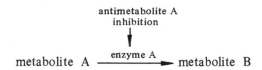

The antimetabolite is similar in structure to a given metabolite, and this characteristic of complementarity allows it to combine with the active site of the enzyme, altering the enzyme-substrate complex dissociation. This mechanism is valid for enzymes in general with the exception of *allosteric* enzymes, so called for having a binding site other than the active site.

Owing to the unusual kinetic and structural characteristics of allosteric enzymes, Monod and co-workers have proposed the following model for them:

1. All are polymers, made up of one or more identical subunits and, therefore, can exist in at least two different conformational states.

2. Each one of the identical subunits has a single catalytic site, specific for the substrate, and a separate allosteric site for each allosteric effector (inhibitor or activator).

3. For each conformational state the catalytic and allosteric sites have equal affinities for their respective ligands.

4. The various conformational states of the enzyme are in mutual dynamic equilibrium.

5. The transition from one state to another occurs with simultaneous alterations in all the identical subunits within a particular molecule.

Based on this model, a new concept of inhibition has emerged recently. It was shown that the enzyme can also be inhibited by chemical substances that have no structural similarity with the substrate. These substances are called *allosteric inhibitors*. They exert their action either by competing directly with activator substances for regulatory sites or by causing conformational changes, which result in decreased affinity for the substrates by catalytic sites. An example of allosteric inhibition called *feedback* or *terminal product inhibition* is represented by the following diagram:

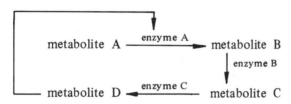

In a series of reactions catalyzed by enzymes only the first enzyme A is inhibited, due to the accumulation of the terminal metabolite D. The interaction of the inhibitor, the metabolite D, with the enzyme A is not necessarily at the same site that interacts with the substrate, that is, with its active site; in fact it usually is with another, the regulatory site, to which the name *allosteric site* is given.

Therefore, the allosteric inhibitor does not need to have any chemical resemblance to the substrate, because the allosteric site and the catalytic site are located in different portions of the enzyme. Interaction of the inhibitory drug with the allosteric site results in an alteration in the conformational state of the enzyme, which adopts a form in which its affinity, at the catalytic site, for the substrate is decreased. Hence, with the purpose of obtaining antagonists of the enzyme A, derivatives of the metabolite D can be synthesized instead of structural analogues of the metabolite A.

4. Examples of Enzyme Inhibitors

Among many other types of drugs that act as enzyme inhibitors may be cited (a) some anticholinesterase agents—inhibitors of acetylcholinesterase; (b) certain antidepressants—inhibitors of monoamine oxidase; (c) a number of diuretics—inhibitors of carbonic anhydrase; (d) the salicylates; (e) several antibiotics.

Pharmacodynamic Agents

Various pharmacodynamic agents inhibit biochemical processes of fundamental importance. Among many others are the following: inhibitors of biosynthesis and metabolism of epinephrine, inhibitors of biosynthesis and metabolism of serotonin, inhibitors of biosynthesis of histamine, inhibitor of biosynthesis of uric acid (see Chapter 2, Section IV. E).

Chemotherapeutic Agents

Penicillins and cephalosporins owe their action to inhibition of transpeptidase, an enzyme catalyzing the cross-linking reaction of linear polymers that comprise the bacterial cellular wall. The action of cycloserine is also on the bacterial wall, but through inhibition of alanine racemase and D-alanyl-D-alanine synthetase, enzymes involved in the formation of dipeptide to complete the pentapeptide side chain of the cellular wall. Antimonials having schistosomicidal action inhibit the phosphofructokinase of the parasites.

Heavy metals, such as Hg^{2+}, Cu^{2+}, Ni^{2+}, Pb^{2+}, Zn^{2+}, Co^{2+}, Cd^{2+}, Mn^{2+}, Mg^{2+}, Ca^{2+}, and Ba^{2+}, can cause inhibition of enzymes, often with highly toxic effects to the human or animal body. The poisoning subsequent to interaction of metallic ions with enzymes results from their complexation with one or more of the following groups existing in any living cell and in most enzymes: $-OH$, $-COOH$, $-PO_3H_2$, $-SH$, $-NH_2$, imidazole ring.

B. Drugs Acting as Suppressors of Gene Function

Many drugs act as suppressors of gene function. Most of them are chemotherapeutic agents, found among antibiotics, fungicides, antimalarials, trypanocides, antineoplastics, antivirals.

Gene function can be suppressed in several steps of protein biosynthesis (Figure 3.5). Suppressors of gene function can act as (a) inhibitors of biosynthesis of nucleic acids; (b) inhibitors of protein synthesis.

1. Inhibitors of Biosynthesis of Nucleic Acids

Many substances show strong activity as inhibitors of biosynthesis of nucleic acids. Most of them are exceedingly toxic, because they lack selectivity. They interact with biochemical processes both of parasite and host. Some of them,

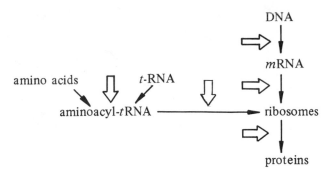

Figure 3.5 **Phases of protein biosynthesis liable to inhibition. The short arrows indicate the possible sites of attack by chemotherapeutic agents, mainly antibiotics**

however, are used as chemotherapeutic agents.

Inhibitors of biosynthesis of nucleic acids can be divided into two groups: (*a*) those that interfere with the biosynthesis of nucleotide precursors; (*b*) those that interfere in the polymerization of nucleotides in nucleic acids.

To the first group belong certain analogues of amino acids, folic acid, purines, and pyrimidines and their respective nucleosides. For instance: azaserine, DON, azotomycin, fluorouracil, azauracil, fluororuridine, idoxuridine, azauridine, cytarabine, mercaptopurine, azathioprine, thioguanine, azaguanine, psicofuranine, aminopterin, methotrexate. They act by metabolic antagonism.

In the second group can be included several antibiotics and other chemotherapeutic agents, such as certain antineoplastics, antibacterials, antimalarials, trypanocides, and schistosomicides, which act (*a*) either by intercalation or apposition in the nucleic acids or (*b*) by inhibition of enzymes involved in nucleic acid synthesis.

Daunomycin, dactinomycin, ethidium, proflavine, quinacrine, chloroquine, lucanthone, hycanthone, and several other chemotherapeutic agents complex intercalation between pairs of DNA bases (Figure 3.1).

Alkylating agents, such as nitrogen mustards, aziridines, methanesulfonate esters, epoxides, mitomycins, and several other substances complex with nucleic acids by apposition, by forming a crossed linkage with the DNA double strand (Figures 3.1 and 3.6).

Rifamycins act by inhibition of RNA synthesis, interfering specifically with RNA polymerase function of sensitive bacterial cells.

Nalidixic acid inhibits selectively DNA synthesis of pathogenic microorganisms.

2. *Inhibitors of Protein Synthesis*

Several chemotherapeutic agents owe their activity to inhibition of protein

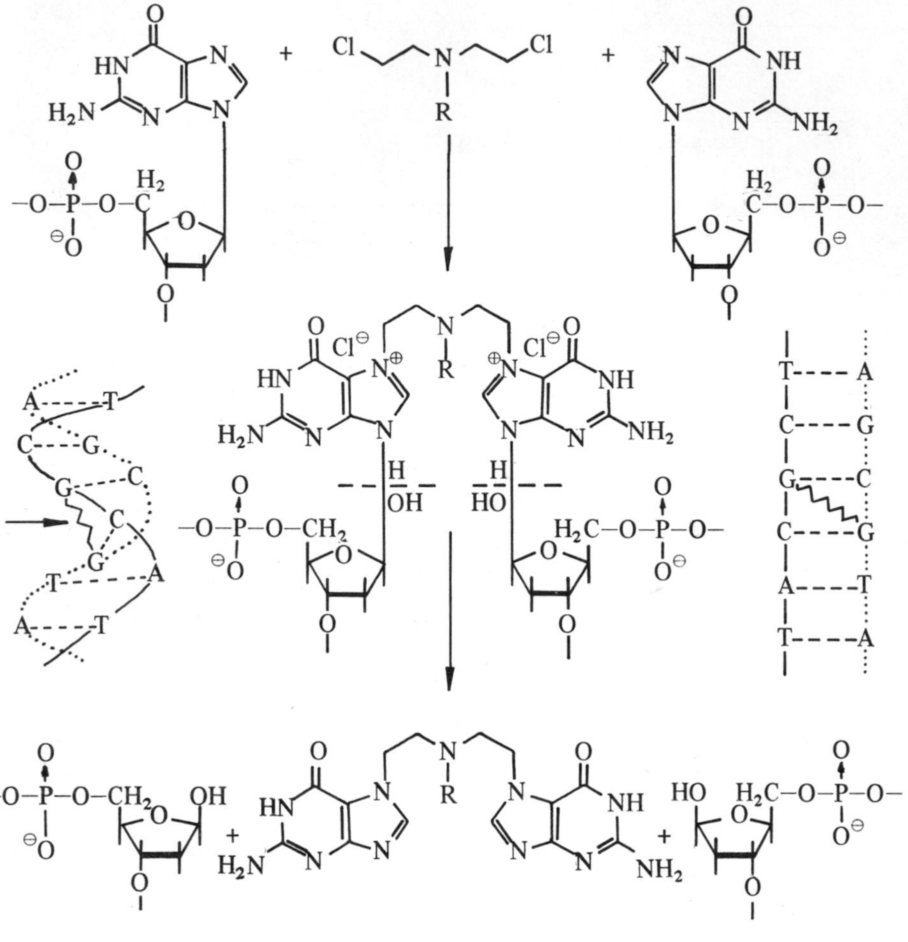

Figure 3.6 Crossed linkage of guanine bases of DNA twin strands by a bifunctional alkylating agent followed by depurination and excision of a *bis*(guanin-7-yl) derivative of the alkylating agent. [After P. D. Lawley and P. Brookes, *J. Mol. Biol.*, **25**, 143 (1967).]

biosynthesis of parasites, interfering in this way with the translation of genetic message. A well-studied case is that of puromycin. Owing to its structural resemblance to the terminal aminoacyl adenosine moiety of *t*RNA, it can interrupt the translation of the genetic code. Unfortunately, it is too toxic for use in therapeutics.

Among many other antibiotics that inhibit protein biosynthesis may be cited streptomycin, neomycin, kanamycin, tetracyclines, chloramphenicol, lincomycin, erythromycin, fusidic acid, and cycloheximide. The site of action of these and other antibiotics is shown in Figure 3.7.

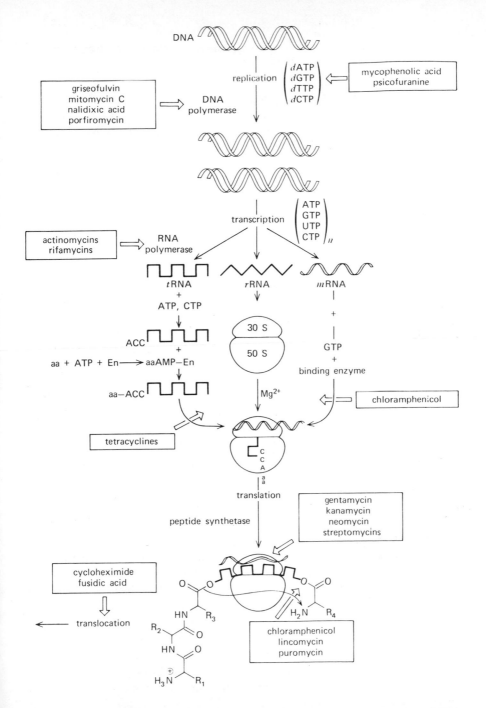

Figure 3.7 Site of action of some antibiotics and other drugs. The arrows indicate inhibition of specific reactions in replication, transcription, translation, and translocation in molecular biology processes.

C. Antimetabolites

The strategy of chemotherapy consists in exploring the morphological and biochemical differences between the invading cell or parasite and the cells of the host. Although cells of all living organisms are morphologically and biochemically very similar, some differences have been found, such as the following:

1. The parasite and the host present clear morphological differences. For instance, bacterial cells have cellular walls, but animal cells do not. This explains the highly selective and nontoxic action of penicillins.

2. The parasite can use essential enzymatic systems not used by the host and vice versa. For example, bacteria sensitive to sulfonamides must synthesize their own folic acid from p-aminobenzoic acid, but mammals receive already preformed folic acid from nutrients.

3. An enzyme may not be identical in all species or even in all tissues of the same species. For instance, phosphofructokinase from schistosomae is 80 times more sensitive to antimonials than the same enzyme from mammals.

4. The host can detoxify the antimetabolite by a route not available to the invading cell. This difference explains the antineoplastic action of mercapto-purine, because some tumors which are deficient in xanthine oxidase cannot detoxify it to thiouric acid, whereas in host cells this oxidation does occur.

5. The invading cell has a hyperactive transport system, which allows it to increase the concentration of certain substances in its interior. Thus, some leukemic cells have this system for folic acid. The concentration of folic acid rises inside the cells, making them more sensitive to methotrexate, an anti-metabolite of this acid.

6. The parasite or invading cell can convert enzymatically the antimetabolite to a lethal form, whereas the host does not have this ability. For this reason drug metabolism in the first case receives the name *lethal synthesis*.

7. The parasite can have a ribosome different from the mammalian ribosome. In fact, bacterial ribosomes sediment at 70 S, being composed of 30 S and 50 S subunits, whereas mammalian ribosomes sediment at 80 S, and they are comprised of 40 S and 60 S subunits, although ribosomes of mammalian mitochondria are similar to bacterial ribosomes. This explains the selective action of some antibiotics, such as streptomycin, kanamycin, neomycin, viomycin, chloramphenicol, erythromycin, oleandomycin, lincomycin, that affect only 70 S ribosomes. In this way they interfere almost exclusively with the translation process in the protein biosynthesis of bacteria. Some other antibiotics—for example, puromycin, tetracyclines, fusidic acid, cycloheximide—affect 70 S as well as 80 S ribosomes and, for this reason, they are not so selective in their toxic action.

1. Classical and Nonclassical Antimetabolites

Classical antimetabolites are those that result from isosteric replacement of the structure of essential metabolites. They act by taking the place of essential metabolites in biochemical processes of living organisms through complexation with the *active* site of enzymes. Examples are allopurinol, fluorouracil, and methotrexate; they are structural analogues, respectively, of hypoxanthine, uracil, and folic acid.

Nonclassical antimetabolites present remote or no structural resemblance to that of essential metabolites. Their action results from complexation with the *allosteric* site of enzymes. Examples are pyrimethamine and cycloguanil. Although bearing only a vague similarity to folic acid, the substrate of dihydrofolate reductase, the two drugs act as inhibitors of this enzyme, and, for this reason, they find use as antimalarials.

2. Mechanism of Action

Metabolic antagonists seem to act in two different ways: either by interfering in the synthesis of the metabolite or by preventing its use. The first is a case of *noncompetitive inhibition*. For instance, administration of bishydroxycoumarin to animals causes the typical syndrome of avitaminosis K_1; this deficiency is overcome by high doses of this vitamin. The second case is one of *competitive inhibition*. The classical example is that of the sulfonamides, which take the place of *p*-aminobenzoic acid in the biosynthesis of the coenzymically active form of folic acid.

3. Examples of Antimetabolites

Many compounds have been synthesized with the aim of obtaining antimetabolites. However, only a few of them have proved to be useful, and these are being used in therapeutics. On the other hand, many other drugs, such as the antihistaminics, a few anticoagulants, several cholinergic agents, certain adrenergic agents, some antibiotics, although having structural similarities to that of metabolites, have not resulted from the application of the antimetabolite concept, notwithstanding the fact that such an interpretation may be given. Neither are antimetabolites the antagonists of pharmacodynamic agents, such as nalorphine, which antagonizes morphine, and bemegride, which is an antagonist of barbiturates.

Among several others, may be cited the following examples of antimetabolites: antagonists of amino acids, antagonists of purines and pyrimidines, and antagonists of vitamins:

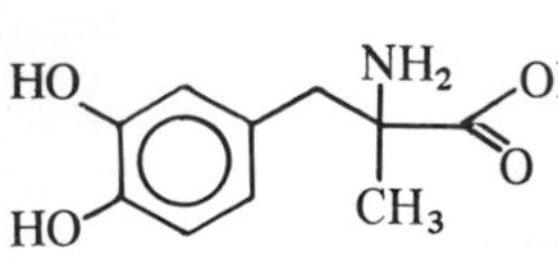

methyldopa
(antagonist
of dopa)

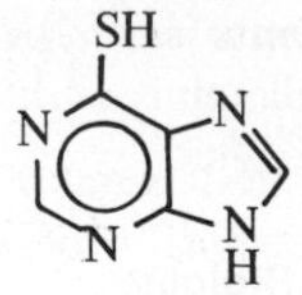

mercaptopurine
(antagonist of
hypoxanthine)

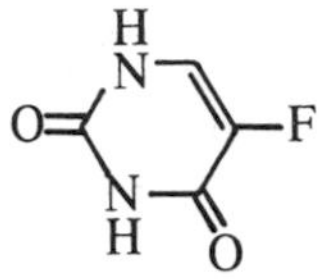

fluorouracil
(antagonist
of uracil)

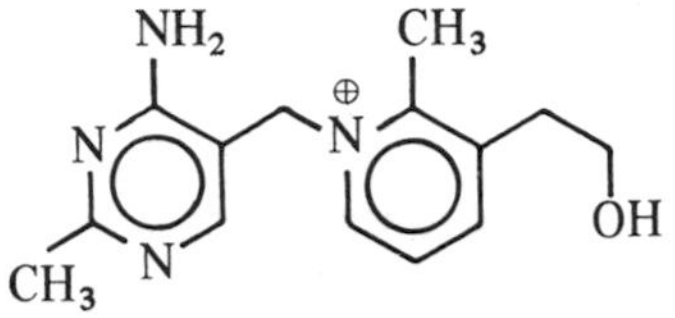

pyrithiamine
(antagonist
of thiamine)

D. Chelating Agents

Chelating agents are those substances which have the property of combining
with a metallic ion by donating a pair of electrons and thus forming annular
compounds, or chelates, usually of five or six members. For instance, oxine is
used as a chemotherapeutic agent, owing to its ability to chelate iron, a metal
essential to the metabolism of certain microorganisms.

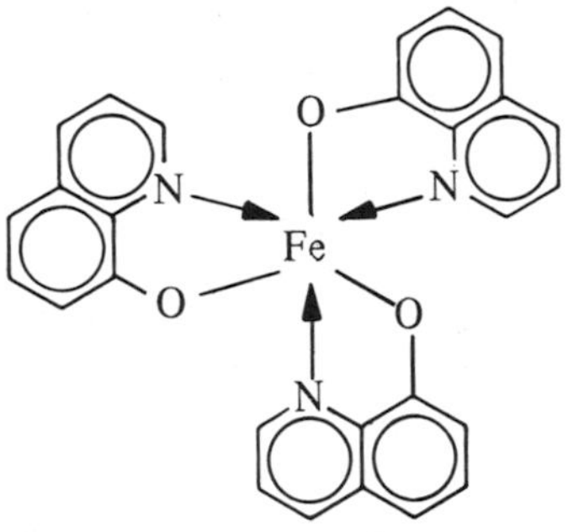

3:1-oxine-ferric chelate
saturated (inactive)

Several other drugs have the property of forming chelates and owe their
biological action partially or totally to this property: tetracyclines, hexachlorophene, salicylic acid, thiouracil, thiosemicarbazones, histamine, epinephrine,
norepinephrine, and isoniazid.

Some chelating agents are used as *antidotes* in poisoning by metallic ions. Dimercaprol, D-penicillamine, desferrioxamine-B, ethylenediaminetetraacetic acid, and *trans*-1,2-diaminocyclohexanetetraacetic acid are some examples.

E. Drugs Acting on Biological Membranes

1. *Action of Drugs on Membranes*

Several drugs act on cellular membranes, modifying their physiological role and producing, in consequence, pharmacological effects. To this group belong some drugs of topical use, such as antiseptics and polyene antibiotics.

Antiseptics

Although not equivalent, the terms antiseptics, sterilizants, disinfectants, and biocides are used as synonyms. Many antiseptics do not kill bacteria but only prevent their multiplication: organic defenses represented by antibodies and phagocytes eliminate the focus of infection. They act primarily on cytoplasmic membrane. Examples are phenols (hexachlorophene, used until recently but now restricted), cationic antiseptics (chlorhexidine), and polypeptide antibiotics (polymyxin, gramicidin S, tyrocidins, colistin).

Polyene Antibiotics

These antibiotics, such as nystatin and amphotericin B, increase membrane permeability. They do not act on bacteria but on fungi (*Candida albicans,* for example), apparently because they have affinity for sterols, which are present in the membranes of fungi and other higher organisms but not in bacterial membranes.

2. *Action of Drugs on Transport Systems*

There are several mechanisms of transport of substances across cellular membranes: passive diffusion, active transport, facilitated diffusion. Some drugs owe their action to interference with one or more of these mechanisms. For instance, insulin facilitates the diffusion of hexoses and amino acids in some tissues. Copper ions decrease the facilitated diffusion of glucose.

F. Nonspecific Action of Drugs

The action of structurally nonspecific drugs, such as biological depressors, a class to which certain hypnotics, the general anesthetics, and the volatile insecticides belong, does not derive from their interaction with specific receptors but results from their physicochemical properties. It seems that their action originates from the accumulation of such drugs at some point of vital importance to the cell, with consequent disorganization of a chain of metabolic processes.

REFERENCES

General

A. Goldstein, L. Aronow, and S. M. Kalman, *Principles of Drug Action*, 2nd ed., Wiley-Interscience, New York, 1974.

A. Korolkovas, *Fundamentos de Farmacologia Molecular: Base para o Planejamento de Farmacos*, 2^a ed. revista, EDART and Universidade de São Paulo, São Paulo, 1974.

A. Korolkovas, *Actual. Chim. Ther.*, 2ème serie, 13 (1974).

A. Albert, *Selective Toxicity*, 5th ed., Wiley, New York, 1973.

B. A. Callingham, *Biochemical Pharmacology*, Wiley, New York, 1973.

C. J. Cavallito, Ed., *Structure-Activity Relationships*, Vol. I, Pergamon, Oxford, 1973.

S. Dikstein, Ed., *Fundamentals of Cell Pharmacology*. Thomas, Springfield, Ill., 1973.

R. M. Featherstone, Ed., *A Guide to Molecular Pharmacology-Toxicology*, Dekker, New York, 1973.

Z. M. Bacq, Ed., *Fundamentals of Biochemical Pharmacology*, Pergamon, London, 1971.

B. B. Brodie and J. R. Gillette, Eds., "Concepts in Biochemical Pharmacology," Part 1, *Handbuch der experimentellen Pharmakologie*, Vol. XXVIII/1, Springer, Berlin, 1971.

D. J. Triggle, *Neurotransmitter-Receptor Interactions*, Academic, London, 1971.

A. Korolkovas, *Essentials of Molecular Pharmacology: Background for Drug Design*, Wiley-Interscience, New York, 1970.

E. J. Ariëns, Ed., *Molecular Pharmacology*, Vol. I, Academic, New York, 1964.

R. B. Barlow, *Introduction to Chemical Pharmacology*, 2nd ed., Methuen, London, 1964.

W. C. Holland, R. L. Klein, and A. H. Briggs, *Introduction to Molecular Pharmacology*, Macmillan, New York, 1964.

Types of Drug Action

A. Felmeister, *J. Pharm. Sci.*, **61**, 151 (1972).

A. Korolkovas, *Rev. Port. Farm.*, **22**, 224 (1972).

A. L. Lehninger, *Bioenergetics: The Molecular Basis of Biological Energy Transformations*, 2nd ed., Benjamin, New York, 1971.

R. E. Sonntag and G. J. Van Wylen, *Introduction to Thermodynamics: Classical and Statistical*, Wiley, New York, 1971.

J. C. Dearden and E. Tomlinson, *J. Pharm. Pharmacol.*, **22**, Suppl. 53S (1970).

T. Higuchi and S. S. Davis, *J. Pharm. Sci.*, **59**, 1376 (1970).

W. Klyne and P. M. Scopes, *Farmaco, Ed. Sci.*, **24**, 533 (1969).

R. J. Pinney and V. Walters, *J. Pharm. Pharmacol.*, **21**, 415 (1969).

A. H. Beckett, *Progr. Drug Res.*, **1**, 455 (1959).

J. Ferguson, *Proc. Roy. Soc. (London)*, Ser. B, **127**, 387 (1939).

Physicochemical Parameters and Pharmacological Action

P. H. Doukas, "The Role of Charge-Transfer Processes in the Action of Bioactive Materials," in E. J. Ariëns, Ed., *Drug Design*, vol. V, Academic, New York, 1975, pp. 133-167.

B. Hetnarski and R. D. O'Brien, *J. Med. Chem.*, **18**, 29 (1975).

A. Leo *et al.*, *J. Med. Chem.*, **18**, 865 (1975).

A. Cammarata and T. M. Bustard, *J. Med. Chem.*, **17**, 981 (1974).

F. Darvas, *J. Med. Chem.*, **17**, 799 (1974).

W. J. Dunn III and C. Hansch, *Chem.-Biol. Interactions*, **9**, 75 (1974).

T. K. Lin *et al.*, *J. Med. Chem.*, **17**, 151, 749, 751 (1974).

A. G. Turner, *Methods in Molecular Orbital Theory*, Prentice-Hall, Englewood Cliffs, N. J., 1974.

A. Korolkovas, *Ciência Cultura*, **25**, 131, 215 (1973).

C. Tanford, *The Hydrophobic Effect: Formation of Micelles and Biological Membranes*, Wiley-Interscience, New York, 1973.

P. N. Craig, *J. Med. Chem.*, **15**, 144 (1972).

American Chemical Society, "Biological Correlations: The Hansch Approach," *Advan. Chem. Ser.*, **114** (1972).

C. Hansch and W. J. Dunn, III, *J. Pharm. Sci.*, **61**, 1 (1972).

V. E. Marquez *et al.*, *J. Med. Chem.*, **15**, 36 (1972).

A. Cammarata, *Annu. Rep. Med. Chem.*, **1970**, 245 (1971).

P. N. Craig, *J. Med. Chem.*, **14**, 680 (1971).

R. Daudel and A. Pullman, Eds., *Aspects de la Chimie Quantique Contemporaine*, Centre National de la Recherche Scientifique, Paris, 1971.

B. Duperray, *Chim. Ther.*, **6**, 305 (1971).

E. R. Garrett, *Progr. Drug Res.*, **15**, 271 (1971).

L. B. Kier, *Molecular Orbital Theory in Drug Research*, Academic, New York, 1971.

A. Leo *et al.*, *Chem. Rev.*, **71**, 525 (1971).

J. W. McFarland, *Progr. Drug Res.*, **15**, 123 (1971).

F. Peradejordi *et al.*, *J. Pharm. Sci.*, **60**, 576 (1971).

J. L. Rabinowitz and R. M. Myerson, Eds., "Absorption Phenomena," *Top. Med. Chem.*, **4** (1971).

G. Swidler, *Handbook of Drug Interactions*, Wiley-Interscience, New York, 1971.

M. S. Tute, *Advan. Drug Res.*, **6**, 1 (1971).

J. G. Wagner, *Biopharmaceutics and Relevant Pharmacokinetics*, Hamilton Press, Hamilton, Ill., 1971.

K. Fukui, *Top. Current Chem.*, **15**, 1 (1970).

L. B. Kier, Ed., *Molecular Orbital Studies in Chemical Pharmacology*, Springer, Berlin, 1970.

J. A. Pople and D. L. Beveridge, *Approximate Molecular Orbital Theory*, McGraw-Hill, New York, 1970.

M. S. Tute, *J. Med. Chem.*, **13**, 48 (1970).

E. I. Eger, II, *et al.*, *Anesthesiology*, **30**, 129, 136 (1969).

C. Hansch, *Accounts Chem. Res.*, **2**, 232 (1969).

C. Hansch, *Farmaco, Ed. Sci.*, **23**, 293 (1968).

A. Cammarata and R. C. Allen, *J. Pharm. Sci.*, **56**, 640 (1967).

N. R. Draper and H. Smith, *Applied Regression Analysis*, Wiley, New York, 1966.

S. M. Free, Jr., and J. W. Wilson, *J. Med. Chem.*, **7**, 395 (1964).

R. W. Taft, Jr., *J. Phys. Chem.*, **64**, 1805 (1960).

P. H. Bell and R. O. Roblin, Jr., *J. Amer. Chem. Soc.,* **64**, 2905 (1942).

L. P. Hammett, *Physical Organic Chemistry,* McGraw-Hill, New York, 1940.

H. Meyer, *Arch. Exp. Pathol. Pharmakol.,* **42**, 109 (1899).

E. Overton, *Vierteljahrsschr. Naturforsch. Ges. Zürich,* **44**, 88 (1899).

Drug Receptors

J. A. Katzenellenbogen, *Annu. Rep. Med. Chem.,* **9**, 222 (1974).

H. P. Rang, Ed., *Drug Receptors,* Macmillan, London, 1973.

A. Albert, *Annu. Rev. Pharmacol.,* **11**, 13 (1971).

E. De Robertis, *Science,* **171**, 963 (1971).

J. C. Meunier *et al., C. R. H. Acad. Sci.,* Ser. D, **274**, 117 (1971).

R. Miledi and L. T. Potter, *Nature (London),* **233**, 599 (1971).

R. Miledi *et al., Nature (London),* **229**, 554 (1971).

H. P. Rang, *Nature (London),* **231**, 91 (1971).

A. S. V. Burgen, *Annu. Rev. Pharmacol.,* **10**, 7 (1970).

J.-P. Changeux *et al., Proc. Nat. Acad. Sci. U.S.A.,* **67**, 1241 (1970).

S. Ehrenpreis, *Progr. Drug Res.,* **14**, 59 (1970).

R. D. O'Brien *et al., Proc. Nat. Acad. Sci. U.S.A.,* **65**, 438 (1970).

R. Porter and M. O'Connor, Eds., *Molecular Properties of Drug Receptors,* Churchill, London, 1970.

Theories of Drug Action

M. Rocha e Silva and F. Fernandes, *Eur. J. Pharmacol.,* **25**, 231 (1974).

D. E. Koshland, Jr., *Pure Appl. Chem.,* **25**, 119 (1971).

N. H. Jurkiewicz *et al., Pharmacology,* **5**, 129 (1971).

J. F. Danielli, J. Moran, and D. J. Triggle, Eds., *Fundamental Concepts in Drug-Receptor Interactions,* Academic, New York, 1970.

H. P. Rang and J. M. Ritter, *Mol. Pharmacol.,* **6**, 357, 383 (1970).

R. W. Noble, *J. Mol. Biol.,* **39**, 479 (1969).

M. Rocha e Silva, *Eur. J. Pharmacol.,* **6**, 294 (1969).

J. Wyman, *J. Amer. Chem. Soc.,* **89**, 2202 (1967).

C. Cennamo, *J. Theor. Biol.,* **21**, 260 (1968).

D. E. Koshland, Jr., and K. E. Neet, *Annu. Rev. Biochem.,* **37**, 359 (1968).

J.-P. Changeux *et al., Proc. Nat. Acad. Sci. U.S.A.,* **57**, 335 (1967).

J. Wyman, *J. Amer. Chem. Soc.,* **89**, 2202 (1967)

W. D. M. Paton and H. P. Rang, *Advan. Drug Res.,* **3**, 57 (1966).

J. M. van Rossum, *Advan. Drug Res.,* **3**, 189 (1966).

B. Belleau, *Advan. Drug Res.,* **2**, 89 (1965).

E. J. Ariëns and A. M. Simonis, *J. Pharm. Pharmacol.,* **16**, 137, 289 (1964).

Mechanism of Drug Action

J. W. Corcoran and F. E. Hahn, Eds., *Mechanism of Action of Antimicrobial and Antitumor Agents,* Springer, Berlin, 1975.

R. N. Lindquist, "The Design of Enzyme Inhibitors: Transition State Analogs," in E. J. Ariëns, Ed., *Drug Design,* vol. V, Academic, New York, 1975, pp. 23-80.

T. J. Bardos, *Top. Current Chem.,* **52**, 63 (1974).

G. H. Haggis *et al., Introduction to Molecular Biology,* 2nd ed., Longmans, London, 1974.

H. Kersten and W. Kersten, *Inhibitors of Nucleic Acid Synthesis,* Springer, Berlin, 1974.

J. B. G. Kwapinski, *Molecular Microbiology,* Wiley, New York, 1974.

R. R. Rando, *Annu. Rep. Med. Chem.,* **9**, 234 (1974).

J. M. Barry and E. M. Barry, *Molecular Biology: An Introduction to Chemical Genetics,* Prentice-Hall, Englewood Cliffs, N.J., 1973.

R. M. Hochster and J. H. Quastel, Eds., *Metabolic Inhibitors: A Comprehensive Treatise,* 4 vols., Academic, New York, 1963, 1972, 1973.

M. K. Jain, *The Bimolecular Lipid Membrane: A System,* Van Nostrand Reinhold, New York, 1972.

A. Korolkovas, *Rev. Bras. Clin. Ter.,* **1**, 729, 769 (1972).

J. H. Biel and L. G. Abood, Eds., *Biogenic Amines and Physiological Membranes in Drug Therapy,* 2 vols., Dekker, New York, 1971.

F. E. Hahn, Ed., "Complexes of Biologically Active Substances with Nucleic Acids and Their Modes of Action," *Progr. Mol. Subcell. Biol.,* **2**, 1-400 (1971).

D. F. H. Wallach and H. Fischer, Eds., *The Dynamic Structure of Cell Membranes,* Springer, Berlin, 1971.

B. N. Lewin, *The Molecular Basis of Gene Expression,* Wiley-Interscience, New York, 1970.

T. Y. Shen, *Angew. Chem., Int. Ed. Engl.,* **9**, 678 (1970).

Th. Bücher and H. Sies, Eds., *Inhibitors: Tools in Cell Research,* Springer, New York, 1969.

A. D. Russell, *Progr. Med. Chem.,* **6**, 135 (1969).

G. Valette and P. Rossignol, *Actual. Pharmacol.,* **22**, 85 (1969).

L. Fowden *et al., Advan. Enzymol.,* **29**, 89 (1967).

J. L. Webb, *Enzyme and Metabolic Inhibitors,* Academic, New York, 1963-1966.

G. H. Hitchings and J. J. Burchall, *Advan. Enzymol.,* **27**, 417 (1965).

N. O. Kaplan and M. Friedkin, *Advan. Chemother.,* **1**, 499 (1964).

R. A. Peters, *Biochemical Lesions and Lethal Synthesis,* Pergamon, Oxford, 1963.

D. W. Wooley, *Progr. Drug Res.,* **2**, 613 (1960).

C. Mentzer, *Actual. Pharmacol.,* **7**, 173 (1954).

D. W. Wooley, *A Study of Antimetabolites,* Wiley, New York, 1952.

C. J. Martin, *Biological Antagonism,* Blakiston, New York, 1951.

R. O. Roblin, Jr., *Chem. Rev.,* **38**, 255 (1946).

PART 2

DRUGS ACTING ON THE CENTRAL NERVOUS SYSTEM

Drugs acting on the central nervous system can either depress or stimulate its functions. These drugs are usually divided into three broad classes:

1. Nonselective central-nervous-system depressants: general anesthetics, hypnotics and sedatives, narcotic analgetics, antipyretic and antirheumatic analgetics

2. Selective modifiers of central nervous system: anticonvulsants, antitussives, psychotherapeutic agents, central intraneural blocking agents

3. Central-nervous-system stimulants

These drugs are studied in Chapters 4 to 12.

REFERENCES

A. Burger, Ed., *Drugs Affecting the Central Nervous System,* Dekker, New York, 1968.

W. S. Root and J. G. Hofmann, Eds., *Physiological Pharmacology,* Vol. I and II, Academic, New York, 1963, 1965.

CHAPTER 4

GENERAL ANESTHETICS

I. INTRODUCTION

General anesthetics are drugs which produce analgesia, loss of consciousness, and muscular relaxation by depressing reversibly the central nervous system.

The word *anesthesia* was coined by Oliver Wendell Holmes, who used it for the first time on November 21, 1846, in a letter to William T. G. Morton. It comes from the Greek and means *without perception* or *insensibility*.

Four stages of anesthesia can be distinguished:

Stage I--*Analgesia.* The patient remains conscious but sleepy, owing to depression of higher cortial centers; this stage is suitable for dental extractions, incisions and drainage of abscesses.

Stage II -- *Delirium* or *excitement.* The patient loses consciousness but is excited and delirious as a result of depression of higher motor centers; however, no operative procedures can be performed during this stage.

Stage III -- *Surgical anesthesia.* The patient progressively loses his reflexes as a consequence of depression of skeletal muscles. This stage is divided into planes numbered 1 to 4, in order of increasing skeletal muscle relaxation and decreasing respiration. At this stage most surgical procedures are performed.

Stage IV -- *Respiratory* or *medullary paralysis.* The patient stops breathing and his blood circulation collapses, owing to a depression of vital functions; at this stage no operative procedures are performed and, therefore, it is never intentionally reached.

II. CLASSIFICATION

General anesthetics are divided into inhalation anesthetics and intravenous anesthetics (Table 4.1).

A. Inhalation Anesthetics

Also called *volatile anesthetics,* inhalation anesthetics may be gases (cyclopropane, ethylene, nitrous oxide) or volatile liquids (chloroform, ether, ethyl chloride, fluroxene, halothane, methoxyflurane, trichloroethylene, vinyl ether).

Some of them form explosive mixtures with air or other gases. They vary greatly in potency, safety, and ability to induce analgesia and muscular relaxation. Based on their chemical structure, volatile liquids may be divided into ethers (ethyl ether, vinyl ether, fluroxene, methoxyflurane) and halogenated hydrocar-

Table 4.1 General Anesthetics Most Commonly Used

Official Name	Proprietary Name	Chemical Name	Structure
cyclopropane*		trimethylene	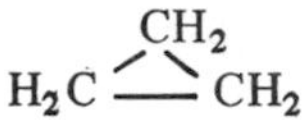 $\begin{array}{c} CH_2 \\ H_2C \diagup\!\!\diagdown CH_2 \end{array}$
ether*		diethyl ether	$C_2H_5OC_2H_5$
ethyl chloride[+]	Kelene	chloroethane	CH_3CH_2Cl
halothane*	Fluothane	2-bromo-2-chloro-1,1,1-trifluoroethane	$CF_3CHClBr$
nitrous oxide*		nitrogen monoxide	N_2O
chloroform		trichloromethane	$CHCl_3$
ethylene		ethene	$H_2C=CH_2$
fluroxene[+]	Fluoromar	2,2,2-trifluoroethyl vinyl ether	$CF_3CH_2OCH=CH_2$
methoxyflurane[+]	Penthrane	2,2-dichloro-1,1-difluoroethyl methyl ether	$Cl_2CHCF_2OCH_3$
enflurane	Ethrane	2-chloro-1,1,2-trifluoroethyldifluoromethyl ether	CHF_2OCF_2CHClF
sevoflurane		2,2,2-trifluoro-1-(trifluoromethyl)ethyl fluoromethyl ether	$FH_2COCH(CF_3)_2$
vinyl ether[+]	Vinethene	divinyl oxide	$CH_2=CHOCH=CH_2$
tribromoethanol	Avertin Ethobrom	2,2,2-tribromoethanol	Br_3CCH_2OH
trichloroethylene[+a]	Chlorylen Trilene Trimar	1,1,2-trichloroethene	$Cl_2C=CHCl$
ketamine[+b]	Ketalar Ketaject	2-(o-chlorophenyl)-2-(methylamino)cyclohexanone	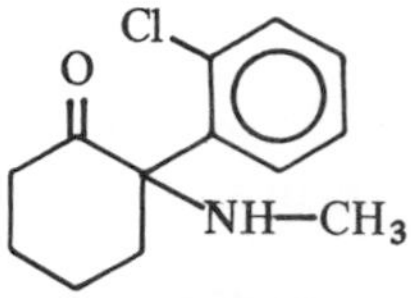

*In *USP XIX* (1975).
[+]In *NF XIV* (1975).
[a] Used as an analgetic.
[b] Used i.v. or i.m.

bons (chloroform, ethyl chloride, halothane, trichloroethylene, halopropane, teflurane and isoflurane).

Cyclopropane

Highly flammable, colorless, with characteristic odor, and explosive when mixed with certain concentrations of air (3.0 to 8.5%) or oxygen (2.5 to 50%). It is the most potent anesthetic gas currently used.

Ether

Colorless, highly volatile, flammable, with pungent odor. In the presence of oxygen, explosive ether peroxides are formed. Explosion is prevented by addition of ethanol, which reacts with peroxides, forming acetaldehyde.

Vinyl Ether

Clear, more volatile than ether, and with a slight ethereal odor. In order to prevent oxidation, phenyl-α-naphthylamine is added. It must be kept in small, tight, light-resistant containers. This drug was rationally designed by Leake and Chen. Based on the hybridization principle, they expected that divinyl ether, that can be considered a combination of ethylene and ether, might have the rapidity of induction of the former and the potency of the latter. Their expectations materialized, but the drug is more flammable and thus more hazardous than ether.

Chloroform

Colorless, sweet-smelling, toxic to the liver. It is synthesized from ethanol by haloform reaction. In the presence of oxygen and light, it forms phosgene gas, which is highly corrosive to lungs, because inside the body it reacts with water, producing hydrochloric acid. Phosgene formation is prevented by the addition of stabilizing agents, such as ethanol.

Halothane

Nonflammable, nonexplosive, and nonirritant. It resulted from research aimed at nonexplosive anesthetics. The presence of three atoms of fluorine imparts to it an extremely high stability. It is the anesthetic of choice for asthmatic patients, because it dilates the bronchioles. Unfortunately repeated administration of halothane may cause liver damage.

B. Intravenous Anesthetics

Intravenous anesthetics are nonexplosive solids. They produce rapid loss of consciousness but insufficient anesthesia and muscular relaxation. The most commonly used are ultrashort-acting barbiturates (thiopental, thiamylal, and methohexital), and ketamine. All are used for basal anesthesia, that is, to achieve a degree of unconsciousness before anesthetic administration. Barbiturates are studied in the next chapters; but those used as intravenous anesthetics are listed in Table 4.2.

Ketamine

Administered intramuscularly or intravenously, this nonbarbiturate anesthetic "induce[s] a cataleptic state in which the patient appears to be awake but is dissociated from the environment, is unresponsive to pain, and has no subsequent memory of the procedure," according to the AMA. It may be used as the sole anesthetic for diagnostic and minor surgical interventions of short duration.

III. MECHANISM OF ACTION

Several theories have been advanced for the mechanism of action of general anesthetics, since their action cannot be explained by a unified theory. Actually, the theories presented only describe the effects caused by these anesthetics, without elucidating how the effects are produced. Since their chemical structure, physicochemical properties, and pharmacological effects are so varied, it is admitted that they depress nonselectively the central nervous system by a physicochemical mechanism; that is, they owe their action to their physicochemical properties and not to the complexation with a pharmacological receptor. In other words, general anesthetics are structurally nonspecific drugs.

Theories of general-anesthetic action may be classified as physical theories and biochemical theories. Physical theories are based mainly on two physicochemical properties of an anesthetic molecule: its polarizability and its volume. The principal physical theories are lipid theory, permeability theory, surface tension or adsorption theory, molecular-size theory, neurophysiological theories, clathrate theory, iceberg theory.

Biochemical theories include inhibition of oxidation theory, interference with mitochondrial ATP formation theory, suppression of ion movements theory, neurophysiological theory.

None of these theories, however, is supported by unquestionable experimental evidence. Several authors suggest that the main effect produced by general anesthetics results from physical interactions, such as those causing conformational changes in macromolecules, and a secondary role is reserved for biochemical changes.

A. Physical Theories

Lipid Theory

Advanced by Meyer (1899) and Overton (1901), it postulates that anesthetic action is directly related to the partition coefficient of the anesthetic agent between olive oil and water: the higher the numerical value of this coefficient, the greater the anesthetic activity of the drug. Recently, by using oil-gas partition coefficient, Eger and coworkers found a better correlation between minimal alveolar concentrations and solubilities of some general anesthetics in olive oil.

The Overton-Meyer theory does not explain, however, the mechanism of action of general anesthetics. It merely expresses the direct parallelism between liposolubility and anesthetic action.

Molecular-Size Theory

On the basis of the fact that xenon produces narcosis, Wulf and Featherstone advanced the hypothesis that anesthetic activity results from the molecular size of the compounds used. They showed that anesthetic activity is related to molecular volume b, which appears in the van der Waals equation

$$\left(P + \frac{a}{V^2}\right)(V - b) = RT$$

The molecular volume b should be greater than that of those substances, such as water, oxygen, and nitrogen, that could normally occupy the lateral space between the lipid and the protein layers of the cellular membrane.

Actually the values of b for such substances are, respectively, water, 3.05; oxygen, 3.18; and nitrogen, 3.91. With a greater b value (nitrous oxide, 4.4; xenon, 5.1; ethylene, 5.7; cyclopropane, 7.5; chloroform, 10.2; ethyl ether, 13.4), when occupying the space between the lipid layers normally occupied by water, oxygen, and nitrogen, general anesthetics would cause alteration in the cellular structure and subsequent depression of function, resulting in anesthesia.

Clathrate Theory

According to Pauling (1961), the important phase of the central nervous system in anesthesia is not the lipid one, as assumed in former theories, but the aqueous one. Considering that certain compounds, such as chloroform and xenon, form *in vitro* microcrystalline hydrates, Pauling postulated that similar crystals, made up by water molecules and called *clathrates*, form in the encephalic fluid. They are stabilized by general anesthetics bound by van der Waals forces to side chains of proteins and other solutes. These microcrystalline hydrates alter the conduction of electrical impulses necessary for maintenance of wakefulness; in consequence narcosis or anesthesia occurs. However, in a recent paper Erlander presented evidence to contradict the clathrate theory. Nevertheless it is interesting to combine Wulf and Featherstone's theory with Pauling's. By building molecular models, both of microcrystalline hydrates and of general anesthetics, it is seen that the latter ones can be easily enclosed

Table 4.2 Sodium Salts of Barbiturates and Thiobarbiturates Used as Intravenous Anesthetics

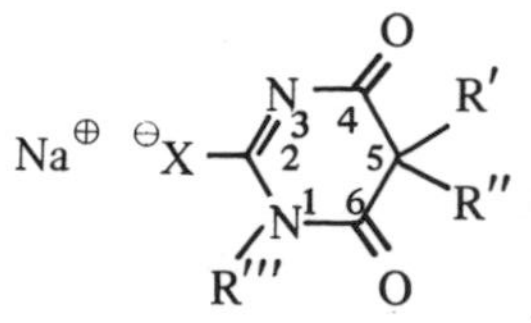

Official Name	Proprietary Name	R'	R''	R'''	X
methohexital*	Brevital Brietal	$-CH_2-CH=CH_2$	$-CH-C\equiv C-CH_2-CH_3$ $\quad\;\;\;\|$ $\quad\;\;CH_3$	$-CH_3$	O
thiamylal[+]	Surital	$-CH_2-CH=CH_2$	$-CH-CH_2-CH_2-CH_3$ $\quad\;\;\|$ $\quad\;CH_3$	$-H$	S
thiopental*[a]	Pentothal	$-CH_2-CH_3$	$-CH-CH_2-CH_2-CH_3$ $\quad\;\;\|$ $\quad\;CH_3$	$-H$	S

*In *USP XIX* (1975).
[+]In *NF XIV* (1975).
[a] Also used as an anticonvulsant (Table 5.2).

The following thiobarbiturates are also marketed: butalital (Baytinal, Transithal, Ullreval), methitural (Diogenal, Neraval), and thialbarbital (Intranarcon, Kemithal).

inside the former ones.

B. Biochemical Theories

Inhibition of Oxidation Theory

Quastel (1952, 1963) has demonstrated that anesthetics suppress brain oxygen uptake *in vitro*. For instance, they inhibit the oxidation of the coenzyme NADH to NAD^+ (nicotinamide adenine dinucleotide, formerly known as diphosphopyridine nucleotide = DPN):

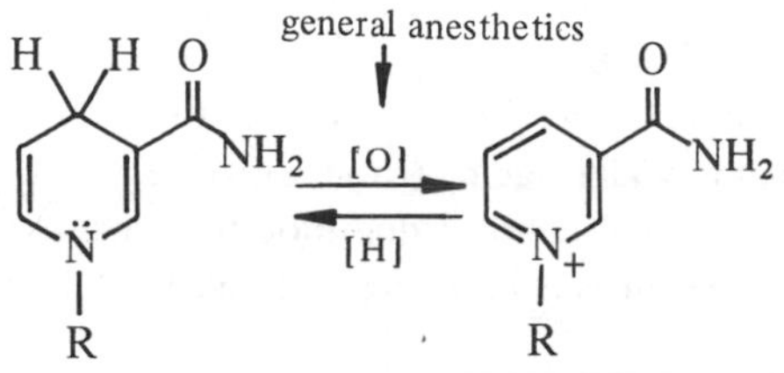

$$\text{NADH} \quad \underset{[H]}{\overset{[O]}{\rightleftharpoons}} \quad NAD^+$$

NADH   NAD⁺
(DPNH)   (DPN⁺)

By preventing this oxidation, general anesthetics depress the function of the citric acid cycle, because NAD^+ is involved in oxidative decarboxylation in the tricarboxylic acid (Kreb's) cycle. Since the biological oxidation of NADH is controlled by the phosphorylation of ADP to ATP, it follows that general anesthetics also inhibit oxidative phosphorylation.

These phenomena, however, may be only the consequences rather than the cause of anesthesia. Very likely diminished oxygen uptake results from the decrease in CNS activity caused by anesthesia. It is also known that dinitrophenol, although being an uncoupler of oxidative phosphorylation, does not produce anesthesia.

IV. ADJUNCTS TO ANESTHETICS

As adjuncts to general anesthesia, some patients receive supplementary drugs usually in preanesthetic medication. This practice fulfills many purposes, among them the following:

1. Reduction of anxiety. The drugs prescribed are sedatives and hypnotics, such as barbiturates (pentobarbital and secobarbital) and nonbarbiturates (ethinamate, glutethimide, chloral hydrate), tranquilizers, such as phenothiazines (promethazine, triflupromazine), and antianxiety agents, such as chlordiaze-

poxide, diazepam, hydroxyzine, and meprobamate.

2. Control of pain. For this purpose strong analgetics are administered (morphine, levorphanol, hydromorphone, oxymorphone, piminodine, pethidine, fentanyl, alphaprodine, methotrimeprazine, pentazocine).

3. Inhibition of salivation. The drugs most commonly used are anticholinergic agents, such as atropine, scopolamine, hyoscyamine.

4. Prevention of nausea and vomiting. This is accomplished by administration of phenothiazine antiemetics (promethazine, chlorpromazine, perphenazine, propiomazine, thiethylperazine), nonphenothiazine antiemetics (trimethobenzamide, dimenhydrinate, diphenhydramine), and the butyrophenone droperidol.

5. Reduction of the amount of general anesthetic by synergism or summation. The drugs used are sedatives and basal anesthetics.

6. Production of skeletal-muscle relaxation. The most commonly used are nondepolarizing blocking agents (tubocurarine, dimethyltubocurarine, gallamine, and pancuronium) and depolarizing blocking agents (succinylcholine, and decamethonium); sometimes other drugs are used, such as hexafluorenium bromide and mephenesin.

7. Production of basal anesthesia. For this purpose the following are used: tribromoethanol, short- or ultrashort-acting barbiturates.

Two mixtures are also widely used: Mepergan (promethazine plus meperidine) and Innovar (fentanyl plus droperidol).

Sometimes as adjuncts to anesthesia other drugs are also used: antiarrhythmic agents (propranolol, procainamide, lidocaine, quinidine), α-adrenergic blocking agents (phentolamine, tolazoline, phenoxybenzamine), vasopressors, analeptics, and narcotic antagonists (nalorphine, levallorphan, naloxone).

REFERENCES

Introduction

F. T. Evans and C. Gray, Eds., *General Anesthesia*, Butterworth, London, 1965.

E. M. Papper and R. J. Kitz, Eds., *Uptake and Distribution of Anesthesthic Agents*, McGraw-Hill, New York, 1963.

History

H. K. Beecher, *Anesthesiology*, 29, 1068 (1968).

T. E. Keys, *The History of Surgical Anesthesia*, Schuman's, New York, 1945.

Classification

A. Cherkin, *Annu. Rev. Pharmacol.*, 9, 259 (1969).

J. W. Dundee, *Clin. Pharmacol. Ther.*, 8, 91 (1967).

C. D. Leake and M. Y. Chen, *Proc. Soc. Exp. Biol. Med.*, 28, 151 (1930).

Mechanism of Action

C. Hansch *et al.*, *J. Med. Chem.*, 18, 546 (1975).

P. Seeman, *Pharmacol. Rev.*, 24, 583 (1972).

D. L. Clark *et al.*, *J. Comp. Physiol. Psychol.*, 68, 315 (1969).

E. I. Eger II *et al.*, *Anesthesiology*, 30, 129, 136 (1969).

S. R. Erlander, *J. Macromol. Sci.-Chem.*, A2, 595 (1968).

Pharmacology Society Symposium, "The Molecular Pharmacology of Anesthesia," *Federation Proc.*, 27, 870-913 (1968).

B. P. Schoenborn and R. M. Featherstone, *Advan. Pharmacol.*, 5, 1 (1967).

J. H. Quastel, *Brit. Med. Bull.*, 21, 49 (1965).

M. A. B. Brazier, *Brit. J. Anaesth.*, 33, 194 (1961).

H. W. Magoun, *Brit. J. Anaesth.*, 33, 183 (1961).

S. L. Miller, *Proc. Nat. Acad. Sci. U.S.A.*, 47, 1515 (1961).

L. Pauling, *Science*, 134, 15 (1961).

R. J. Wulf and R. M. Featherstone, *Anesthesiology*, 18, 97 (1957).

L. J. Mullins, *Chem. Rev.*, 54, 289 (1954).

J. H. Quastel, *Curr. Res. Anesth. Analg.*, 31, 151 (1952).

M. Michaelis and J. H. Quastel, *Biochem. J.*, 35, 518 (1941).

E. Overton, *Studien über die Narkose*, Fischer, Jena, 1901.

H. H. Meyer, *Arch. Exp. Pathol. Pharmakol.*, 42, 109 (1899).

CHAPTER 5

HYPNOTICS AND SEDATIVES

I. INTRODUCTION

Hypnotics and sedatives are nonselective or general depressants of the central nervous system; they are used to reduce restlessness and emotional tension and to induce either sleep or sedation.

Sedative-hypnotics are widely and increasingly consumed all over the world. It is estimated that an average American takes 33 tablets of barbiturates per year. Sedation, for instance, is sought in one or more of the following situations: emotional strain, chronic tension, hypertension, potentiation of analgetics, control of convulsions, adjuncts to anesthesia, narcoanalysis. Hypnotics, on the other hand, are prescribed to overcome insomnia of several types; in many cases insomnia results from unsolved problems.

The difference between hypnotic and sedative action depends on dosage: a higher dosage causes a hypnotic effect, whereas a small dosage produces only sedation. In high doses some of these drugs are used to induce surgical anesthesia or as basal anesthetics.

The most common adverse reactions are drowsiness, lethargy, and hangover. Coma and even death, caused by depression of the vital medullary centers of the brain, result from overdosage. Prolonged use, even in therapeutic doses, can cause psychic and physical dependence. Abrupt withdrawal results sometimes in a severe abstinence syndrome, characterized by convulsions and delirium; coma and death may occur.

Poisoning is treated by removal of the drug from the stomach and maintenance of adequate respiration and circulation. Emetics should not be used in comatose patients.

II. CLASSIFICATION

Hypnotics and sedatives vary widely in chemical structure. Useful agents are found in many different chemical classes. Those currently used belong to the following groups: bromides, alcohols, aldehydes and derivatives, carbamates, acyclic ureides, barbiturates, cyclic amides and imides, benzodiazepines (Tables 5.1 and 5.2).

Table 5.1 Nonbarbiturate Hypnotics and Sedatives

Official Name	Proprietary Name	Chemical Name	Structure
ethchlorvynol[+]	Placidyl Serenesil	1-chloro-3-ethyl-1-penten-4-yn-3-ol	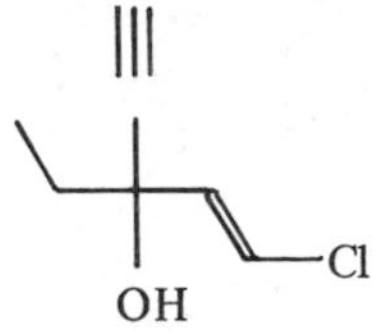
chloral hydrate*	Noctec Somnos	trichloroacetaldehyde monohydrate	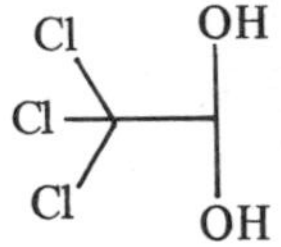
chloral betaine[+]	Beta-Chlor	complex of chloral hydrate and betaine	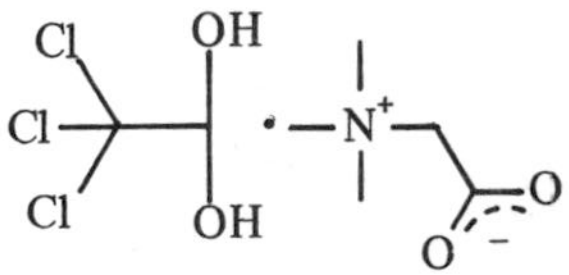
triclofos	Triclos	2,2,2-trichloroethyl dihydrogen phosphate	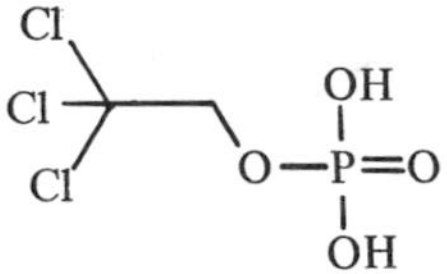
paraldehyde*	Paral	2,4,6-trimethyl-s-trioxane	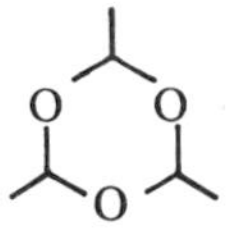

95

ethinamate[+]	Valmid	1-ethynylcyclohexanol carbamate 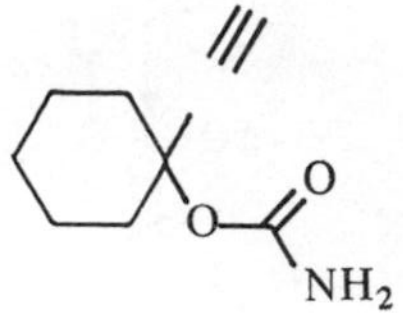
glutethimide[+]	Doriden	2-ethyl-2-phenylglutarimide 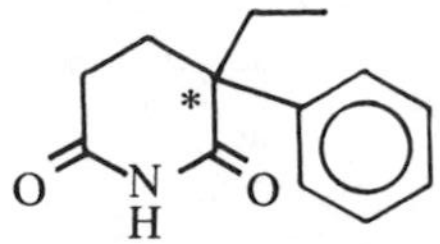
methyprylon[+]	Noludar	3,3-diethyl-5-methyl-2,4-piperidinedione 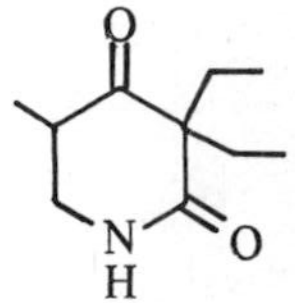
methaqualone[+]	Parest Quaalude Sopor	2-methyl-3-o-tolyl-4(3H)-quinazolinone
thalidomide	Contergan Kevadon	3-phthalimidopiperidine-2,6-dione 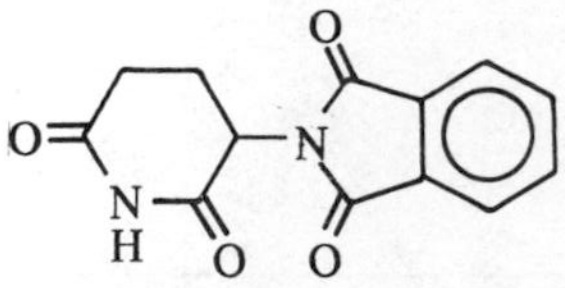

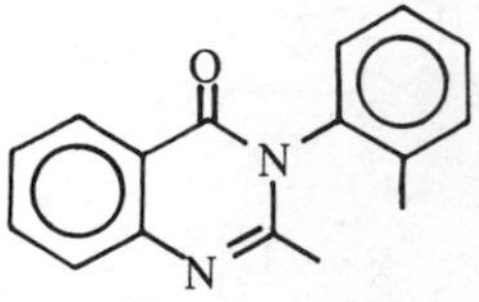

*In *USP XIX* (1975).
[+] In *NF XIV* (1975).

Table 5.2 Barbiturates Used as Hypnotics and Sedatives

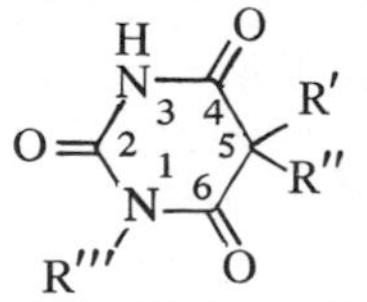

Official Name	Proprietary Name	R'	R''	R'''
Prolonged effect (6 or more hours)				
barbital	Veronal	$-C_2H_5$	$-C_2H_5$	$-H$
mephobarbital[+]	Mebaral	$-C_2H_5$	$-C_6H_5$	$-CH_3$
phenobarbital[*]	Luminal	$-C_2H_5$	$-C_6H_5$	$-H$
Intermediate effect (3-6 hours)				
amobarbital[*][+]	Amytal	$-C_2H_5$	$-CH_2CH_2CH(CH_3)_2$	$-H$
aprobarbital	Alurate	$-CH_2CH=CH_2$	$-CH(CH_3)_2$	$-H$
butabarbital[+]	Bubartal Butisol	$-C_2H_5$	$-CH-C_2H_5$ $\quad\quad$ CH_3	$-H$
butalbital	Sandoptal	$-CH_2CH=CH_2$	$-CH_2CH(CH_3)_2$	$-H$
probarbital	Ipral	$-C_2H_5$	$-CH(CH_3)_2$	$-H$
talbutal[+]	Lotusate	$-CH_2CH=CH_2$	$-CH-C_2H_5$ $\quad\quad$ CH_3	$-H$
Short effect (less than 3 hours)				
pentobarbital[*][+]	Nembutal	$-C_2H_5$	$-CH-C_3H_7$ $\quad\quad$ CH_3	$-H$
secobarbital[*]	Seconal	$-CH_2CH=CH_2$	$-CH-C_3H_7$ $\quad\quad$ CH_3	$-H$
hexobarbital[+]	Cyclonal Dorico Evipal Narcosan Privenal	$-CH_3$	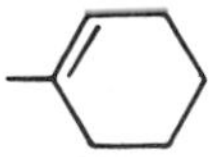	$-CH_3$

[*]In *USP XIX* (1975).
[+]In *NF XIV* (1975).

A. Bromides

These inorganic sedatives were widely used during the last century. The salts mainly prescribed were NaBr, KBr, NH_4Br, $CaBr_2 \cdot 2H_2O$, $SrBr_2 \cdot 6H_2O$. Now these drugs are rarely used, because they tend to accumulate in the body and cause severe intoxication called bromism, characterized by dermatitis, gastro-intestinal disorders, and mental disturbance. Bromism is treated with sodium or ammonium chloride.

B. Alcohols

Several alcohols exert hypnotic action. Certain structure-activity relationships are found in them: (*a*) hypnotic activity increases with increasing chain length until *n*-hexanol or *n*-octanol; (*b*) unsaturation enhances both activity and toxicity; (*c*) tertiary alcohols are more active than secondary ones, and secondary more than primary; (*d*) branching results in greater depression; (*e*) introduction of another hydroxyl group tends to decrease toxicity as well as activity; (*f*) replacement of a hydrogen by a halogen enhances activity.

The first member of the family, methanol, is not used as a hypnotic and sedative because it causes blindness. Ethanol in many forms (beer, wine, brandy, whisky) has been used for centuries, but since chronic alcoholism is rapidly developed and only large doses are effective, it is not favored as a hypnotic. New hypnotic alcohols are clomethiazole (Distaneurin) and 2-methyl-1-phenyl-3-butyne-1,2-diol (Centalun). Older ones are amylene hydrate, chlorobutanol (Chloretone), and meparfynol (Dormison).

Ethchlorvynol

Colorless to yellow liquid which darkens on exposure to light and air. A pungent odor. Excessive intake may cause physical dependence. The hypnotic dose is 500 mg; the sedative dose, 100 to 200 mg.

C. Aldehydes and Derivatives

Those most widely used are chloral hydrate and paraldehyde, as well as some latent forms of chloral hydrate: chloral betaine, chloralodol (Dormex), chloral-osium (=glucochloralium), cloretate (Clorets), dichloralphenazone (Samilclor), mecloralurea (Heraldium), petrichloral (Petriclor), triclofos.

Chloral Hydrate

The oldest of the hypnotic agents, it was introduced on the basis of the mistaken belief that *in vivo* it would slowly undergo the haloform reaction, releasing chloroform, whose anesthetic properties were already known. Later on, it was found that in the body it is largely reduced to trichloroethanol, to which most of its hypnotic action may be ascribed.

Chloral is an oily liquid which, by hydration, yields chloral hydrate, a colorless or white crystalline substance, with penetrating odor and bitter caustic taste, being quite irritating to the skin and mucous membrane. Chloral hydrate is very soluble in water and in ethanol. Its depressant action is increased synergistically by alcohol. In the old days (1870) of Chicago, Mickey Finn used to add chloral hydrate to cocktails; this mixture, known as "knockout drops" or "Mickey Finn," is a very potent depressant, although the hemiacetal $Cl_3CCH(OH)OCH_2CH_3$ (chloral ethanolate) thus formed has no greater hypnotic activity than the hydrate.

Paraldehyde

Colorless liquid, with an unpleasant taste, used by oral and rectal routes. It imparts a potent odor to a patient's breath soon after administration. Its use is limited largely to alcoholics and psychotics. Its preparation consists in condensation of 3 molecules of acetaldehyde, for it is a cyclic acetal of acetaldehyde. During storage, paraldehyde is easily oxidized to glacial acetic acid.

D. Carbamates

Certain urethanes have been introduced as sedative-hypnotics: mepentamate (Psicosedine), hexapropymate (Merinax), hydroxyphenamate (Listica), mebutamate (Capla), emylcamate (Nuncitat), meprobamate (Equanil), nisobamate, oxanamide (Quiactin), procimate (Equipax), tybamate (Solacen).

E. Acyclic Ureides

Acyclic ureides are derivatives of urea and monocarboxylic acids. Their general formula is $R-CONHCONH_2$. They differ from cyclic ureides, which are derivatives of urea and dicarboxylic acids; the prototype of cyclic ureides is barbituric acid.

Several drugs of this class are marketed: acetylcarbromal (Sedamyl), apronal (Sedormid), bromisovalum (Bromural), capuride (Pacinox), carbromal (Adaline), ectylurea (Levanil). They exert a weak depressant effect and are used primarily as daytime sedatives.

F. Barbiturates

To date, about 2500 barbiturates have been synthesized, and roughly 30 are marketed. The most widely used are listed in Table 5.2. Others are (*a*) prolonged effect: metharbital (Gemonil), febarbamate (Getril); (*b*) intermediate effect: allobarbital (Dial), butallylonal (Pernoston), buthetal (Neonal), nealbarbital (Censedal), propallylonal (Noctal), vinbarbital (Delvinal), vinylbital (Byconox); (*c*) short effect: cyclobarbital (Phanodorm), cyclopentobarbital (Cyclopal), heptabarb (Medomin), hexethal (Orthal).

1. *Physicochemical Properties*

Barbiturates are colorless, crystalline, solid compounds, with melting point range between 96 and 205°C. They have acidic character, which in the case of barbituric acid itself results from three factors: (*a*) symmetrical and conjugated cyclic system of enol form; (*b*) active methylene between two carbonyl groups; (*c*) presence of a diiminocarbonyl type of system in the tautomeric forms seen on the next page.

The trienol form may be considered as 2,4,6-trihydroxypyrimidine. X-ray diffraction studies indicate that the triketo tautomer is the predominant form, at least in the crystalline state.

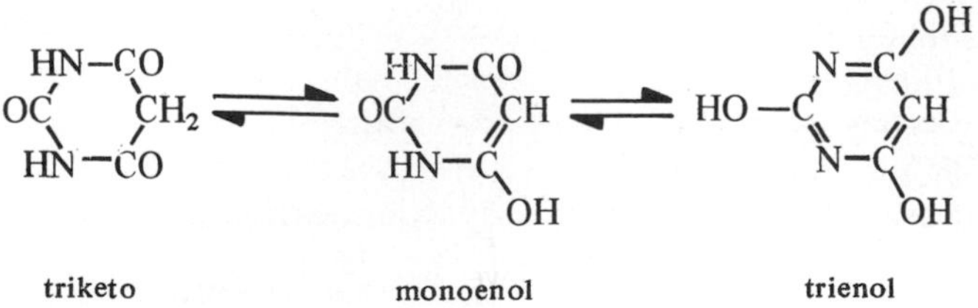

| triketo | monoenol | trienol |

In barbiturates disubstituted at position 5 and trisubstituted at 1 and 5, the acidic character is imparted by the group $-CONHCO-$, owing to the absence of a hydrogen atom at 5:

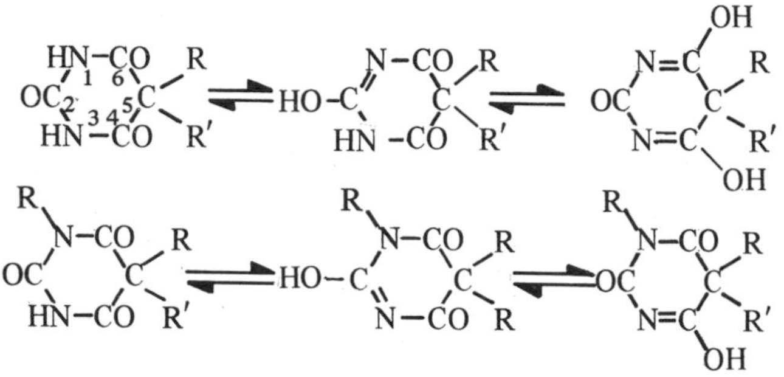

As free acids, barbiturates are sparingly soluble in water but freely soluble in organic solvents. For this reason they are often converted to sodium salts, which are water-soluble. In the presence of acids, sodium salts of barbiturates precipitate as free barbiturate acids; to prevent this, when used in the injectable form, potassium carbonate buffer is added.

2. *Action and Metabolism*

Hypnotic and sedative effects produced by barbiturates are usually ascribed to their action at the level of the thalamus and ascending reticular formation, which interferes with the transmission of nervous impulses to the cortex. The usual dose varies from 30 to 600 mg.

Since monoamine oxidase inhibitors potentiate the depressant action of the barbiturates, their concomitant use is not recommended.

Especially as sodium salts, barbiturates are completely absorbed from the gastrointestinal tract; for the parenteral route, soluble sodium salts are necessarily used. They are uniformly distributed in all tissues, reaching higher concentrations in the liver and kidneys. Their elimination is principally through the kidneys, in the following forms: unchanged, partially oxidized at the side chain, partially conjugated.

The metabolism of barbiturates, which occurs in the liver, proceeds through one or more of the following pathways: (*a*) desulfuration of 2-thiobarbiturates; (*b*) loss of any *N*-alkyl groups; (*c*) oxidation of groups attached to 5; (*d*) hydrolytic opening of the barbituric acid ring. In the metabolism, lipid-soluble barbiturates are transformed into more polar metabolites that can be excreted. The resulting metabolites can be inactive, or as active as, or more active than, the parent barbiturate. For instance, phenobarbital is hydroxylated in the para position of the phenyl ring, giving an inactive compound. Thiopental is oxidized to an alcohol, to a carboxylic acid, and to pentobarbital, which is a longer-acting barbiturate.

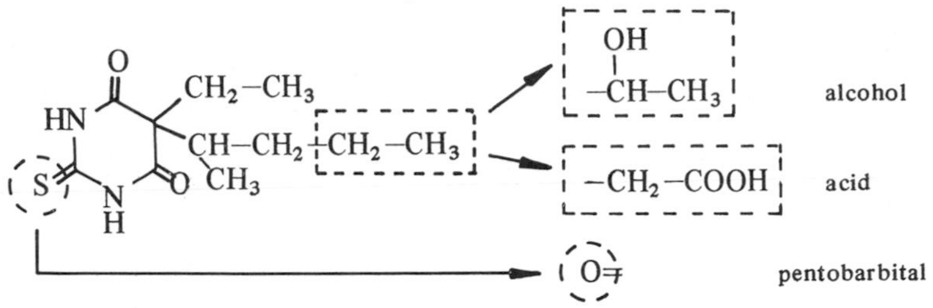

$$\text{alcohol} \quad -CH-CH_3 \text{ (OH)}$$

$$\text{acid} \quad -CH_2-COOH$$

$$\text{pentobarbital} \quad O$$

thiopental

3. *Structure-Activity Relationships*

Fischer observed that several hypnotics contained quaternary carbons. This observation prompted him to synthesize 5,5-disubstituted barbiturates. Several hundreds of these compounds have since been synthesized. Attempts have been made to establish structure-activity relationships, in order to bring some order into the chaos of this field. Tatum divided barbiturates into four groups according to their duration of effect. Now these groups comprise the following barbiturates:

1. Prolonged duration of effect (4 to 12 hours): barbital, mephobarbital, metharbital, phenobarbital
2. Intermediate duration of effect (2 to 8 hours): amobarbital, aprobarbital, butabarbital, probarbital, talbutal, vinbarbital
3. Short duration of effect (up to 3 hours): cyclobarbital, heptabarb, pentobarbital, secobarbital
4. Ultrashort duration of effect (less than 3 hours): thiamylal, thiopental, hexobarbital, methohexital

Barbiturates with prolonged duration of effect are used mainly for the treatment of epilepsy and to maintain day-long sedation in states of anxiety and tension. Those with intermediate and short duration of effect are used chiefly as hypnotics and sedatives, for the treatment of insomnia and for preanesthesia sedation. Those having ultrashort duration of effect are used principally as intravenous anesthetic agents for basal anesthesia.

Some general rules have been found in the study of structure-activity relationships of barbiturates:

1. Duration of effect depends mainly on the substituents at position 5 which confer lipid solubility. The effect increases until the total number of carbon atoms of both substituents reaches 8; further lengthening of the side chain results in convulsant or inactive products. To be more specific: (*a*) ultrashort effect: long chain at 5 and atom of S instead of O at 2; (*b*) short effect: long chain at 5, atom of O at 2; (*c*) intermediate effect: shorter, less branched chain at 5, atom of O at 2; (*d*) long effect: phenyl group or short and saturated chain at 5, atom of O at 2.

2. Alkyl groups attached to 1 and 3 shorten effect and give rise to stimulant properties.

3. Methylation of one N atom increases affinity for lipids and has a tendency to decrease effect.

4. Atom of S at 2 shortens onset of action, owing to very rapid uptake into the central nervous system, and shortens duration of action, owing to rapid redistribution into fatty tissue.

5. Phenyl group at C-5 imparts anticonvulsant properties.

Results of recent molecular orbital calculations performed by Pullman and coworkers with the PCILO method on barbital, phenobarbital, hexobarbital, and amobarbital agree with available experimental data from X-ray crystallography as to the conformations of these molecules. They have shown that substituents at C-5 exert a negligible influence on the charge distribution in the barbiturate ring. They concluded, therefore, that it is likely that the pharmacological activity of barbiturates depends on the electronic properties of the barbituric ring system, perhaps in its ability to form a hydrogen bond with adenine. As to substituents at C-5, they may be responsible for the transport of the drugs to their sites of action and may be involved in an adequate contact with those sites.

G. Cyclic Amides and Imides

Glutethimide

Structurally related to barbiturates, it is a white powder, soluble in ethanol but almost insoluble in water. It is used for induction of sleep. Cases of circulatory failure and death have occurred when glutethimide was ingested with alcohol and other depressants. Its margin of safety is less than that of barbiturates. The usual dose is 125 to 250 mg for sedation and 500 mg to 1 g for hypnosis.

Methaqualone

Used for patients who cannot tolerate barbiturates, this quinazolinone is contraindicated in pregnant patients and in persons with suicidal tendencies. It has fallen into disrepute because of habituation in young people.

Thalidomide

Introduced in 1957 but since it causes phocomelia, a teratogenic effect characterized by the absence of arms and legs, it is strongly contraindicated during pregnancy and is no longer generally available. It has been used for the treatment of transitory cutaneous reaction in leprosy.

H. Benzodiazepines

Several drugs of this type have been introduced: chlordiazepoxide, clozapine, diazepam, nitrazepam, oxazepam, flurazepam, nimetazepam (Hypnon), and perlapine. Although having hypnotic and sedative action, they are mostly used as antianxiety drugs, formerly called "minor tranquilizers." However, nitrazepam is widely used as a hypnotic, and flurazepam was recently introduced as a hypnotic. (See Chapter 10 for names and structures.)

I. Mixtures

Several different categories of mixtures of hypnotics are marketed. Usually they contain one or more barbiturates, with various other agents: antispasmodics, antihistaminics, antiemetics, sympathomimetics, analeptics. According to the American Medical Association, "the use of these mixtures is not advisable."

III. MECHANISM OF ACTION

Some biochemical mechanisms have been proposed in explanation of the action of hypnotics and sedatives, such as specific inhibition of respiratory enzymes and uncoupling of oxidative phosphorylation. Such proposed mechanisms, however, are generally considered to be inadequate.

Hypnotics and sedatives are structurally nonspecific drugs. Their action probably does not result from their attachment to specific receptor sites but from their physicochemical properties. It is very likely that, by modifying the dielectric constant and structure of water surrounding biopolymers, they induce conformational changes in some macromolecules connected with an important physiological role. This is quite possible, because most of these drugs (bromides and a few other agents are exceptions), although of very different chemical structures, present two common features: (*a*) a group that can be involved in hydrogen bonding; (*b*) groups that can lower the dielectric constant of water. In other words, hypnotics and sedatives might act by mechanisms suggested by Koshland's induced-fit and Belleau's macromolecular perturbation theories (see Chapter 3, Sections IV D and E).

Recently, Hansch and coworkers tried to correlate hypnotic activity with hydrophobicity. Although they found good correlation for some drugs, this correlation does not hold for all hypnotics. Kyogoku and Yu have found that barbital binds selectively to the adenine moiety of FAD and NAD. There is increasing evidence that barbiturates, owing to their structural similarity to thymine, exert their effects by interacting selectively, through hydrogen bonding, with the adenine moiety of many macromolecules, such as FAD and NADH, involved in important biochemical processes.

At the present time, it is generally accepted that hypnotics act by interfering with functions of the reticular activating system, either by stimulating the sleep center or by inhibiting the function of the arousal center.

REFERENCES

General

F. Kagan, *et al.*, Eds., *Hypnotics: Methods of Development and Evaluation*, Halsted, New York, 1975.
E. D. Weitzman, *Advances in Sleep Research*, 2 vols., Halsted, New York, 1974-1975.
K. W. Wheeler, *J. Med. Chem.*, **6**, 1 (1963).

Introduction

I. Oswald, *Pharmacol. Rev.*, **20**, 273 (1968).

Classification

B. Pullman *et al., J. Theor. Biol.*, **35**, 375 (1972).
E. M. Sellers *et al., Clin. Pharmacol. Ther.*, **13**, 37, 50 (1972).
A. J. Mandell *et al., Biol. Psychiatry*, **1**, 13 (1969).
M. T. Bush and E. Sanders, *Annu. Rev. Pharmacol.*, **7**, 57 (1967).

Mechanism of Action

R. J. H. Davies and N. Davidson, *Biopolymers*, **10**, 21 (1971).
Y. Kyogoku and B. S. Yu, *Chem.-Biol. Interactions*, **2**, 117 (1970).
Y. Kyogoku *et al., Nature (London)*, **218**, 69 (1968).

CHAPTER 6

ANTICONVULSANTS

I. INTRODUCTION

Anticonvulsants are drugs which selectively depress the central nervous system. These drugs are used mainly for suppression of epileptic seizures without impairment of the central nervous system and without depression of respiration. They are effective in 75 to 80% of patients.

Epilepsy, from the Greek word meaning seizure, is defined as a recurrent, paroxysmal cerebral dysrhythmia which is characterized by abnormal and excessive electroencephalographic (EEG) discharge and a disturbance or loss of consciousness. It may or may not be associated with body movements. Epilepsy is a widespread disease. It is estimated that 0.5 to 1% of the world's population suffer from epileptic seizures.

According to John Hughlings Jackson, the father of modern concepts of epilepsy, epileptic seizures are caused by "occasional, sudden, excessive, rapid and local electrical discharges of gray matter." This almost century-old proposal was later substantiated by the electroencephalograph.

Notwithstanding the fact that new classifications for epileptic seizures have been proposed and that, according to the characteristic seizure type and EEG of the patients, more than a dozen forms of epilepsy may have been distinguished, it is usual to consider only the following three principal types of epilepsy: (a) *grand mal*—the most frequent form, in which the seizures last from 2 to 5 minutes, being characterized by sudden loss of consciousness, tonic and clonic convulsions of all muscles, urinary incontinence; (b) *petit mal*—the seizures last from 5 to 30 seconds, being characterized by brief attacks of unconsciousness; this type occurs often in children of high or at least medium intelligence; it starts at the ages of 4 to 8 and rarely continues after 15; (c) *psychomotor seizures*—they last from 2 to 3 minutes and are characterized by attacks without convulsions.

Several hypotheses have been advanced to explain the cause of epileptic discharges, but none is completely satisfactory. It is known, however, that congenital defects, head trauma, hypoxia at birth, concussion or fracture of the skull, abcess, neoplasm, inflammatory vascular changes subsequent to many infectious diseases, intrathecal administration of some drugs, and certain psychotropics are some of the many factors that can give rise to epileptic seizures.

Since a patient may need to take anticonvulsants all his life, the toxicity of various drugs should be considered. Most anticonvulsants produce adverse effects, such as damage to the bone marrow, liver, and kidneys, severe dyscrasias, gastrointestinal disturbance, drowsiness, alopecia, and nephropathies. Paradoxically, some anticonvulsants used for one type of epilepsy may aggravate or precipitate seizures of another type.

There are examples of antagonism between anticonvulsants and other drugs and sometimes between different anticonvulsants. For instance, phenobarbital antagonizes phenytoin by speeding its metabolism; reserpine antagonizes phenytoin.

Potentiation of anticonvulsants is achieved by several drugs: barbiturates by monoamine oxidase inhibitors; phenytoin by disulfiram, isoniazid, and amino-salicyclic acid.

II. CLASSIFICATION

Anticonvulsant agents are found in several different chemical classes: bromides, barbiturates, hydantoins, oxazolidinediones, succinimides, acylureides, benzo-diazepines, and miscellaneous. Most of them can be represented by a common fundamental structure (Table 6.1).

A. Bromides

Now mainly of historical interest, bromides were introduced as antiepileptic drugs on a false premise. It was thought in the last century that epilepsy was related to the practice of masculine masturbation called onanism, after Onan (Genesis 38:6-9), and that bromides had anaphrodisiac action. The salts mostly used were sodium bromide and potassium bromide. They have moderate effect on grand mal, but even so they are recommended when other drugs are ineffective.

Intoxication by bromides (bromism) is treated with sodium chloride in great amount of ammonium chloride and a thiazide saluretic.

B. Barbiturates

Structures and other details concerning barbiturates were presented in Chapter 5. Phenobarbital, mephobarbital, and metharbital are used as anticonvulsants. Even at nonsedating doses, these barbiturates protect against epileptic seizures. Their anticonvulsant activity, therefore, is not a consequence of sedation, but it results from another mechanism.

Table 6.1 Major Classes of Anticonvulsants

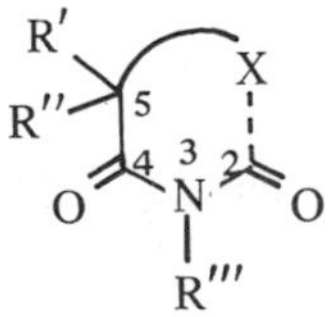

Official Name	Proprietary Name	X	Chemical Name
Barbiturates		$-CO-NH-$	
phenobarbital*	Luminal		5-ethyl-5-phenylbarbituric acid
mephobarbital[+]	Mebaral		5-ethyl-1-methyl-5-phenyl-barbituric acid
metharbital	Gemonil		5,5-diethyl-1-methylbarbituric acid
Hydantoins		$-NH-$	
phenytoin*	Dilantin		5,5-diphenylimidazolidine-2,4-dione
ethotoin	Peganone		3-ethyl-5-phenylhydantoin
mephenytoin[+]	Mesantoin		3-methyl-5-ethyl-5-phenyl-hydantoin
albutoin	Euprax		3-allyl-5-isobutyl-2-thio-hydantoin
Oxazolidinediones		$-O-$	
trimethadione*	Tridione		3,5,5-trimethyloxazolidine-2,4-dione
paramethadione*	Paradione		5-ethyl-3,5-dimethyloxa-zolidine-2,4-dione
allomethadione	Dimedion		3-allyl-5-methyloxazoli-dine-2,4-dione
Succinimides		$-CH_2-$	
phensuximide[+]	Milontin		*N*-methyl-2-phenylsuccinimide
methsuximide[+]	Celontin		*N*,2-dimethyl-2-phenylsuc-cinimide
ethosuximide*	Zarontin		2-ethyl-2-methylsuccinimide
Acylureides		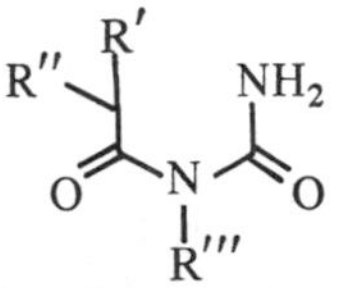	
phenacemide[+]	Phenurone		phenylacetylurea
ethylphenacemide	Pheneturide		phenylethylacetylurea
chlorphenacemide	Comitiadon		α-chlorophenylacetylurea

*In *USP XIX* (1975).
[+]In *NF XIV* (1975).

Phenobarbital

It is the most widely used anticonvulsant, being the drug of choice in most types of epilepsy but especially in grand mal. The usual dosage is 120 to 200 mg daily in divided doses. It is also used as the sodium salt.

C. Hydantoins

These drugs are free acids; they are insoluble in water. But strong base converts them to useful salts.

Phenytoin

White powder, practically insoluble in water, and slightly soluble in ethanol. It is used in free form and as the sodium salt primarily for control of grand mal, alone or together with phenobarbital. A common serious side effect is gingival hyperplasia, a reaction which seldom occurs with mephenytoin and apparently never with ethotoin. Sometimes hirsutism and hyperactivity occur, principally in the young.

D. Oxazolidinediones

Trimethadione

It is a white, granular, crystalline substance, with a slight camphoraceous odor, soluble in water and ethanol. Used principally for the treatment of petit mal, especially in refractory cases, because of its severe side effects, some of them fatal.

Paramethadione

It is an oily liquid, slightly soluble in water but freely soluble in ethanol. Introduced in 1947, it has the same applications and exerts similar effects as trimethadione but may cause fewer side effects in certain patients.

E. Succinimides

Phensuximide

A crystalline solid, slightly soluble in water, freely soluble in ethanol. It is the safest of the succinimides, used to control petit mal seizures.

Ethosuximide

It is the drug of choice for petit mal. The usual dosage for adults is 500 mg daily; for children under 6 years, 250 mg. Recently morsuximide (Morfolep), a derivative of phensuximide, was introduced in Hungary.

F. Acylureides

Phenacemide

An odorless and tasteless white crystaline solid. Structurally speaking, it may be considered an open-chain analogue of hydantoins. It has a broad spectrum of action, being effective in epilepsy of the three types. But since it produces potentially fatal adverse reactions, it should be used only as a drug of last resort, for the treatment of psychomotor epilepsy refractory to other drugs.

Chlorphenacemide

It is a less toxic chloro derivative of phenacemide.

G. Benzodiazepines

These drugs are used mainly as antianxiety agents; therefore, their structures and physicochemical properties are given in Chapter 11.

Diazepam

Used principally as an antianxiety agent, this drug has shown strong anticonvulsant properties. According to the AMA, for *status epilepticus* it is now regarded as the drug of choice.

Nitrazepam

The same use as diazepam. In myoclonic seizures it is more effective than diazepam.

Oxazepam

A metabolite of diazepam, it has shown activity in psychomotor seizures.

Prazepam

It is a long-acting potent anticonvulsant with slight sedative action.

H. Miscellaneous

Structures and names of some of these agents are given in Table 6.2. The following ones are also used: paraldehyde, lidocaine, quinacrine, barbexaclone (Maliasin, an equimolecular complex between propylhexedrine and phenobarbital) and sodium di-*n*-propylacetate (Depakin).

III. MECHANISM OF ACTION

Although various mechanisms of anticonvulsant action have been proposed, none

Table 6.2 Miscellaneous Anticonvulsants

Official Name	Proprietary Name	Chemical Name	Structure
primidone*	Mysoline	5-ethyldihydro-5-phenyl-4,6-(1H,5H)-pyrimidinedione	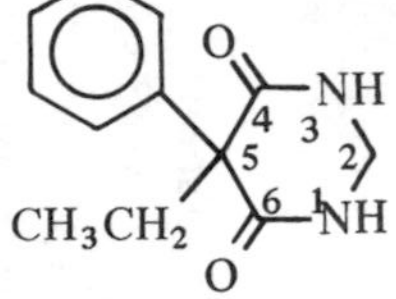
acetazolamide	Diamox	5-acetamido-1,3,4-thiadiazole-2-sulfonamide	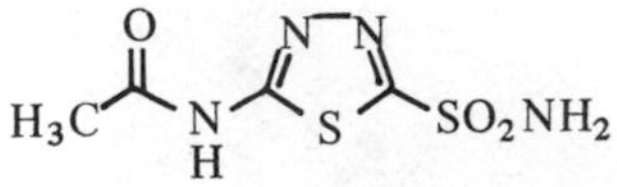
sulthiame	Conadil Contravul Elisal Ospolot	N-(4'-sulfamoylphenyl)-1,4-buthanesultham	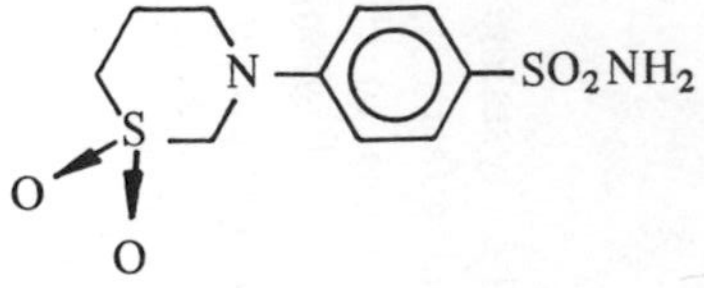
carbamazepine*[a]	Tegretol	5H-di-benzo[b,f]-azepine-5-carboxamide	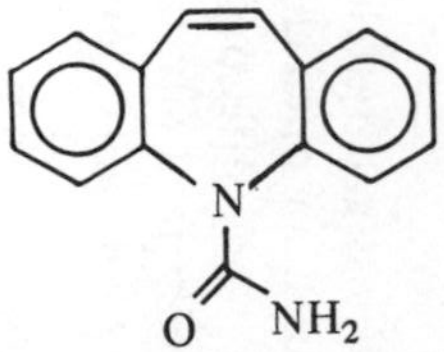

*In *USP XIX* (1975).

[a] Also used for trigeminal neuralgia.

enjoys general acceptance. Recent evidence, however, favors the hypothesis that anticonvulsants, similar to what was seen of general anesthetics and sedative-hypnotics, are structurally nonspecific drugs: that is, they owe their action to physicochemical properties and not to complexation with specific receptors. Let us examine briefly some theories that have been advanced.

Interference in Cholinergic Transmission

Consideration that many anticonvulsants inhibit the biosynthesis of acetylcholine led to the hypothesis that they act by this mechanism. However, experimental evidence does not support extrapolation of this theory to all anticonvulsants. For instance, atropine and other anticholinergics are ineffective as anticonvulsants. Furthermore, the correlation found between antiepileptic activity and acetylcholinesterase activity is not significant.

Increase of Concentration of Biogenic Amines

It is based on the fact that various anticonvulsants, by nonspecific depression of CNS function, increase the levels of serotonin in the brain and that certain monoamine oxidase inhibitors have anticonvulsant activity. This theory cannot be applied, however, to all types of anticonvulsant drugs.

Involvement of γ-Aminobutyric Acid

This theory claims that convulsions result from reduction of γ-aminobutyric acid (GABA) concentration, because it was found that convulsions disappear after administration of GABA. There is no unquestionable evidence, however, that anticonvulsants act by interfering in some way with the metabolism, concentration, or action of GABA.

Carbonic Anhydrase Inhibition

Since certain inhibitors of carbonic anhydrase exert antiepileptic effect, it has been postulated that anticonvulsants owe their action to the decrease of cerebral respiration and an increased concentration of carbon dioxide, which decreases nerve conduction, in consequence of carbonic anhydrase inhibition. However, it is likely that the action of these drugs results from systemic acidotic effects rather than carbonic anhydrase inhibition.

Macromolecular Conformational Changes

On the basis of the mode of action of lipid-soluble depressants and consideration of Koshland's induced-fit and Belleau's macromolecular perturbation theories, it has been proposed that anticonvulsants act through induction of conformational changes in oxidative systems essential to brain respiration.

The difference in activity of anticonvulsants—some being active in grand mal, others in petit mal, and so on—might be explained by differences in physicochemical properties of the drugs. Thus, peculiar structural features would direct some agents to be localized more extensively in one area, and another group of drugs in another one.

REFERENCES

General

M. J. Eadie, *Drugs,* 8, 386 (1974).

H. Kutt and S. Louis, *Drugs,* 4, 227, 256 (1972).

D. M. Woodbury *et al.,* Eds., *Antiepileptic Drugs,* Raven, New York, 1972.

S. Livingston, *Drug Therapy for Epilepsy: Anticonvulsant Drugs–Usage, Metabolism and Untoward Reactions,* Thomas, Springfield, Ill., 1966.

S. J. Hopkins, *Manuf. Chemist,* 36(2), 54 (1965).

A. Spinks and W. S. Waring, *Prog. Med. Chem.,* 3, 261 (1963).

W. J. Close and M. A. Spielman, *J. Med. Chem.,* 5, 1 (1961).

Introduction

H. H. Jasper, A. A. Ward, Jr., and A. Pope, *Basic Mechanisms of the Epilepsies,* Little, Brown, Boston, 1969.

C. A. Marsan, *Epilepsia,* 6, 275 (1965).

H. Gastaut *et al., Epilepsia,* 5, 297 (1964).

History

G. Chen *et al., Epilepsia,* 4, 66 (1963).

J. Y. Bogue and H. C. Carrington, *Brit. J. Pharmacol. Chemother.,* 8, 230 (1953).

Classification

E. H. P. Young, *Rep. Progr. Appl. Chem.,* 51, 176 (1966).

J. G. Millichap, *Postgrad. Med.,* 37, 22 (1965).

Mechanism of Action

L. H. Sternbach *et al., J. Med. Chem.,* 17, 374 (1974).

N. Camerman and A. Camerman, *Mol. Pharmacol.,* 7, 406 (1971).

A. Camerman and N. Camerman, *Science,* 168, 1457 (1970).

I. J. Wilk, *J. Chem. Educ.,* 34, 199 (1957).

CHAPTER 7

ANALGETICS: NARCOTIC AGENTS

I. INTRODUCTION

Analgetics are selective central-nervous-system depressants used to relieve pain without impairing consciousness. They act by raising the threshold of pain perception. The term *analgesia* comes from the Greek word meaning "without pain."

Based on analgetic potency and on differences in the development of dependence and tolerance, until recently analgetics were commonly divided into *narcotic* (or strong) analgetics and *nonnarcotic* (or weak) analgetics. Now this classification is obsolete. Nevertheless, the American Medical Association divides analgetics into two classes: (*a*) strong analgetics—subdivided into narcotics and nonnarcotics; (*b*) mild analgetics. Drugs of the first class are used to alleviate severe pain, whereas agents of the second class are used to treat mild to moderate pain.

Even in therapeutic doses, narcotic analgetics can cause respiratory depression (especially in elderly and debilitated patients), constipation, vomiting, nausea, cardiovascular disturbances, and several other untoward effects, such as drowsiness, mental clouding, and changes in mood. Chronic administration of these drugs can produce *tolerance* and psychic and physical *dependence,* formerly called *addiction.*

Abrupt withdrawal of narcotics from addicts or from patients subjected to prolonged treatment may cause a characteristic *abstinence syndrome.* Several hypotheses have been advanced to explain these phenomena. According to Goldstein and Goldstein, narcotics act by inhibiting an enzyme whose synthesis is repressed by the end product. The enzyme inhibition leads to a deficiency of the final product. For a certain pharmacological response, larger and larger doses are required, thus explaining tolerance and the psychic and physical dependence caused by narcotics. On the other side, interruption of normal biosynthetic pathways may result in high concentrations of the enzymes involved in the process and also of the intermediates accumulated near the site of inhibition. This phenomenon, coupled with the temporary acceleration of the normal biosynthetic process as a result of abrupt withdrawal of narcotics, gives rise to the abstinence syndrome.

To assess the dependence liability of chemical substances, Seevers and Deneau elaborated a procedure, now standard, based on the proven assumption that any drug capable of suppressing completely all, or nearly all, specific symptoms of morphine abstinence also creates physical dependence if administered for a long enough period of time. This test is useful also in the search for nalorphine-like, specific morphine antagonists. The test is performed in Rhesus monkeys.

II. CLASSIFICATION

Although having different chemical structures, centrally acting analgetics present, according to Jacobson and coworkers, the following common features: (a) a quaternary carbon atom, (b) a phenyl (or isostere) ring attached to this carbon atom, (c) a tertiary amino group, two saturated carbon atoms removed, and (d) a phenolic hydroxyl group in m position relative to the quaternary carbon atom attachment if the tertiary nitrogen is a part of a six-membered ring. Some compounds, such as benzimidazoles, tetrahydroisoquinolines, and dithienyl-butenylamines, which exhibit strong analgetic activity in test animals, do not present all these features, but they have not yet been introduced into thera-peutics. Those currently used in medicine as such, or in the *in vivo* form, have the following common structure, where Ar is an aromatic ring, Am an amino group, and X is C or N:

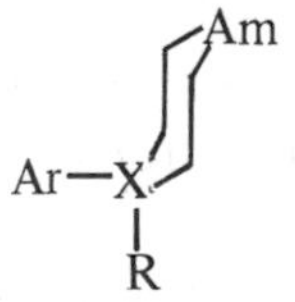

This means that most analgetics are characterized by the presence of the γ-phenyl-*N*-methylpiperidine moiety.

Analgetics may be divided into the following classes: morphine and derivatives, morphinans, benzomorphans, phenylpiperidines, diphenylpropylamines, pheno-thiazines.

Narcotic antagonists are closely related chemically to morphine.

As to metabolism, the major routes of practically all these analgetics is *N*-dealkylation.

A. Morphine and Relatives

Morphine is one of the 25 or so alkaloids isolated from the unripe seed capsules of the poppy, *Papaver somniferum.* It was first isolated from opium in 1803. From the same source, other alkaloids of clinical importance were isolated years later: codeine by Robiquet in 1832, papaverine by Merck in 1848. All are derivatives of benzylisoquinoline. In 1925 Robinson proposed a structure for morphine which was confirmed in 1952 by the total synthesis of this alkaloid by Gates and Tschudi and later by Elad and Ginsburg, besides other investigators. More recently its stereochemistry and absolute configuration have been elucidated (Figure 7.1).

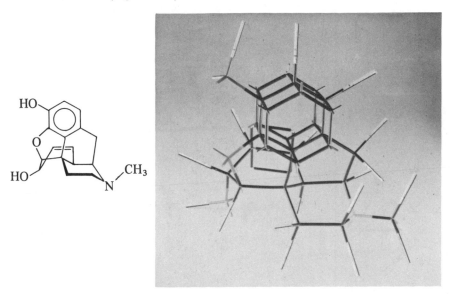

Figure 7.1 Absolute configuration of (−)-morphine.

Morphine occurs in opium in concentration varying from 5 to 20%. Various processes are available for extraction, but the final step is generally the precipitation of morphine from an acid solution by an excess of ammonia, followed by recrystallization from boiling ethanol.

Not only are morphine and its derivatives used as analgetics, but also a mixture of the total alkaloids of opium marketed as Pantopon, and *paregoric* (camphorated opium tincture). Now almost obsolete are *laudanum* (tincture of opium) and *Dover's powder* (powder of ipecac and opium).

The main derivatives of morphine are listed in Table 7.1. Most of them are used as salts, which occur as water-soluble, white, or practically white, crystalline powders. Also marketed are oxycodone and dihydrocodeine (Paracodin).

Morphine

The natural substance is (−)-morphine. As the free alkaloid it is an odorless,

Table 7.1 Morphine and Derivatives

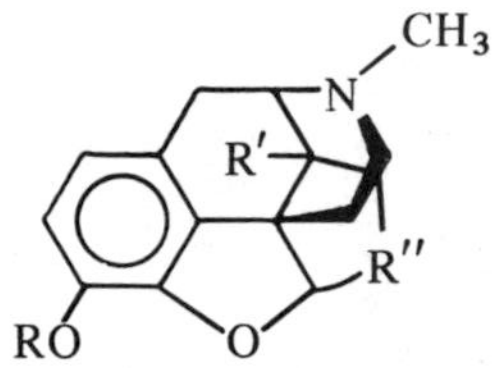

Official Name	Proprietary Name	Chemical Name	R	R'	R''
morphine*			H	H	
codeine*+			CH₃	H	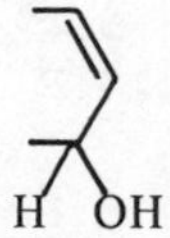
ethylmorphine	Dionin		CH₂CH₃	H	
hydromorphone+	Dilaudid	dihydromorphinone	H	H	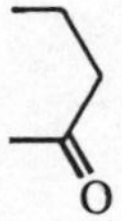

hydrocodone[+]	Dicodid	dihydrocodeinone	CH$_3$	H	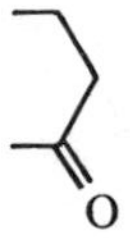
oxymorphone[+]	Numorphan	(–)-14-hydroxydihy-dromorphinone	H	OH	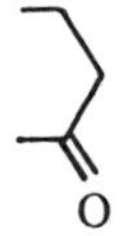
heroin		diacetylmorphine	CH$_3$CO	H	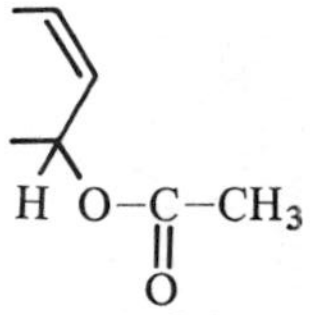

*In *USP XIX* (1975).
[+] In *NF XIV* (1975).

117

white, crystalline substance with a bitter taste. It is practically insoluble in water but soluble in alkaline solution because of its phenolic hydroxyl group. It forms readily water-soluble salts with most acids. The preferred forms used in medicine are the sulfate and hydrochloride. These salts are prepared by neutralizing a hot aqueous suspension of morphine with diluted acids. The usual dosage is 2 to 20 mg, intravenously or subcutaneously; oral administration is not recommended. Morphine is detoxified in the liver by conjugation, largely at the 3-phenolic hydroxy group.

Codeine

Since it occurs in opium only in small amount, usually between 0.2 and 0.08%, codeine is prepared by methylation of the phenolic hydroxyl group of morphine. It is a colorless, efflorescent, and light-sensitive crystalline substance, slightly soluble in water and freely soluble in ethanol. By treatment with acids, salts are formed; the most used are sulfate and phosphate. Codeine is used both as analgetic and antitussive.

Heroin

Its analgetic ability is greater than morphine's, but it has intense dependency liability. For this reason it is forbidden in most countries.

Ethorphine

It is 1000 to 80,000 times more potent than morphine, depending on the test system used. Owing to its high potency, it is only used in the capture of large wild animals.

B. Morphinans

Levorphanol, NF XIV, (Levo-dromoran), (−)-3-hydroxy-*N*-methylmorphinan. Used as tartrate salt, sparingly water-soluble, colorless crystals. Its dextrorotatory optical antipode, called dextrorphan, has antitussive properties but it is not used in medicine. The racemic mixture, known as racemorphan, is marketed in Germany under the name of Citarin. The usual dosage of levorphanol is 2 to 3 mg by injection or orally.

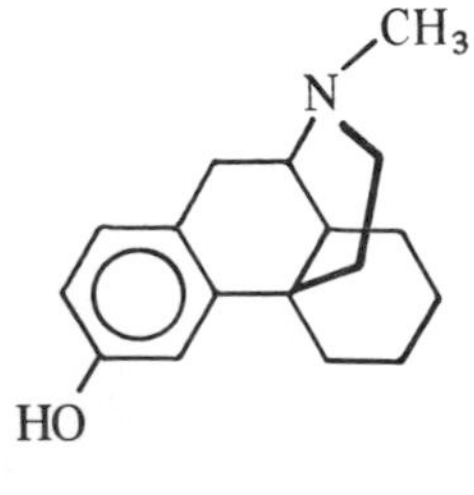

C. Benzomorphans

Pentazocine, NF XIV, (Talwin), *cis*-2-dimethylallyl-5,9-dimethyl-2'-hydroxy-6,7-benzomorphan. Used as the lactate salt of the racemic mixture. Its analgetic and respiratory depressant activity results almost exclusively from the (−)-isomer. It was synthesized by Archer *et al.* in 1964 as part of a program to achieve analgetics devoid of dependence liability. It is administered parenterally. Unfortunately, it is addictive.

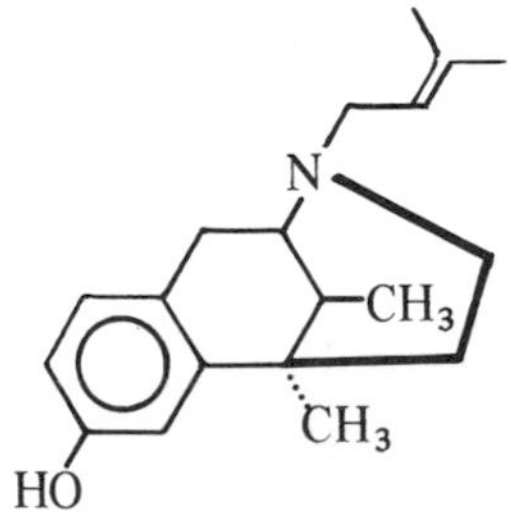

D. Phenylpiperidines

Derivatives of 4-phenylpiperidines were the first synthetic strong analgetics. They were synthesized in Germany in 1941. Although apparently not related structurally to morphine, actually they represent a close resemblance to this molecule, with a central quaternary carbon atom, an ethylene chain, an amino group, and an aromatic ring. Some of them are marketed as hydrochlorides or other salts, which occur usually as water-soluble, white, crystalline powders. Those most commonly used in medicine are listed in Table 7.2. Also marketed is phenopiridine (Operidine).

Meperidine

It is used as the hydrochloride. It is one of the most commonly used analgetic substitutes for morphine. Meperidine exerts several pharmacologic effects: analgetic, spasmolytic, local anesthetic, and mild antihistaminic. This multiple activity is explained by its structural resemblance to, respectively, morphine, atropine, cocaine, and histamine. The usual dosage is 50 to 150 mg, by intramuscular, intravenous, oral, or subcutaneous routes.

Fentanyl

Used as citrate, a colorless crystalline powder, it is slightly soluble in water. It was introduced in 1964. As an analgetic, it is about 50 times more active than morphine. In association with droperidol, it is marketed as Innovar, being extensively used in neuroleptoanalgesia.

Bezitramide

Recently introduced. It is more potent than dextromoramide (Table 7.3). It has

Table 7.2 Phenylpiperidine Derivatives

Official Name	Proprietary Name	Chemical Name	Structure
meperidine*	Demerol	ethyl 1-methyl-4-phenyl-4-piperidinecarboxylate	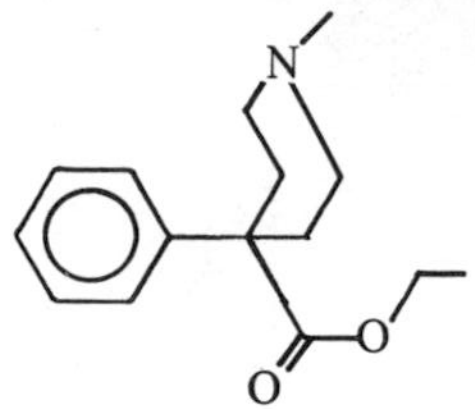
alphaprodine[+]	Nisentil	(±)-1,3-dimethyl-4-phenyl-4-propionoxypiperidine	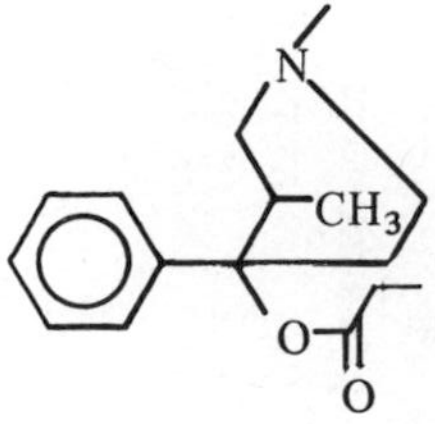
anileridine[+]	Leritine	ethyl 1-(p-amino phenethyl)-4-phenyl-4-piperidinecarboxylate	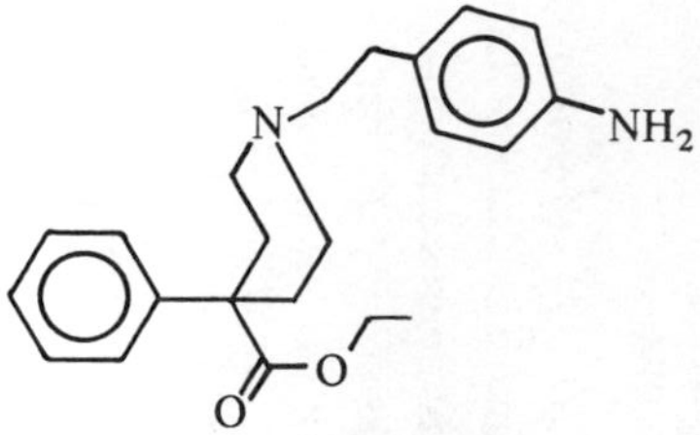
ethoheptazine	Zactane	ethyl hexahydro-1-methyl-4-phenyl-1H-azepine-4-carboxylate	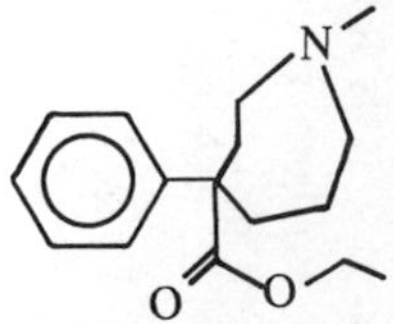

piminodine[+]	Alvodine	ethyl 4-phenyl-1-[3-(phenylamino)propyl] piperidine-4-carboxylate	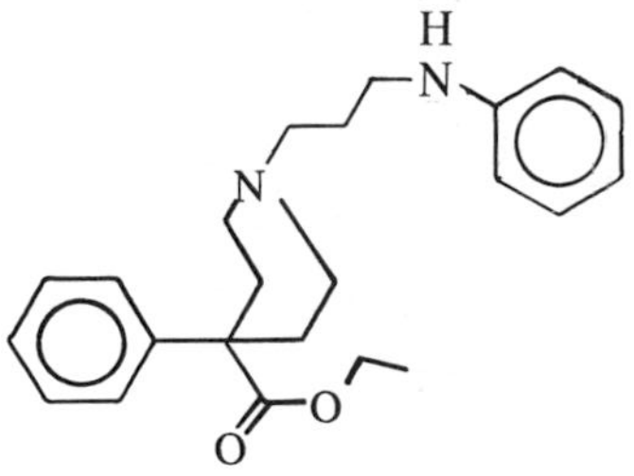
fentanyl*	Sublimaze	1-phenethyl-4-*N*-propionylanilinopiperidine	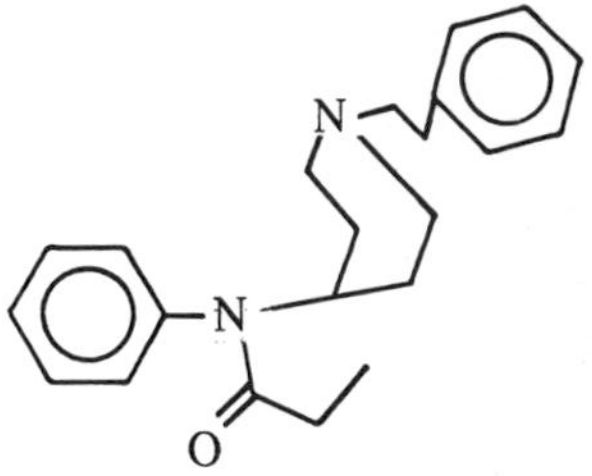
bezitramide	Burgodin	1-(3-cyano-3,3-diphenylpropyl)-4-(2-oxo-3-propionyl-1-benzimidazolinyl)piperidine	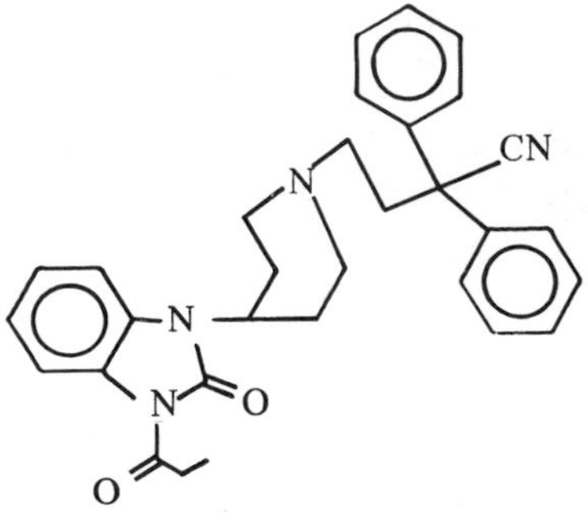

*In *USP XIX* (1975).
[+]In *NF XIV* (1975).

Table 7.3 Diphenylpropylamine Derivatives

Official Name	Proprietary Name	Chemical Name	Structure[a]
methadone*	Dolophine	(±)-6-(dimethylamino)4,4-diphenyl-3-heptanone	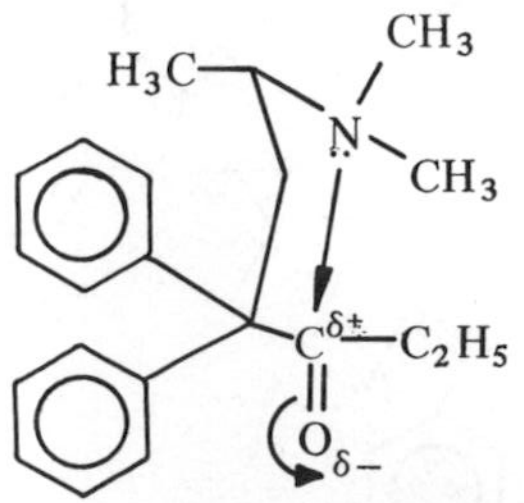
dextromoramide	Palfium	(+)-4-[2-methyl-4-oxo-3,3-diphenyl-4-(1-pyrrolidinyl)-butyl] morpholine	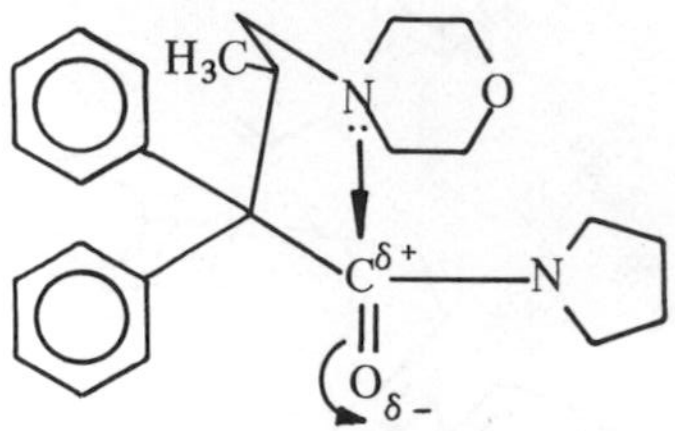
propoxyphene*+	Darvon	(2S,3R)-(+)-4-(dimethylamino)-3-methyl-1,2-diphenyl-2-butanol propionate (ester)	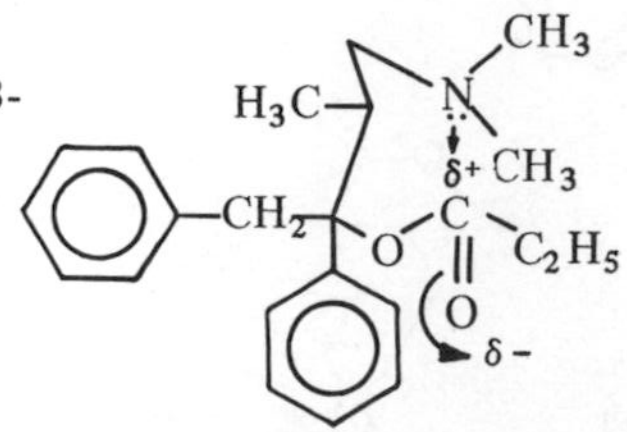

*In *USP XIX* (1975).
+ In *NF XIV* (1975).
[a] Structures may also be drawn so as to resemble those of morphine and Table 7.2.

prolonged action and is useful for the treatment of severe chronic pain.

E. Diphenylpropylamines

Analgetics having this structure are optically active. Those commonly used are listed in Table 7.3. They are white, or colorless, crystalline powders which are soluble in water. Another one is dipipanone (Pipadone).

Methadone

It exists as a racemic mixture, with the (−)-isomer more active than the (+)-isomer. Methadone is more active than morphine but its toxicity is also greater. It is used for relief of many types of pain and also in addict treatment, as a narcotic substitute, because it prevents or completely alleviates a morphine abstinence syndrome. The usual dosage is 2.5 to 10 mg, administered by intramuscular, subcutaneous, or oral routes.

Propoxyphene

It is the (+)-isomer, used as hydrochloride. It is a pure analgetic, whereas its (−)-isomer is a pure antitussive.

Dextromoramide

Used as tartrate. The (−)-isomer is practically inactive, and the racemic mixture has half the activity of dextromoramide, which is more active than morphine, methadone, or meperidine.

F. Phenothiazines

Methotrimeprazine, NF XIV, (Levoprome), (−)-10-[3-(dimethylamino)-2-methylpropyl]-2-methoxyphenothiazine. As a phenothiazine analgetic used in medicine, it is closely related to chlorpromazine.

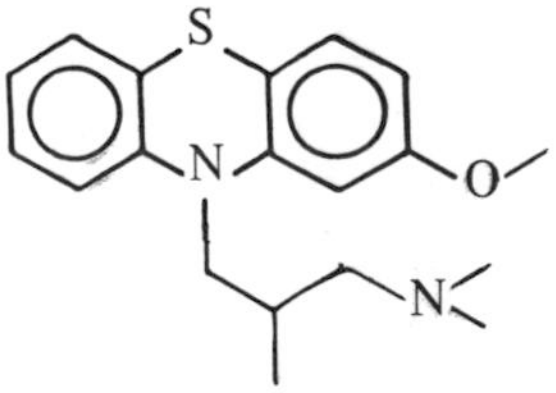

It is about one-half as potent as morphine, but it produces neither physical nor psychic dependence. However, since it causes marked postural hypotension, its use is restricted to nonambulatory patients. The usual dosage is 20 mg by intramuscular injection every 4 to 6 hours. It is sold without narcotic prescription.

III. MECHANISM OF ACTION

The major pharmacological action of analgetics is on the cerebrum and medulla of the central nervous system. Smooth muscle and glandular secretions of the respiratory and gastrointestinal tracts are their main target. The precise mechanism of action is not known. However, it seems that they interfere with the pain impulses in the neighborhood of the thalamus, through complexation with specific receptor sites. These receptors are perhaps not static and rigid structures but dynamic entities which can both induce conformational changes in the molecules with which they interact and undergo themselves a similar alteration. In other words, analgetic receptors can be either static or dynamic.

Recently, an endogenous material named endorphin, with morphine-like activity, was isolated from several vertebrates. It is a polypeptide of 7 to 8 amino acids with molecular weight 1000. Its distribution in different brain regions is similar to that of the opiate receptor.

A. Static Receptor

Taking into consideration that analgetics derived from, or analogous to, morphine have in common the N-methyl-γ-phenylpiperidino moiety, Beckett and Casy proposed the receptor topography shown in Figure 7.2. There three sites are essential:

1. A flat structure that allows a linkage with the aromatic ring of the drug through van der Waals-type of forces.
2. An anionic site able to associate with the positively charged basic center of the drug.
3. A suitably oriented cavity in order to accommodate the $-CH_2CH_2-$ portion (relative to carbons 15 and 16) projecting from the piperidine ring that lies in front of the plane containing the aromatic ring and the basic center.

B. Dynamic Receptor

The attractive hypothesis of Beckett and Casy, which admits an essentially inflexible receptor site, had to be modified when it was found that certain analgetics, although possessing high activity, did not present an exactly complementary structure to the proposed receptor area. For this and other reasons and on the basis of the fact that macromolecules may undergo conformational changes, Portoghese postulated that the formation of different complexes of narcotic analgetics with receptors can, in many cases, involve different modes of interaction instead of only one type of drug-receptor complexation (Figure 7.3).

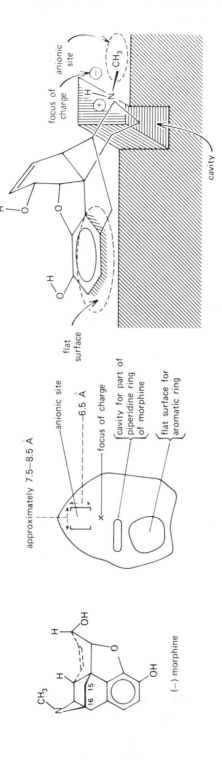

Figure 7.2 Receptor site of analgetics derived from and analogous to morphine. [After A. H. Beckett and A. F. Casy, *Prog. Med. Chem.,* 4, 171 (1965)].

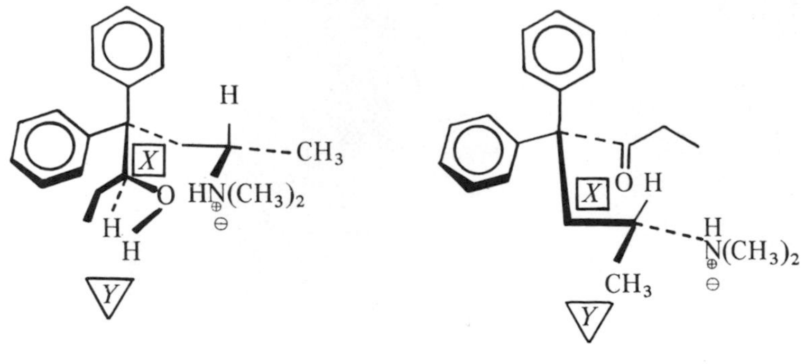

Figure 7.3 One possible mechanism whereby different polar groups in analgetic molecules may cause inversion in the configurational selectivity of analgetic receptors. Hydrogen-bonding proton donor and acceptor dipoles are noted by *X* in the square and by *Y* in the triangle, respectively. The anionic site is represented by ⊖. Left, 3S:6S-methadol; right, R-methadone. [Reprinted from P. S. Portoghese, *J. Med. Chem.*, 9, 609 (1965), by permission of the American Chemical Society and the author.]

IV. NARCOTIC ANTAGONISTS

Narcotic antagonists are drugs that prevent or abolish excessive respiratory depression caused by the administration of morphine and related compounds. They act by competing for the same receptor sites of analgetics, with which they are structurally related, the only difference being the moiety linked to the amine nitrogen atom: in antagonists usually it is an allyl group (Table 7.4).

The phenomenon of narcotic antagonism was first observed in 1915, by Pohl, in *N*-allylnorcodeine. Years later, the antagonistic effect of nalorphine was described. This discovery led eventually to the introduction of other antagonists. Those now available have similar actions, the most potent being naloxone. A newly introduced one is nalbuphine. An older one is cyclazocine.

Narcotic antagonists are also used to test for narcotic dependence. For instance, nalorphine causes an increase in pupil size of those who are narcotic dependents and a decrease in nonaddicts.

Nalorphine

Used as the hydrochloride, which occurs as white, water-soluble crystals that darken when exposed to light and air. It antagonizes not only morphine, of which it is a derivative, but also meperidine, methadone, and levorphanol. However, its antagonistic activity toward other respiratory depressants, such as general anesthetics and barbiturates, is weak. For this reason in barbiturate

Table 7.4 Narcotic Antagonists

Official Name	Proprietary Name	Chemical Name	Structure
nalorphine[+]	Nalline	N-allylnor-morphine	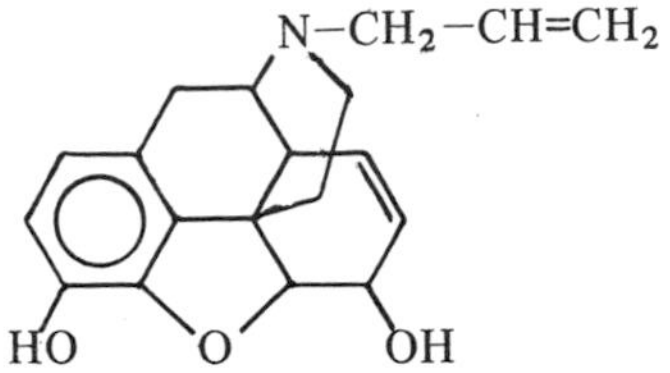
levallorphan[+]	Lorfan	(−)-N-allyl-3-hydroxymor-phinan	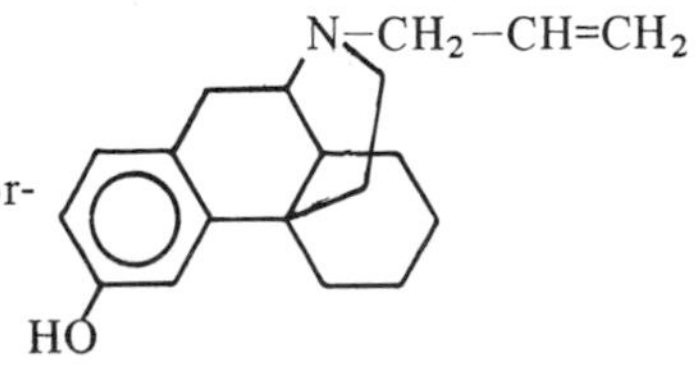
naloxone*	Narcan	N-allyl-14-hydroxynordi-hydromorphi-none	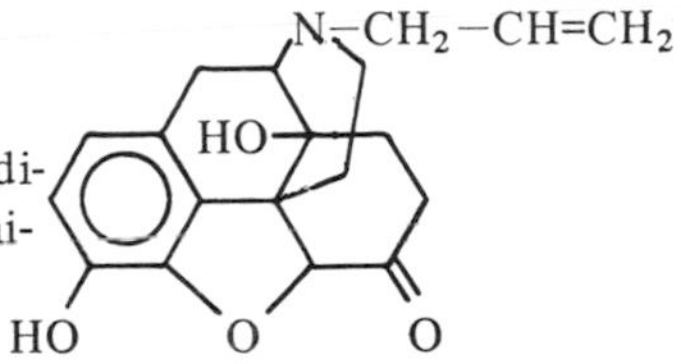

*In *USP XIX* (1975).
[+]In *NF XIV* (1975).

intoxication it deepens, rather than alleviates, depression. The routes of administration are intravenous, intramuscular, or subcutaneous. The usual dosage is 5 to 10 mg initially, repeated at 10- to 15-minute intervals if necessary, up to a maximum of three doses.

Levallorphan

Used as the tartrate, an odorless, water-soluble, white crystalline powder. It is a morphinan derivative. Its action is similar to that of nalorphine, but it is more potent. It is administered intravenously, 1 to 2 mg initially.

Naloxone

It is more potent than nalorphine or levallorphan. It has no hypnoanalgetic, psychotomimetic, miotic, or respiratory depressant actions. Naloxone antagonizes the effects of narcotics and the depression produced by pentazocine and other antagonists. The usual dosage, intravenously, is 0.4 mg, repeated at 2- to 3-minute intervals.

REFERENCES

General

D. H. Clouet, *Narcotic Drugs: Biochemical Pharmacology,* Plenum, New York, 1971.

A. Herz and H. J. Teschemacher, *Advan. Drug Res.,* 6, 79 (1971).

A. F. Casy, *Progr. Med. Chem.,* 7, 229 (1970).

G. Thuillier *et al., Chim. Ther.,* 5, 79 (1970).

G. deStevens, Ed., *Analgetics,* Academic, New York, 1965.

Introduction

M. Green, *Annu. Rev. Biochem.,* 39, 821 (1970).

A. Goldstein and D. B. Goldstein, *Proc. Assoc. Research Nervous Mental Diseases,* 46, 265 (1968).

M. H. Seevers, *J. Amer. Med. Ass.,* 206, 1263 (1968).

L. F. Small *et al., Studies on Drug Addiction,* Supplement 138 to the Public Health Reports, U.S. Government Printing Office, Washington, D.C., 1938.

Classification

E. Shefter, *J. Med. Chem.,* 17, 1037 (1974).

H. O. J. Collier, *Advan. Pharmacol. Chemother.,* 7, 333 (1970).

G. deStevens, *Pure Appl. Chem.,* 19, 89 (1969).

N. E. Eddy and E. L. May, *Synthetic Analgesics,* Part IIB, "6,7-Benzomorphans," Pergamon, Oxford, 1966, pp. 113-182.

M. Gates, *Sci. Amer.,* 215(5), 131 (1966).

J. Hellerbach, O. Schnider, H. Besendorf, and B. Pellmont, *Synthetic Analgesics,* Part IIA, "Morphinans," Pergamon, Oxford, 1966, pp. 1-112.

D. Ginsburg, *The Opium Alkaloids,* Interscience, New York, 1962.

P. A. J. Janssen, *Synthetic Analgesics,* Part I, "Diphenylpropylamines," Pergamon, New York, 1960.

Mechanism of Action

J. J. Kaufman and W. S. Koski, "Physicochemical, Quantum Chemical, and Other Theoretical Techniques for the Understanding of the Mechanism of Action of CNS Agents: Psychoactive Drugs, Narcotics, and Narcotic Antagonists and Anesthetics," in E. J. Ariëns, Ed., *Drug Design,* vol. V, Academic, New York, 1975, pp. 251-340.

G. H. Loew and D. S. Berkowitz, *J. Med. Chem.,* 18, 656 (1975).

J. L. Marx, *Science,* 189, 708 (1975).

G. W. Pasternak and S. H. Snyder, *Mol. Pharmacol.*, **11**, 478 (1975).

L. J. Saethre *et al.*, *Mol. Pharmacol.*, **11**, 492 (1975).

J. J. Kaufman *et al.*, *Int. J. Quantum Chem.*, Quantum Biology Symp. no. 1, 289 (1974).

L. Jung *et al.*, *Chim. Ther.*, 6, 341 (1971).

J. W. Lewis *et al.*, *Annu. Rev. Pharmacol.*, **11**, 241 (1971).

L. Jung and H. Lami, *Chim. Ther.*, **5**, 391 (1970).

R. Haller, *Arzneim.-Forsch.*, **23**, 608 (1968).

P. S. Portoghese, *J. Pharm. Sci.*, **55**, 865 (1966).

A. H. Beckett and A. F. Casy, *Progr. Med. Chem.*, **4**, 171 (1965).

Narcotic Antagonists

H. F. Fraser and L. S. Harris, *Annu. Rev. Pharmacol.*, 7, 277 (1967).

W. R. Martin, *Pharmacol. Rev.*, **19**, 463 (1967).

CHAPTER 8

ANALGETICS: ANTIPYRETIC AND ANTIRHEUMATIC AGENTS

I. INTRODUCTION

Antipyretic and antirheumatic agents are drugs used for treatment of connective-tissue diseases, such as rheumatoid arthritis, rheumatic fever, osteoarthritis, psoriatic arthritis, palindromic rheumatism, Reiter's syndrome, lupus erythematosus, and ankylosing spondylitis. These illnesses are second only to cardiovascular diseases in the number of people affected: in the United States about 61% of the population.

The etiological factors in most chronic inflammatory diseases are still unknown. Several mediators of inflammation, however, are already known, such as histamine, serotonin, kinins, hyaluronic acid depolymerizers, acetylcholine, epinephrine, prostaglandins, antigen-antibody complexes, and lysosomal enzymes. The causative factors, however, remain obscure.

Certain antipyretic and antirheumatic agents are also used in gout, a disease caused by an excess of uric acid occurring as sharp irritating needles in the body and manifested also by recurring attacks of acute gouty arthritis, consisting of painful inflammation of joints.

II. CLASSIFICATION

Agents studied in this chapter have a great variety of chemical structures.

A. Salicylates

Salicylates were among the first drugs of this class and still are the most commonly used (Table 8.1). They act on the heat-regulating centers of the hypothalamus and exert antipyretic effects in feverish patients but none on normal body temperature. Salicylic acid has antipyretic and antirheumatic

Table 8.1 Salicylates Most Widely Used

Official Name	Proprietary Name	Chemical Name	Structure
sodium salicylate[+]		sodium salicylate	
salicylamide	Salicin	o-hydroxybenzamide	
aspirin[*][+] (acetylsalicylic acid)	Aspro	acetylsalicylic acid	
aluminum aspirin	Aluminum dulcet	hydroxybis- (salicylato)- aluminum diacetate	

*In *USP XIX* (1975).
[+]In *NF XIV* (1975).

activities, but it is too toxic to be used as such or in the form of salts. Some less irritating and more palatable derivatives have been prepared by one of four ways: (*a*) alteration of the carboxyl group through formation of salts, esters, or amides; (*b*) substitution on the hydroxyl group; (*c*) modification on both functional groups; (*d*) introduction of another hydroxyl group or a different group into the phenyl ring and, in some cases, alterations of one or more of the three functional groups (see Table 2.7).

Old derivatives of the first type have little value as antipyretics. They are mostly of topical application, being used mainly as counterirritants or rubefacients, because they are well absorbed through the skin (see Chapter 42, Section I.D). Examples of rubefacients are methyl salicylate and diethylamine salicylate (Algesal). The following are used as analgetics: sodium salicylate, choline salicylate (Arthropan), morpholine salicylate (Deposal), ethanolamine salicylate (Salusal), calcium acetyl salicylate and salicylamide (Salicin).

The second type of salicylic acid derivatives is represented by acetylsalicylic acid, or aspirin, one of the most widely used drugs. It has analgetic, antipyretic, and antirheumatic actions; it is administered orally in powder, capsule, or tablet or rectally in suppository form.

Derivatives of the third type are the larger group. They resulted from the application of the salol principle (see Chapter 2). *In vivo* they are hydrolyzed to aspirin, the active compound. Examples are salts of aspirin [aluminum aspirin, calcium acetylsalicylate (Calcium Aspirin, Dispril), calcium aspirin-urea complex (Calurin), tetrahydrate magnesium acetylsalicylate (Fyracyl)], carbethyl salicylate (Sal-Ethyl Carbonate), phenyl salicylate (Salol), ethenzamide (Etocil, Lucamide), acetylsalicylic anhydride (Pircan), benorylate (a hybridization product of aspirin with acetaminophen). Recèntly, two aspirin derivatives that can be administered by intravenous injection and are less toxic than the parent compound have been introduced: lysine acetylsalicylate (Aspegic) and acetyl-salicylic acid-sodium glycerophosphate (Ivepirine).

Representatives of the fourth type are sodium gentisate (Gentasol), diace-tylpirocatechic acid (Movilene), 4-hydroxyphthalic acid (Eupirine), carsalam, chlortenoxazine, and flufenisal.

Salicylates are not devoid of untoward effects. They produce gastro-intestinal disturbances, such as dyspepsia, nausea, vomiting, occult bleeding. Large doses may cause salicylism and serious acid-base imbalance. Salicylate poisoning is extremely dangerous in children, owing to the profound physio-logic alteration it produces, which may even result in death.

Acetylsalicylic Acid

The official name in the United States is aspirin, which is still the proprietary name in other countries. It is the drug produced in the largest tonnage. It consists in slightly water-soluble, white crystals or white, crystalline powder. It should be kept dry, because in the presence of moisture it hydrolyzes into salicylic and acetic acids; decomposition is detected by the appearance of a violet color when the product is treated with a ferric chloride solution.

Aspirin is the prototype of analgetic-antipyretics; it is the mild analgetic of choice for treatment of headache, neuralgia, myalgia, and other pains. It is absorbed mostly unchanged, but *in vivo* it undergoes mainly hydrolysis and conjugation.

B. *p*-Aminophenol Derivatives

Among them are phenacetin and acetaminophen. Both were introduced during the last century as a result of a search for substitutes for acetanilid. Although having analgetic-antipyretic properties, acetanilid and phenacetin produce methemoglobin and, therefore, they are no longer recommended. Both phenacetin and acetanilid are metabolized to acetaminophen, which is the active drug. The metabolic fate of these substances is shown in Figure 8.1.

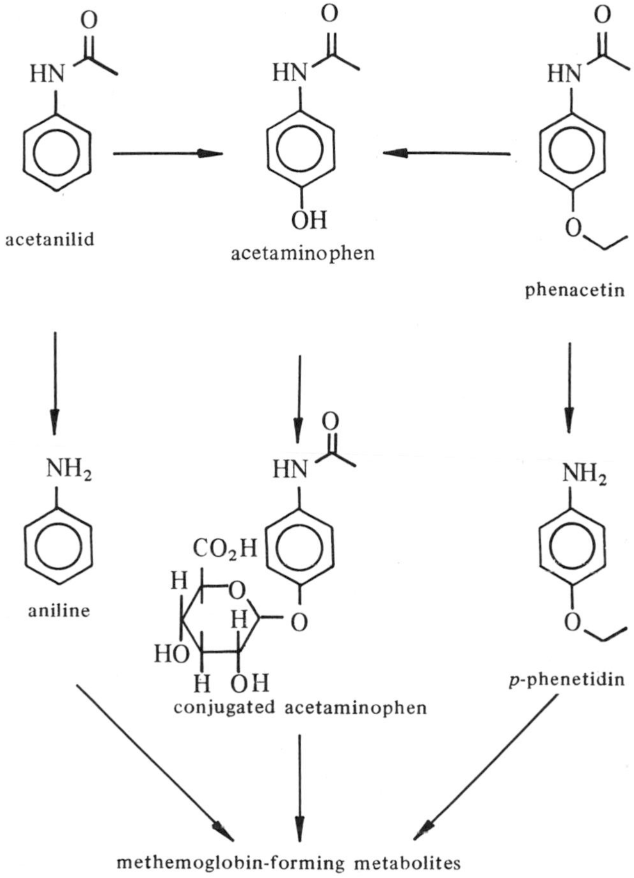

Figure 8.1 Metabolic fate of *p*-aminophenols.

A new drug of this group is chloracetadol, a hybrid association of acetaminophen with chloralium.

Acetaminophen, USP, (Tempra, Tylenol), 4-acetamidophenol (Figure 8.1), has analgetic and antipyretic activities. It is a safer drug than acetanilid or phenacetin.

C. Pyrazolone Derivatives

This group (Table 8.2) comprises derivatives of 5-pyrazolone, as well as derivatives of 3,5-pyrazolidinedione.

Table 8.2 Pyrazolone Derivatives

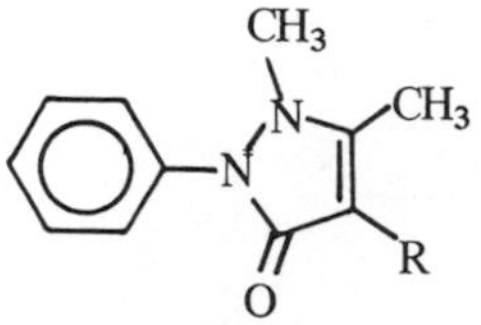

5-pyrazolone 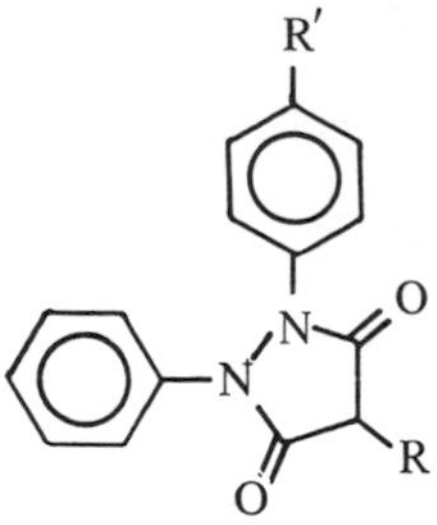   3,5-pyrazolidinedione

Official Name	Proprietary Name	Chemical Name	R	R'
5-pyrazolone derivatives				
dipyrone	Dimethone Narone Pyral	sodium [N-(1,5-dimethyl-3-oxo-2-phenylpyrazolin-4-yl)-N-methylamino] methanesulfonate monohydrate	$\underset{\displaystyle \mathrm{CH_2SO_3^-Na^+}}{\overset{\displaystyle \mathrm{N\text{-}CH_3}}{\mid}}$	
3,5-pyrazolidinedione derivatives				
phenylbutazone*	Butazolidin	4-butyl-1,2-diphenyl-3,5-pyrazolidinedione	-C$_4$H$_9$-n	H
oxyphenbutazone[+]	Tandearil	4-butyl-1-(p-hydroxyphenyl)-2-phenyl-3,5-pyrazolidinedione	-C$_4$H$_9$-n	OH
sulfinpyrazone	Anturane	1,2-diphenyl-4-[2-(phenylsulfinyl)ethyl]-3,5-pyrazolidinedione	$\underset{\displaystyle \mathrm{C_6H_5\text{-}S{\rightarrow}O}}{\overset{\displaystyle -CH_2CH_2}{\mid}}$	H

*In *USP XIX* (1975).
[+]In *NF XIV* (1975).

1. Derivatives of 5-Pyrazolone

The most widely used drugs of this group are dipyrone, antipyrine, and aminopyrine. Others have also found clinical application, especially in Europe.

Derivatives of 5-pyrazolone are prone to cause fatal agranulocytosis and other blood dyscrasias. While antipyrine is not usually associated with agranulocytosis, other drugs are more effective. Since less toxic and equally effective analgetic-antipyretic and antirheumatic agents are available, the use of 5-pyrazolone derivatives is only justifiable as a last resort to reduce fever when safer measures have failed. In the United States their use has been discontinued. They occur as water-soluble, colorless, odorless, white powders.

2. Derivatives of 3,5-Pyrazolidinedione

The most widely used are listed in Table 8.2. Derivatives of 3,5-pyrazolidinedione exert antiinflammatory, antipyretic, and analgetic effects, but they "are often more effective in ankylosing spondylitis and gout than in acute rheumatoid arthritis and other arthropathies," according to the American Medical Association. Unfortunately, these drugs are toxic; the most serious side effect is bone-marrow depression. For these reasons, they are contraindicated to many patients: children and senile people; those with gastrointestinal lesions or with hepatic, renal, or cardiovascular disease; those who are allergic to drugs or who have blood dyscrasias.

D. Arylacetic Acid Derivatives

The most commonly used drug of this group is indomethacin (Indocin). However, several related drugs are available.

Indomethacin, NF (Indocin), 1-(p-chlorobenzoyl)-5-methoxy-2-methylindole-3-acetic acid. It is a practically water-insoluble, pale yellow to yellow-tan crystalline powder, being sensitive to sunlight and alkaline solutions. It is used as an antiinflammatory analgetic in rheumatoid arthritis, osteoarthritis, ankylosing spondylitis, and occasionally in gout. Its side effects are frequent and sometimes serious: leukopenia, hemolytic anemia, aplastic anemia, purpura, and thrombocytopenia. For this reason it should not be used as an antipyretic or analgetic. This drug is contraindicated in infants and children, pregnant women, nursing mothers, and patients with gastrointestinal lesions:

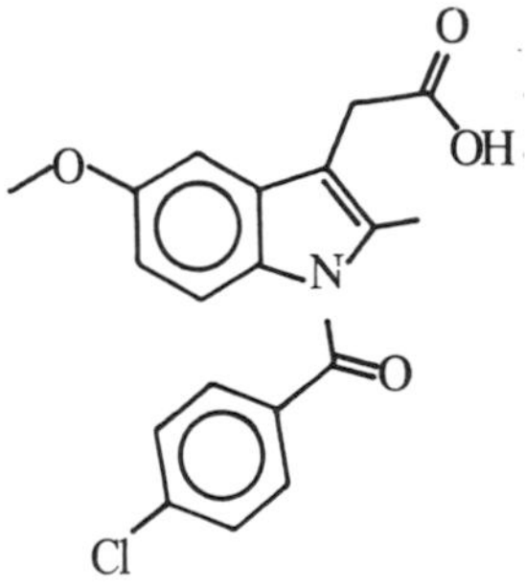

E. Adrenocortical Steroids

They are used to relieve pain and to control inflammation in rheumatoid arthritis. Considering that their adverse reactions are numerous (ulcerogenic effects, osteoporosis, spontaneous fractures, psychoses, cataracts, hypertension), they are drugs of second choice and should be reserved for those patients with moderately severe rheumatoid arthritis who do not respond to other antirheumatic agents. Adrenocortical steroids with antiinflammatory action most widely used are listed in Table 8.3. For systemic use, corticoids of choice are

Table 8.3   Antiinflammatory Adrenocortical Steroids

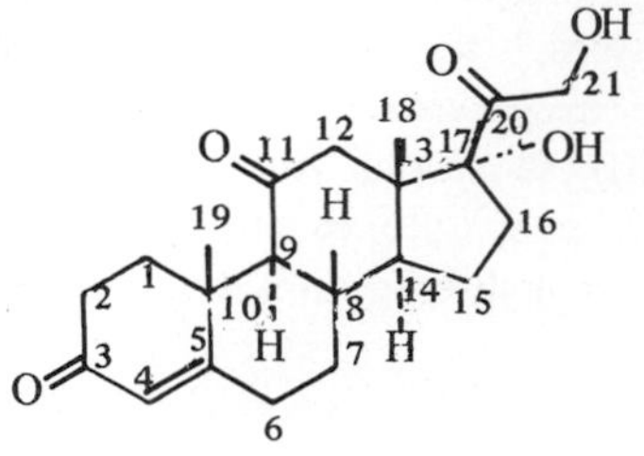

cortisone

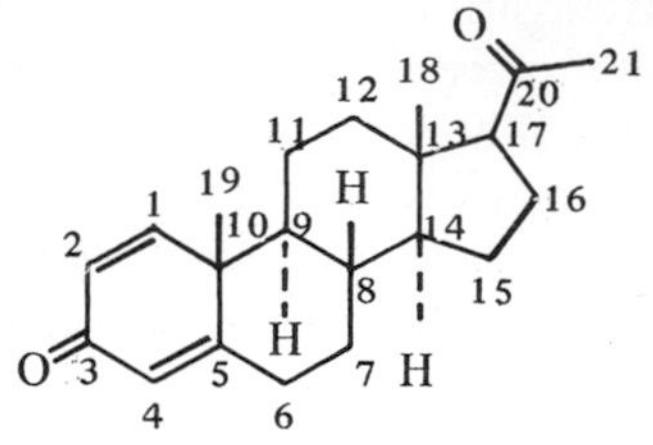

pregna-1,4-diene-3,20-dione

Official Name	Proprietary Name	Chemical Name
cortisone*	Cortone	17-hydroxy-11-dehydrocorticosterone
dexamethasone*	Decadron Deronil Hexadrol	9α-fluoro-11β,17,21-trihydroxy-16α-methylpregna-1,4-diene-3,20-dione
hydrocortisone*	Cortef Cortril Hydrocortone	11β,17,21-trihydroxypregn-4-ene-3,20-dione
prednisone*	Delta Deltasone Meticorten	17,21-dihydroxypregna-1,4-diene-3,11,20-trione
prednisolone*	Delta cortef Hydeltra Meticortilone	11β,17,21-trihydroxypregna-1,4-diene-3,20-dione
triamcinolone acetonide*[+]	Kenalog Aristocort Acetonide	9-fluoro-11β,16α,17,21-tetrahydroxypregna-1,4-diene-3,20-dione cyclic 16,17-acetal with acetone
triamcinolone hexacetonide*		9-fluoro-11β,16α,17,21-tetrahydroxypregna-1,4-diene-3,20-dione cyclic 16,17-acetal with acetone 21-(3,3-dimethylbutyrate)

*In *USP XIX* (1975).
[+]In *NF XIV* (1975).

prednisone and prednisolone. However, for topical application, especially for the treatment of eczema and psoriasis, several other drugs of this class are available.

Most of adrenocortical steroids occur as white or practically white, odorless, crystalline powders that are insoluble or practically insoluble in water but soluble in organic solvents; some derivatives, however, are freely soluble in water. The doses vary widely, from 10 to 100 mg daily.

Some adrenocortical steroids are used only in the free form. Others are used also as derivatives. For instance, prednisolone is used as such and also as acetate, sodium phosphate, succinate, butyl acetate. Hydrocortisone occurs as such and as acetate, sodium acetate, sodium phosphate, cypionate, 21-dimethyl aminoacetate. Triamcinolone is marketed also as acetonide, and dexamethasone as 21-isonicotinate (Auxiloson). Fluocortolone occurs both as hexanoate (Ficoid) and pivalate.

F. Anthranilic Acids

Older members of this class are mefenamic acid (Ponstel), flufenamic acid (Arlef), niflumic acid (Nifluril), and glaphenine (Glifan, Glifanan). Some new analogues became recently available: aclantate, clonixeril, clonixin, diclofenac, floctafenine, meclofenamic acid, and niflumic acid morpholinoethyl ester.

Mefenamic Acid

Chemically, it is N-(2,3-xylyl)-anthranilic acid. It is water-insoluble, colorless, crystalline powder, with antipyretic and antiinflammatory activities, but since it may cause serious adverse reactions, other mild analgetics are preferred.

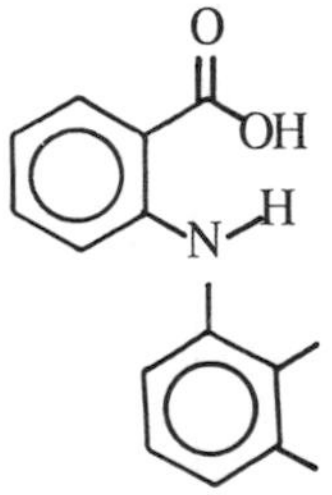

mefenamic acid

gold sodium thiomalate, *USP*

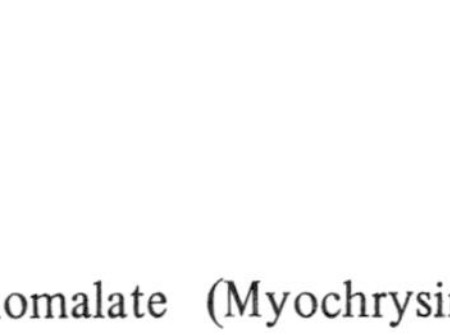

aurothioglucose, *USP*

G. Gold Compounds

Those most commonly used are gold sodium thiomalate (Myochrysine), aurothioglucose (Solganal), and aurothioglycanide (Lauron).

H. Immunosuppressants

According to the American Medical Association, immunosuppressants are being used experimentally in patients with rheumatoid arthritis who have progressive bone and gout deterioration despite adequate treatment with other drugs, as well as in patients with psoriatic arthritis or systemic lupus erythematosus. Those mostly used are azathioprine (Imuran), chlorambucil, cyclophosphamide, and methotrexate (Antineoplastic Agents, Chapter 34).

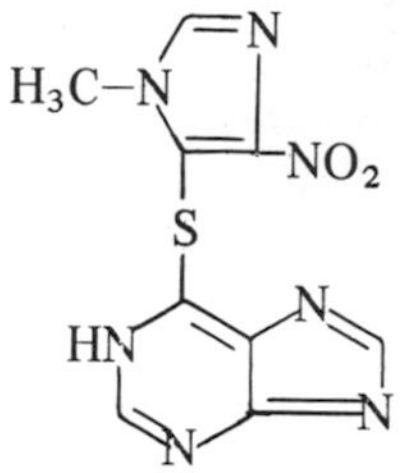

azathioprine, *USP*

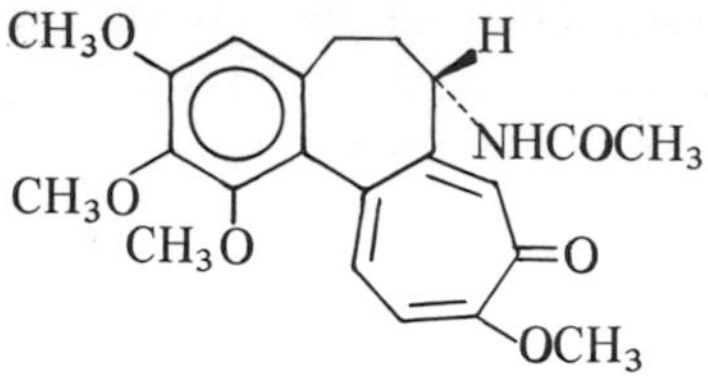

colchicine, *USP*

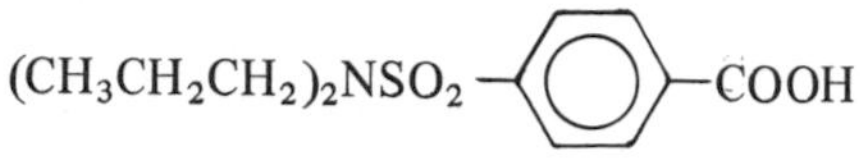

probenecid, *USP*

I. Miscellaneous

Many other drugs, some recently introduced, not included in the preceding chapters have antipyretic or antirheumatic activities. Certain ones are: allopurinol, benzydamine (Dorinamin), bufexamac (Droxaryl), chloroquine, colchicine, corticotropine, etebenecid (Longacid), griseofulvin, hydroxychloroquine, ibuprofen (Brufen), metofoline, orotic acid (Dioron), namoxyrate (Namol), naphthypramide, penicillamine, phenamol, probenecid (Benemid), proxazole (Toness).

J. Mixtures

Several mixtures of analgetics, antipyretics, antirheumatics, and other drugs are commercially available. The reasoning behind the formulating of most of these mixtures is irrational.

III. MECHANISM OF ACTION

The mechanism of action of several drugs studied in this chapter is well known; for most of them it is only hypothesized.

Well known is the mechanism of action of uricosuric agents, such as allopurinol, probenecid, and sulfinpyrazone, used in the treatment of gout.

Owing to its structural resemblance to hypoxanthine and xanthine, allopurinol and its metabolite alloxanthine inhibit xanthine oxidase, the enzyme involved in the biosynthesis of uric acid, which is responsible for gout, and thus prevent the formation of this acid (see Chapter 2, Section IV.E.1, for structures).

Probenecid, a drug designed to depress tubular secretion of penicillin, inhibits tubular reabsorption of uric acid and in this way increases its renal excretion. The same mechanism is operative in sulfinpyrazone.

The action of colchicine against attack of acute gouty arthritis may be related to its property of being a mitotic poison. It may relieve pain by disrupting microtubules in leucocytes and thus preventing phagocytosis and consequent inflammation.

As to other agents studied in this chapter, several modes of action have been proposed. Recently, it has been suggested that acidic antiinflammatory agents, such as aspirin and indomethacin, owe their effects to inhibition of the synthesis or release of prostaglandins. Another hypothesis claims that antirheumatic drugs act by dislodging certain small peptides from their binding sites. These free peptides can then protect connective tissues against chronic inflammatory processes.

Although no universal agreement has been reached as to how antiinflammatory drugs act, experimental evidence supports the following mechanisms proposed by Whitehouse:

1. Steroids, because they are lipophilic, can function as nuclear regulators of cellular metabolism.

2. Salicylates and some other agents (indomethacin, fenamic acids, phenylbutazone, gold salts) uncouple oxidative phosphorylation, which is represented by the following equation:

$$NADH + H + \tfrac{1}{2}O_2 + 3ADP + 3P_i \rightarrow NAD + H_2O + 3ATP$$

3. Since they are uncouplers of oxidative phosporylation, they inhibit the synthesis of ATP.

4. Salicylates and probably some other agents may chelate an ion, such as copper or zinc, which may be important for an enzyme.

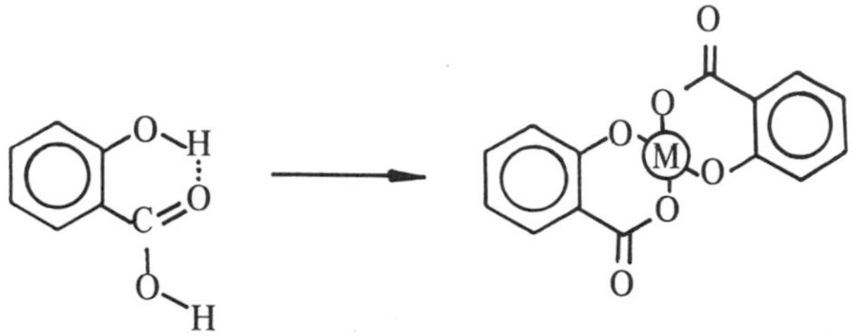

This chelation may result in the uncoupling of oxidative phosphorylation and, in consequence, interference with the metabolic processes and the movement of molecules into the inflamed joint, with the subsequent slowing of the inflamatory process.

5. The acidic antiinflammatory agents (salicylates, 4-isobutylphenylacetic acid, phenylbutazone, indomethacin, and fenamic acids) compete, in the anionic form, with pyridoxal-5-phosphate for the coenzyme-binding site on the enzyme apoprotein, probably a lysine ϵ-amino group:

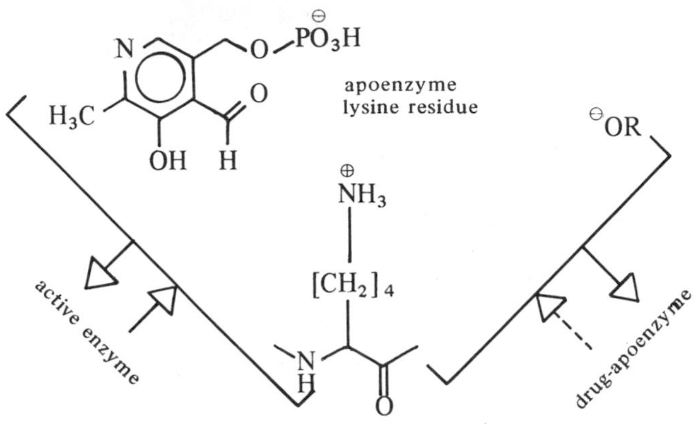

Kier and Whitehouse have found that certain antiinflammatory agents (fenamic acids, cortisol, phenylbutazone, indomethacin, salicylates), as well as some inflammagenic amines (histamine, serotonin), present very similar distances (4.4 to 6.17 Å) between atoms supposedly involved in the interaction with receptors. For this reason they advanced the hypothesis that both groups of substances act on the same receptor. By having distances very close to those found in inflammagenic amines, antiinflammatory drugs might displace histamine and/or serotonin from the receptor site. However, they do not act as competitive antagonists but as false feedback inhibitors to prevent the biogenesis of those amines, if indeed the biosynthesis of histamine and/or serotonin is autoregulated by either allosteric or end-product inhibition.

According to Vane, salicylates inhibit prostaglandin synthetase, the enzyme which is involved at an early stage in the biosynthesis of prostaglandins.

REFERENCES

General

R. J. Perper, Ed, "Mechanisms of Tissue Injury with Reference to Rheumatoid Arthritis," *Ann. N. Y. Acad. Sci.,* **256,** 1-450 (1975).

R. A. Scherrer and M. W. Whitehouse, Eds., *Antiinflammatory Agents: Chemistry and Pharmacology,* 2-Vols., Academic, New York, 1974.

H. E. Paulus and M. W. Whitehouse, "Drugs for Chronic Inflammatory Disease," in A. A. Rubin, Ed., *Search for New Drugs,* Dekker, New York, 1972, pp. 1-114.

M. Rocha e Silva and J. Garcia Leme, *Chemical Mediators of the Acute Inflammatory Reaction,* Pergamon, Oxford, 1972.

T. Y. Shen, *Angew. Chem., Int. Ed. Engl.,* **11,** 460 (1972).

W. Müller *et al.,* Eds., Colloquia Geigy, *Rheumatoid Arthritis,* Academic, London, 1971.

R. H. Ferguson and J. W. Worthington, *Ann. Intern. Med.,* **73,** 109 (1970).

W. Murray and S. Piliero, *Annu. Rev. Pharmacol.,* **10,** 171 (1970).

Classification

H. Terada, *J. Med. Chem.,* **17,** 330 (1974).

K. J. Doebel, *Pure Appl. Chem.,* **19,** 49 (1969).

R. W. Rundles *et al., Annu. Rev. Pharmacol.,* **9,** 345 (1969).

A. K. Done, *Clin. Toxicol.,* **1,** 4, 451 (1968).

D. J. McCarty, Jr., and G. McLaughlin, *Top. Med. Chem.,* **2,** 217 (1968).

I. M. Krakoff, *Clin. Pharmacol. Ther.,* **8,** 427 (1967).

J. H. Shelley, *Clin. Pharmacol. Ther.,* **8,** 427 (1967).

A. B. Gutman, *Advan. Pharmacol.,* **4,** 91 (1966).

M. J. H. Smith and P. K. Smith, Eds., *The Salicylates: A Critical Bibliographic Review,* Interscience, New York, 1966.

M. W. Whitehouse, *Progr. Drug Res.,* **8,** 321 (1965).

L. H. Sarett *et al., Progr. Drug Res.,* **5,** 11 (1963).

H. K. von Rechenberg, *Butazolidin,* Arnold, London, 1962.

P. K. Smith, *Acetophenetidin: A Critical Bibliographic Review,* Interscience, New York, 1958.

L. A. Greenberg, *Antipyrine: A Critical Bibliographical Review,* Hillhouse, New Haven, Conn., 1950.

Mechanism of Action

E. G. McQueen, *Drugs,* **6,** 104 (1973).

J. Flower *et al., Nature New Biology,* **238,** 104 (1972).

H. Levitan and J. L. Barker, *Science,* **176,** 1423 (1972).

G. Kaley and R. Weiner, *Ann. N.Y. Acad. Sci.,* **180,** 338 (1971).

J. R. Vane, *Nature New Biology,* **231,** 232 (1971).

M. W. Whitehouse, *Pure Appl. Chem.,* **19,** 35 (1969).

L. B. Kier, *J. Med. Chem.,* **11,** 915 (1968).

L. B. Kier and M. W. Whitehouse, *J. Pharm. Pharmacol.,* **20,** 793 (1968).

M. W. Whitehouse, *Biochem. Pharmacol.,* Special Suppl. to vol. 17, 293 (1968).

CHAPTER 9

ANTITUSSIVES

I. INTRODUCTION

Antitussives are agents which aid in decreasing the frequency of coughing.

Cough is a physiologic protective reflex partially under voluntary control, and its function is to expel irritating substances or excess secretions from the respiratory tract. Several factors may be involved in cough etiology: irritative, allergic, infectious, vascular, neoplastic, and psychogenic are the most common. The act of coughing consists of three main phases: (*a*) a sharp, deep inspiration ending with closure of the glottis; (*b*) a contraction of the thoracic and abdominal muscles followed by bronchoconstriction, resulting in a marked but momentary increase in intrathoracic pressure; (*c*) the opening of glottis and concomitant rapid and speedy expulsion of air and irritating material. Cough is regulated by the cough center, which is located in the medulla oblongata. But cough can also result from mechanical, chemical, or other types of stimulation of the nerve endings of the respiratory tract.

Antitussives can act either by raising the threshold of the cough center or by reducing the number of impulses transmitted to the cough center from the peripheral receptors; some antitussives can act by both mechanisms. Accordingly, antitussives are divided into two groups: centrally acting antitussives and peripherally acting antitussives. The first group includes opium alkaloids, their semisynthetic derivatives and synthetic antitussives. The second group is comprised of demulcents, expectorants, mucolytic agents, and other agents, which, in general, are not used invidivually but incorporated into antitussive pharmaceutical preparations.

As to adverse reactions, centrally acting antitussives may produce nausea, constipation, dizziness, somnolence, vomiting, and other disturbances. They do not cause such a deep respiratory depression as narcotic analgetics. Usually they do not induce dependence because of the short period of use. Peripherally acting antitussives produce also untoward effects, but they are mild and less frequent.

II. CLASSIFICATION

Antitussives may be conveniently divided into two groups: peripherally acting agents and centrally acting agents. The opium alkaloids (Chapter 7) possess antitussive activity. Codeine is the antitussive of choice. The structures and names of others commonly used appear in Table 9.1. The following are also marketed: caramiphen (Panparnit), dimethoxanate (Cothera), hydrocodone (Dicodid), pipazethate (Theratus).

A. Peripherally Acting Antitussives

According to their way of exerting antitussive action, these agents can be grouped into the following classes: demulcents, expectorants, mucolytics, and inhibitors of peripheral cough receptors.

1. *Demulcents*

Demulcents are substances that act by protectively coating the irritated mucosa, by stimulating the production of saliva, and very likely by exerting a mild local anesthetic action. They are used mostly as vehicles for the more specific antitussives. Among demulcents are found acacia, glycerin, honey, and licorice. These and other demulcents are also used as dermatological agents (see Chapter 42).

2. *Expectorants*

Expectorants are agents that modify the output and viscosity of the respiratory tract fluid which is then more easily removed. According to their mechanism of action they can be divided into (*a*) sedative expectorants—they act by stimulating gastric reflexes through stomach irritation: ammonium chloride, ammonium carbonate, glyceryl guaiacolate, potassium guaiacolsulfonate, syrup of ipecac, iodinated glycerin (Organidin), potassium iodide, hydroiodic acid syrup; (*b*) stimulant expectorants—they act by stimulating secretory cells of the respiratory tract: volatile organic oils (such as terpin hydrate), terebene, tolu balsam, creosote, chloroform, saponins, and glycosides.

3. *Mucolytics*

Mucolytic agents have been useful as antitussives, because in irritative states there is an increase in concentration of mucoproteins and mucopolysaccharides, the main constituents of normal respiratory-tract secretions. Among these agents, used as aerosol inhalants, may be cited some detergents, tyloxapol (Alevaire), pimetine, bromhexine (Bisolvon), acetylcysteine (Mucomyst), mesna (Mistrabon), sobrerol (Sobrepin), methylcysteine (Visclair), iodide, Ascoxal, and

Table 9.1 Antitussives Commonly Used

Official Name	Proprietary Name	Chemical Name	Structure
noscapine	Nectadon	(−)-narcotine	
dextromethorphan[+]	Romilar	(+)-3-methoxy-17-methyl-9α,13α,14α-morphinan	
levopropoxyphene[+]	Novrad	(−)-α-4-(dimethyl-amino)-3-methyl-1,2-diphenyl-2-butanol propionate	
carbetapentane	Toclase	2-[2-(diethylami-no)ethoxy]ethyl 1-phenylcyclopen-tanecarboxylate	
chlophedanol	ULO	1-o-chlorophenyl-1-phenyl-3-dimeth-ylaminopropan-1-ol	

144

Table 9.1 Antitussives Commonly Used (continued)

Official Name	Proprietary Name	Chemical Name	Structure
benzonatate[+]	Tessalon Ventussin	2,5,8,11,14,17,20, 23,26-nonaoxa-octacosan-28-yl p-(butylamino)-benzoate	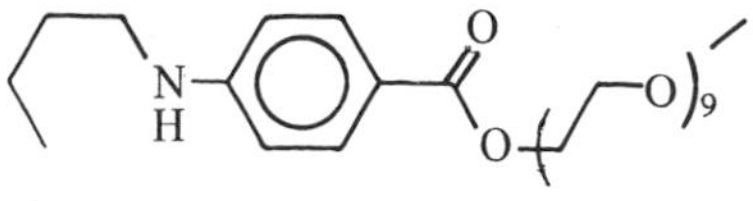
glyceryl guaiacolate[+]	Robitussin	3-(o-methoxyphen-oxy)-1,2-propane-diol	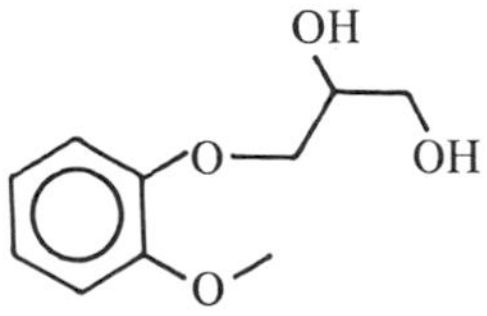
terpin hydrate[+]		terpinol	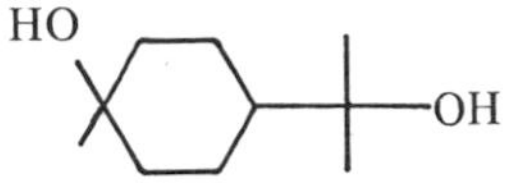

[+]In *NF XIV* (1975).

some enzymes: pancreatic dornase (Dornavac), trypsin, chymotrypsin.

4. Inhibitors of Peripheral Cough Receptors

In this class are included several local anesthetics or agents related to these drugs. The prototype is benzonatate, but oxalamine (Bredon), Libexin, and related compounds act in the same way.

Benzonatate

Chemically related to the local anesthetic tetracaine, it is a colorless or pale-yellow oil, insoluble in water but soluble in most organic solvents. It was introduced in 1956, as a result of a planned search for more specific anti-tussive agents based on the rationale that they could be found among compounds having a local anesthetic moiety combined with a substance having selective affinity for myelin. Its antitussive action is essentially peripheral, at the level of pulmonar receptors.

B. Centrally Acting Antitussives

Drugs acting by depression of the cough center are divided into two classes: (1) opium alkaloids and semisynthetic derivatives; (2) synthetic antitussives.

1. Opium Alkaloids and Derivatives

Both morphine and codeine, studied in Chapter 7, have antitussive activity. Codeine is considered the antitussive of choice. Other alkaloids of opium used as antitussives also seen in Chapter 7 are hydrocodone, ethylmorphine, and hydromorphone. New derivatives of opium alkaloids with similar activity are 3-methyl-*N*-methylmorphinan and nicocodine (Lyopect). In Europe, glauvent, an aporphinic alkaloid, is also marketed as an antitussive.

Hydrocodone

Used as bitartrate, a fine, white, water-soluble crystalline powder. Its action is similar to that of codeine, but it produces less untoward effects and has greater dependence liability. It is prepared from codeine or from dihydrothebaine. The usual dose is 5 to 10 mg three or four times daily.

Noscapine

It is extracted from opium, of which it constitutes 0.75 to 9%. It is a fine, water-insoluble, crystalline powder. Formerly known as narcotine, it contains two asymmetric centers (C-1 and C-9), allowing two possible diastereoisomeric pairs (designated α and β), whose absolute configurations were determined by Ohta *et al.* The natural isomer is (−)-noscapine. It does not produce dependence and is free of untoward effects peculiar to narcotic antitussives. The usual dosage is 15 to 30 mg.

2. Synthetic Antitussives

They were developed through exploration of side effects of analgetics, sympathomimetics, spasmolytics, antihistaminics, and phenothiazine depressant agents. In general they are used as salts, especially hydrochlorides, which occur as water-soluble, white or colorless crystalline powders.

Dextromethorphan

Used as hydrobromide, it is a sparingly water-soluble crystalline powder with a faint odor. It is a derivative of morphinan, being the optical antipode of levorphanol, an analgetic and antitussive agent. Since dextromethorphan is devoid of analgetic, constipative, addictive, and other properties of levorphanol, this pair of isomers is a good example showing the importance of stereochemistry in drug action. The dosage varies from 15 to 30 mg one to four times daily.

Chlophedanol

It was introduced in 1960 as a result of research with the objective of discovering new antispasmodics. It is almost as potent as codeine, the usual dosage being 25 mg three to four times daily.

Many other synthetic antitussives have been marketed in the U. S. and other countries. Several have been recently introduced. Time will tell if these become important.

REFERENCES

General

F. P. Doyle and M. D. Mehta, *Advan. Drug Res.,* 1, 107 (1964).
C. I. Chappel and C. von Seemann, *Progr. Med. Chem.,* 3, 89 (1963).

Introduction

C. L. Winek, *New Engl. J. Med.,* 280, 840 (1969).

Classification

N. B. Eddy, H. Friebel, K.-J. Hahn, and H. Halbach, *Codeine and Its Alternates for Pain and Cough Relief,* World Health Organization, Genève, 1970.
E. B. Shaw, *J. Amer. Med. Assoc.* 208, 1493 (1969).
M. Ohta *et al., Chem. Pharm. Bull. (Tokyo),* 12, 1080 (1964).

10

PSYCHOTHERAPEUTIC AGENTS

I. INTRODUCTION

Psychotherapeutic agents are selective modifiers of the central nervous system which are used for the treatment of psychiatric disorders. They are also called *psychoactive* or *psychotropic* agents and include those drugs which either depress or stimulate selectively mental activity. These drugs that affect the mind have given origin to a new branch of pharmacology, *psychopharmacology*, and have dramatically decreased the need for beds in mental hospitals.

Many theories have been advanced to explain the chemical pathology of mental and emotional disorders, but none is of general acceptance. One of the most recent relates them to cerebral amines, which coordinate and regulate biochemical, physiological, psychological, and clinical phenomena. According to Dewhurst, proponent of this theory, those amines are either stimulant or depressant, and the effects they produce are mediated by two specific cerebral receptors: a disturbance in their normal mechanism causes functional psychoses. Since the etiology of functional psychoses is not yet determined, agents used in the treatment of these disorders are not curative: they only alleviate their symptoms by mechanisms which are not completely elucidated.

As to the sites of action of these drugs, it is already known that they are located in the hypothalamus, brainstem, and probably other subcortical parts of the brain involved in the coordination of emotional behavior.

Many authors classify all psychotropic drugs, according to their predominant clinical use, in the following six groups: antipsychotic agents, antianxiety agents, antidepressant drugs, hallucinogenic agents, muscle relaxants, and analeptics. The first four groups are studied in this chapter, the fifth, in Chapter 11, and the sixth, in Chapter 12.

II. ANTIPSYCHOTIC AGENTS

Antipsychotic agents, known also as *neuroleptics,* and formerly called *major tranquilizers,* not only produce calm in severely disturbed psychiatric patients but also relieve them of the symptoms of their diseases. However, contrary to the effect caused by hypnotics and sedatives, they do not cloud consciousness or

148

depress vital centers. For this reason, they are used for prolonged treatment of acute and chronic schizophrenia. These drugs are characterized by antipsychotic activity, failure to produce deep coma and anesthesia even at large doses, production of effects on the extrapyramidal system, and lack of psychic or physical dependence.

They are used in the treatment of patients with psychotic disorganization of thought and behavior, and in relief of severe emotional tension. In other words, their main application is the therapy of functional psychoses, especially schizophrenia. However, they are not curative, their action being primarily palliative, since the causative factor of functional psychoses is unknown.

Most antipsychotic agents have also antiemetic, sympatholytic, and α-adrenergic blocking actions. All produce closure of the eyelid. None is exempt from untoward effects. Since some of these drugs produce autonomic blockade in varying degrees, a vasopressor such as norepinephrine should be administered if circulatory collapse occurs. However, epinephrine never should be used in these cases, because in a patient with partial adrenergic blockade it may lead to a dangerous fall in the blood pressure.

Because neuroleptic drugs potentiate the action of other central-nervous-system depressants, extreme care should be taken if a need arises to use them concurrently with alcohol, hypnotics such as barbiturates, narcotic analgetics, or general anesthetics.

A. Classification

Neuroleptic drugs can be suitably divided into the following classes: phenothiazine derivatives, thioxanthene derivatives, *Rauwolfia* alkaloids, butyrophenone derivatives, and miscellaneous antipsychotic agents.

1. *Phenothiazine Derivatives*

Although many others are marketed, the most commonly used phenothiazine derivatives are listed in Table 10.1. Their side chains vary considerably, but most of them present one of the following three: (*a*) dimethylaminopropyl moiety (chlorpromazine, promazine, triflupromazine); (*b*) piperidyl moiety (mesoridazine, thioridazine); (*c*) piperazinyl moiety (acetophenazine, butaperazine, carphenazine, fluphenazine, perphenazine, prochlorperazine, trifluoperazine, thiopropazate).

From studies of structure-activity relationship it has been concluded that in the phenothiazine derivatives and analogues the structural features connected with high antipsychotic potency are the following: (*a*) a tricyclic ring system with a six- or seven-membered central ring; (*b*) a chain of three atoms between the central ring and a terminal amino group; and (*c*) an electron-attracting atom or group, such as chloro, methoxy, or trifluoromethyl, at the meta position relative to the atom of the central ring attached to the side chain.

Table 10.1 Phenothiazine Derivatives

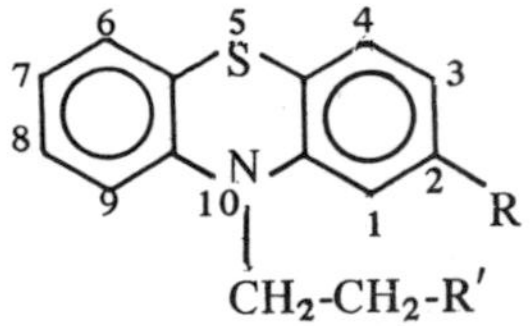

$$\text{CH}_2\text{-CH}_2\text{-R}'$$

Official Name	Proprietary Name	Chemical Name	R	R'
chlorpromazine*	Thorazine	2-chloro-10-[3-(dimethylamino)-propyl]phenothiazine	Cl	$-\text{CH}_2\text{N}(\text{CH}_3)_2$
promazine[+]	Sparine	10-[3-(dimethylamino)propyl]phenothiazine	H	$-\text{CH}_2\text{N}(\text{CH}_3)_2$
triflupromazine[+]	Vesprin	2-trifluoromethyl-10-[3-(dimethylamino)propyl]phenothiazine	CF_3	$-\text{CH}_2\text{N}(\text{CH}_3)_2$
prochlorperazine*	Compazine	2-chloro-10-[3-(4-methyl-1-piperazinyl)propyl]phenothiazine	Cl	$-\text{CH}_2\text{-N}\overbrace{}\text{N-CH}_3$
trifluoperazine[+]	Sterlazine	2-trifluoromethyl-10-[3-(4-methyl-1-piperazinyl)propyl]phenothiazine	CF_3	$-\text{CH}_2\text{N}\overbrace{}\text{N-CH}_3$
perphenazine[+]	Trilafon	2-chloro-10-{3-[4-(2-hydroxyethyl)-1-piperazinyl]propyl}phenothiazine	Cl	$-\text{CH}_2\text{N}\overbrace{}\text{NCH}_2\text{CH}_2\text{OH}$
fluphenazine*	Permitil Prolixin	2-trifluoromethyl-10-{3[4-(2-hydroxyethyl)-1-piperazinyl]propyl}phenothiazine	CF_3	$-\text{CH}_2\text{N}\overbrace{}\text{NCH}_2\text{CH}_2\text{OH}$
acetophenazine[+]	Tindal	2-acetyl-10-{3[4-(2-hydroxyethyl)-1-piperazinyl]-propyl}phenothiazine	COCH_3	$-\text{CH}_2\text{-N}\overbrace{}\text{NCH}_2\text{CH}_2\text{OH}$

butaperazine	Repoise	2-butyryl-10-[3-(4-methyl-1-piperazinyl)propyl] phenothiazine	COC_3H_7	 $-CH_2N\!\!\diagdown\!\!\diagup NCH_3$ (piperazine)
carphenazine[+]	Proketazine	2-propionyl-10-{3-[4-(2-hydroxyethyl)-1-piperazinyl] propyl} phenothiazine	COC_2H_5	 $-CH_2N\!\!\diagdown\!\!\diagup NCH_2CH_2OH$ (piperazine)
thiopropazate	Dartal	2-chloro-10-{3-[4-(2-acetoxyethyl)-piperazinyl] propyl} phenothiazine	Cl	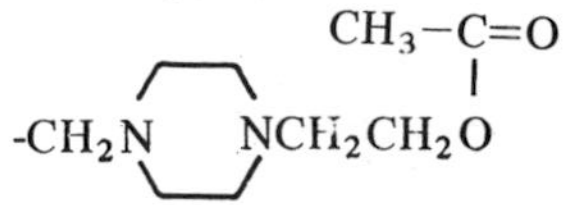 $-CH_2N\!\!\diagdown\!\!\diagup NCH_2CH_2O-C(CH_3)=O$ (piperazine)
piperacetazine[+]	Quide	2-acetyl-10-{3-[4-(2-hydroxyethyl)-piperidino] propyl} phenothiazine	$COCH_3$	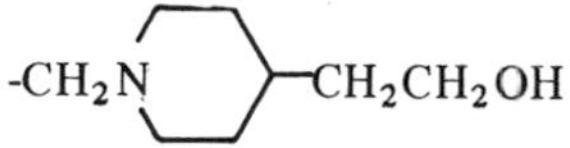 $-CH_2N\!\!\diagdown\!\!\diagup CH_2CH_2OH$ (piperidine)
thioridazine*	Mellaril	2-methylthio-10-[2-(1-methyl-2-piperidyl)ethyl] phenothiazine	$S-CH_3$	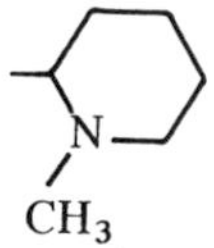 piperidine, $N-CH_3$
mesoridazine	Serentil	2-methylsulfinyl-10-[2-(1-methyl-2-piperidyl)ethyl] phenothiazine	$S\!\rightarrow\!O$ $-CH_3$	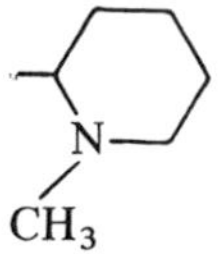 piperidine, $N-CH_3$

*In *USP XIX* (1975).
[+]In *NF XIV* (1975).

As free bases these drugs are insoluble in water. For this reason, they are used generally as salts, especially hydrochlorides, which occur as water-soluble, white, crystalline powders. Besides neuroleptic action, some of them exert also antiemetic effect or potentiate effects of analgetics, sedatives and anesthetics.

Phenothiazines are not exempt from untoward effects. The most serious are extrapyramidal reactions, ascribed to the piperazine moiety, and pigmentary retinopathy, due to $-S-CH_3$, moiety which substitutes Cl in thioridazine and mesoridazine. Other adverse reactions are jaundice, dermatitis, agranulocytosis, convulsions, skin and eye changes.

In vivo phenothiazines undergo extensive metabolism. For instance, the phenothiazine ring is sulfoxidized and hydroxylated, mainly in positions 7 and 3. The side chain is mono- or didemethylated and also *N*-oxidized. The metabolites are excreted to a large extent as glucuronides.

Chlorpromazine

Used both as hydrochloride and as free base. Owing to its sedative properties, it is widely used to reduce combative behavior in psychotic patients and to control anxiety and tension in the same patients after hospitalization. Larger doses, especially in children, can give rise to extrapyramidal symptoms, which are controlled by reduction of dose and administration of an antiparkinsonism drug. Another common side effect is orthostatic hypotension, when the drug is given by parenteral route.

Chlorpromazine potentiates other depressants, such as barbiturates, morphine-like analgetics, and general anesthetics. The usual dosage by intramuscular, intravenous, or oral routes is 30 mg to 1 g daily in divided doses.

2. Thioxanthene Derivatives

Structurally related to phenothiazines, they resulted from isosteric replacement in chlorpromazine and analogues.

The most commonly used are listed in Table 10.2. Their pharmacological actions are very similar to those of phenothiazine derivatives. The following are also marketed: clopenthixol (Sordinol), flupenthixol (Fluanxol).

Chlorprothixene

This isostere of chlorpromazine was introduced in 1961. It is effective in treatment of schizophrenia and produces less extrapyramidal symptoms than most phenothiazines. The usual dosage, by oral route, is 30 to 600 mg daily.

Thiothixene

Structurally related to chlorprothixene, it was introduced in 1967. Used in

Table 10.2 Thioxanthene Derivatives

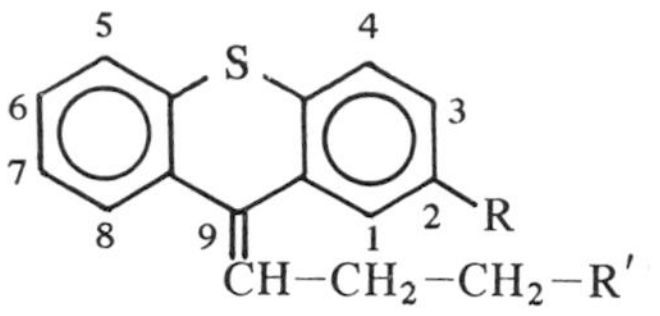

Official Name	Proprietary Name	Chemical Name	R	R'
chlorprothixene[+]	Taractan	cis-2-chloro-9-(3-dimethylaminopropylidine)thioxanthene	$-Cl$	$-N(CH_3)_2$
thiothixene[+]	Navane	cis-N,N-dimethyl-9-[3-(4-methyl-1-piperazinyl)propylidene]thioxanthene-2-sulfonamide	$-SN(CH_3)_2$ (with O above and O below)	$-N\overbrace{}N-CH_3$

[+]In *NF XIV* (1975).

treatment of hospitalized patients with chronic schizophrenia. Larger doses produce extrapyramidal reactions.

3. Rauwolfia Alkaloids

Those mostly used are listed in Table 10.3. As antipsychotic agents they are not as effective as phenothiazines, being administered only to patients who cannot tolerate these drugs. *Rauwolfia* alkaloids are also used as antihypertensive agents. The most potent is reserpine.

A semisynthetic derivative, metoserpate (Pacitran), which is the methyl ether of methyl 18-epireserpate, has been tried clinically with good results.

Rauwolfia alkaloids cause depletion of catecholamines and serotonin, an ability which is not shared by phenothiazines and butyrophenones. Some authors believe that this depletion is responsible for their pharmacological action, either antipsychotic or sedative.

These drugs cause side effects of mental depression, parkinsonism, allergic reactions, and other minor symptoms.

Stereochemistry plays an important role in pharmacological actions of these alkaloids. The natural reserpine is (−)-isomer. It has six asymmetric centers, making, therefore, possible the existence of 64 isomers. (+)-Reserpine and several other isomers have been synthesized, but none manifests appreciable antipsychotic or antihypertensive properties. Alkaline hydrolysis of reserpine

Table 10.3 *Rauwolfia* Alkaloids

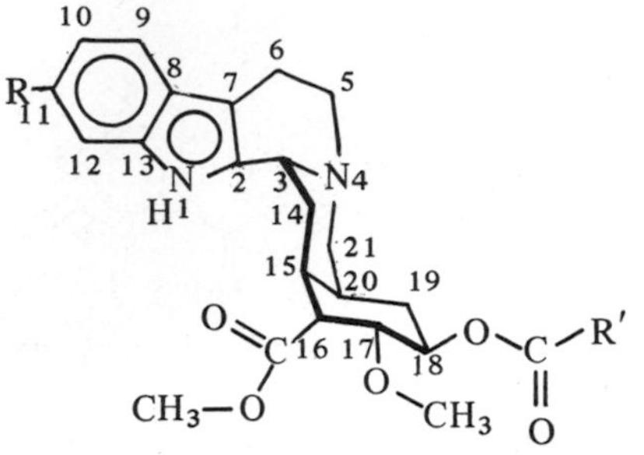

Official Name	Proprietary Name	R	R'
reserpine*	Reserpoid Serpasil	CH_3O-	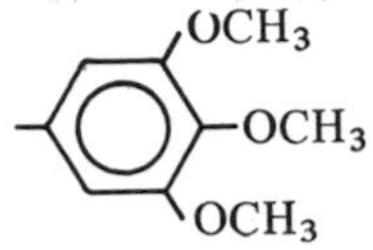
deserpidine	Harmonyl	$H-$	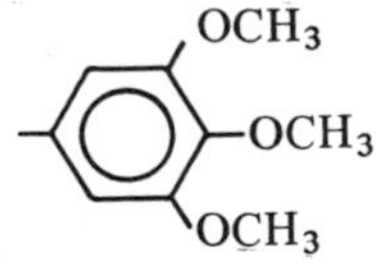
syrosingopine	Singoserp	CH_3O-	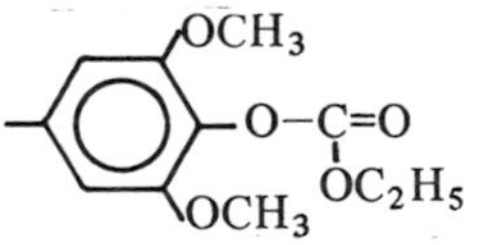
rescinnamine	Moderil	CH_3O-	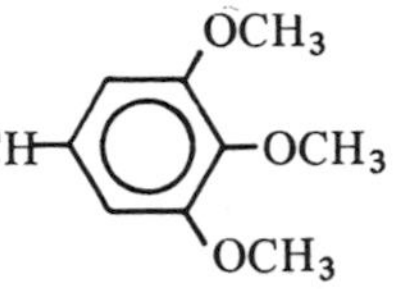

*In *USP XIX* (1975).

gives reserpic acid, 3,4,5-trimethoxybenzoic acid, and methanol.

Reserpine

Extracted from various species of *Rauwolfia*, it occurs as white to light-yellow crystalline powder, which is practically insoluble in water. It decomposes on exposure to light or by oxidation, especially in solution. To prevent this decomposition, which is followed by loss of activity, several stabilizers are used: sodium metabisulfite, urethan, and nordihydroguaiaretic acid; they are useful, however, only if the reserpine solution is kept from light. In high doses reserpine may cause extrapyramidal reactions.

4. Butyrophenone Derivatives

Antipsychotic agents of this class most widely used are in Table 10.4. All those listed are derived from 4-phenylpiperidin-4-ol. For high activity it is essential

Table 10.4 Butyrophenone Derivatives

$$F-\bigcirc-\overset{O}{\underset{||}{C}}-CH_2-CH_2-CH_2-R$$

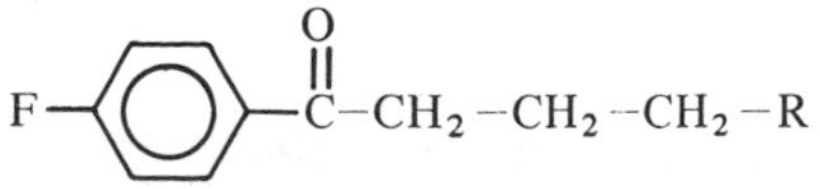

Official Name	Proprietary Name	Chemical Name	R
haloperidol*	Haldol Serenase	4-[4-(p-chlorophenyl)-4-hydroxypiperidino]-4'-fluorobutyrophenone	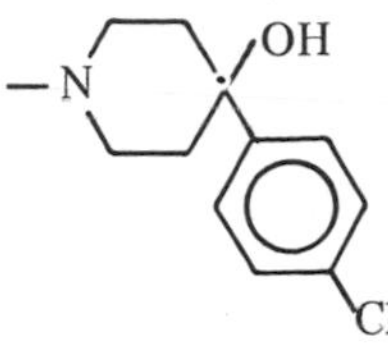
droperidol[+]	Inapsine	1-{1-[3-(p-fluoroben-zoyl)propyl]-1,2,3,6-tetrahydro-4-pyridyl}-2-benzimidazolinone	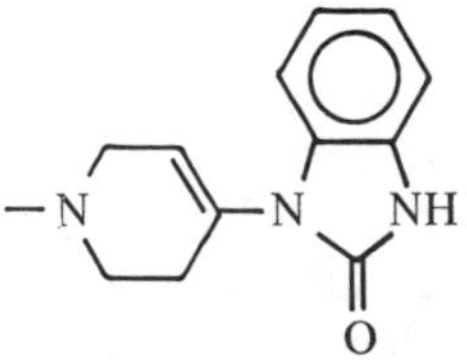
trifluperidol	Psicoperidol Psychoperidol Triperidol	1'-fluoro-4-[4-hydroxy-4-(α,α,α-trifluoro-m-tolyl)piperidino]-butyrophenone	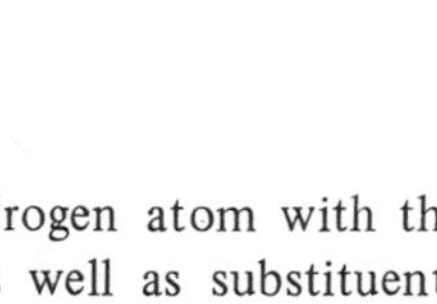

*In *USP XIX* (1975).
[+] In *NF XIV* (1975).

to have a trimethylene side chain linking the basic nitrogen atom with the benzoyl group. The fluorine atom in para position as well as substituents CF_3, Cl or CH_3 on the other aromatic ring increase potency. The tertiary alcohol function may be replaced by a tertiary amide group, as in droperidol.

Their pharmacological action is similar to that of phenothiazines. Among the side effects they produce most frequently are extrapyramidal symptoms, such as restlessness, akathisia, and dystonia, especially with large doses.

The representative of this class is haloperidol. Another widely used butyrophenone, especially in neuroleptanalgesia, is droperidol. A congener of derivatives of the class in pimozide (Orap), whose antipsychotic action is extremely long.

5. Miscellaneous Antipsychotic Agents

Although chemists who synthesized these compounds had not the intention, some of these agents are structurally related to reserpine. For instance, benzquinamide, molindone, oxypertine, phenoharmane, and tetrabenazine may be considered products of dissociation of the reserpine molecule, since they incorporate parts of its structural features.

Other drugs of this group are flurothyl (Indoklon), tetrahydropalmatine, and hydroxyzine (Table 20.4).

B.   Mechanism of Action

Antipsychotic agents are selective depressors of the central nervous system. Their central sites of action are located in the hypothalamus, brainstem, and very likely other subcortical regions of the brain involved in the coordination of emotional behavior. But the mechanism of action of these agents is not known with certainty.

Considering that neuroleptic drugs are extremely hydrophobic substances, it is very likely that at least some of their actions may be ascribed to their physicochemical properties. For instance, owing to their tendency to accumulate at interfaces and cell membranes, they can stabilize biological membranes, alter their permeability, and interfere with neural transmission. Such is especially the case with phenothiazines. They interact with some components of many membranes, including those present in nerves, muscles, and junctions, and thus influence biogenic amine transport.

Another physicochemical mechanism of action proposed for phenothiazines is that they act as free radicals. A third one is based on the evidence that some of them act as electron donors in charge transfer complexes with many macromolecules, including DNA. This hypothesis, however, cannot be applied to all phenothiazine neuroleptics, because a number of these drugs are poor electron donors.

Since minor alterations in their chemical structures result in pronounced changes on their pharmacological activity, antipsychotic agents are most certainly structurally specific drugs; that is, they produce their effects by interacting with specific receptors. In the case of phenothiazines those receptors might be flavoproteins, for there is evidence that they interact with succinic dehydrogenase, $NADH_2$, cytochrome-c-reductase and D-amino acid oxidase.

Janssen has called attention to the fact that all potent neuroleptics have two structural features in common which he considered essential for high antipsychotic activity:

1. A straight chain of three carbon atoms linking the basic ring nitrogen with a carbon, nitrogen, or oxygen atom, this atom being part of one of the following moieties: benzoyl group, 2-phenothiazine or a thixanthene-tricyclic system, phenoxypropyl side chain, 2-phenylpent-2-ene side chain, cyclohexane ring.

2. A six-membered basic heterocyclic ring, such as piperazine or piperidine, substituted in positions 1 and 4; the best substituents in position 4 are phenyl, anilino, methyl, or hydroxyethyl moieties.

According to Janssen, therefore, not withstanding their diversity of structure, all potent neuroleptics, because they have those two chemical features in common, act at the molecular level by the same mechanism, by complexing with specific receptors.

Concerning their nature, recent evidence points out that they are catecholaminergic or dopaminergic receptors. Thus, Horn and Snyder showed that structures of dopamine and chlorpromazine are partially superimposable and, based on this fact, suggested that the neuroleptic drugs owe their effects to a selective interaction with catecholaminergic receptors. Further experimental evidence justifies the postulate that all antipsychotic agents act by specifically blocking postsynaptic dopaminergic receptors. In consequence of the resulting functional deficiency of this neurotransmitter, a feedback mechanism is activated to stimulate biosynthesis of dopamine.

III. ANTIANXIETY AGENTS

Antianxiety agents, formerly called *minor tranquilizers,* and known also as *anxiolytics* and *tensiolytics,* are used to control neuroses and stress. In large doses they may be helpful in the treatment of severe psychomotor excitability, such as *delirium tremens.* They have shown usefulness in certain symptoms of toxic psychoses.

Several hypotheses have tried to explain the cause of anxiety. Most of them postulate that catecholamines, norepinephrine, and epinephrine are implicated, but the real etiology of emotional disturbances is not yet determined.

Several types of tests are available for screening potential antianxiety agents. Measurement of muscle-relaxant effects, behavioral tests, and observation of neurophysiological effects are the most widely used.

Untoward effects accompany the therapeutic action of antianxiety agents. Drowsiness is the most common. Others are more rare: ataxia, dizziness, headache, dryness of the mouth, blood dyscrasias, jaundice. Prolonged use of large doses may cause physical and psychic dependence. Massive doses may result in coma and death but less frequently than with barbiturates. One curious untoward effect is stimulation of appetite, with consequent gain in weight.

A. Classification

According to their chemical structure, antianxiety agents can be divided into three classes: propanediol carbamates and related compounds, benzodiazepines, and miscellaneous compounds. The most effective are benzodiazepines.

1. Propanediol Carbamates and Related Compounds

Those mostly used are listed in Table 10.5. Their potencies are about the same.

Table 10.5 Propanediol Carbamates and Related Antianxiety Agents

Official Name	Proprietary Name	Chemical Name	Structure
meprobamate***+**	Equanil Miltown	2-methyl-2-propyl-1,3-propanediol dicarbamate	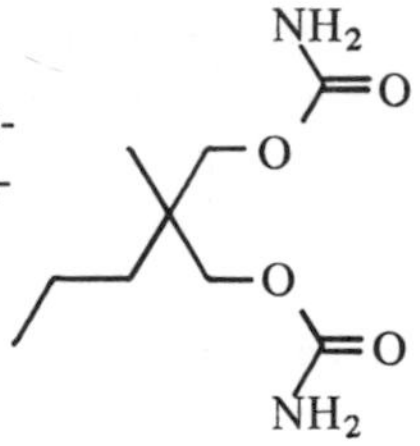
tybamate	Solacen Tybatran	mono-*N*-butyl-2-methyl-2-propyl-1,3-propanediol dicarbamate	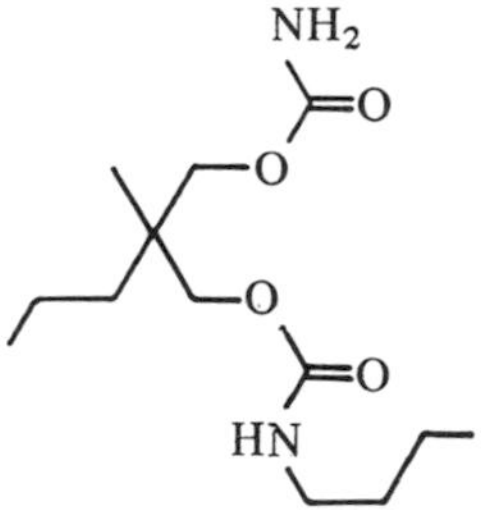
phenaglycodol	Ultran	2-(*p*-chlorophenyl)-3-methyl-2,3-butanediol	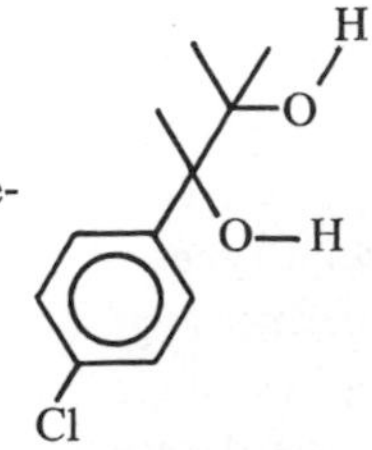

*In *USP XIX* (1975).
+In *NF XIV* (1975).

In vivo most of them are converted to inactive metabolites. Ethinamate, mainly used as hypnotic, has also antianxiety activity and belongs to this class.

These drugs usually occur as white or colorless crystals, with bitter taste, and are practically insoluble in water.

Other members of this class are: carisoprodol (Rela, Soma), cyclarbamate (Calmalone), emylcamate (Nuncital), mebutamate (Capla), oxyfenamate (Listica), phenprobamate (Gamaquil), tolboxane (Clarmil).

Meprobamate

It is the prototype of this class of agents, being one of the most widely used. Its potency is comparable to that of barbiturates but less than that of benzodiazepines. Used to treat psychosomatic and musculoskeletal disorders, it is also used as a sedative and hypnotic. Meprobamate is a relatively safe drug. Nevertheless, prolonged use of large doses may cause physical and psychic dependence and withdrawal syndrome subsequent to abrupt discontinuance. The usual dosage, by oral route, is 400 mg three or four times daily. It is also useful in the prevention of petit mal seizures and in the treatment of alcoholism. Meprobamate has the property of inducing the hepatic microsomal enzymes and thus accelerates the metabolism of a number of drugs.

Tybamate

As an antianxiety agent it is more effective than meprobamate and produces less untoward effects. It should not be administered to patients inclined to drug abuse. The usual dosage is 750 mg to 2 g daily. It is not recommended for children.

2. Benzodiazepines

They are the drugs of choice for treatment of anxiety. Table 10.6 lists some of those available, but the most widely used as anxiolytics are chlordiazepoxide, diazepam, and oxazepam. Several others, however, are being marketed: chlorazepate (Tranxene), clonazepam, demethdiazepam, demoxepam, lorazepam, medazepam (Nobrium), prazepam, tetrazepam. Many more are under animal or human evaluation. Benzodiazepines are also effective as muscle relaxants and for treatment of chronic alcoholics.

For greatest antianxiety potency, benzodiazepines should present the following structural features: (*a*) a methyl group attached to the nitrogen atom in the 1 position; (*b*) an electron-withdrawing group, such as Cl, NO_2, or CF_3, in the 7 position; (*c*) a phenyl group (or a phenyl group with an electronegative substituent, such as F, in the ortho position) in the 5 position.

Benzodiazepines are metabolized *in vivo*, by hydrolysis, hydroxylation, dealkylation, reduction, and conjugation to either active or inactive compounds. Diazepam, for instance, yields the equally active oxazepam as one of its

Table 10.6 Benzodiazepines

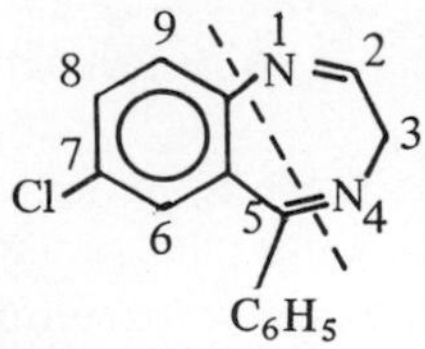

Official Name	Proprietary Name	Chemical Name	Structure
chlordiazepoxide*[+]	Librium	7-chloro-2-(methylamin-o)-5-phenyl-3H-1,4-ben-zodiazepine-4-oxide	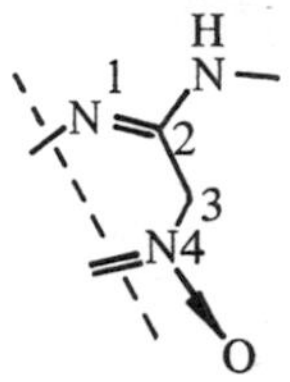
oxazepam[+]	Serax	7-chloro-1,3-dihydro-3-hydroxy-5-phenyl-2H-1,4-benzodiaze-pin-2-one	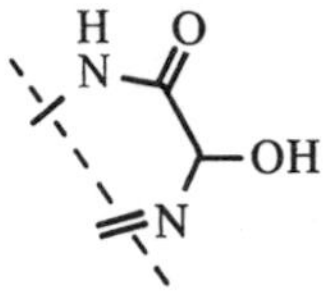
diazepam*	Valium	7-chloro-1,3-dihydro-1-methyl-5-phenyl-2H-1,4-benzodiazepin-2-one	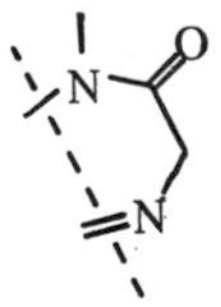
nitrazepam	Mogadon	1,3-dihydro-7-nitro-5-phenyl-2H-1,4-benzo-diazepin-2-one	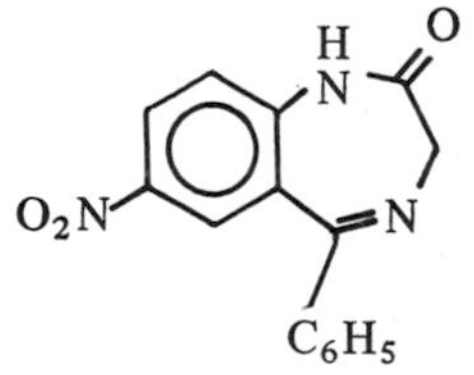
flurazepam[+a]	Dalmane	7-chloro-1-(2-diethyla-minoethyl)-5-(2-fluoro-phenyl)-1,3-dihydro-2H-1,4-benzodiazepin-2-one	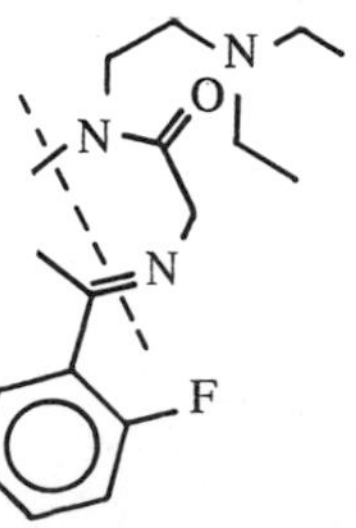

*In *USP XIX* (1975).
[+]In *NF XIV* (1975).
[a]Marketed in the United States as hypnotic.

several metabolites. Usually, after undergoing metabolic disposition, the resulting compounds are excreted as glucuronides.

Chlordiazepoxide

Introduced in 1960 as antianxiety agent, it is used as the hydrochloride, a highly water-soluble, hygroscopic, light-sensitive, colorless crystalline substance. It is rapidly absorbed, having a half-life of about 24 hours. Two of its metabolites— a lactam and a N-demethylated derivative—are pharmacologically active. Chlordiazepoxide is less potent than diazepam but produces less drowsiness. The usual dosage is 15 to 40 mg daily.

Diazepam

Introduced in 1964, it is more effective than chlordiazepoxide in the treatment of anxiety and tension. It is also helpful in combatting withdrawal symptoms in chronic alcoholics but should be used with caution. It has shown effectiveness in certain types of epilepsy. Other recommended uses are as premedication before surgery, hypnotic, sedative, muscle relaxant. In labor it has many advantages over other drugs. The usual dosage is 5 to 40 mg daily in divided doses. One of its metabolites is oxazepam, which is also used as antianxiety agent.

3. Miscellaneous Antianxiety Agents

These drugs are used especially as adjuncts in the treatment of anxiety and musculoskeletal disorders, but the effects ascribed to them may be due only to sedative action. The most widely used drug of this class is hydroxyzine (NF), either as hydrochloride (Atarax) or as pamoate (Vistaril Pamoate). Chemically, it is 1-(p-chlorobenzhydryl)-4-[2-(2-hydroxyethoxy)ethyl] piperazine.

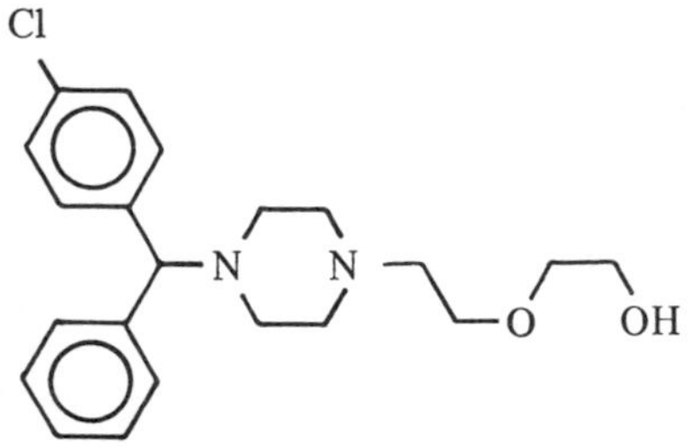

Besides hydroxyzine, many other drugs of this class are available. Among them, the following: buclizine (Softran), Cenestil, chlormezanone (Trancopal), difencloxazine (Olympax), Diphenazin, etodroxyzine (Drimyl), fluoresone (Bripadon), mephenoxalone (Trepidone), oxanamide (Quiactin), pinethanate

(Sycotrol), trimetozine (Trioxazin), Valmane, valnoctamide (Nirvanil).

Owing to untoward toxic effects, presently the following drugs of this class are seldom used: amphenidone, azacyclonol, benactyzine, and captodiame.

B. Mechanism of Action

The predominant idea, supported by experimental evidence, is that antianxiety agents act on catecholamine pathways; this is supported by the fact that benzodiazepines and barbiturates decrease the turnover of norepinephrine, serotonin, and other biogenic amines in the brain. The decreased turnover of norepinephrine and serotonin may be at least partially responsible for some of the pharmacological and clinical effects of antianxiety drugs. However, how they act at the molecular level is not known.

IV. ANTIDEPRESSANT DRUGS

Antidepressant drugs are those agents used to restore mentally depressed patients to an improved mental status. They are useful especially in depression and in depressive symptoms and, to a certain extent, in the treatment of depressive phases of certain types of schizophrenia. They decrease the intensity of the patient's symptoms, reduce his tendency to suicide, accelerate the rate of his improvement, and promote his mental well-being.

Depression accompanies many physical, mental, and emotional disorders. According to their etiology, depressive symptoms can be *reactive* or *endogenous.* The first ones are precipitated by a shocking event in the patient's life. The second ones are more often related to involutional changes after the age of forty and they may occur in cycles.

Various theories have been advanced to explain the biochemical causes of affective disorders. The most accepted is the *catecholamine hypothesis.* It postulates that depression results from a deficiency of catecholamines, especially norepinephrine, at functionally important central catecholamine receptors, whereas mania is caused by an excess of catecholamines in the brain. An alternative theory is the *serotonin hypothesis,* which ascribes depression and mania to deficiency and excess, respectively, of serotonin and not of catecholamines. According to a more recent hypothesis both depression and mania result from a central indoleamine defficiency, but in depression the adrenergic activity is diminished, whereas in mania it is increased.

A. Classification

The main classes of antidepressant drugs are tricyclic compounds, MAO inhibitors, lithium and strontium salts, and miscellaneous antidepressant agents.

1. *Tricyclic Compounds*

They are chemically similar to phenothiazine antipsychotic agents, and, as in the case of those drugs, their antidepressant activity is related to structure, depending essentially on the tricyclic nucleus, the lateral chain, and the nature of basic amino group. There are, however, some differences between tricyclic antidepressants and phenothiazine antipsychotics.

1. The central ring of the tricyclic system of antidepressants is usually constituted of seven or eight atoms, which confers to them a more angled or twisted conformation than is found in phenothiazine antipsychotics.

2. Although the side chain in antidepressants is usually composed of three carbon atoms, as it is a requirement in potent antipsychotics, strong activity is also present in some compounds in which this chain has only two carbon atoms.

3. The amino group in antidepressants is usually secondary, and not tertiary as in antipsychotics.

Tricyclic compounds are the most widely used drugs for the treatment of depressed patients, being more effective and less dangerous than MAO inhibitors. They are useful in treating endogenous and exogenous depressions. They are contraindicated, however, in patients with angina pectoris, congestive heart failure, and paroxysmal tachycardia and should not be administered with, or soon after, MAO inhibitors, since this mixture can cause a toxic atropine-like reaction, which may be fatal. An interval of at least two weeks should be kept between discontinuation of one type of these drugs and the administration of the other type. Untoward effects are common and varied: dryness of the mouth, blurred vision, constipation, orthostatic hypotension.

In vivo tricyclic compounds undergo metabolism, yielding demethylated, hydroxylated, and conjugated metabolites. For instance, imipramine forms several different metabolites, including desipramine, which accumulates in the body and is responsible for the observed pharmacological action. By the same process, amitriptyline is *N*-demethylated to nortriptyline.

Tricyclic andipressants are generally used as hydrochlorides, which occur as water-soluble, white or colorless, crystalline powders.

Table 10.7 lists drugs of this class that are most widely used. However, many other tricyclic antidepressants are available. Examples: azaphen, carpipramine (Defekton), clomipramine (Anafranil), iprindole (Galatur, Prondol), trimipramine (Surmontil).

Imipramine

It is the prototype of tricyclic antidepressant compounds. The initial dosage is 100 to 150 mg daily in divided doses.

2. *MAO Inhibitors*

Monoamine oxidase (EC.1.4.3.4), more commonly called MAO, plays the

Table 10.7 Tricyclic Antidepressants

Official Name	Proprietary Name	Chemical Name	Structure
imipramine*	Tofranil	5-[3-(dimethylamino)-propyl]-10,11-dihy-dro-5H-dibenz[b,f]-azepine	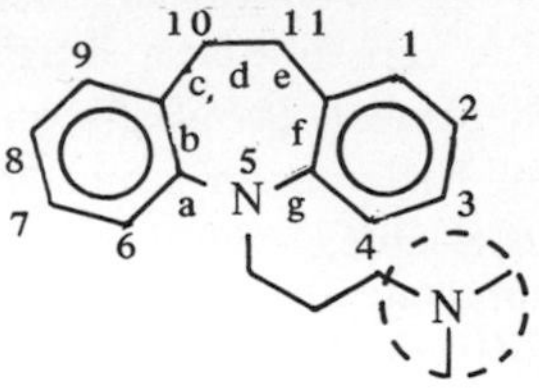
desipramine[+]	Norpramin Pertofrane	5-(3-methylaminopro-pyl)-10,11-dihydro-5H-dibenz-[b,f]azepine	
amitriptyline*	Elavil	5-(3-dimethylamino-propylidene-10,11-dihydro-5H-dibenzo-[a,b]cycloheptane	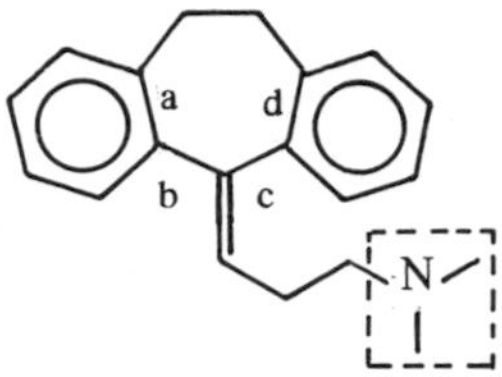
nortriptyline[+]	Aventyl	5-(3-methylaminopropyli-dene)-10,11-dihydro-5H-dibenzo[a,d]cycloheptane	
protriptyline[+]	Vivactil	5-(3-methylaminopro-pyl)-5H-dibenzo[a,d]-cycloheptene	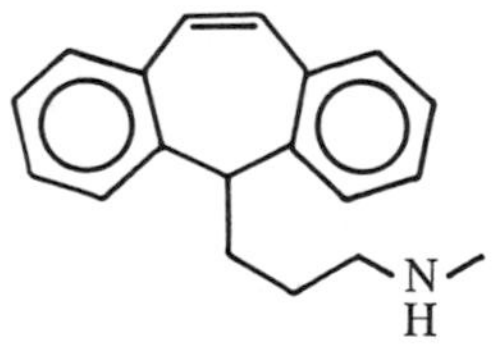
doxepin	Sinequan	N,N-dimethyl-3-(di-benz-[b,e]oxepin-11-(6H)-ylidene)pro-pylamine	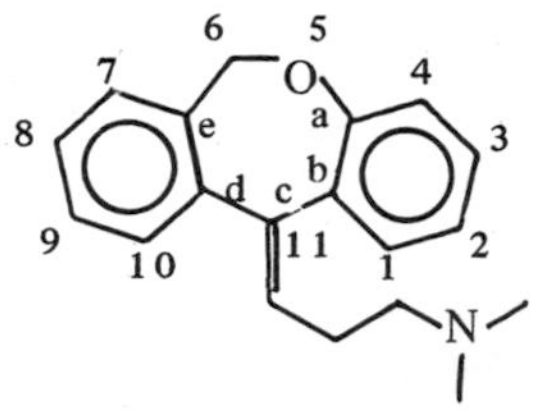

*In *USP XIX* (1975).
[+]In *NF XIV* (1975).

physiological role of oxidatively deaminating primary and secondary amines to aldehydes, ammonium, and hydrogen peroxide. It metabolizes, therefore, catecholamines, although in this process another enzyme, cathecol-O-methyltrans-ferase, plays a more important role.

The structure of MAO varies according to organs, tissues, and species from which it is extracted. It follows that it is not a single enzyme, but a group of enzymes, to which the name isoenzymes is given. For instance, in the human brain there occur at least four different molecular forms of MAO, with different substrate specificities. Hence it is not wise to extrapolate the structure activity relationships of one group of MAO inhibitors to another group.

Inhibitors of MAO are found in different chemical classes of substances: hydrazines, hydrazides, cyclopropylamines, indolealkylamines, carbolines, pyri-dines, and others. They present, however, the common property of preventing the oxidative deamination of the biogenic amines, such as epinephrine, nor-epinephrine, serotonin, tryptamine, tyramine, and dopamine. In this way they increase the concentration of these central excitatory amines in the body, in-cluding that of serotonin and norepinephrine in the brain. Their action, however, is not restricted to MAO inhibition. They likewise affect other susceptible enzymes.

Although there are many compounds that inhibit MAO, the most widely used for the treatment of depression are those listed in Table 10.8.

Most MAO inhibitors are hydrazine or hydrazide derivatives. Hydrazine moiety is highly reactive and may form a strong linkage with MAO, inhibiting this enzyme for up to five days. The onset of action of hydrazine derivatives is slow: 2 to 3 weeks; of tranylcypromine, more rapid.

The MAO inhibitors are used mainly in the treatment of mental depression and certain anxiety states. They appear to be most effective in endogenous depressions. As a group these drugs are less effective and produce more serious untoward effects than the tricyclic antidepressants. For these reasons, they are drugs of second choice and should be used only in patients who were already treated with them or with those in whom tricyclic compounds have been not effective.

Among the adverse reactions caused by these drugs are dryness of the mouth, excessive perspiration, dizziness, blurred vision, nausea, constipation, postural hypotension, hepatitis, leukopenia.

Patients receiving anti-MAO drugs must avoid the ingestion of cheese, broad beans, pickled herring, yeast extracts, chicken liver, canned figs, chocolate, alcoholic beverages, and certain other foods which contain amines, such as tyramine. They must also avoid the concomitant use of other antidepressant drugs, such as amitriptyline, and several other types of drugs (alcohol, ampheta-mines, anesthetics, antihistaminics, antihypertensives, metaraminol, methyldopa, morphine, muscle relaxants, meperidine, pressor agents, reserpine, sympatho-mimetics). Such substances may precipitate hypertensive crises resulting in death. Hypertensive crises are treated with phentolamine.

Table 10.8   MAO Inhibitors

Official Name	Proprietary Name	Chemical Name	Structure
phenelzine[+]	Nardil	β-phenylethylhydrazine	
isocarboxazid[+]	Marplan	5-methyl-3-isoxazole-carboxylic acid 2-benzylhydrazide	
nialamide	Niamid	isonicotinic acid 2-[2-(benzylcarbamoyl)-ethyl]hydrazide	
tranylcypromine[+]	Parnate	(±)-*trans*-2-phenylcyclo-propylamine	
pargyline[+a]	Eutonyl	*N*-methyl-*N*-(2-propyn-yl)benzylamine	

[+]In *NF XIV* (1975).
[a]Used as an antihypertensive.

3. Lithium and Strontium Salts

Lithium salts (carbonate, acetate), as well as strontium carbonate, are the most recent acquisition of antidepressive therapy. They are effective primarily for control of mood in patients with manic-depressive psychoses. They are not recommended for other psychiatric disorders. Patients treated with these salts should be maintained under close medical observation, because of the toxic effects the salts produce.

4. Miscellaneous Antidepressant Agents

This class can be divided in several groups:

1. β-Adrenergic blocking agents: alprenolol, bunolol, pindolol, practolol,

propranolol

2. Amphetamines: amphetamine, dextroamphetamine, methamphetamine, methylphenidate

3. Antiemetics: metoclopramide, sulpiride

4. Bicyclic antidepressants: amedalin, clodazone, daledalin, 5-hydroxytryptophan, Impavido, Iometraline, quipazine, talopram, talsupram, tetramisole, thiazesim, trazodone (Trittico)

5. GABA and related compounds: GABA, 4-hydroxybutyric acid, muscimol, valtrate

6. Heterocyclic antidepressants: chlormezanone, 1,2-diisopropylurazole, etifoxine, fipexide (Vigilor), mepiprazole, nelopam, nuciterine, sydnophen, toprilidine, xylazine

7. Hypnotics and anticonvulsants: beclamide, capuride, glutethimide, methaqualone, methyprylon

8. Miscellaneous antidepressants: Gordazyl, α-methyltyrosine, tofenacin (Elamol).

B. Mechanism of Action

The exact mechanism of action of antidepressant drugs is not known, since the etiology of depressive disorders is unknown. However, it is generally accepted that the pharmacological and clinical effects produced by these drugs result from their actions on central monoaminergic, especially catecholaminergic, pathways. In other words, they affect the biosynthesis or the storage of certain biogenic amines, such as norepinephrine and serotonin.

There is evidence that tricyclic antidepressants and MAO inhibitors increase the levels of one or more of these amines at neuronal receptor sites in the central nervous system. As to lithium salts, it has been suggested that they decrease the level of epinephrine or serotonin at the same receptor sites.

Considering that the chemical structures of some MAO inhibitors (Table 10.8) are reminiscent of the structure of amphetamine, a catecholaminergic agent, it is very likely that, besides acting by inhibition of MAO, those drugs also compete with catecholamines for the same receptor sites. This dual mechanism is cooperative toward an increase in the level of catecholamines in the central nervous system.

V. HALLUCINOGENIC AGENTS

Hallucinogenic agents, also called *psychotomimetics, psychosometics, psychotogenics, psychodysleptics, psychedelics,* and *mysthicomimetics,* although of scant therapeutic application, present great practical interest, because they produce psychoses, some of them intense, and they have large illegal consumption, constituting a serious problem in some countries.

To a certain extent some of these agents are used in medicine to produce model psychoses as aids in psychotherapy. Such is the case with mescaline, psilocybin, and LSD. Another licit use is in the investigation of the relationship among mind, brain, and biochemistry, with the purpose of elucidating the etiology of mental diseases, such as schizophrenia.

Self-administration of these drugs is extremely dangerous. Severe reactions, such as acute panic, chronic anxiety, and depressive states have been observed. Under the effect of these drugs, crimes and suicides have been committed. Since the illicit use of psychotomimetic agents is increasing, several national and supranational agencies are joining efforts and promoting massive campaigns to curb this trend.

A. Classification

Presently several compounds are known which cause psychoses. However, the true hallucinogens belong to the following chemical classes (Figure 10.1): indolealkylamines (LSD = diethylamide of (+)-lysergic acid, psilocybin, psilocin, N,N-dimethyltryptamines), phenylalkylamines (mescaline and derivatives), and substituted amphetamines (the best known is DOM = 2,5-dimethoxy-4-methylamphetamine).

From this classification we have excluded tetrahydrocannabinols, extracted from *Cannabis sativa,* commonly called marijuana, because they are euphoriants and not truly hallucinogenics. Also we exclude the anticholinergics, antihistaminics, and others with psychomimetic activity whose mechanism of action is not related to antagonism with serotonin.

B. Mechanism of Action

Since the pharmacological and physiological bases of hallucinogenic potency has not been yet elucidated, it is not possible to propose a general theory of hallucinogenesis. However, long ago there was observed the antagonism between LSD, an extremely potent hallucinogenic agent, and serotonin, a phenomenon which recent investigations have confirmed. Also known is the crossed tolerance among LSD, psilocybin, and mescaline. This led to the hypothesis that hallucinogenic agents act by competing with serotonin for the same central receptors.

Two lines of reasoning are used in support of this hypothesis. On one side, there are structural similarities among the several classes of the most potent hallucinogenic agents. All of them have, or by partial cyclization of the side chain may assume, an indolealkylamine or, better, tryptamine conformation, which makes them total structural analogues of serotonin or partial structural analogues of LSD (Figure 10.1). There is no evidence to suggest that such structures are

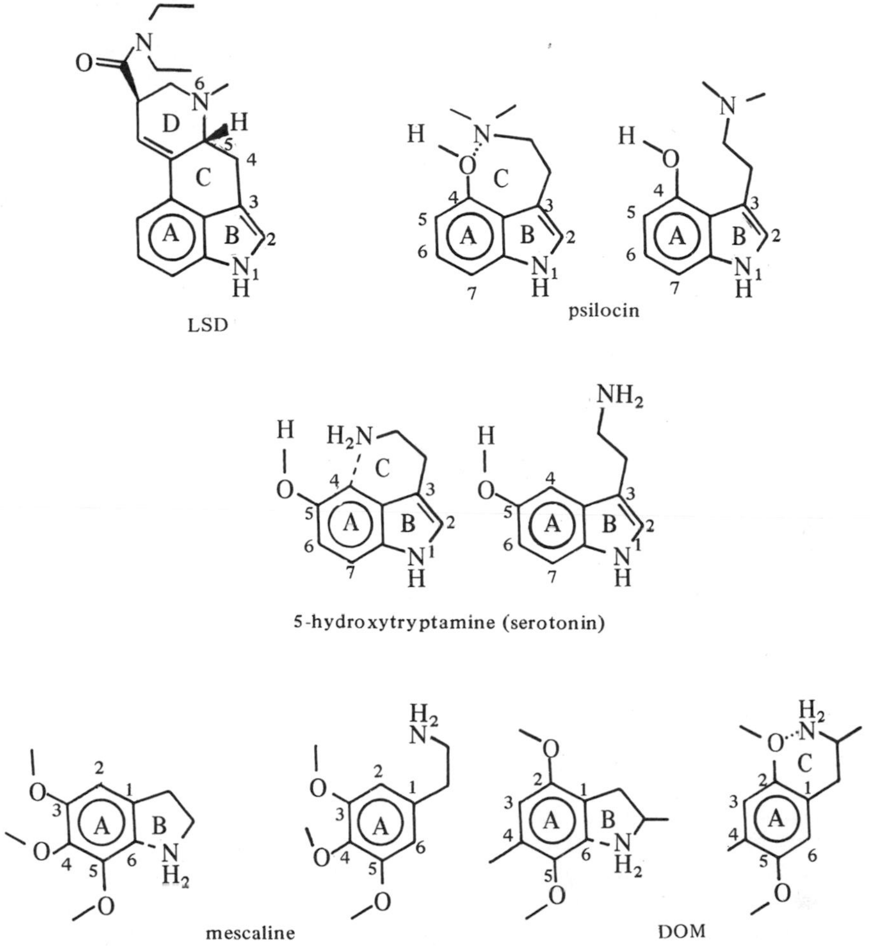

Figure 10.1 The most important hallucinogenic agents. Note how the common structural feature present in all of them relates to serotonin.

actually present in solution or at receptors. Molecular orbital calculations, on the other side, indicate that hallucinogenic agents, because they have adequate HOMO, can act as electron donors in charge transfer complexes.

The potency of hallucinogenic agents is, therefore, in part determined by the ability (*a*) to assume a structure which is partially analogous to that of LSD and (*b*) to donate electrons.

REFERENCES

Introduction

Z. Giora, *Psychopathology: A Cognitive View*, Halsted, New York, 1975.

J. R. Cooper *et al.*, *The Biochemical Basis of Neuropharmacology*, 2nd ed., Oxford University Press, London, 1974.

J. J. Kaufman and E. Kerman, *Int. J. Quantum Chem.*, Quantum Biology Symp. n° 1, 259 (1974).

A. Korolkovas, *Rev. Bras. Anestesiol.*, **24**, 343 (1974).

W. G. Klopfer and M. R. Reid, *Problems in Psychotherapy: An Eclectic Approach*, Halsted, New York, 1974.

S. H. Snyder, *Madness and the Brain*, McGraw-Hill, New York, 1974.

S. H. Snyder *et al.*, *Science*, **184**, 1243 (1974).

E. Usdin and S. H. Snyder, Eds., *Frontiers in Catecholamine Research*, Pergamon, New York, 1974.

L. E. Hollister, *Clin. Pharmacol. Ther.*, **13**, 803 (1972).

S. H. Snyder, Ed., *Perspectives in Neuropharmacology*, Oxford University Press, New York, 1974.

R. H. Rech and K. E. Moore, Eds., *An Introduction to Psychopharmacology*, Raven, New York, 1971.

H. Weil-Malherbe and S. I. Szara, *The Biochemistry of Functional and Experimental Psychoses*, Thomas, Springfield, Ill., 1971.

W. G. Clark and J. del Giudice, Eds., *Principles of Psychopharmacology*, Academic, New York, 1970.

H. E. Himwich and H. Alpers, *Annu. Rev. Pharmacol.*, **10**, 313 (1970).

T. A. Ban, *Psychopharmacology*, Williams and Wilkins, Baltimore, 1969.

F. T. von Brucke, O. Hornykiewicz, and E. B. Sigg, *The Pharmacology of Psychotherapeutic Drugs*, Springer, Berlin, 1969.

A. Cerletti and F. J. Bove, *The Present Status of Psychotropic Drugs*, Excerpta Medica Foundation, Amsterdam, 1969.

D. F. Klein and J. M. Davis, *Diagnosis and Drug Treatment of Psychiatric Disorders*, Williams and Wilkins, Baltimore, 1969.

M. Gordon, Ed., *Psychopharmacological Agents*, 3 vols., Academic, New York, 1964, 1967, 1974.

Antipsychotic Agents

I. S. Forrest *et al.*, Eds., *Phenothiazines and Structurally Related Drugs*, Raven Press, New York, 1974.

M. H. J. Koch, *Mol. Pharmacol.*, **10**, 425 (1974).

P. A. J. Janssen, "Structure-Activity Relationships (SAR) and Drug Design as Illustrated with Neuroleptic Agents," in C. J. Cavallito, Ed., *Structure-Activity Relationships*, Vol. 1, Pergamon, Oxford, 1973, pp. 37-73.

C. D. Wise and L. Stein, *Science*, **181**, 344 (1973).

C. L. Zirkle and C. Kaiser, *Annu. Rep. Med. Chem.*, **8**, 1, 11 (1973).

J. J. Kaufman and E. Kerman, *Int. J. Quantum Chem.*, **6**, 319 (1972).

J. J. Kaufman and A. A. Manian, *Int. J. Quantum Chem.*, **6**, 375 (1972).

P. Seeman, *Pharmacol. Rev.*, **24**, 583 (1972).

A. S. Horn and S. H. Snyder, *Proc. Nat. Acad. Sci. U.S.A.*, **68**, 2325 (1971).

L. Stein and C. D. Wise, *Science*, **171**, 1032 (1971).

S. Gabay and S. R. Harris, *Top. Med. Chem.*, **3**, 57 (1970).

H. E. Lehmann and T. A. Ban, Eds., *The Thioxanthenes*, Karger, Basel, 1969.

P. A. J. Janssen, *Farm. Rev.*, **65**, 272 (1966).

Antianxiety Agents

E. Costa and P. Greengard, Eds., *Mechanism of Action of Benzodiazepines*, Raven, New York, 1975.

M. Protiva, *Farmaco. Ed. Sci.*, **28**, 58 (1973).

M. Protiva, *Med. Actual. (Drugs of Today)*, **9**, 199 (1973).

C. D. Wise *et al.*, *Science*, **177**, 180 (1972).

B. J. Ludwig and J. R. Potterfield, *Advan. Pharmacol. Chemother.*, **9**, 173 (1971).

L. H. Sternbach, *Angew. Chem., Int. Ed. Engl.*, **10**, 34 (1971).

G. A. Archer and L. H. Sternbach, *Chem. Rev.*, **68**, 747 (1968).

R. C. Elderfield, *Heterocyclic Compounds*, Vol. IX, Wiley, New York, 1967, pp. 332-361.

G. Zbinden and L. O. Randall, *Advan. Pharmacol.*, **5**, 213 (1967).

Antidepressant Drugs

J. Mendels, Ed., *Psychobiology of Depression*, Halsted, New York, 1975.

J. Becker, *Depression: Theory and Research*, Halsted, New York, 1974.

H. S. Akiskal and W. T. McKinney, Jr., *Science*, **182**, 20 (1973).

B. J. Carroll, *Clin. Pharmacol. Ther.*, **12**, 743 (1972).

B. T. Ho, *J. Pharm. Sci.*, **61**, 821 (1972).

J. M. Davis and W. E. Fann, *Annu. Rev. Pharmacol.*, **11**, 285 (1971).

B. T. Ho and W. M. McIsaac, Eds., *Brain Chemistry in Mental Disease*, Plenum, New York, 1971.

A. I. Salama *et al.*, *J. Pharmacol. Exp. Ther.*, **178**, 474 (1971).

F. Sulser and E. Sanders-Bush, *Annu. Rev. Pharmacol.*, **11**, 209 (1971).

G. G. S. Collins *et al.*, *Nature (London)*, **225**, 817 (1970).

S. Gershon, *Clin. Pharmacol. Ther.*, **11**, 168 (1970).

D. Glick, Ed., *Analysis of Biogenic Amines and Their Related Enzymes*, Wiley-Interscience, New York, 1970.

R. Kapeller-Adler, *Amine Oxidases and Methods for Their Study*, Wiley-Interscience, New York, 1970.

E. Marley and B. Blackwell, *Advan. Pharmacol. Chemother.*, **8**, 185 (1970).

R. A. Maxwell *et al.*, *J. Pharmocol. Exp. Ther.*, **173**, 158 (1970).

W. G. Dewhurst, *Nature (London)*, **218**, 1130 (1968).

S. Garattini and M. Dukes, *Antidepressant Drugs,* Excerpta Medica Foundation, Amsterdam, 1967.

J. O. Cole and J. R. Wittenborn, Eds., *Pharmacology of Depression,* Thomas, Springfield, Ill., 1966.

B. Belleau and J. Moran, *Ann. N. Y. Acad. Sci.,* **107**, 822 (1963).

Hallucinogenic Agents

R. K. Siegel and L. J. West, Eds., *Hallucinations: Behavior, Experience, and Theory,* Wiley, New York, 1975.

F. A. B. Aldous *et al., J. Med. Chem.,* **17**, 1100 (1974).

B. Pullman *et al., J. Med. Chem.,* **17**, 439 (1974).

R. W. Baker *et al., Mol. Pharmacol.,* **9**, 23 (1973).

A. Korolkovas, *Rev. Paul. Med.,* **81**, 43 (1973).

R. Mechoulam, Ed., *Marijuana,* Academic, New York, 1973.

G. K. Aghajanian, *Annu. Rev. Pharmacol.,* **12**, 157 (1972).

P. Brawley and J. D. Duffield, *Pharmacol. Rev.,* **24**, 31 (1972).

F. C. Brown, *Hallucinogenic Drugs,* Thomas, Springfield, Ill., 1972.

E. Campaigne and D. R. Knapp, *J. Pharm. Sci.,* **60**, 809 (1971).

S. Cohen, *Progr. Drug Res.,* **15**, 68 (1971).

W. Keup, Ed., *Origin and Mechanisms of Hallucinations,* Plenum, New York, 1971.

Marijuana and Health: A Report to the Congress, National Institute of Health, 1971.

C. H. Chothia and P. J. Pauling, *Proc. Nat. Acad. Sci. U.S.A,* **65**, 477 (1970).

D. H. Efron, Ed., *Psychotomimetic Drugs,* Raven, New York, 1970.

R. Mechoulam, *Science,* **168**, 1159 (1970).

B. Weiss and V. Laties, *Annu. Rev. Pharmacol.,* **9**, 297 (1969).

L. B. Kier, *J. Pharm. Sci.,* **57**, 1188 (1968).

N. Giarman and D. K. Freedman, *Pharmacol. Rev.,* **17**, 1 (1965).

CHAPTER 11

CENTRAL INTRANEURAL BLOCKING AGENTS

I. INTRODUCTION

Central intraneuronal blocking agents are psychotherapeutic drugs which diminish the sklcletal-muscle tone and involuntary movement. They comprise two categories of drugs:

1. Centrally acting skeletal-muscle relaxants—those drugs that selectively depress the central nervous system which controls muscle tone. They are used to promote relaxation of skeletal-muscle spasms and in the treatment of tetanus and certain orthopedic procedures.

2. Antiparkinsonism drugs—those compounds with anticholinergic or other pharmacodynamic activity that are used primarily in the treatment of Parkinson's disease, which affects more than 1 million people in the United States alone.

II. MUSCLE RELAXANTS

Most of the centrally acting skeletal-muscle relaxants exhibit also other activities and for this reason are used for the relief of several conditions. Thus, some are hypnotics and sedatives, whereas others have shown usefulness in the treatment of anxiety and tension.

Adverse but transient reactions of muscle relaxants are drowsiness, flushing, weakness, dizziness, blurred vision, lethargy, and lassitude. Large doses occasionally produce nausea, vomiting, abdominal distress, heartburn. Activities that require motor coordination, sound judgment, and mental alertness should be avoided by patients treated with these drugs.

A. Classification

Muscle relaxants may be grouped in three classes: mephenesin and related drugs, heterocycles and their derivatives, and miscellaneous agents. Available data do not allow reliable comparison among these agents regarding their therapeutic value, effectiveness, and safety.

1. *Mephenesin and Related Drugs*

Drugs of this class most widely used are listed in Table 11.1. Others are: guaifenesine (Dilyn), styramate (Sinaxar), fenyramidol (Analexin).

Table 11.1 Mephenesin and Related Drugs

Official Name	Proprietary Name	Chemical Name	Structure
mephenesin	Myanesin Tolserol	3-(o-tolyloxy)-1, 2-propanediol	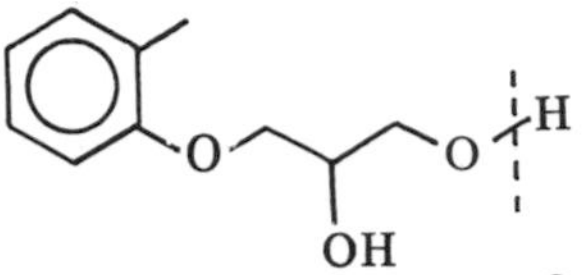
mephenesin carbamate	Tolseram	2-hydroxy-3-o-tolyloxypropyl carbamate	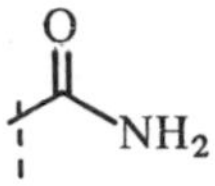
methocarbamol[+]	Robaxin	3-(o-methoxyphen-oxy)-1,2-propane-diol 1-carbamate	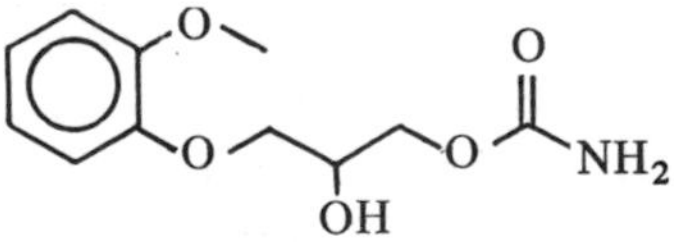
chlorphenesine carbamate	Maolate	3-p-chlorophenoxy-2-hydroxypropyl carbamate	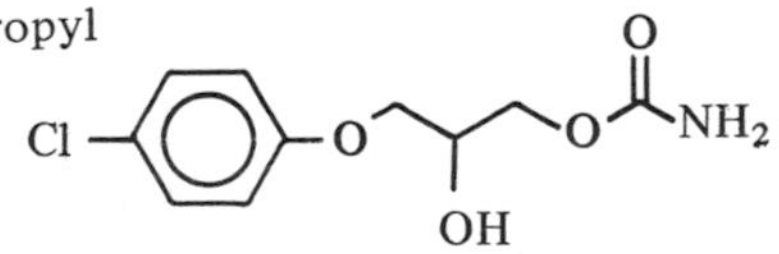

[+]In *NF XIV* (1975).

Mephenesin

It is the prototype of this class; it occurs as a sparingly soluble-in-water, odorless, white powder, with bitter taste. It is rapidly absorbed, and it is uniformly distributed throughout most tissues but reaches a higher concentration in brain. The muscle relaxation it produces is brief. In order to avoid the gastrointestinal disturbances it produces, mephenesin should be given after meals or with milk. The usual dosage by oral route is 1 to 3 g three to five times daily.

Mephenesin Carbamate

Introduced in 1954, it has the same actions and potency as, but longer duration of action than, the parent drug.

2. *Heterocycles and Their Derivatives*

Besides those listed in Table 11.2, the following are used: metaxalone (Zorane),

Table 11.2 Heterocycles and Their Derivatives

Official Name	Proprietary Name	Chemical Name	Structure
chlorzoxazone	Paraflex	5-chloro-2-benzoxazolinone	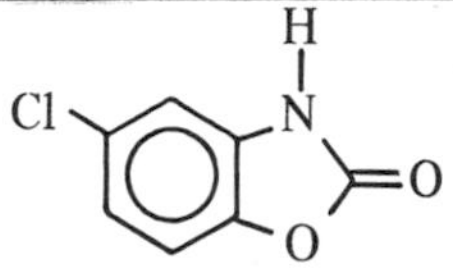
metaxalone	Skelaxin	5-(3,5-dimethyl phenoxymethyl)-2-oxazolidinone	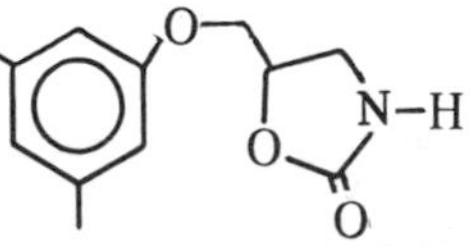

Quiloflex, 2-(γ-methoxypropylaminomethyl)-benzo-1,4-dioxan, marketed in Europe, orphenadrine, which has also antiparkinsonism activity (Table 11.4), mephenoxalone and chlormezanone, both used also as antianxiety agents. A new type of skeletal-muscle relaxant, recently introduced, is dantrolene (Dantrium).

3. *Miscellaneous Agents*

This class is represented presently by chlordiazepoxide, diazepam, idrocilamide (Brolitene), midetone (Mydocalm), nefopam, tetrazepam (Myolastan), and thiocolchicoside (Colcamyl, Coltrax). The first two were studied in Chapter 10 (Table 10.6), because their main application is as antianxiety agents. The last two are marketed only in Europe.

B. Mechanism of Action

Skeletal-muscle relaxants just studied act centrally on the brain and spinal cord. There are, however, muscle relaxants which act *peripherally* at the neuromuscular junction of voluntary muscle; these drugs, called curaremimetics, are studied in Chapter 16.

The exact way centrally acting skeletal-muscle relaxants exert their effects is still poorly understood. It is generally believed that mephenesin and related compounds, as well as the heterocycles and their derivatives, act by blocking or retarding transmission of nervous impulses in internuncial synapses within the spinal cord and in the brain stem, thalamus, and basal ganglia.

Those drugs with muscle-relaxant properties which also exhibit other pharmacological actions, such as hypnotics, sedatives, antipsychotics, anxiolytics, act by mechanisms peculiar to them as already seen in preceding chapters.

III. ANTIPARKINSONISM DRUGS

Parkinson's disease is named after James Parkinson, who first described it in 1817. Various types of Parkinson's disease are known: idiopathic paralysis agitans, postencephalitic and arteriosclerotic parkinsonism. It is characterized by tremor, rigidity, akinesia or bradykinesia, and postural instability. Treatment of this disease involves not only drug therapy but also physiotherapy, exercises, activities, and psychological support. Drugs seen in this section only alleviate the symptoms, allowing a number of patients to live normal lives.

Assays used to disclose useful antiparkinsonism drugs are numerous. Some are based on blockade of tremors of central origin induced in animals by chemical agents, such as Tremorine and Oxotremorine. Others are based on the ability of potential agents to antagonize the activating effects of physostigmine and similar drugs on the electroencephalogram of laboratory animals. A third group of assays involves the induction of tremors in primates by localized lesions in the midbrain and blocking them by potential antiparkinsonism drugs.

Drugs used in Parkinson's disease are not highly toxic. Usually they may be administered for long periods of time. Nevertheless, they cause some minor untoward effects, such as dryness of the mouth, constipation, blurred vision, and gastrointestinal disturbances.

Doses should be adjusted individually. Since Parkinson's disease requires continuous medication, from time to time it is necessary to change drugs and adjust gradually the dosage of the new drug.

A. Classification

Antiparkinsonism drugs presently available can be divided into four classes: central anticholinergics, antihistaminics, phenothiazines, and miscellaneous.

1. *Central Anticholinergics*

They are tertiary amines and thus are able to cross the blood-brain barrier in the undissociated form in order to exert central effects. Quaternary ammonium compounds with anticholinergic action have difficulty in crossing this barrier and, therefore, are ineffective as antiparkinsonism agents. Atropine and scopolamine were the only effective antiparkinsonism drugs for many years, but they have been supplanted by synthetic analogues listed in Table 11.3. For control of neuroleptic-induced parkinsonism, dexetimide (Tremblex) was recently introduced.

Since, in the form of bases, central anticholinergics are insoluble, they are generally used as hydrochlorides which are white, crystalline, slightly water-soluble substances. These drugs are all structurally related, exert similar actions, and are used for the same purpose. The prototype is trihexyphenidyl.

Table 11.3   Anticholinergics Used in Parkinsonism

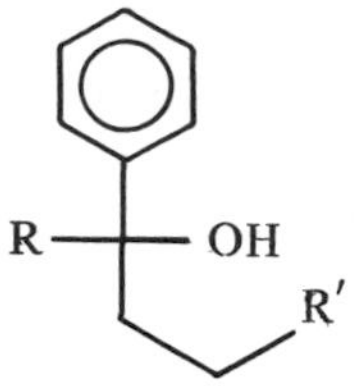

Official Name	Proprietary Name	Chemical Name	R	R'
trihexyphenidyl*	Artane Pipanol Temin	α-cyclohexyl-α-phenyl-1-piperi-dinepropanol	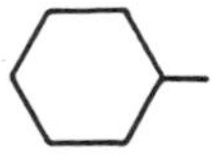	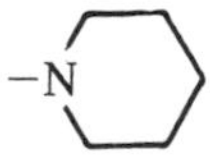
biperiden[+]	Akineton	α-5-norbornen-2-yl-α-phenyl-1-piperi-dinepropanol	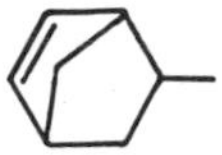	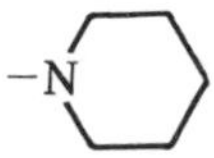
cycrimine[+]	Pagitane	α-cyclopentyl-α-phenyl-1-piperi-dinopropanol	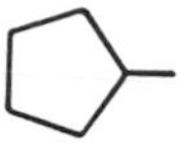	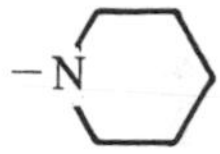
procyclidine[+]	Kemadrin	α-cyclopentyl-α-phenyl-1-pyrroli-dinepropanol	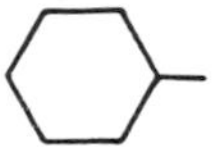	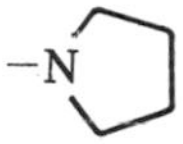
benztropine*	Cogentin	3α-(diphenylmeth-oxy)tropane	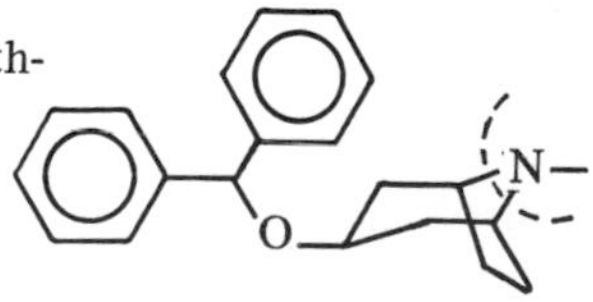	
ethybenzatropine	Ponalid	3-(diphenylmeth-oxy)-8-ethylnor-tropane		

*In *USP XIX* (1975).
[+]In *NF XIV* (1975).

Trihexyphenidyl

It is often used as drug of first choice, frequently as an adjunct to levodopa, to treat idiopathic and postencephalitic parkinsonism and to relieve drug-induced extrapyramidal reactions.

Benztropine

Used as mesylate. It was designed rationally by the process of hybridization or association of mixed moieties (tropine base of atropine and benzohydryloxy moiety of diphenhydramine) with the purpose of obtaining a drug having simultaneously anticholinergic and antihistaminic properties. Benztropine does exert these two activities, as well as local anesthetic effect as a result of its structural resemblance to cocaine.

Benztropine is mainly used as an adjunct to other drugs in patients who do not respond properly, or have become tolerant, to other agents. Because of its prolonged action, it is used as bedtime medication. It has a cumulative action, which may cause toxic reactions. Benztropine is prepared by reaction between diphenyldiazomethane and tropine followed by salification with methanesulfonic acid.

Table 11.4 Antihistaminics Used in Parkinsonism

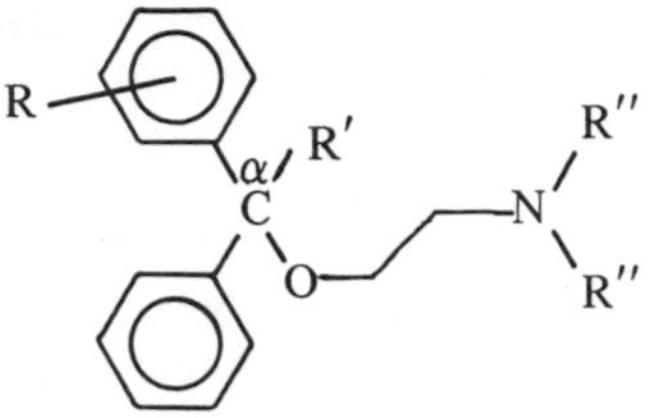

Official Name	Proprietary Name	Chemical Name	R	R'	R''
diphenhydramine*	Benadryl	2-(diphenylmeth-oxy)-N,N-dimethyl-ethylamine		H	CH$_3$
orphenadrine[+]	Norflex	N,N-dimethyl-2-[o-methyl-α-phenyl-benzyl)oxy]ethyl-amine	o-CH$_3$	H	CH$_3$
chlorphenoxamine[+]	Phenoxene	2-(p-chloro-α-meth-yl-α-phenylbenzyl-oxy)-N,N-dimethyl-ethylamine	p-Cl	CH$_3$	CH$_3$
clofenetamine	Keithon	2-(p-chloro-α-meth-yl-α-phenylbenzyl-oxy)triethylamine	p-Cl	CH$_3$	C$_2$H$_5$
phenindamine[+]	Thephorin	2,3,4,9-tetrahydro-2-methyl-9-phenyl-1H-indeno[2,1-c]pyridine	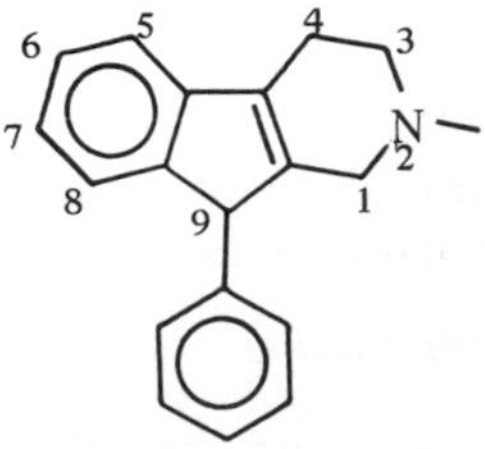		

*In *USP XIX* (1975).
[+]In *NF XIV* (1975).

2. Antihistaminics

Those used in Parkinson's disease appear in Table 11.4. Other antihistaminics are not effective (Chapter 20). Introduced beginning in 1960, these tertiary amines with central effects are not as effective as anticholinergics in alleviating parkinsonism symptoms but cause less undesirable atropinelike effects. They are mainly used, in the form of hydrochlorides, as adjuncts to trihexyphenidyl or its congeners and to control promptly severe extrapyramidal reactions to neuroleptic agents.

3. Phenothiazines

Phenothiazines used as antiparkinsonism drugs have central and peripheral anticholinergic and antihistaminic actions (Table 11.5). They serve as adjuncts to

Table 11.5   Phenothiazines and Thioxanthenes Used in Parkinsonism

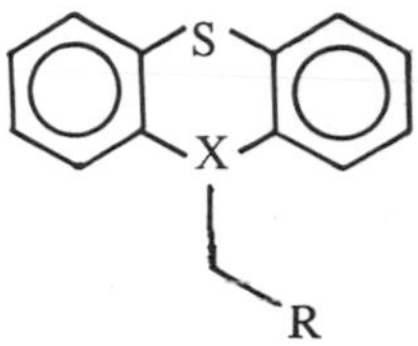

Official Name	Proprietary Name	Chemical Name	X	R
ethopropazine* (profenamine)	Parsidol	10-(2-diethylamino-propyl)phenothiazine	N	$-CH(CH_3)N(C_2H_5)_2$
methixene	Tremaril Tremonil Trest	9-(N-methyl-3-pip-cridylmethyl)thioxanthene	C	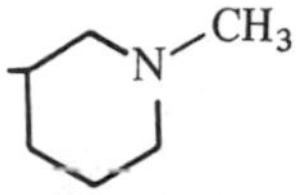 (piperidine ring) $N-CH_3$

*In *USP XIX* (1975).

trihexyphenidyl or its congeners. Untoward effects are many, including agranulocytosis. Large doses in long-term treatment may exacerbate parkinsonian symptoms. Their administration with antihistaminics is not recommended. They are used as hydrochlorides.

4. Miscellaneous Drugs

Those most widely used are listed in Table 11.6. Besides these, the following two, elantrine and piroheptine, structurally related to tricyclic antidepressants, are sometimes used.

Levodopa

It is the immediate precursor of dopamine in the biosynthesis of catecholamines

Table 11.6 Miscellaneous Drugs Used in Parkinsonism

Official Name	Proprietary Name	Chemical Name	Structure
caramiphen	Panparnit Parpanit	2-diethylaminoethyl 1-phenylcyclopentane-1-carboxylate	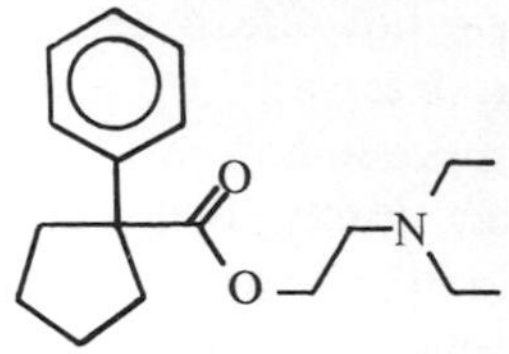
phenglutarimide	Aturbal Aturban Aturbane	2,2-diethylamino-ethyl-2-phenyl-glutarimide	
levodopa*	Dopar Larodopa Levopa	(–)-3-(3,4-dihydroxy-phenyl-L-alanine	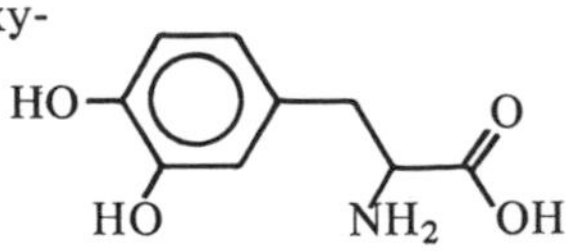
amantadine[+]	Symmetrel	1-adamantanamine	

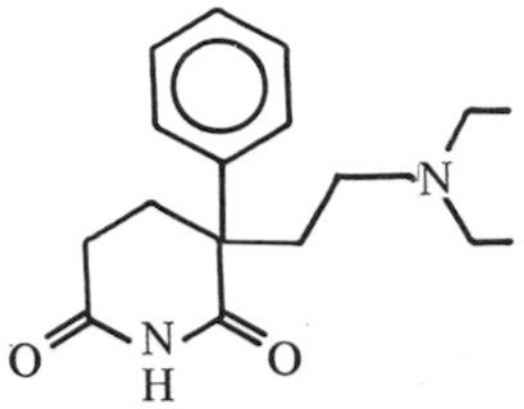

*In *USP XIX* (1975).
[+]In *NF XIV* (1975).

(Chapter 19). Its introduction in therapy is very recent and was prompted by the observation that parkinsonian patients present a low level of dopamine in the basal ganglia and substantia nigra. Levodopa appears to be the drug of choice in idiopathic or postencephalitic parkinsonism. According to the American Medical Association, it "is the most effective agent presently available for treating parkinsonism." Because levodopa is inactivated by peripheral enzymatic action, large doses are usually required to attain therapeutic results. However, large doses cause numerous and troublesome adverse reactions, especially gastrointestinal, cardiovascular, and psychiatric disorders, besides neurologic symptoms, in nearly 100% of patients.

With the purpose of achieving the same efficiency with lower doses, the concomitant administration of levodopa- and dopa-decarboxylase inhibitors is presently being investigated. These inhibitors are hydrazines or hydrazides structurally related to levodopa. Two are already marketed: benserazide (in combination with levodopa = Madopar) and carbidopa (Sinemet). Another drug used in combination with levodopa is fusaric acid. Decarboxylase inhibitors exert their action in peripheral tissues, inhibiting the conversion of levodopa to dopamine, but they do not affect this biochemical reaction in the central nervous system. Clinical trials have shown that combination of levodopa with decarboxylase inhibitors above-mentioned results in a faster, but not greater, therapeutic effect.

Levodopa should not be administered simultaneously with MAO inhibitors, because this combination may result in hypertensive crises. The same caution should be kept toward antipsychotics and antianxiety agents and some antihypertensive drugs, such as methyldopa and *Rauwolfia* alkaloids.

The usual dosage of levodopa by oral route ranges from 300 mg to 1 g, administered three to seven times daily with food in order to reduce gastrointestinal disturbances.

Apomorphine, NF

This morphine derivative, administered subcutaneously, is a very effective and safe emetic, acting in 10 to 15 minutes. Recent experiments indicate that it may also become a useful agent in Parkinson's disease, because it relieves tremors in patients. Since part of its structure resembles that of dopamine, it acts as a competitive antagonist of neuroleptics at the dopaminergic receptors:

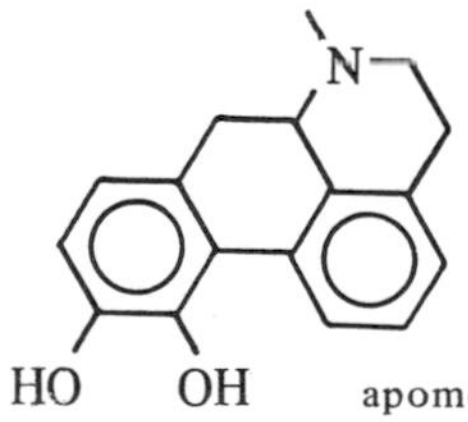

HO   OH   apomorphine   HO   OH   dopamine

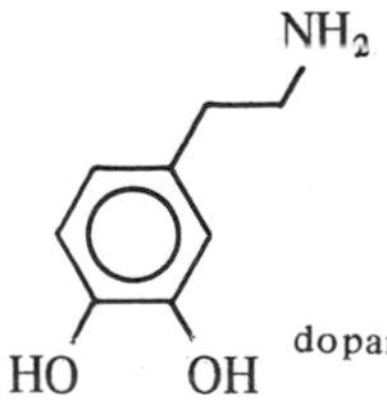

Amantadine, NF

Used as hydrochloride. Its effectiveness is of the same order of the standard antiparkinsonism drugs, especially anticholinergics, but it is not as high as that of levodopa. Amantadine is also used as prophylactic agent against various strains of Asian influenza.

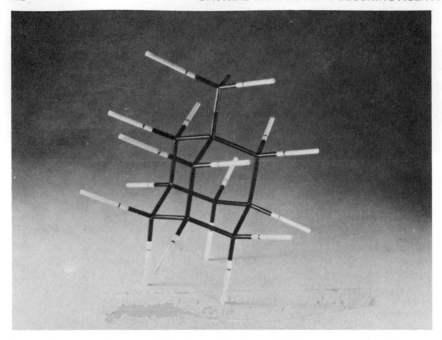

Molecular model of amantadine

B. Mechanism of Action

Antiparkinsonism drugs known thus far act by two different mechanisms, thus summarized by Vernier:

1. Decrease of extrapyramidal facilitation
2. Increase of extrapyramidal inhibition

The first mechanism is operative in most anticholinergics which block central acetylcholine muscarinic transmission: scopolamine, atropine, benztropine, trihexyphenidyl, and others.

The second mechanism explains the action of other types of drugs, which restore deficient central adrenergic transmission, mainly dopamine. Increase of extrapyramidal inhibition is accomplished in several ways: (*a*) supply of a precursor—such as the action of levodopa; (*b*) release of a local transmitter—thus acts amantadine; (*c*) increase in transmitter synthesis—this is another mode of action of amantadine; (*d*) prolongation of transmitter availability and block of dopamine uptake—antihistaminics and anticholinergics, respectively, act by these mechanisms: (*e*) action as transmitter—it is the case with apomorphine.

REFERENCES

Muscle Relaxants

J. Malatray, *Prod. Probl. Pharm.,* **23**, 425 (1968).

E. J. Pribyl, *Med. Chem.,* **6**, 246 (1963).

C. D. Lunsford *et al., J. Amer. Chem. Soc.,* **82**, 1166 (1960).

F. M. Berger, *Pharmacol. Rev.,* **1**, 243 (1949).

Antiparkinsonism Drugs

J. Marks, *The Treatment of Parkinsonism with L-Dopa,* American Elsevier, New York, 1974.

J. Presthus, *Med. Actual. (Drugs of Today),* **10**, 315 (1974).

J. R. Bianchine *et al., Clin. Pharmacol. Ther.,* **13**, 584 (1972).

B. Boshes, *Advan. Intern. Med.,* **18**, 219 (1972).

C. Mawdsley *et al., Clin. Pharmacol. Ther.,* **13**, 575 (1972).

R. M. Pinder, *Progr. Med. Chem.,* **9**, 191 (1972).

A. Barbeau, *Lancet,* **I**, 395 (1971).

Symposium, on Levodopa in Parkinson's Disease, *Clin. Pharmacol. Ther.,* **12**, 317-416 (1971).

V. G. Vernier, *Annu. Rep. Med. Chem.,* **1970**, 42 (1971).

A. Barbeau and F. H. McDowell, *Levodopa and Parkinsonism,* Davis, Philadelphia, 1970.

D. B. Calne, *Clin. Pharmacol. Ter.,* **11**, 789 (1970).

F. H. McDowell *et al., Ann. Intern. Med.,* **72**, 29 (1970).

O. Hornykiewicz, *Pharmacol. Rev.,* **18**, 925 (1966).

CHAPTER **12**

CENTRAL-NERVOUS-SYSTEM STIMULANTS

I. INTRODUCTION

Central-nervous-system (CNS) stimulants are drugs that exert their action through nonselective excitation of the central nervous system. Some drugs of this group produce a powerful action, whereas others cause a weak effect. Two general mechanisms are involved in CNS stimulation: selective blockade of neuronal inhibition and direct neuronal excitation. In the first mechanism two processes have been described: (*a*) the blockade of postsynaptic inhibition; (*b*) the blockade of presynaptic inhibition. Strychnine acts by the first mechanism. Picrotoxin is an example of a drug that acts by the second mechanism. These two drugs, strychnine and picrotoxin, although clinically used for many years, are obsolete; they are now used almost exclusively in experimental investigations. Other CNS stimulants do not block inhibitory processes. The stimulant effects they produce result, therefore, from enhancement of synaptic excitation.

Until recently, it was customary to divide central-nervous-system stimulants into two fundamental classes: *analeptics* per se and *psychoanaleptics.* Today, however, more is known about psychoanaleptics; for this reason, they are being studied either as antidepressant drugs or as antianxiety agents, according to their clinical actions, as was mentioned in Chapter 10. As to analeptics, from the Greek word which means restorative, this name is now considered synonymous with *general stimulants,* which do not include cerebral or psychic stimulants.

Central-nervous-system stimulants are used for a variety of purposes: treatment of depressive states, maintenance of wakefulness or alertness, restoration of consciousness, recuperation of normal reflexes, restoration of respiration or blood pressure.

Analeptics may cause nausea, vomiting, cardiac arrythmias, convulsions, and other side effects if taken in higher doses. Euphoria, confusion, agitation, and hallucinations are some of the delayed psychotic reactions.

II. CLASSIFICATION

A rough pharmacological division of stimulants distinguishes three classes: central stimulants, respiratory stimulants, and convulsant stimulants. Taking into account their main site of action, however, stimulants may be more properly divided into two main groups: general stimulants and psychic stimulants. The first are rarely used, whereas use of the second is wide and increasing.

A. General Stimulants

Also called analeptics, general stimulants comprise most of the central-nervous-system stimulants. Their clinical use is now limited. Contrary to the opinion formerly accepted, they do not display any useful action as specific antagonists to hypnotics and sedatives. For treatment of respiratory depression associated with drug-induced coma, supportive therapy (maintenance of adequate respiratory exchange and electrolyte balance by use of dialysis) is currently the regime of choice. Analeptics are given only as adjunctive agents.

Intoxication by narcotics is currently treated with narcotic antagonists (nalorphine, levallorphan) and not with general stimulants.

Analeptics are useless in many conditions, some for which they were prescribed years ago: carbon dioxide intoxication, carbon monoxide poisoning, anoxia, respiratory failure caused by drowning, pulmonary insufficiency resulting from chronic lung disease, electric shock. According to the American Medical Association, "they are also ineffective for treating cardiac arrest, depression caused by an overdosage of central nervous system stimulants, and pulmonary insufficiency caused by chronic lung diseases (e.g., emphysema)."

Their primary site of action is in the medulla, where in therapeutic doses they stimulate the depressed centers in order to restore respiratory activity and increase muscle tone. However, greater doses cause stimulation of the midbrain, cortex, and/or spinal cord. Some of these drugs act by a reflex mechanism.

According to their site of action, general stimulants may be divided into three groups: medullary stimulants, spinal-cord stimulants, and reflex stimulants.

1. Medullary Stimulants

These drugs stimulate the medullary centers by acting directly on them. Since in high doses they produce convulsions, these analeptics are also called *convulsant stimulants*. In this group is included bemegride (Megimide), which is no longer marketed in the United States. Others, such as nikethamide, pentetrazol, and picrotoxin, once very popular, have seen a sharp decline in their use. The most currently used are ethamivan and doxapram, which are relatively new drugs.

According to their chemical structures, medullary stimulants may be divided into the following groups: nikethamide and derivatives, tetrazoles, and miscellaneous compounds. Those mostly used are listed in Table 12.1. Several other

Table 12.1 Medullary Stimulants

Official Name	Proprietary Name	Chemical Name	Structure
carbon dioxide*		carbon dioxide	CO_2
nikethamide	Coramine Nikorin	N,N-diethylnic- otinamide	
ethamivan[+]	Emivan	N,N-diethylvanil- lamide	
pentylenetetrazol (pentetrazol)	Metrazol	6,7,8,9-tetrahydro- 5H-tetrazoloazepine	
doxapram[+]	Dopram	1-ethyl-3,3-diphen- yl-4-(2-morphol- inoethyl)-2-pyr- rolidinone	
picrotoxin[a]			

*In *USP XIX* (1975).
[+]In *NF XIV* (1975).
[a]See text.

agents are, however, available. Among them are the following: bemegride (Megimide), dimefline (Remeflin), and flurothyl (Indoklon). Camphor and its derivatives administered internally have enjoyed wide popularity as central stimulants. Presently, their use for this purpose is considered unsound. Nevertheless they are properly used as mild antiseptics, analgetics and antipruritics.

a Nikethamide and Derivatives

Of this group, members of which have in common the diethylamide or similar moiety, nikethamide and ethamivan are the most widely used. Others, however, are marketed: nikethamide chlormethylate (Inicardio), camphotamide (Camphramine, Tonicorine), phthalic tetraethyldiamide (Coretonin), 3,5-dimethylisoxazole-4-carboxylic acid diethylamide (Cycliton), 3-ethoxy-4-hydroxybenzoic acid diethylamide (Anacardiol), dimorpholamine (Théraleptique, Therapline), pretcamide (Micoren), cotinine (Scotine).

Nikethamide

It occurs either as a slightly viscous liquid or as crystalline solid, because the melting point is 22 to 24°. It is miscible with water, ether, chloroform, and ethanol. Once widely used as a respiratory and cardiac stimulant, this pyridine derivative is now considered by the American Medical Association as of no value in drug-induced coma and may be dangerous, because the margin between the analeptic and convulsant dose is narrow. Therefore, its use is inadvisable.

Ethamivan

It is a colorless crystalline powder, which is sparingly soluble in water. Its use is inadvisable, for the same reasons pointed out in the case of nikethamide.

b. Tetrazoles

Although tetrazole is pharmacologically inactive, alkylation of positions 1 or 5 results in convulsant and analeptic compounds. The one clinically used is pentetrazol, called pentylenetetrazol in the United States.

Pentylenetetrazol

The international name is pentetrazol. It occurs as water-soluble, odorless white crystals which should be kept out of light. Years ago it was extensively used for treatment of apnea and drug-induced coma, psychoses, and senility. Now its use as an analeptic is inadvisable, as an antipsychotic it is obsolete, and as a geriatric it is not recommended. It is still used in the screening of new potential anticonvulsants.

c. Miscellaneous Compounds

This group consists of drugs of varied structures. Besides those listed in Table 12.1, it should be mentioned that several others are marketed outside the United States; among them the following ones: 2-amino-4-methylpyridine (Ascensil, Pirid), diethadione (Persisten, Ledosten), amiphenazole (Daptazole, Dizol), tacrine (Romotal), ecinamine (Gilutensin), 2-piperidinomethylcyclohexanone (Karion, Spiractin), meclofenoxate (Lucidril).

Carbon Dioxide

Because of its role as the natural stimulant of the respiratory center, it is used at the end of general anesthesia, in patients requiring mechanical hyperventilation in open-heart surgery. It is ineffective if the respiratory center is depressed. Furthermore, when inhaled concentrations reach 10%, initial stimulation may be followed by depression of respiration.

Picrotoxin

It is the active principle extracted from the seed of *Anamirta cocculus* or *Cocculus indicus.* By stimulating the respiratory center, it exerts long-lasting, highly potent respiratory action. Now it is considered obsolete, and its use is inadvisable, because it is ineffective in hastening arousal and can cause severe convulsions. Picrotoxin is a molecular addition compound of two substances, picrotoxinin and picrotin; only the former is active.

2. Spinal-Cord Stimulants

The only one of interest is strychnine. This alkaloid, which is extracted from seeds of *Strychnos nux-vomica,* was first isolated by the French pharmacists Pelletier and Caventou in 1818, and its structure, proposed by Robinson in 1947, was confirmed by total synthesis by Woodward and coworkers:

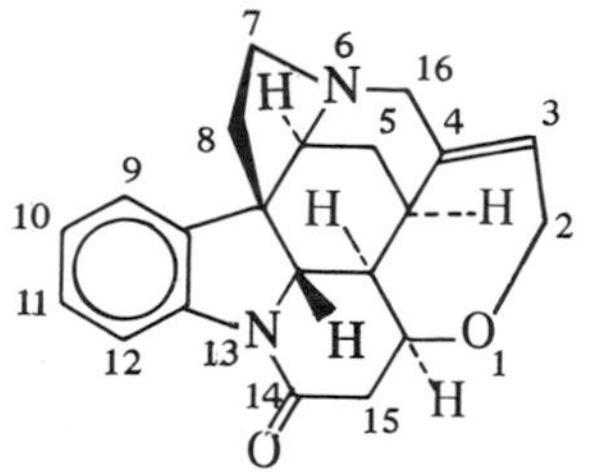

strychnine

For some time since Strecker introduced it in clinical use in 1861, strychnine has been used as bitter or "muscular tonic," but now it has no rational indication in medicine, except as a rodenticide. However, it is widely used as an experimental tool in pharmacological and physiological studies of the central nervous system, because it is a powerful convulsant, producing stimulation of all parts of the CNS. Accidental strychnine poisoning is treated by prevention of convulsions and support of respiration; the drug of choice is a short-acting barbiturate. A new drug, echinopsine, has properties similar to those of strychnine.

3. Reflex Stimulants

These drugs produce stimulation of the medullary centers by a reflex mechanism. They act on various peripheral sites, including the carotid sinus, and the impulses thus generated stimulate reflexly the medullary centers. Drugs of this group are seldom used, and most of them are considered either obsolete or dangerous.

Aromatic ammonia spirit is the only official preparation of this group. The usual dosage is 2 ml diluted in water. Irritation of the nose and throat gives rise to reflex stimulation.

B. Psychic Stimulants

Also called *cerebral stimulants* and *psychic energizers,* they include xanthine derivatives, phenalkylamine derivatives, and miscellaneous drugs. However, some authors place in this class also the antidepressant drugs, such as tricyclic compounds and monoamine oxidase inhibitors, that have already been studied in Chapter 10.

In the past, psychic stimulants were given to overcome the "hangover" effects that occur during the arousal period in patients with drug-induced coma. "However, their administration for this purpose is neither advisable nor logical," according to the American Medical Association. Nevertheless, they are useful to heighten or improve the mood.

1. Xanthine Derivatives

Xanthine derivatives used as cerebral stimulants are also called *psychomotor stimulants.* They stimulate the cerebral cortex and the medullary centers, besides other parts of the CNS system.

The xanthines most widely used in medicine are caffeine, theophylline, and theobromine. In small doses they are taken especially as coffee, tea, cocoa, and chocolate drinks, in order to enhance mental alertness and wakefulness, to lessen fatigue, and to produce diuresis. Excessive amounts, however, cause insomnia and restlessness in some people. Larger doses of caffeine and its salts were once used to stimulate the respiratory center, but now they are not recommended as respiratory stimulants. Aminophylline (Table 12.2) is used as a coronary dilator and antiasthmatic.

Xanthine derivatives are wrongly called alkaloids by some authors. By Elderfield's definition, however, alkaloids are nitrogenated substances with basic character. Theophylline and theobromine, although nitrogenous substances, have no basic but rather acidic character. Therefore, they are actually acids and not alkaloids. Caffeine forms acid salts but only at extremely low pH.

The following structures represent the parent structure xanthine, two of its anionic structures, and the structures of caffeine, theophylline, and theobromine:

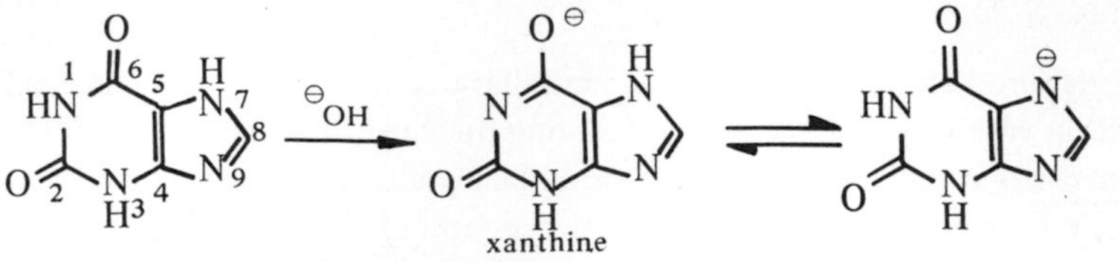

xanthine

1,3,7—CH$_3$ = caffeine   1,3—CH$_3$ = theophylline   3,7—CH$_3$ = theobromine

Of the three xanthines, caffeine is the most potent cerebral stimulant; theophylline follows next, whereas theobromine is almost devoid of stimulant properties. Theophylline produces more diuresis than theobromine and the latter more than caffeine.

Table 12.2 Xanthine Derivatives

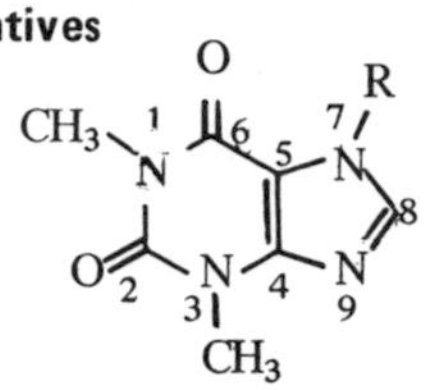

Official Name	Proprietary Name	Chemical Name	R
caffeine*		1,3,7-trimethylxan-thine	—CH$_3$
theophylline*+^{a}		1,3-dimethylxanthine	—H
aminophylline*^{b}	Diaphyllin Tefamin	theophylline ethylenediamine	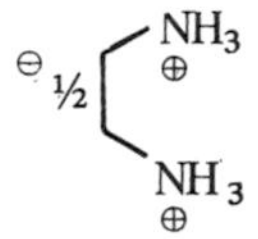
dimethazan		7-(2-dimethylamino-ethyl)theophylline	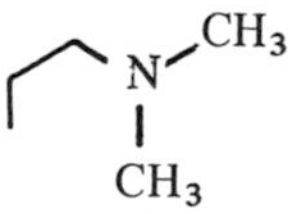
fenetylline	Captagon	7-[2-(1-methylphen-ethylamino)ethyl] theophylline	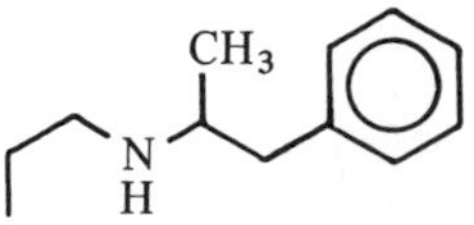

*In *USP XIX* (1975).
+In *NF XIV* (1975).
^{a}Used as a pharmaceutical necessity for meralluride injection, smooth-muscle re-laxant, and diuretic.
^{b}Used as a brochiolar (smooth-muscle) relaxant and seldom as a diuretic.

Xanthines mostly used as psychic stimulants are listed in Table 12.2. The following ones are also marketed: bamifylline (Trentadil), diprophylline (Cordalin), proxyphylline (Spasmolysin). Those used as diuretics are studied in Chapter 24. Xanthines may be given by injection, or orally in the form of black coffee.

As a fully methylated purine, caffeine may affect DNA. It cannot, though, be incorporated into DNA, because its methyl group at the 7 position prevents its forming a stable bond to deoxyribose at the 9 position. However, it may interfere in the genetic code in several ways: (a) alteration of the normal base ratios in the DNA precursor pool; (b) intercalation between base pairs of DNA; (c) prevention of repair of DNA strands damaged by radiation or other means. These mechanisms and also the inhibition of DNA polymerase caused by caffeine have been invoked to explain mutagenic and teratogenic effects ascribed to this trimethylxanthine. However, for centuries man has been drinking coffee, tea, mate, guaraná, cocoa, and other beverages that contain 1 to 4% caffeine without apparent harmful effects.

Caffeine

It occurs as a water-soluble white powder. It can be extracted form *Coffea arabica* or prepared by methylation of either theobromine or theophylline. It is used as both psychic stimulant and muscle stimulant. Often caffeine is combined with analgetics, especially aspirin, phenacetin, and acetanilid. The usual dosage is 60 to 200 mg. Larger doses cause insomnia, restlesness, excitement, tachycardia, diuresis, and other untoward effects. Caffeine is used also in combination with sodium benzoate and as citrated caffeine. With benzoic and citric acids caffeine does not form salts but charge transfer complexes, acting as an electron acceptor.

Nonofficial preparations are many. Examples are caffeine N-acetyltryptophanate (Adrifane), used in France; caffeine + nicotinic acid (Cosaldon), indicated mainly as a vasodilator in the treatment of cerebral sclerosis.

Fenetylline

Formerly called amfetylin. it is used as the hydrochloride, a white, crystalline powder, which occurs in two crystal forms; in therapy it is used as the racemic mixture. Fenetylline resulted from a rational design based on hybridization or molecular association of mixed moieties, both having psychic stimulating activity: theophylline and amphetamine. It stimulates the hypocampus, hypothalamus, and reticular formation.

2. Phenalkylamine Derivatives

This class is composed of sympathomimetic amphetamines and related drugs. However, contrary to what happens with catecholamines, they easily cross the blood-brain barrier, and this explains their central effects. They stimulate the normal cerebrospinal axis, lessen the degree of central depression caused by various drugs, and stimulate the medullar respiratory center. Therefore, these substances are not only potent CNS stimulants, but they also exert cardiovascular, hyperthermic, and anorexigenic activities. They show usefulness in treatment of mild depression, narcolepsy, and disturbed hyperkinetic children, and they are also largely used in managing obesity, a problem with at least 30% of the present adult population of the United States. Unfortunately, they manifest a tendency to cause psychic dependence. Although recommended by some authors, their use for fatigue is unjustifiable. Also, phenethylamines elevate the blood pressure. They are, therefore, potentially dangerous to cardiovascular patients.

The most widely used phenalkylamine derivatives as central stimulants and anorexiants are listed in Table 12.3. All present the β-phenethylamine skeleton

Table 12.3   Amphetamines and Related Drugs

Official Name	Proprietary Name	Chemical Name	Structure
amphetamine	Benzedrine	($\pm$)-1-phenyl-2-aminopropane	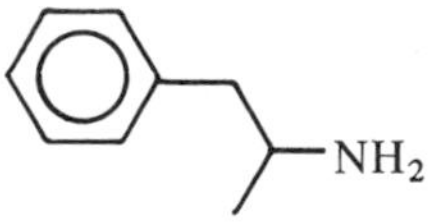
dextroamphetamine* (dexamphetamine)	Dexedrine	(+)-1-methyl-phenethylamine	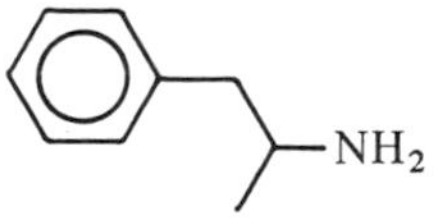
phenylpropanola-mine[+]	Propadrine	($\pm$)-1-phenyl-2-amino-1-propanol	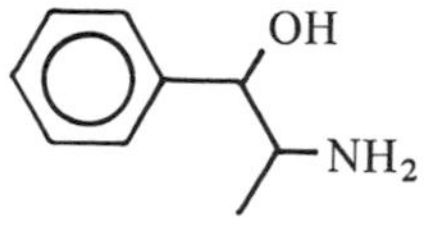
methamphetamine	Desoxyn Drinalfa Methedrine	(+)-N,1-dimethyl phenethylamine	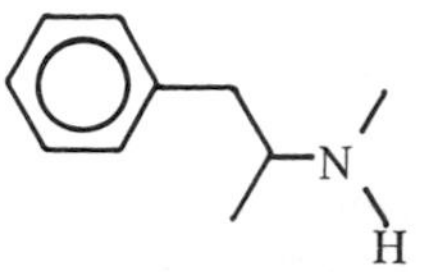
methylphenidate*	Ritalin	methyl α-phenyl-2-piperidine-acetate	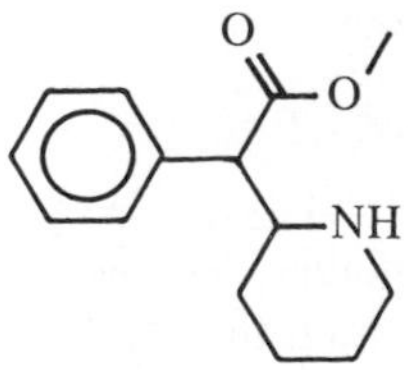

benzphetamine	Didrex	(+)-N-benzyl-N, 1-dimethyl-phenethylamine	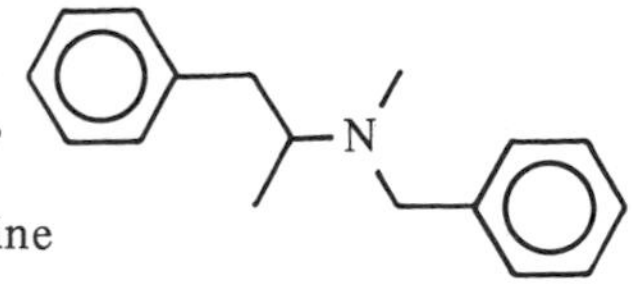
phentermine	Ionamine Wilpo	1-phenyl-2-methyl-2-amino-propane	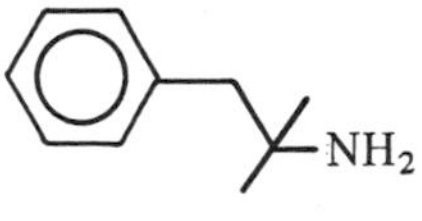
chlorphentermine	Pre-Sate	1-(4-chlorophen-yl)-2-methyl-2-aminopro-pane	
phenmetrazine[*]	Preludin	(±)-3-methyl-2-phenylmor-pholine	
phendimetrazine	Plegine	(+)-3,4-dime-thyl-2-phenyl-morpholine	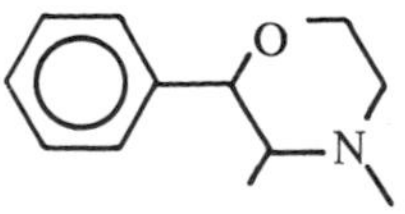
diethylpropion[+] (amfepramone)	Tenuate Tepanil	2-(diethylam-ino)propio-phenone	

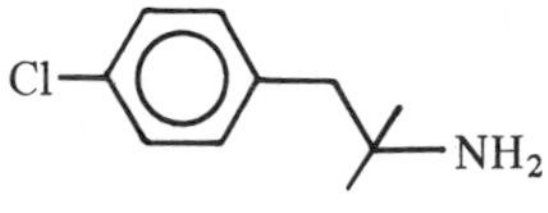

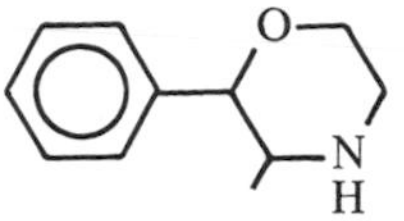

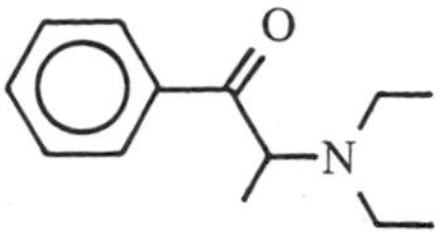

*In *USP XIX* (1975).
[+]In *NF XIV* (1975).

with a α-methyl group, $C_6H_5-\overset{\underset{|}{C}}{C}-C-N$, which is crucial to most of their pharmacological and biochemical activities. Thus, an α-methyl group is very important, because it prevents oxidation of the amino group by MAO and assists in drug transport mechanisms. Furthermore, the right position for the side-chain methyl group is α; shifting of this group to the β position results in a compound devoid of all stimulant and anorexic properties.

Phenylalkylamine derivatives are usually marketed as salts: hydrochloride, sulfate, phosphate, tannate, tartrate. For instance, dextroamphetamine is available in the form of hydrochloride, phosphate, sulfate and tannate; phentermine, as hydrochloride and resin.

Several mixtures containing anorexiants with sedatives, anorexiants with sedatives or bulk-producing agents and vitamins, anorexiants with thyroid preparations and sedatives, and miscellaneous anorexiant mixtures are available.

Amphetamine

This racemic mixture occurs as a colorless liquid. Since solids are easier to handle, its sulfate and phosphate salts are used. Amphetamine has a more marked effect on the cardiovascular system and less effect on the CNS than the (+)-isomer, dextroamphetamine. Amphetamine is not recommended as an anorexiant for this reason.

Dextroamphetamine

Introduced in 1944, this (+)-isomer of amphetamine has a greater psychic stimulating activity than the racemic mixture. Of the amphetamines, this drug is the most effective anorexiant, provided that a program of caloric restriction is followed during its administration and afterwards. However, it may cause psychic dependence, besides other untoward effects. Dextroamphetamine is contraindicated in patients under treatment with MAO inhibitors. The usual dosage is 2.5 to 10 mg three times daily.

Methylphenidate

Introduced in 1956, it is structurally related to amphetamine, having part of its side chain forming a piperidine ring; it can be considered, therefore, a modified isostere of amphetamine. It is a mild cortical stimulant similar to amphetamine and more potent than caffeine. It is being used in the management of hyperkinetic children, but it may cause a deleterious effect in those with psychopathic personalities.

3. Miscellaneous Compounds

Most drugs of this group are being investigated. Some are structurally related to amphetamine, and they can be considered its highly substituted isosteres. It is the case with pemoline (Cylert), tozalinone (Stimsen), and fenozolone (Ordinator), that are oxazolidinone derivatives:

$R = R' = H$ pemoline
$R = R' = CH_3$ tozalinone
$R = H; R' = C_2H_5$ fenozolone

It has been claimed that magnesium pemoline facilitates memory and learning, but recent experiments proved that it does not show this effect.

Another subgroup promoted for the same purpose consists of dimethylaminoethanol salts and esters. Among the salts are the following: tartrate (Atrol), pantothenate (Psicoline), N-acetylglutamate (deanol aceglumate, Otrun), acetamidobenzoate (deanol acetamidobenzoate, Deaner), xenyrate (namoxyrate, namol xenyrate), hemisuccinate (Tonibrol), pyroglutamate (Debrumyl), and pyrisuccinodimaleate (pyrisuccideanol dimaleate, Stivane). The derivative is the phosphoric acid ester (Panclar).

According to the American Medical Association, "the effectiveness of deanol acetamidobenzoate (Deaner) as an antidepressant or in the management of children with behavioral and learning problems has not been demonstrated conclusively."

Other drugs marketed as adjuvants in problems of learning and memory are arginine acetylasparaginate (Drusil), dexoxadrol, dioxadrol (Rydar), encyprate, gamfexine, sodium hexacyclonate (Repriscal), iprindole (Prondol), levoglutamide (Acutil), levoxadrol, mefexamide (Timodyne), meclophenoxate (Lucidril), nicametate (Euclidan), piracetam (Nootropil), pyritinol, scotophobin, tiazesim (Altinil).

A recent acquisition is mazindol (Sanorex), an imidazole-isoindole anorexiant chemically unrelated to phenalkylamines.

Two drugs are promoted as stimulants of sexual activity: 5,6-dihydroxytryptamine and fenclonine.

III. MECHANISM OF ACTION

The mode of action of most drugs studied in this chapter is not known. But evidence already available has permitted one to propose the possible mechanism of action of some of them.

A. General Stimulants

Owing to varied chemical structures, these drugs have different modes of action, which are poorly understood. For instance, the mechanism of action of nikethamide and derivatives is unknown. As to pentetrazol, its action results from stimulation of excitatory and inhibitory neurons. Picrotoxin blocks presynaptic inhibition, which is antagonized by γ-aminobutyric acid. Finally, strychnine acts by interfering with central inhibitory processes, that is, by blocking selectively postsynaptic inhibition.

B. Central Stimulants

Methylxanthines, especially theophylline, owe their action to competitive inhibition of cyclic nucleotide phosphodiesterase, an enzyme that catalyzes the conversion of cyclic 3',5'-adenosinemonophosphate (3',5'-AMP) to 5'-adenosinemonophosphate (5'-AMP), as shown in Figure 12.1. As a result, the concentration

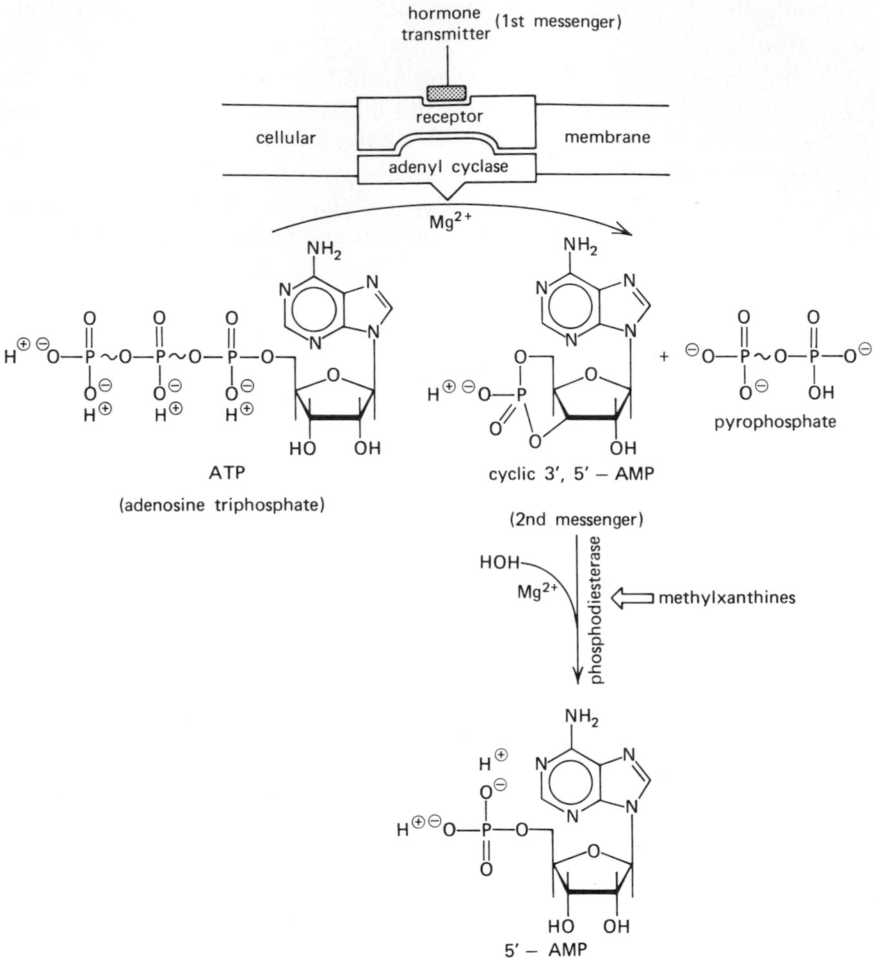

Figure 12.1 Site and mechanism of action of methylxanthines.

of cyclic $3',5'$-AMP is increased in many tissues. The same effect is produced by catecholamines but by a different mechanism: stimulation of adenyl cyclase to convert ATP to cyclic $3',5'$-AMP. Cyclic $3',5'$-AMP plays a critical role in promoting glycogenolysis. Therefore, an increase in cyclic $3',5'$-AMP may cause psychic stimulation observed by administration of methylxanthines by increasing the availability of glucose to the brain.

As to amphetamines and related sympathomimetic drugs, it is known that they owe their peripheral action to the release of catecholamines, mainly norepinephrine, from the storage granules at nervous sympathetic terminals.

Their central actions, however, are not perfectly understood. Several theories have been proposed, but none is of general acceptance. One claims that, since they cross the blood-brain barrier and they are attacked by MAO at a slower rate than other sympathomimetic amines, phenalkylamines could act on serotonin receptors in the brain. Anyway, their effects result from cortical stimulation and possible stimulation of the reticular activating system. There is also evidence that anorexigenic phenalkylamine derivatives may act through stimulation of the lateral nuclei or feeding center of the hypothalamus.

The mechanism of action of dimethylaminoethanol salts and derivatives is suggested to be related to acetylcholine biosynthesis, which proceeds by the following pathways:

$$HO-CH_2-\underset{\underset{NH_2}{|}}{CH}-COOH \xrightarrow[-CO_2]{} HO-CH_2-CH_2-NH_2 \longrightarrow$$

serine   1-ethanolamine

$$CH_3-S-CH_2-CH_2-\underset{\underset{NH_2}{|}}{CH}-COOH \qquad HS-CH_2-CH_2-\underset{\underset{NH_2}{|}}{CH}-COOH$$

methionine   homocysteine

$$HO-CH_2-CH_2-\overset{+}{N}(CH_3)_3 \xrightarrow[\underset{ATP}{choline\ acetylase}]{acetyl\text{-}CoA \quad CoA} CH_3-\overset{\overset{O}{\|}}{C}-O-CH_2-CH_2-\overset{+}{N}(CH_3)_3$$

choline   acetylcholine

Since dimethylaminoethanol $HO-CH_2-CH_2-N(CH_3)_2$ is structurally related to choline, salts and derivatives of the former might act as precursors of acetylcholine biosynthesis that cross the blood-brain barrier better than choline. However, there is no evidence that the abnormal conditions for which these drugs have been prescribed result from acetylcholine deficiency in the brain or that the effects produced by them are a consequence of an increased content of this chemical transmitter.

REFERENCES

Introduction

J. M. van Rossum, *Int. Rev. Neurobiol.*, **12**, 307 (1970).
J. Cole, *J. Amer. Med. Assoc.*, **190**, 448 (1964).
F. Hahn, *Pharmacol. Rev.*, **12**, 447 (1960).

Classification

W. B. Essman, "Drugs Affecting Facilitation of Learning and Memory," in A. A. Rubin, Ed., *Search of New Drugs*, Dekker, New York, 1972, pp. 385-406.

K. Rickels *et al.*, *Clin. Pharmacol. Ther.*, **13**, 595 (1972).

E. J. Fjordungstad, *Chemical Transfer of Learned Information*, North-Holland, Amsterdam, 1971.

K. Rickels *et al.*, *Clin. Pharmacol. Ther.*, **11**, 698 (1971).

O. Vinar *et al.*, Eds., *Advances in Mnemopsychology*, North-Holland, Amsterdam, 1971.

W. L. Byrne, Ed., *Molecular Approaches to Learning and Memory*, Academic, New York, 1970.

E. Costa and S. Garattini, Eds., *Amphetamines and Related Compounds*, Raven, New York, 1970.

B. Weiss and V. G. Laties, *Pharmacol. Rev.*, **14**, 1 (1962).

Mechanism of Action

F. Sulser and E. Sanders-Bush, *Annu. Rev. Pharmacol.*, **11**, 209 (1971).

E. W. Sutherland *et al.*, *Circulation*, **37**, 279 (1968).

PART 3

DRUGS STIMULATING OR BLOCKING THE PERIPHERAL NERVOUS SYSTEM

Drugs acting on the peripheral nervous system can either stimulate it or block it. With the exception of local anesthetics, all these drugs act by altering the transmission of impulses between synapses or between neuroeffector junctions. They comprise the following classes of drugs:

1. Cholinergic and anticholinergic agents
2. Adrenergic stimulants, adrenergic blocking agents, and inhibitors of catecholamine biosynthesis and metabolism
3. Histamine and antihistamine agents
4. Local anesthetics

These drugs are studied in Chapters 13 to 21.

REFERENCES

A. Korolkovas, *Fundamentos de Farmacologia Molecular: Base para o Planejamento de Fármacos,* EDART and Universidade de São Paulo, São Paulo, 1974.

D. J. Triggle, *Neurotransmitter-Receptor Interactions,* Academic, London, 1971.

F. M. Abboud, *Advan. Intern. Med.,* 15, 17 (1969).

D. J. Triggle, *Chemical Aspects of the Autonomic Nervous System,* Academic, New York, 1968.

A. Burger, Ed., *Drugs Affecting the Peripheral Nervous System,* Dekker, New York, 1967.

C. B. Ferry, *Annu. Rev. Pharmacol.,* 7, 185 (1967).

A. G. Karczman, *Annu. Rev. Pharmacol,* 7, 241 (1967).

W. S. Root and F. G. Hofmann, Eds., *Physiological Pharmacology,* Vols. III and IV, Academic, New York, 1967.

CHAPTER 13

CONSIDERATIONS INVOLVING THE AUTONOMIC NERVOUS SYSTEM

I. INTRODUCTION

In order to understand the action of drugs discussed in the next chapters, it is necessary to recall some basic concepts related to anatomy and physiology of the nervous system.

The nervous system is comprised of two fundamental parts (Table 13.1):

1. Central nervous system
2. Peripheral nervous system

Table 13.1 Anatomy of Nervous System

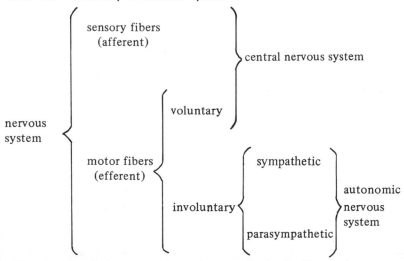

The peripheral nervous system is comprised of afferent (or sensory) and efferent (or motor) neurons of the autonomic and somatic nervous system.

II. ANATOMY

The autonomic nervous system, also called vegetative, visceral, automatic, or involuntary nervous system, innervates almost all tissues of the body, with the exception of skeletal muscle. It consists of nerves, ganglia, and plexuses; it helps to control the so-called vegetative functions, such as arterial pressure, gastrointestinal motility and secretion, body temperature, urinary output, and several other bodily functions. Its activities are controlled mainly by the hypothalamus, which also contains control centers for several other involuntary activities of the body not under autonomic-nervous-system regulation.

Autonomic impulses are transmitted to the body through the sympathetic and parasympathetic systems, the two divisions of the autonomic nervous system. The first division is a thoracolumbar outflow, the second one a craniosacral outflow. The sympathetic system is much more widely distributed in the body, and its fibers ramify more extensively than the parasympathetic system. Both systems, however, are composed of preganglionic and postganglionic fibers. In the sympathetic system the preganglionic fibers are short and the postganglionic, long. The contrary occurs in the parasympathetic system. At sympathetic nerve endings norepinephrine is the neurohumoral transmitter. At ganglia and parasympathetic nerve endings acetylcholine is the chemical mediator of impulses. In other words, norepinephrine is liberated by the postganglionic sympathetic fibers; acetylcholine is liberated by all preganglionic autonomic fibers, all postganglionic parasympathetic fibers, and a few post-ganglionic sympathetic fibers. Owing to their endogenous origin and chemical structures, these mediators—norepinephrine and acetylcholine—are called biogenic amines (Figure 13.1).

The somatic nervous system innervates skeletal muscle. It is composed of (a) *sensory fibers* carrying impulses from the external stimulus and (b) *motor fibers* innervating voluntary muscle cells. The neuroeffector junction between motor nerves and skeletal muscle is called neuromuscular junction.

III. PHYSIOLOGY

The autonomic nervous system regulates activities which are not under voluntary control, such as breathing, digestion, glandular secretion, circulation. It maintains the constancy of the internal environment of the organism, and, therefore, it is of fundamental importance to the well-being of the organism. The two branches of the autonomic nervous system are complementary, acting synergistically, although their activities produce generally opposite effects. (Table 13.2). Thus, depression of one system results in effects reminiscent of stimulation of the other system, and vice versa.

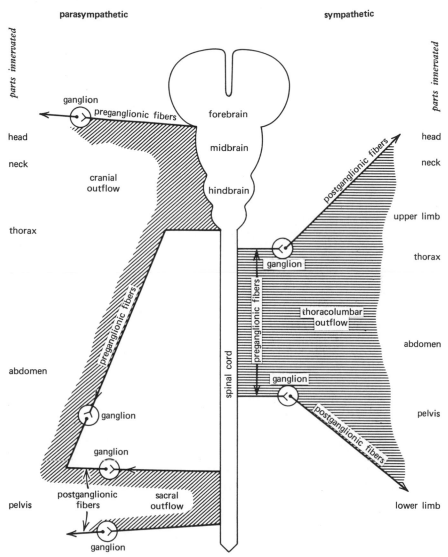

Figure 13.1 Sympathetic and parasympathetic nervous system.

The somatic nervous system, through its sensory or afferent fibers, carries sensory impulses from the skin and special sense organs, and, through its motor efferent limb, controls movement and posture by regulating skeletal muscle activity.

Table 13.2 Actions of the Autonomic Nerves on Various Effectors

Effector	Response to Sympathetic Nerves	Response to Parasympathetic Nerves	Nature of Responses
Eye:			
Pupil	Dilation	Constriction	Opposed
Iris:			
Radial muscles	Contraction		
Circular muscles		Contraction	
Accommodation		Near vision	
Ciliary muscle		Contraction	
Tarsal muscle	Contraction		
Orbital muscle	Contraction		
Nictitating membrane			
(cat, etc.)	Contraction		
Glands:			
Sweat	Secretion[a]		
Salivary	Secretion	Secretion	Parallel
Lacrimal		Secretion	
Respiratory tract		Secretion	
Gastrointestinal tract		Secretion	
Piloerectors	Contraction		
Bronchioles	Relaxation	Contraction	Opposed
Heart:			
Sinus nodal rhythm	Acceleration	Slowing	Opposed
AV node refractory period	Reduced	Increased	Opposed
Atrial conduction rate	Increased	Increased	Parallel
Atrial contraction force	Increased	Decreased	Opposed
Ventricular contraction force	Increased	Decreased (?)	Opposed
Blood vessels:			
Muscle	Dilation		
Coronary	Dilation	Constriction	Opposed
Skin	Constriction		
Viscera	Constriction		
Salivary gland	Constriction	Dilation	Opposed
Erectile tissue	Constriction	Dilation	Opposed
Gastrointestinal tract:			
Muscle wall	Relaxation	Contraction	Opposed
Sphincters:	Contraction	Relaxation	Opposed
Cardiac		Relaxation[b]	
Ileocecal	Contraction		
Spleen	Contraction		
Urinary bladder:			
Fundus	Relaxation	Contraction	Opposed
Trigone and sphincter	Contraction	Relaxation	Opposed

Uterus:
  Nonpregnant
    Cat   Relaxation
    Human   Contraction
  Pregnant   Contraction
Liver   Glycogenolysis

Source. Reproduced from J. R. DiPalma, Ed., *Drill's Pharmacology in Medicine,* 3rd ed., McGraw-Hill, New York, 1965.
[a]Cholinergic fibers
[b]Adrenergic fibers.

REFERENCES

J. Axelrod, *Science,* 173, 598 (1971).

U. S. von Euler, *Science,* 173, 202 (1971).

G. B. Koelle, *Anesthesiology,* 29, 643 (1968).

J. Glowinski and R. J. Baldessarini, *Pharmacol. Rev.,* 18, 1201 (1966).

L. Decsi, *Progr. Drug Res.,* 8, 53 (1965).

J. H. Burn, *The Autonomic Nervous System: For Students of Physiology and of Pharmacology,* Davis, Phildelphia, 1963.

CHAPTER 14

CHEMICAL TRANSMITTERS

I. INTRODUCTION

It is now generally accepted that nerve impulse consists of the movement of ions across nerve membrane, being therefore, in essence, an electrical event. The responses elicited in tissues and organs by nerve impulses result from liberation of specific chemical substances. These substances are called *chemical transmitters*. The main ones are acetylcholine and norepinephrine. The first is involved in cholinergic transmission; the second, in adrenergic transmission.

II. CHOLINERGIC TRANSMISSION

Cholinergic transmission is so called because the chemical transmitter involved is acetylcholine. Four different types of nerves use acetylcholine as chemical transmitter and, for this reason, they are called cholinergic. Cholinergic fibers are (*a*) voluntary; (*b*) preganglionic parasympathetic; (*c*) postganglionic parasympathetic; (*d*) preganglionic sympathetic.

Acetylcholine produces two types of effects: nicotinic and muscarinic. The nicotinic effects are analogous to those produced by nicotine, that is, on ganglia and the motor end plate: stimulus and increase in tone of skeletal muscles. These effects are blocked by tetraethylammonium ions. The muscarinic effects are similar to those produced by muscarine and pilocarpine, that is, on the postganglionic parasympathetic receptors: cardiac inhibition, peripheral vasodilation, pupil contraction, increase of salivation and glandular secretion, contractions and peristaltic action of gastrointestinal and urinary tracts. Such effects are blocked by atropine. Therefore it is assumed that for acetylcholine there are two types of receptors: nicotinic and muscarinic. Besides interacting with these receptors, acetylcholine combines with acetylcholinesterase during the process of its hydrolysis.

As far as the diversity of effects produced by acetylcholine is concerned, a hypothesis that has been advanced states that all these effects result from the different conformations of the molecule of this chemical transmitter. Thus, by interacting with muscarinic receptors in its muscarinic conformation, acetylcholine causes muscarinic effects. Interaction of acetylcholine in nicotinic conformation with nicotinic receptors is responsible for nicotinic effects.

Through molecular orbital calculations, interatomic distances in the preferred conformations of acetylcholine and its analogues muscarine and nicotine were determined (Table 14.1).

Table 14.1 Interatomic Distances in the Preferred Conformations of Acetylcholine, Muscarine, and Nicotine (in Angstroms)

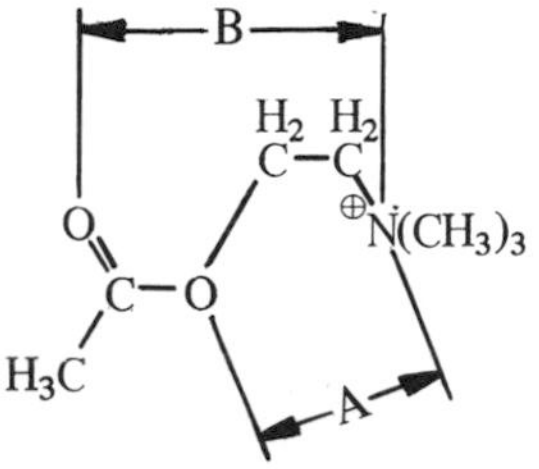

acetylcholine 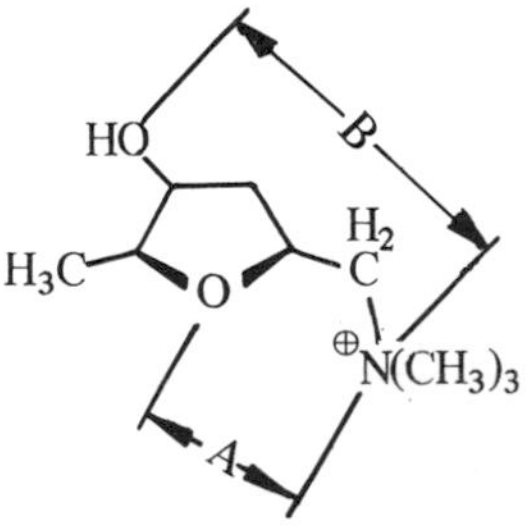   muscarine

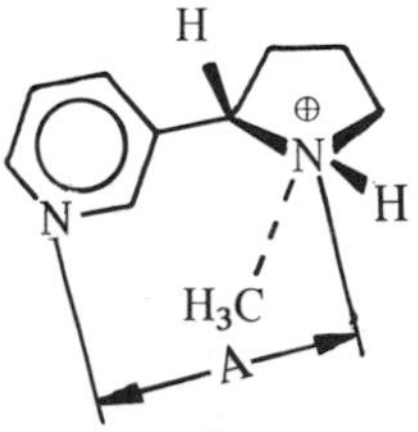

nicotine

	Acetylcholine		Muscarine		Nicotine
	A	B	A	B	A
Found in crystalline state	3.26	5.4	3.07	···	4.4
Calculated by EHT method:					
"Muscarinic" conformation of acetylcholine	3.29	4.9-5.4	3.07	5.7	
"Nicotinic" conformation of acetylcholine	3.85	4.93	···	···	4.36-4.76
Calculated by PCILO method	3.02	4.93	2.91	···	4.3-4.7

Recent evidence favors the hypothesis that acetylcholine exists in only one preferred conformation, the completely extended trans form. It is in the trans form that acetylcholine interacts with its receptors, as found by using the conformationally rigid acetylcholine analogues shown below. The trans isomer is highly active, whereas the cis isomer is inactive:

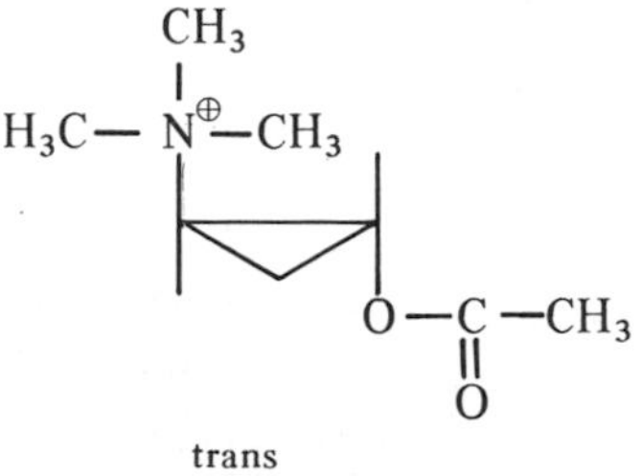

trans 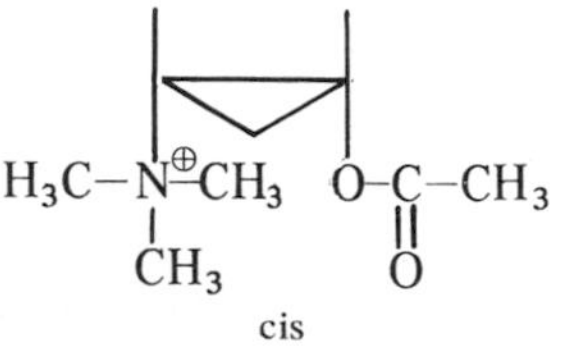   cis

The dual action of acetylcholine is explained thus: in its preferred conformation

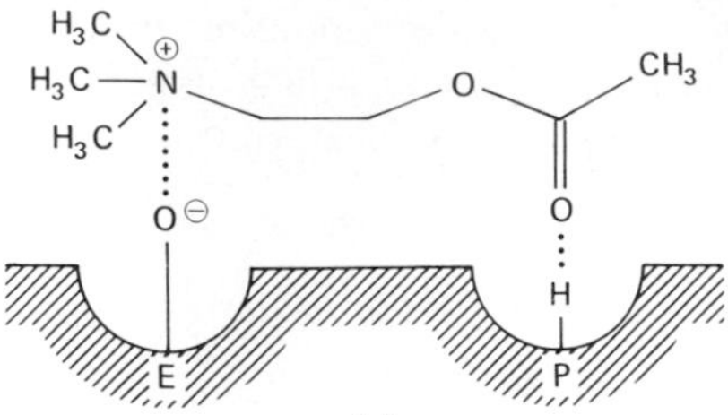

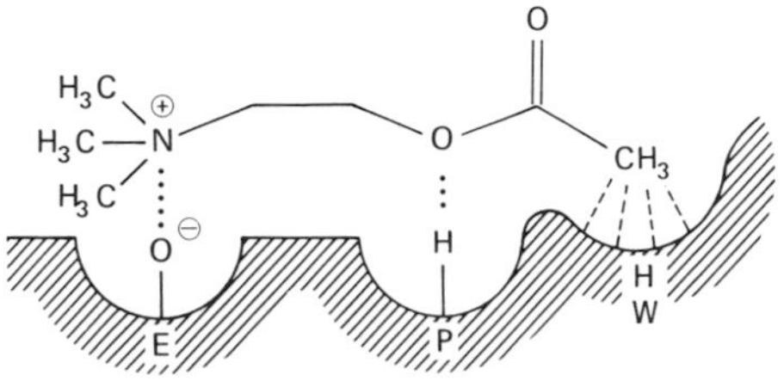

Figure 14.1 Interaction of acetylcholine with (a) nicotinic receptor, through electrostatic attraction E with the quaternary ammonium and hydrogen bonding P with the carbonyl oxygen; (b) muscarinic receptor, through electrostatic attraction E with the quaternary ammonium, hydrogen bonding P with the ester oxygen, and hydrophobic H and van der Waals W interactions with the methyl group. [According to C. H. Chothia, *Nature (London)*, 225, 36 (1970)].

acetylcholine interacts with both receptors but on different sides: with the nicotinic receptor by the carbonyl side and with muscarinic receptor by the methyl side (Figure 14.1).

Several authors have advanced the idea that the acetylcholine receptor is identical to the active site of acetylcholinesterase. Now it is clear that this cannot be so, because between both entities many differences were found as to localization, chemical nature, and chemical reactivity.

The action of acetylcholine is short-lived, because it is rapidly hydrolyzed by acetylcholinesterase.

III. ADRENERGIC TRANSMISSION

Adrenergic transmission received its name because the chemical transmitter involved was thought to be a substance formerly called adrenaline. Now it is known that the main chemical transmitter is norepinephrine (noradrenaline); the lesser ones are dopamine and epinephrine (the international name for adrenaline). These transmitters are called catecholamines, owing to the presence of a catechol nucleus. Adrenergic fibers occur only in the postganglionic sympathetic region of the ANS.

Since epinephrine exerts two types of effects, Ahlquist postulated that there are at least two classes of receptors for it: α and β. However, now the existence of two β receptors, β_1 and β_2, is postulated. Some authors claim that there are other adrenergic receptors, δ and γ.

Alpha adrenergic receptors are involved mainly in smooth-muscle stimulus. Beta receptors are associated with smooth-muscle inhibition, including intestine, and myocardial stimulus: β_1 receptors are located in the heart and small gut, and β_2 receptors, in the bronchus and vascular bed. Activation of the former causes increase of lipolysis and, of the latter, increase of glycogenolysis.

Molecular orbital calculations using the EHT method show that norepinephrine can exist in three main conformations, (b) being the preferred one:

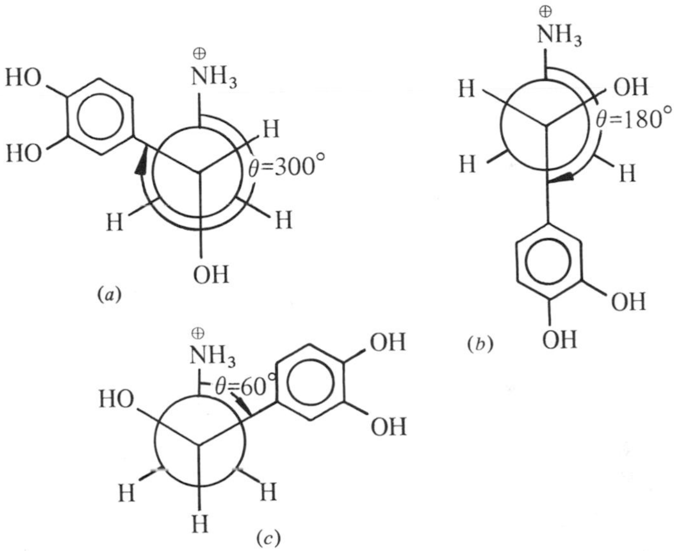

However, by using the INDO method, it is seen that the preferred conformation in solution is (c), in which the θ angle is 60°, and the least stable is (a), in which the same angle is 300°.

From results using the PCILO method, Pullman et al. concluded that in norepinephrine and related sympathomimetic phenethylamines (dopamine, epinephrine, ephedrine, norephedrine, tyramine, phenylpropanolamine, phenethylamine, amphetamine) there are no very marked preferences between folded and extended conformations. External conditions determine, according to these authors, the conformation of these molecules. They also calculated the interatomic distances in the stable conformations of some of these sympathomimetic amines (Figure 14.2). In the extended and folded forms, respectively, other distances in angstroms are (a) between alcoholic O atom and the ring plane: 0 and 0.67 in norepinephrine, 0 and 0.66 in norephedrine; (b) between the quaternary N and the ring plane: -1.32 and -2,16 in norepinephrine, -1.24 and -2,14 in norephedrine.

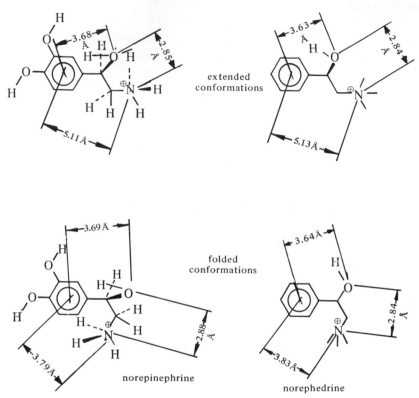

Figure 14.2 Interatomic distances in two sympathomimetic amines (norepinephrine and norephedrine) in their extended and folded forms. [As calculated by Pullman *et al., Mol. Pharmacol.,* **7,** 397 (1971).]

REFERENCES

Introduction

P. A. Shore, *Annu. Rev. Pharmacol.,* **12,** 209 (1972).

J. H. Biel and L. G. Abood, Eds., *Biogenic Amines and Physiological Membranes in Drug Therapy,* 2 vols., Dekker, New York, 1971.

G. Burnstock, *Pharmacol. Rev.,* **21,** 247 (1969).

F. E. Bloom and N. J. Giarman, *Annu. Rev. Pharmacol.,* 8, 229 (1968).

Cholinergic Transmission

G. H. Cocolas *et al., J. Med. Chem.,* 17, 938 (1974).

C. Hebb, *Physiol. Rev.,* 52, 918 (1972).

J. A. Izquierdo, *Progr. Drug Res.,* 16, 334 (1972).

A. Saran and G. Govil, *J. Theor. Biol.,* 37, 181 (1972).

R. W. Baker *et al., Nature (London),* **230,** 439 (1971).

D. L. Beveridge and R. L. Radna, *J. Amer. Chem. Soc.,* 93, 3759 (1971).

J.-P. Changeux *et al., Mol. Pharmacol.,* 7, 538 (1971).

R. Miledi and L. T. Potter, *Nature (London),* **233,** 599 (1971).

R. Miledi *et al., Nature (London),* **229,** 554 (1971).

B. Pullman *et al., Mol. Pharmacol.,* 7, 397 (1971).

J. R. Smythies, *Eur. J. Pharmacol.,* 14, 268 (1971).

W. H. Beers and E. Reich, *Nature (London),* **228,** 917 (1970).

C. H. Chothia, *Nature (London),* **225,** 36 (1970).

L. B. Kier, *Mol. Pharmacol.,* 4, 70 (1968).

L. B. Kier, *Mol. Pharmacol.,* 3, 487 (1967).

Adrenergic Transmission

B. Pullman *et al., Int. J. Quantum Chem.:* Quantum Biology Symp. no. 1, 93 (1974).

B. Pullman *et al., J. Med. Chem.,* 15, 17 (1972).

D. J. Triggle, *Annu. Rev. Pharmacol.,* 12, 185 (1972).

J. Axelrod, *Science,* **173,** 598 (1971).

U. S. von Euler, *Science,* 173, 202 (1971).

N. Kirshner *et al., Advan. Drug Res.,* 6, 121 (1971).

P. B. Molinoff and J. Axelrod, *Annu. Rev. Pharmacol.,* **40,** 465 (1971).

L. Pedersen *et al., J. Pharm. Pharmacol.,* 23, 216 (1971).

J. R. Smythies *et al., Nature (London),* **231,** 185 (1971).

D. Bieger *et al., Eur. J. Pharmacol.,* 9, 156 (1970).

H. J. Schümann and G. Kroneberg, Eds., *New Aspects of Storage and Release Mechanism of Catecholamines,* Springer, Berlin, 1970.

L. B. Kier, *J. Pharm. Pharmacol.,* 21, 93 (1968).

CHAPTER 15

CHOLINERGIC AGENTS

I. INTRODUCTION

Cholinergic agents are drugs that either directly or indirectly produce effects similar to those elicited by acetylcholine.

Acetylcholine is synthesized in nerve tissue, cholinergic synapses, and intestinal wall by the routes seen in Chapter 12, Section III.B. It is very likely that acetylcholine is bound to enzymes liberated by nerve impulses in order to stimulate cholinergic receptors, as shown in Figure 15.1.

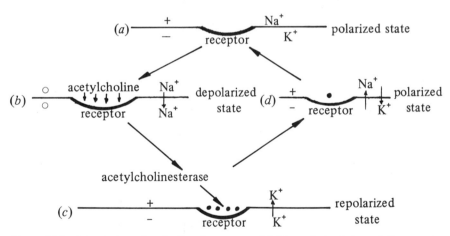

Figure 15.1 Stimulation of a cholinergic receptor. (*a*) The polarized state of the receptor results from the different concentrations of Na$^+$ outside the cell and K$^+$ inside the cell. (*b*) The nerve impulse liberates acetylcholine from storage vesicles. The movement of ions across the membrane eventually leads to a balance of charges, causing receptor stimulation to occur. (*c*) The repolarized state is achieved by hydrolysis of acetylcholine in excess by acetylcholinesterase. (*d*) The polarized state results from the return of the ions to the resting state by active transport.

It is theoretically possible to achieve cholinergic effects in one of the three following ways:

1. Prevention of acetylcholine biosynthesis. No substance is known which

prevents synthesis of acetylcholine by acting on acetylcholine-CoA. However, hemicholinium acts on the choline carrier mechanism and thus causes a depletion of acetylcholine.

2. Inhibition of acetylcholinesterase. Several drugs act by this mechanism, with the result that acetylcholine accumulates at the sites of cholinergic transmission.

3. Stimulation of specific receptor sites. Many cholinergic agents used clinically act this way.

Cholinergic agents, therefore, comprise two classes: cholinomimetics and anticholinesterases. The first produce effects similar to those which result from stimulation of postganglionic parasympathetic nerves; they act directly on effector cells innervated by the parasympathetic division of the ANS. The second cause cholinergic effects indirectly, by inhibiting acetylcholinesterase, the enzyme that hydrolyzes acetylcholine.

Cholinergic agents are used especially to treat gastrointestinal and urinary-bladder disorders. Some are used for treatment of glaucoma and myasthenia gravis.

Cholinomimetics are administered topically, orally, or subcutaneously. The intravenous route is not recommended, because it increases their toxicity besides causing loss of selectivity of action.

Among adverse reactions, cholinergics produce some minor side effects, such as miosis, sweating, excessive salivation, gastrointestinal disorders, bradycardia, and fall in blood pressure. They are contraindicated in patients with mechanical intestinal or urinary obstruction. Anticholinesterases in large doses may cause death as a result of respiratory failure.

II. CLASSIFICATION

According to their mode of action, cholinergic agents may be divided into two main classes: direct cholinergic agents and indirect cholinergic agents. Pilocarpine and cholinesterase reactivators are also included in this chapter.

A. Direct Cholinergic Agents

Direct cholinergic agents, also called cholinomimetics and parasympathomimetics, are drugs that, owing to their resemblance to acetylcholine in both chemical structure and distances between their polar groups as well as in the charge distribution, exert an action which is analogous to that of this chemical transmitter. Those most widely used are listed in Table 15.1. Others are oxapropanium (Dilvasine) and furtrethonium (Furmethide).

Some of them show muscarinic actions; others exhibit nicotinic actions. Muscarinic actions are similar to those produced by the alkaloid muscarine,

Table 15.1 Direct Cholinergic Agents

Official Name	Proprietary Name	Chemical Name	Structure
bethanecol[+]	Urecholine	carbamylmethylcholine	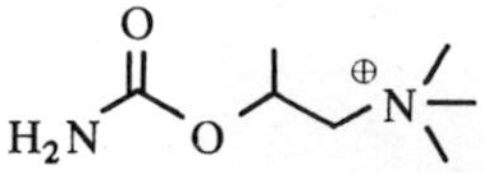
carbachol[*a]	Doryl	choline carbamate	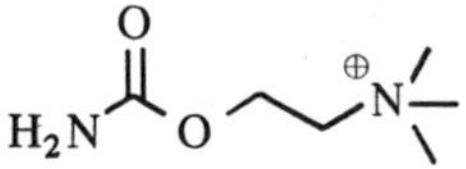
methacholine[+]	Mecholyl	acetyl-β-methylcholine	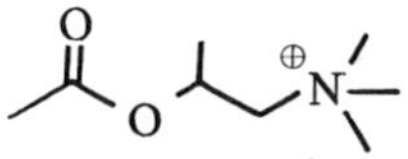

*In *USP XIX* (1975).
[+] In *NF XIV* (1975).
[a] Ophthalmic use.

Table 15.2 Indirect Cholinergic Agents

Official Name	Proprietary Name	Chemical Name	Structure
neostigmine*	Prostigmine	(m-hydroxyphenyl)trimethyl-ammonium dimethylcarbamate	[structure]
pyridostigmine*	Mestinon	3-hydroxy-1-methylpyridinium dimethylcarbamate	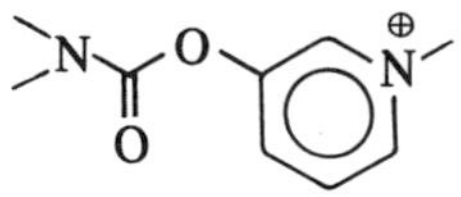 see Figure 2.1
demecarium+	Humorsol	(m-hydroxyphenyl)trimethyl-ammonium decamethylenebis-[methylcarbamate]	see Figure 2.1
physostigmine* (eserine)			see Figure 2.1
ambenonium chloride+	Mytelase	[oxalylbis(iminoethylene)]bis-[(o-chlorobenzyl)diethyl-ammonium]chloride	see Figure 2.1
echothiophate iodide* (ecothiopate)	Phospholine	S-(2-dimethylaminoethyl)-O,O-diethylphosphoro-thionate methiodide	[structure]
isoflurophate*	Floropryl	diisopropyl fluorophosphate	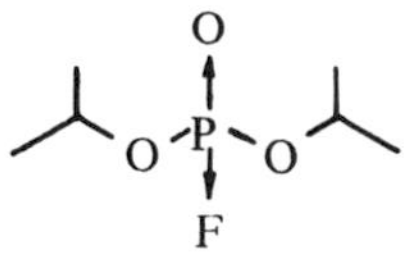 [structure]

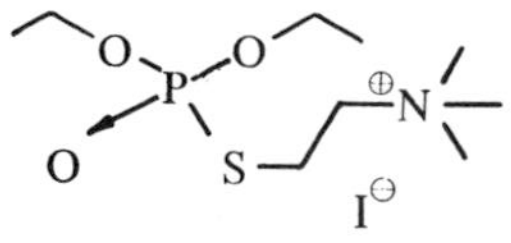

*In USP XIX (1975).
+In NF XIV (1975).

215

present in the mushroom *Amanita muscarina* and not used clinically: peripheral vasodilation, cardiac inhibition, increased salivation and outflow of most secretory glands, contraction of the pupils, and contractions and peristaltic

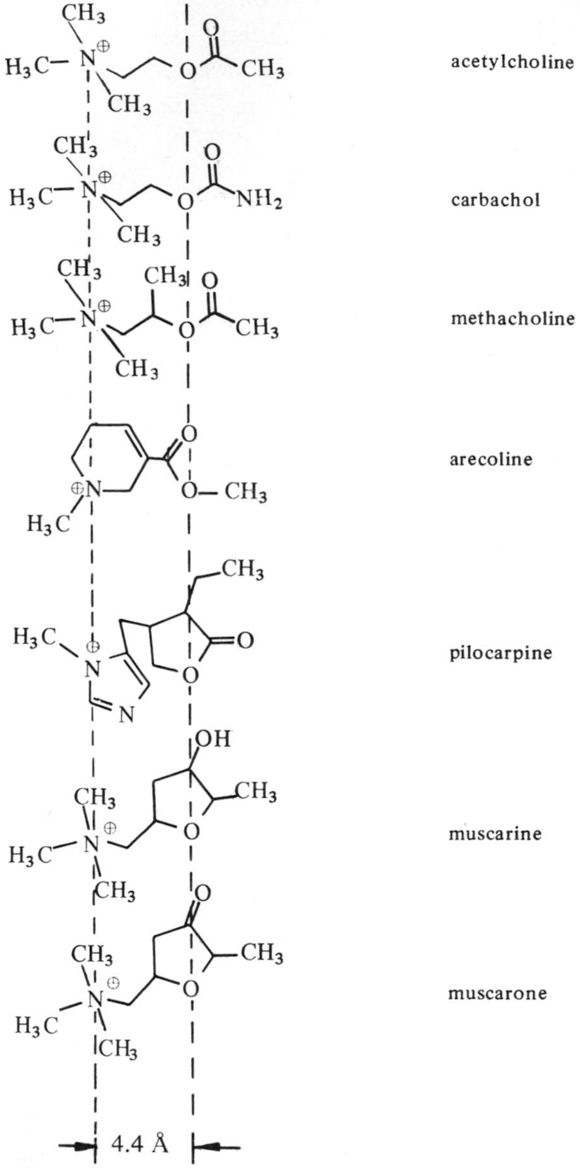

Figure 15.2 Similarities in muscarinic agents. [Adapted from W. H. Beers and E. Reich, *Nature (London)*, **228**, **917** (1970).]

action of the urinary and gastrointestinal tracts. Nicotinic activity is similar to that caused by nicotine, an alkaloid present in the leaves of tobacco, *Nicotiana tabacum,* but not used clinically: stimulation and tonus of skeletal muscles.

Cholinomimetics that manifest muscarinic effects usually contain, attached to the quaternary nitrogen, a chain of five atoms. Thus arose the *five-atom-chain rule* (Figure 15.2). This rule, however, does not apply to those agents which exert nicotinic effects.

Cholinomimetics are structurally related to acetylcholine. They result from systematic isosteric replacement of certain atoms or moieties in the molecule of this chemical transmitter. All are simple onium salts, having the general formula $R\overset{\oplus}{N}(CH_3)_3$.

In this class there are many substances, but just a few are used in therapeutics. Acetylcholine chloride itself, because it is rapidly hydrolyzed, is used only topically in certain dermatological disorders. Other agents which are more resistant to hydrolysis by acetylcholinesterase find more extensive clinical usefulness.

Bethanechol Chloride

It occurs as a water-soluble, white, crystalline solid. As the safest of the choline esters, it is the preferred one for stimulation of the gastrointestinal tract and urinary bladder. Useful in the treatment of atony, the dosage is 5 to 30 mg three or four times daily.

B. Indirect Cholinergic Agents

Indirect cholinergic agents, also called anticholinesterases, are drugs that inhibit the action of acetylcholinesterase; in this way they prevent this enzyme from hydrolyzing acetylcholine; in consequence, this chemical transmitter is accumulated at the sites of cholinergic transmission. They act peripherally, at ganglionic synapses, and at the neuromuscular junction of skeletal muscle.

Anticholinesterases most widely used are listed in Table 15.2. Besides these, however, many others are marketed, such as benzpyrinium (Stigmonine), distigmine (Ubretid), hexadistigmine (BC40), hexastigmine (Bladan), Myastenol, pyrophos (Phosarbin), and tetrastigmine (Nifos).

Some authors divide anticholinesterase agents into *reversible* and *irreversible.* In the first group are placed edrophonium, demecarium, benzpyrinium, ambenonium, physostigmine, neostigmine, and pyridostigmine, The second group is comprised of isoflurophate, echothiophate, and organophosphorus insecticides. This division, however, is not justified. On the one side, the action of physostigmine, neostigmine, and pyridostigmine is not really reversible, because they do not dissociate from acetylcholinesterase but are hydrolyzed the same way as acetylcholine, although more slowly. On the other side, although organophosphorus insecticides phosphorylate covalently the hydroxyl oxygen of serine which constitutes part of the active site of acetylcholinesterase, they do not react with this site in a truly irreversible way, as will be seen later.

Structure-activity relationship studies have shown that for anticholinesterase activity the following structural features are important: (*a*) a substituted amino group; (*b*) the *N,N*-dimethylcarbamate moiety.

Organophosphorus compounds are represented by the general formula

$$R_1 - \overset{\overset{\displaystyle O}{\uparrow}}{\underset{\underset{\displaystyle R_2}{|}}{P}} - X \quad ,$$

in which R_1 and R_2 may be alkyl, alkoxy, aryloxy, amido, mercaptan, or other groups and X is a halo, cyano, carboxyl, phosphono oxy, phenoxy, thiophenoxy, thiocyanate, or other moiety. Some compounds of this group are insecticides (Parathion, Paraoxon, EPN, Malathion, Schradan) and others, nerve gases (tabun, sarin, soman).

Some anticholinesterase agents are short-acting (neostigmine, physostigmine), whereas other ones are long-acting (demecarium, echothiophate, isoflurophate).

Neostigmine

Used in the form of methylsulfate, it is a water-soluble, odorless, white, crystalline powder with a bitter taste. It is hygroscopic; for this reason it is used in injection and not in tablet form. Neostigmine is effective in myasthenia gravis and atony of the gastrointestinal tract and the urinary bladder. Overdosage can cause a cholinergic crisis.

Neostigmine is also used as the bromide in glaucoma and myasthenia gravis. The usual dosage, by subcutaneous or intramuscular routes, is 0.5 to 1 mg.

Physostigmine

Used as the free alkaloid and also in the form of salicylate and sulfate; the latter is less deliquescent than the former. Both salts are water soluble, odorless, white powders, which turn red on prolonged exposure to air and light; by washing with ethanol the red coloration is removed. Aqueous solutions also decompose to a red product in the following sequence: hydrolysis to methylcarbamic acid and eserinol which is readily oxidized to rubreserine, the red substance. This decomposition is prevented by addition of sulfite or ascorbic acid. Physostigmine is used in the treatment of glaucoma.

Demecarium

Used as the bromide, a water-soluble, slightly hygroscopic powder. It may be considered as resulting from the association of two molecules of prostigmine. Used as a long-acting agent, it is applied by direct instillation into the conjunctival sac in cases of primary open-angle glaucoma when short-acting anticholinesterases have shown ineffectiveness. Miosis lasts 3 to 10 days.

Echothiophate

Used as the iodide, a water-soluble, white, crystalline, hygroscopic solid. It has a prolonged action; miosis lasts several days to three weeks. Its main application, topically, is in the treatment of primary open-angle glaucoma when short-acting miotics are ineffective. Prolonged administration may cause iris cysts. It is instilled directly into the conjunctival sac.

Isoflurophate

It is a colorless liquid, slightly soluble in water and freely in ethanol. Since it is absorbed through the intact epidermis and mucous tissues, it should be handled carefully. Its action and use are similar to those of echothiophate. Miosis may last two to four weeks. It undergoes rapid hydrolysis; for this reason patients should not touch the eyes or lids with the dropper or transfer tears or water into the container.

C. Pilocarpine

Pilocarpine

Recorded in *USP XIX* (1975), this alkaloid is used as the hydrochloride and nitrate. Both occur as water-soluble crystals, but the former is hygroscopic and light-sensitive, whereas the latter lacks these disadvantages. Pilocarpine is the drug of choice in initial and maintenance therapy of primary open-angle glaucoma. It is applied by local instillation into the conjunctival sac. See formula in Figure 15.2.

D. Cholinesterase Reactivators

Several substances have the ability to reactivate cholinesterase, and some are used as organophosphate antidotes. But the only one in *USP XIX* (1975) is pralidoxime chloride.

Pralidoxime

Marketed as Protopam, its chemical name is 2-formyl-1-methylpyridinium oxime, and its chloride occurs as a water-soluble, white, crystalline powder. It should be administered orally or by injection as soon as possible after poisoning by a cholinesterase inhibitor.

III. MECHANISM OF ACTION

The mechanism of cholinergic agents depends on the class to which they belong.

A. Cholinominetics

Cholinomimetics, owing to their similarity to acetylcholine, act by complexing with receptors of the chemical transmitter.

Several hypotheses have been put forward to explain how complexation takes place with either muscarinic or nicotinic receptors.

1. *Muscarinic Receptors*

The essential structure of muscarinic agents is a quaternary ammonium group and a methyl group. Muscarinic agents are characterized by the presence, in general, of a chain of five atoms attached to the quaternary nitrogen (Figure 15.2). In the complexation of both muscarinic agonists and muscarinic antagonists with the muscarinic receptor the following structural factors are involved fundamentally: (*a*) a quaternary ammonium group or its equivalent; (*b*) one pair of unshared electrons than can participate in hydrogen bonding. The distance between the quaternary ammonium group and the pair of electrons is 4.4 Å. The interaction is reinforced by a suitably placed alkyl group, corresponding to the methyl group of the acetyl group of acetylcholine, that can participate in hydrophobic interactions.

Interaction of a muscarinic agent with its receptor would be similar to that shown in Figure 14.1 for complexation between acetylcholine and the muscarinic receptor.

2. *Nicotinic Receptors*

The essential structure of nicotinic agents is a quaternary ammonium group and a carbonyl group. All these agents, although having different structures and physicochemical properties, present two common characteristics: a cationic center and a potential acceptor able to form hydrogen bonding 5.9 Å apart (Figure 15.3). Other structural characteristics only assist in the binding of these substances to adjacent sites of the receptor, and thus they either increase of decrease the intensity of action.

Complexation of nicotinic agonists and antagonists with the nicotinic receptor involves (*a*) an electrostatic interaction with the alkylammonium moiety; (*b*) a hydrogen bonding that depends on the acceptor group of the drug and which is formed at 5.9 Å from the center of the positive charge.

On the basis of results obtained with autoradiographic fixation methods of radioactive curariform or cholinomimetic molecules in the end plates of mice diaphragms, Waser proposed a hypothetical model of a very suggestive nicotinic

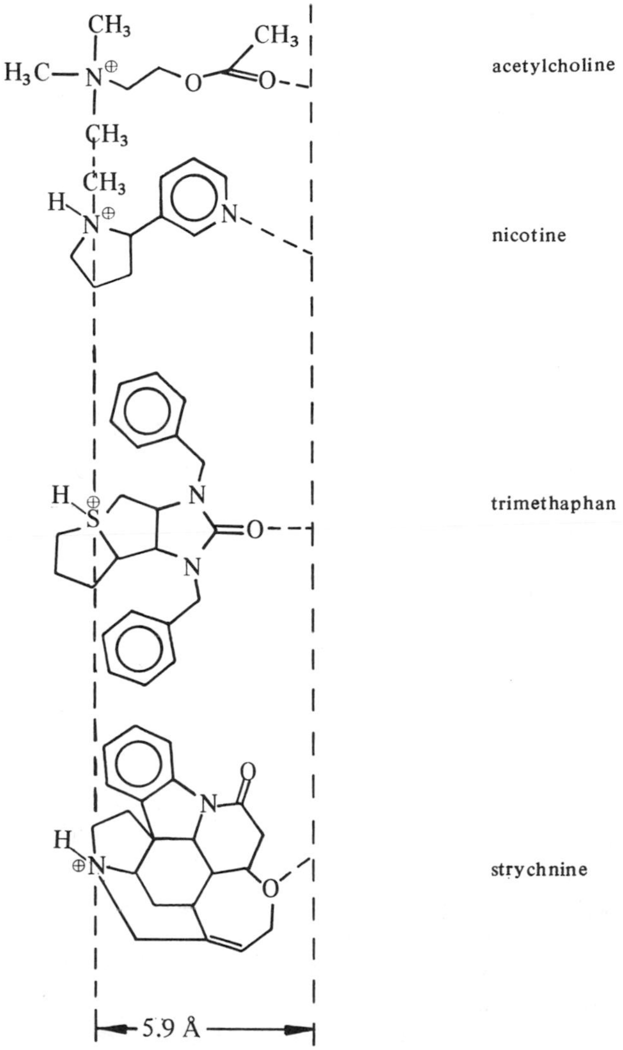

Figure 15.3 Common characteristics found in nicotinic agents. The distance 5.9 Å separates the center of the positively charged atom from the center of van der Waals forces of the atom capable of forming hydrogen bonding as acceptor. [After Beers and Reich, *Nature (London)*, **228**, 917 (1970).]

receptor (Figure 15.4). According to his hypothesis, acetylcholine and other cholinomimetics, when depolarizing the synaptic membrane of a motor plate, produce an opening of the pores and the consequent exchange through them of K^+ and Na^+ ions. The cholinergic blocking agents, being made up of bulky

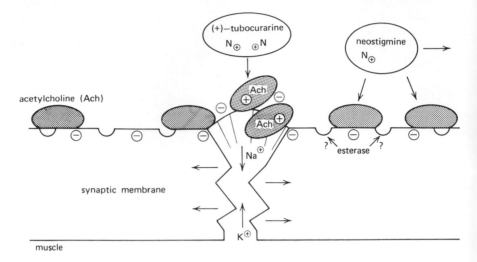

Figure 15.4 Schematic representation of the nicotinic receptor area of end plates, after P. G. Waser, *Actual. Pharmacol.*, **16, 169 (1963).**

molecules and having great chemical affinity for specific groups of the receptor to which they are attracted by electrostatic forces, close the pores and thus prevent the aforementioned ion exchange (Figure 15.5). Cholinergic blocking agents can be removed from their binding to the receptor site, but this demands a high concentration of cholinomimetics.

In accordance with Waser and as shown in Figure 15.5, the nicotinic receptor is a pore in the postsynaptic membrane. The rim of this pore is formed by anionic sites that attract the quaternary nitrogen of acetylcholine. Inside the lumen of the pore are esteratic sites to which the ester group of acetylcholine is attracted. Compounds with groups bulky enough to close the opening of the pore, such as (+)-tubocurarine, would act as antagonists to the actions of acetylcholine at the neuromuscular synapses in skeletal muscles, because they would prevent not only the access of acetylcholine but also the inward and outward flow of ions. In the specific case of bisquaternary compounds, they might bridge the orifice of the pore and deform it, producing agonistic or antagonistic effects.

B. Anticholinesterases

Anticholinesterases act as enzyme inhibitors. They increase the concentration of acetylcholine by inhibiting the hydrolysis of this chemical transmitter by acetylcholinesterase. In this interaction, an enzyme-drug complex is formed bound by several forces, including the covalent bond. In the case of physostigmine and analogues, the covalent bond may be formed which is not so strong as that formed with demecarium and organophosphorus compounds. For this

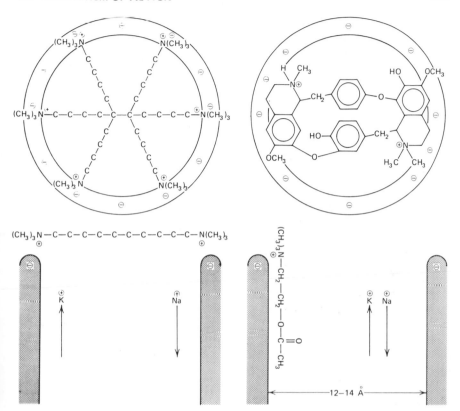

Figure 15.5 Nicotinic receptor, according to P. G. Waser, *in* D. Bovet *et al.*, Eds., *Curare and Curare-like Agents,* Elsevier, Amsterdam, 1959, p. 227.

reason, although in both cases the inhibition is reversible, with organophosphorus compounds the binding lasts much longer, in some cases several weeks, because the Ser-O-P bond, owing to its greater energy, is hydrolyzed much more slowly than Ser-O-C bond. Therefore, physostigmine and derivatives are known as short-acting anticholinesterases, and organophosphorus compounds and demecarium, as long-acting anticholinesterases.

C. Pilocarpine

Pilocarpine acts by several mechanisms. The main one is direct action on autonomic effector cells. But it also stimulates ganglia through complexation with receptors, the same way as cholinomimetics (Figure 14.1).

D. Cholinesterase Reactivators

Organophosphorus insecticides react with the esteratic site of acetylcholinesterase and inhibit it strongly, although not irreversibly. However, regeneration of this site by spontaneous hydrolysis of the complex is very slow. In cases of poisoning by these substances, a cholinesterase reactivator should be administered as soon as possible, because death may occur in 5 minutes to 24 hours, depending on dose, route of administration, organophosphorus agent, and other factors. The mechanism of reactivation is shown in Chapter 2, Section IV.E.4.

REFERENCES

General

A. Korolkovas, *Ciencia Cultura,* **25**, 1136 (1973).

J. Cheymol, *Prod. Probl. Pharm.,* **24**, 6 (1969).

G. B. Koelle, Ed., "Cholinesterases and Anticholinesterase Agents," *Handbuch der Experimentellen Pharmakologie,* Vol. XV, Springer, Berlin, 1963.

Classification

S. Kang, *Int. J. Quantum Chem.,* Quantum Biology Symp. No. 1, 109 (1974).

T. Namba, *Amer. J. Med.,* **50**, 475 (1971).

W. H. Beers and E. Reich, *Nature (London),* **228**, 917 (1970).

J. W. Miller and J. E. Lewis, *Annu. Rev. Pharmacol.,* **9**, 147 (1969).

E. J. Ariëns and A. M. Simonis, *Ann. N.Y. Acad. Sci.,* **144**, 842 (1967).

Z. Votava, *Annu. Rev. Pharmacol.,* **7**, 223 (1967).

I. M. Leopold and E. Keates, *Clin. Pharmacol. Ther.,* **6**, 130, 262 (1965).

Mechanism of Action

G. H. Cocolas *et al., J. Med. Chem.,* **17**, 938 (1974).

R. W. Baker *et al., Nature (London),* **230**, 439 (1971).

G. R. Hillman and H. G. Mautner, *Biochemistry,* **9**, 2633 (1970).

S. Ehrenpreis *et al., Pharmacol. Rev.,* **21**, 131 (1969).

L. B. Kier, *Mol. Pharmacol.,* **3**, 487 (1967).

M. Martin-Smith *et al., J. Pharm. Pharmacol.,* **19**, 561 (1967).

M. Wurzel, *Ann. N.Y. Acad. Sci.,* **144**, 694 (1967).

A. Bebbington and R. Brimblecombe, *Advan. Drug. Res.,* **2**, 143 (1965).

P. G. Waser, *Pharmacol. Rev.,* **13**, 465 (1961).

CHAPTER 16

ANTICHOLINERGIC AGENTS

I. INTRODUCTION

Anticholinergics, or cholinergic blocking agents, are drugs that block the activity resulting from acetylcholine. They may act at different sites, such as (*a*) at the postganglionic terminations of the parasympathetic nervous system—these are called antimuscarinics; (*b*) at the ganglia of both sympathetic and parasympathetic nervous system they are known as ganglionic blocking agents; and (*c*) at the neuromuscular junctions of the voluntary nervous system—these are called neuromuscular blocking agents.

II. ANTIMUSCARINICS

Antimuscarinics, also called parasympatholytics, cholinolytics, anticholinergics, atropinics, and parasympathetic blockers, inhibit the action of acetylcholine on postganglionic cholinergic nerves and smooth muscles, producing the following effects: mydriatic, cycloplegic, antispasmodic, and antisecretory.

These drugs are valuable, widely used agents, employed mainly as mydriatics, cycloplegics, antispasmodics, and antiulcer agents.

Antimuscarinics produce the following side effects: blurred vision, constipation, dryness of the mouth, and urine retention.

As a matter of convenience, myotropic agents are also included in this chapter. They do not act on cholinergic transmission but directly on smooth-muscle fiber. Although not being parasympatholytics, they exert spasmolytic or antispasmodic action. The prototype of these drugs is papaverine. For this reason they are also called papaverinics. Drugs with papaverinic action are found among several chemical groups.

Furthermore, a great number of antimuscarinics manifest both atropinic and papaverinic actions. They, therefore, are considered antispasmodics with double effect.

A. Classification

Antimuscarinics present the general structure called *spasmolytic formula*, in

$$R-COO-(CH_2)_n-N\!\!<$$

which R is an anionic group linked to the basic N through the bridge -COO- (or an isosteric group) and the chain $(CH_2)_n$, in which n is in most cases 2 or sometimes 3. The optimal distance between the $-N\!\!<$ and $-\overset{\overset{O}{\|}}{C}-$ groups is about 5 Å. These drugs are, therefore, very similar to cholinergic agents (Figure 15.2).

Most antimuscarinics are tertiary bases. However, some are quaternary ammonium compounds, but these substances do not readily cross the brain-blood barrier and, therefore, do not stimulate the central nervous system.

They belong to different chemical classes—amino alcohols, derivatives of amino alcohols (esters, ethers, carbamates), amines, diamines, amino amides, and other groups. Those most widely used are listed in Table 16.1. Many others, however, are marketed.

The principal *myotropic agents,* included in this chapter as a matter of convenience because of their spasmolytic action, are papaverine and its relatives (Table 16.2). Although there are many more direct-acting myotropics available, of various chemical types, the antimuscarinics are more important.

B. Mechanism of Action

Antimuscarinics are competitive or surmountable agonists of acetylcholine and other muscarinic agents. They compete for the same receptor sites—namely, exocrine glands and smooth and cardiac muscle—which are affected by muscarine. A representation of this competitive antagonism is Figure 3.2.

Myotropic agents are noncompetitive antagonists of cholinergics. They act as physiological antagonists, by interacting with various smooth-muscle stimulants.

III. GANGLIONIC BLOCKING AGENTS

Ganglionic blocking agents act at the ganglionic synapses of both the sympathetic and parasympathetic nervous system. They block the transmission of nerve impulses through the autonomic ganglia, by occupying receptor sites and by stabilizing the postsynaptic membranes against the actions of acetylcholine.

These agents find some use as antihypertensive drugs. Some of the untoward effects of these drugs are dizziness or syncope from orthostatic hypotension, impotence, constipation, urinary retention, dryness of the mouth, loss of visual accomodation. These undesirable effects caused the obsolescence of the ganglionic blocking agents.

A. Classification

Ganglionic blocking agents can be conveniently classified in the following four

Table 16.1 Official Antimuscarinics

Official Name	Proprietary Name	Chemical Name	Structure
atropine*	Ekomine Metanite Mintussin Tropin	(±)-tropine tropate	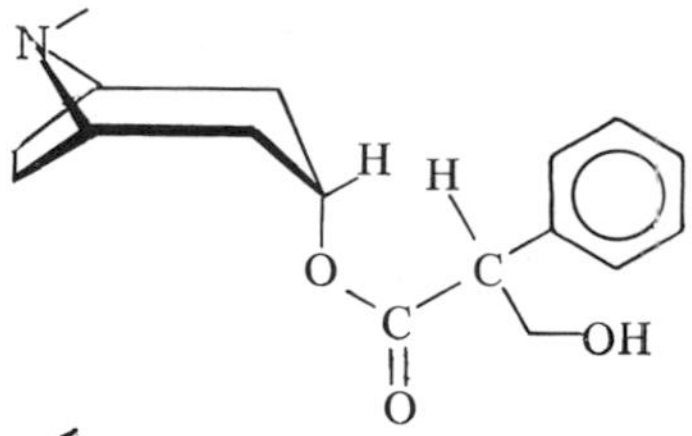
scopolamine*	Buscopan Buscopin	6β,7β-epoxy-3α-tropanyl S(–)-tropate	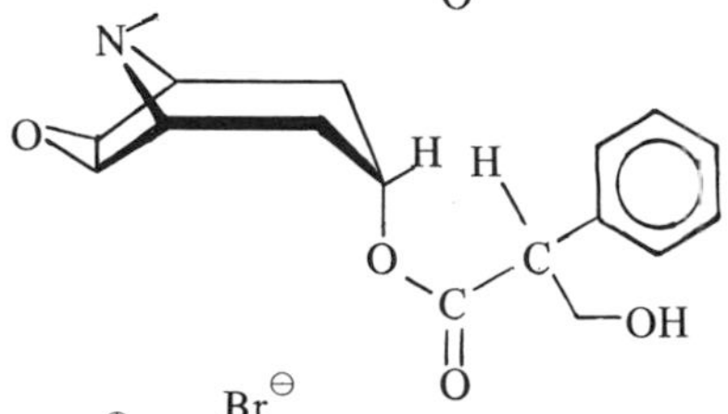
methscopolamine bromide[+]	Methine Pamine	N-methylscopolammonium bromide	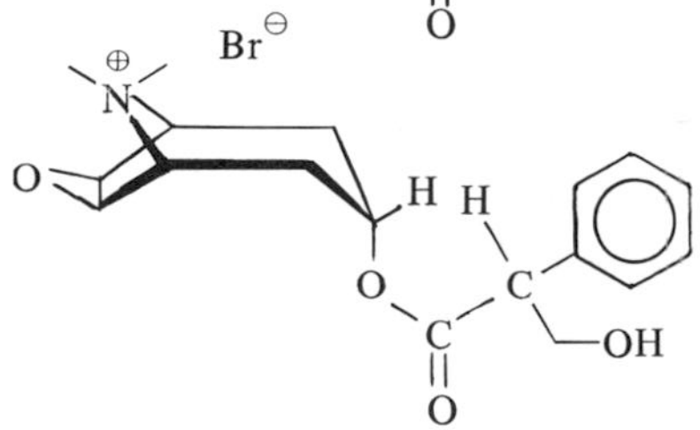
homatropine*	Mesopan Novatropine	tropine mandelate	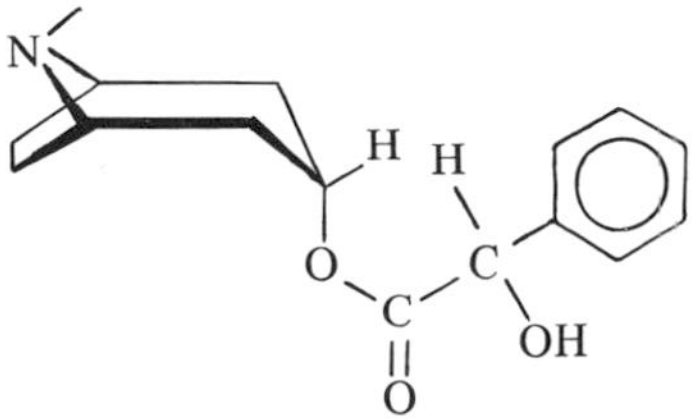

227

Table 16.1 Official Antimuscarinics (continued)

Official Name	Proprietary Name	Chemical Name	Structure
hyoscyamine[+]	Levsin	3α-tropanyl S(−)tropate	same as that of atropine
eucatropine*	Euphtalmine	1,2,2,6-tetramethyl-4β-piperidyl mandelate	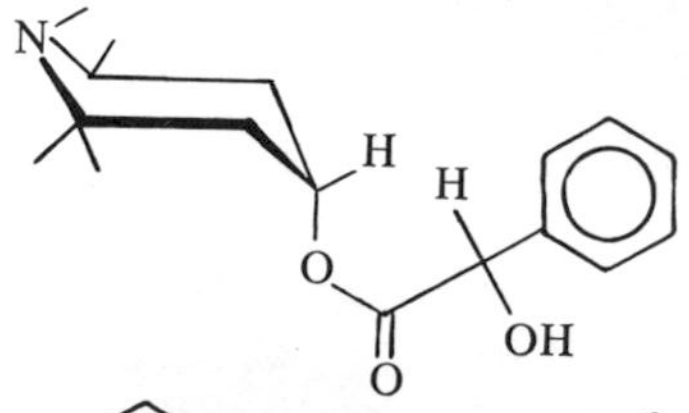
dicyclomine* (dicycloverine)	Bentyl	2-(diethylamino)ethyl[bicyclo-hexyl]-1-carboxylate	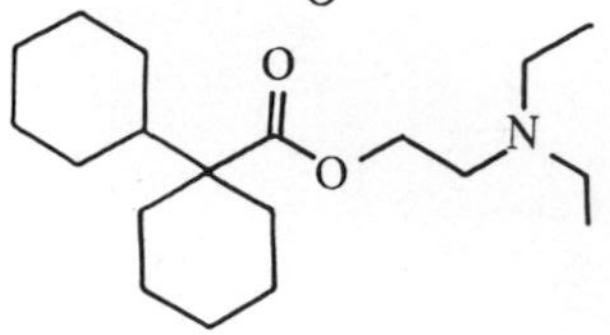
cyclopentolate*	Cyclogyl	2-(dimethylamino)ethyl 1-hydr-oxy-α-phenylcyclopentaneacetate	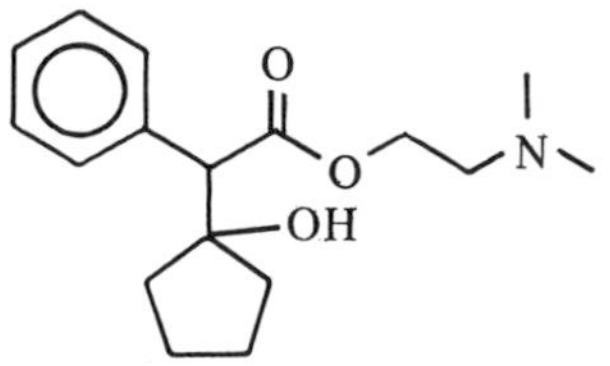
glycopyrrolate[+] (glycopyrronium)	Robinul	N-methyl-3-pyridinyl-α-cyclopent-ylmandelate methobromide	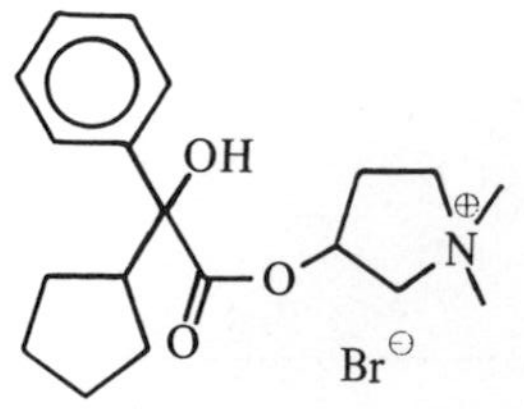

methantheline bromide[+]	Banthine	2-diethylaminoethyl 9-xanthene-carboxylate methobromide 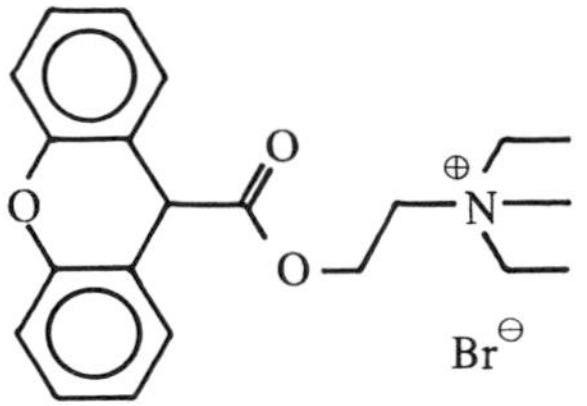
propantheline bromide*	Probanthine	2-(diisopropylamino)ethyl xanthene-9-carboxylate metho-bromide 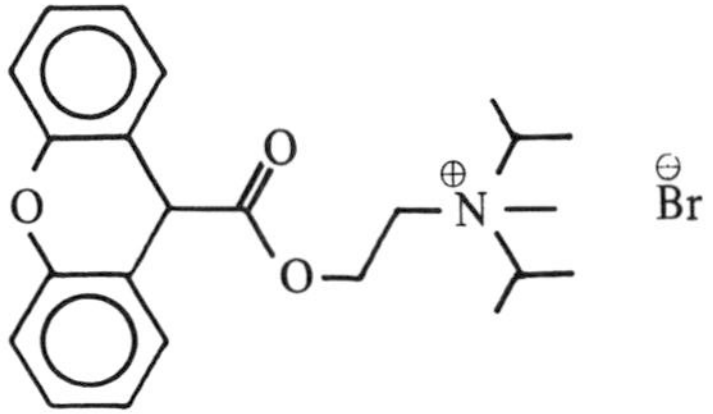
tropicamide*	Mydriacyl	N-ethyl-2-phenyl-N-(4-pyridylmethyl)-hydracrylamide 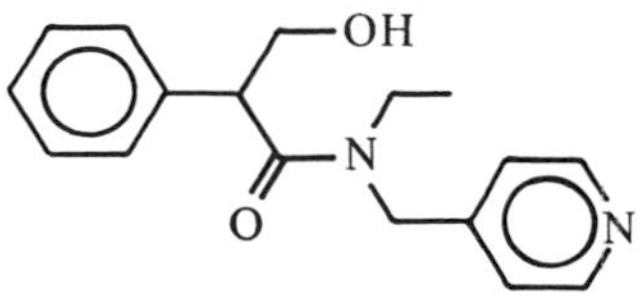

*In *USP XIX* (1975).
[+]In *NF XIV* (1975).

Table 16.2 Derivatives of Papaverine with Antispasmodic Action

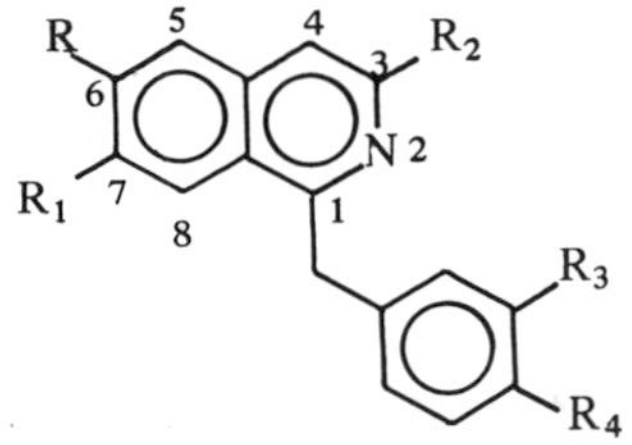

Official Name	Proprietary Name	R	R_1	R_2	R_3	R_4
papaverine[+]	Cardoverina Dispamil Pavabid	$-OCH_3$	$-OCH_3$	$-H$	$-OCH_3$	$-OCH_3$
ethaverine	Barbonin Isovex	$-OC_2H_5$	$-OC_2H_5$	$-H$	$-OC_2H_5$	$-OC_2H_5$
	Eupaverin	$-OCH_3$	$-OCH_3$	$-C_2H_5$	$-H$	$-H$
dimoxyline	Paveril	$-OCH_3$	$-OCH_3$	$-CH_3$	$-OCH_3$	$-OC_2H_5$

[+]In *NF XIV* (1975).

Table 16.3 Ganglionic Blocking Agents

Official Name	Proprietary Name	Chemical Name	Structure
trimethidinium methosulfate	Camphidonium Euprex Ostensin	(+)-3-(3-dimethylaminopropyl)-1,3,8,8-tetramethyl-3-azabicyclo-[3.2.1] octane dimethosulfate	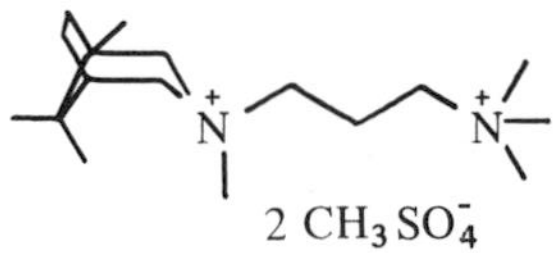
mecamylamine[+]	Inversine	N,2,3,3-tetramethyl-2-norbornan-amine	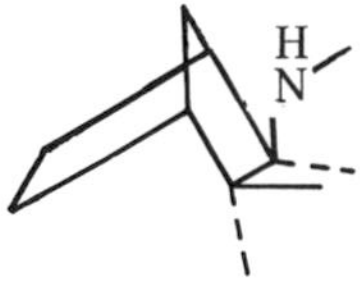
trimethaphan camsylate*	Arfonad	(+)-1,3-dibenzyldecahydro-2-oxoimidazo-[4,5-c] thieno [1,2-a]-thiolium 2-oxo-10-bornanesulfonate (1:1)	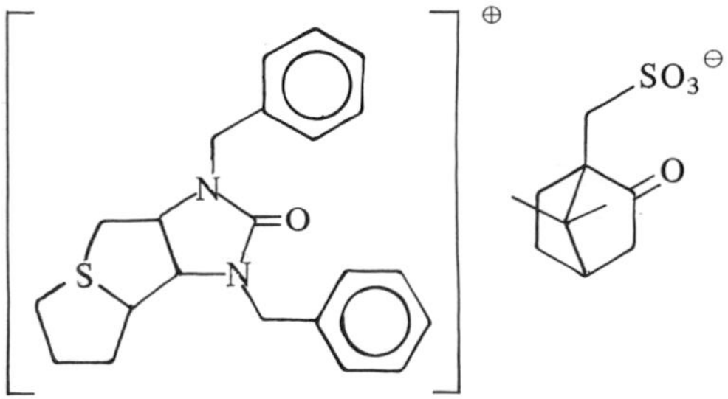

*In *USP XIX* (1975).
[+] In *NF XIV* (1975).

major groups:

    1. Monoquaternary ammonium compounds: tetraethylammonium
    2. Bisquaternary ammonium compounds: (*a*) symmetrical substances: penta-methonium, pentolinium, hexamethonium, azamethonium, Planium; (*b*) asymmetrical substances: trimethidinium, chlorisondamine, pentacynium
    3. Secondary and tertiary amines: mecamylamine, pempidine
    4. Miscellaneous compounds: trimethaphan

Three ganglionic blocking agents are listed in Table 16.3. Several others are marketed: chlorisondamine chloride (Ecolid), hexamethonium bromide (Bistrium), pempidine (Pempidil), pentamethonium bromide (Lytensin), pentolinium tartrate (Ansolysen).

B. Mechanism of Action

As to their mode of action, ganglionic blocking agents are divided into three groups:

    1. Depolarizing ganglionic blocking agents. Actually they are ganglionic stimulants, producing a nicotinic effect, but are not used clinically.
    2. Nondepolarizing competitive ganglionic blocking agents. They compete with acetylcholine for the receptor sites but do not produce transmission of nervous impulses. Examples are tetraethylammonium salts, hexamethonium, azamethonium, trimethaphan, and mecamylamine (which, besides having a competitive action, has also a noncompetitive one—it is a dual antagonist).
    3. Nondepolarizing noncompetitive ganglionic blocking agents. The effect they produce does not result from interaction with the receptor site of acetylcholine but with the noncompetitive receptors of the ganglia. Examples are chlorisondamine and trimethidinium.

Several of these drugs have two quaternary nitrogens separated by a chain of six atoms; hence this chain is called *ganglioplegic distance.* Various hypotheses have been advanced about the interaction of ganglionic blocking agents with receptor sites. According to Barlow, bisonium salts bind to the receptor at two points. More recently, in order to explain the action of ganglionic blocking agents as antagonists of acetylcholine, Goldstein *et al.* postulated that there are two different receptors for acetylcholine: one in the ganglion cell and the other in the muscle end plate. These receptors differ essentially in the distance between anionic sites. The ganglionic site interacts strongly with hexamethonium; hence this drug is a ganglionic blocker, because it prevents the receptor from responding to acetylcholine. Decamethonium, however, is too long to fit the ganglionic receptor; it acts as a neuromuscular blocking agent by preventing complexation of acetylcholine with the muscle end plate receptor (Figure 16.1).

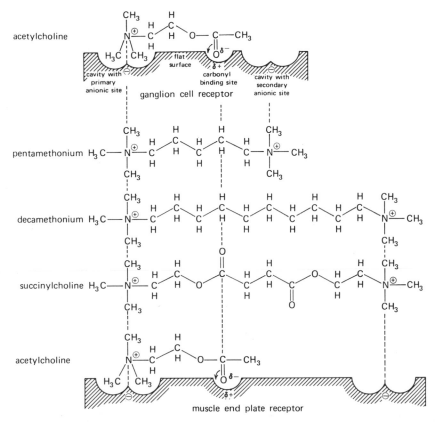

Figure 16.1 Hypothetical acetylcholine receptors in ganglion cell (above) and in muscle end plate (below). (After Goldstein *et al.*, *Principles of Drug Action*, Harper & Row, New York, 1968.)

IV. NEUROMUSCULAR BLOCKING AGENTS

Neuromuscular blocking agents are drugs that bring about voluntary-muscle relaxation and have some points in common with some ganglionic blocking agents. Since their activity is similar to that of curare, they are also called curariform or curarimimetic drugs. They may be depolarizing or nondepolarizing agents.

These drugs are used mainly as adjuncts to anesthesia, in order either to produce prolonged muscular relaxation during surgery and other type of treatment or to facilitate endotracheal endoscopy or intubation.

Depolarizing agents exert an evanescent action at the end plate, causing a sustained depolarizing effect that results in muscle paralysis. The effect is enhanced by anticholinesterases.

Nondepolarizing agents act at the myoneural end plate, causing total paralysis of the muscle fibers that lasts for about 10 to 20 minutes.

Neuromuscular blocking agents, with the exception of succinylcholine, do not undergo metabolism, being excreted unchanged by the kidneys.

A. Classification

Neuromuscular blocking agents can be classified into two groups:

1. Depolarizing agents: succinylcholine (suxamethonium, Anectine), decamethonium (*NF,* Figure 16.1) (Syncurine), suxethonium (Brevidil E), carbolonium (Imbretil)

2. Nondepolarizing agents: tubocurarine (*USP*) (Tubarine), dimethyltubocurarine (Metubine), gallamine (*USP*) (Flaxedil), benzoquinonium (Mytolon), hexafluronium (Mylaxen), pancuronium (Pavulon), piprocurarium (Brévicurarine), dacuronium, decadonium, diadonium, dimethylphaeanthine

Most of these drugs are used as chlorides, bromides, or iodides.

Succinylcholine (USP)

Used as the chloride, a water-soluble, odorless, crystalline substance. It can be considered a product of conjunction of two molecules of acetylcholine (Figure 16.1). It is used to cause brief (2 to 4 minutes) muscle relaxation, because it is soon hydrolyzed by both acetylcholinesterase and pseudocholinesterase. Among other side effects, it produces bradycardia, sustained rise in intraocular pressure, arrhythmias.

Tubocurarine (USP)

Marketed as the chloride, a water-soluble, white or yellowish-white to gray or light tan odorless, crystalline powder. It is used in tetanus and in surgery to produce skeletal muscle relaxation, which lasts about 40 minutes. Unlike succcinylcholine, which is readily metabolized, tubocurarine is a very dangerous drug; its improper administration may be fatal. An antidote is edrophonium chloride (*USP*).

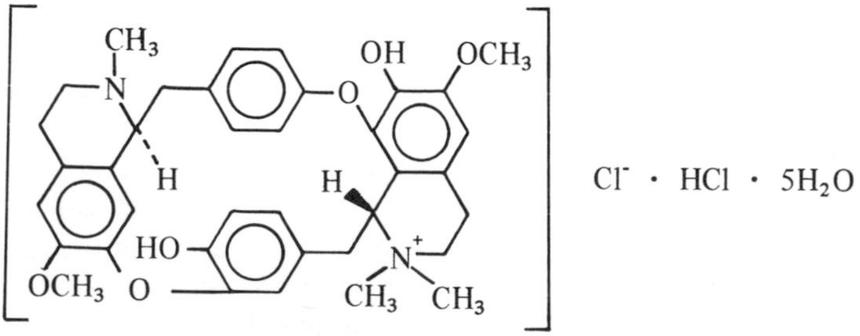

$Cl^- \cdot HCl \cdot 5H_2O$

(+)-tubocurarine chloride hydrochloride pentahydrate
(tubocurarine chloride)

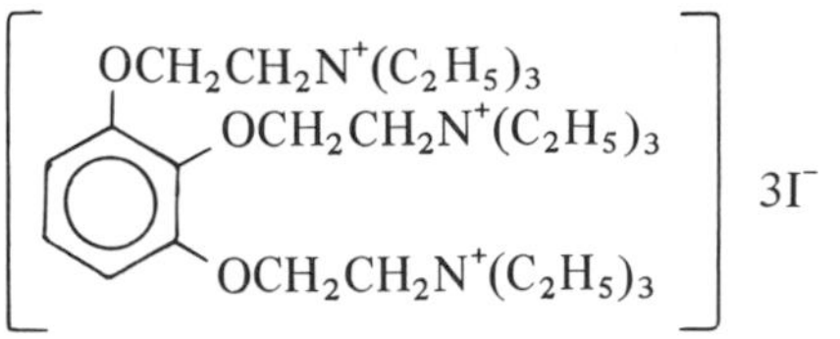

$$
\left[\begin{array}{c} OCH_2CH_2N^+(C_2H_5)_3 \\ OCH_2CH_2N^+(C_2H_5)_3 \\ OCH_2CH_2N^+(C_2H_5)_3 \end{array} \right] 3I^-
\qquad
\left[\begin{array}{c} C_2H_5 \\ HO \quad N^+(CH_3)_2 \end{array} \right] Cl^-
$$

[v-phenenyltris(oxyethylene)]-
tris[triethylammonium] triiodide
(gallamine triethiodide)

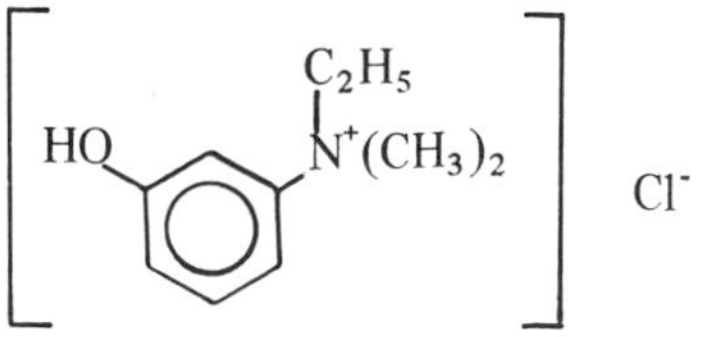

ethyl(3-hydroxyphenyl)dimethyl-
ammonium chloride
(edrophonium chloride)

B. Mechanism of Action

According to their mode of action, neuromuscular blocking agents are classified into three types:

1. Depolarizing blocking agents. They cause depolarization of the membrane of the muscle end plate, similar to that produced by acetylcholine itself, owing to its nicotinic effect, at ganglia and neuromuscular junctions; examples are decamethonium and succinylcholine.

2. Competitive blocking agents. It is thought that they compete with acetylcholine for the receptor site at the myoneural end plate but are unable to effect the depolarization characteristic of the natural neuroeffector; examples are tubocurarine, dimethyltubocurarine, gallamine, laudexium, benzoquinonium, decadonium, and diadonium.

3. Mixed blocking agents. These act by both of the aforementioned mechanisms; examples are decamethonium and succinylcholine, which, although usually considered depolarizing agents, act also through a competitive mechanism.

Considering that many of the compounds that manifest curarizing activity are characterized by a chain of 10 atoms between the two quaternary nitrogens, as in the case of (+)-tubocurarine, succinylcholine, and decamethonium, this chain is known as the curarizing distance (*C-10 structure*) and it was thought to be equivalent to about 14 Å. Recent crystallographic studies by Pauling and Pechter have shown, however, that this distance is about 10.6 Å. Likewise, there are curarimimetics in which the distance between the two quaternary nitrogens is about 20 Å, and the chain that separates them has 16 atoms (*C-16 structure*).

Complexation of these drugs with their receptors is depicted in Figure 15.5.

Koelle has proposed that the protein containing a cholinergic receptor has a tetrameric structure. A similar, but more detailed, representation was given by De Robertis to his *proteolipid* receptor. According to this author, this receptor is a proteolipid (a hydrophobic protein) located on the external surface of the postsynaptic membrane, very likely constituted by four subunits (Figure 16.2).

According to De Robertis, the ionophoric part of the proteolipid receptor molecules, which are arranged in parallel, can form the wall of a pore that, at

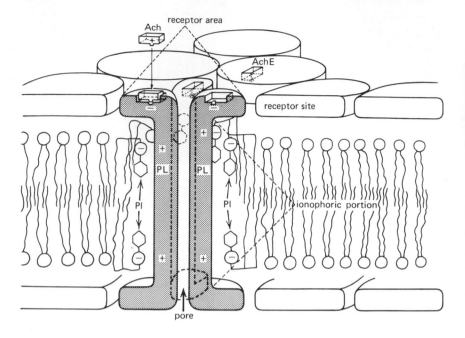

Figure 16.2 Schematic representation of the possible macromolecular organization of the postsynaptic membrane at a receptor site. The wall of a pore is constituted by the ionophoric parts of four proteolipid molecules PL, arranged in parallel, which traverse the membrance, attached to phosphatidyl inositol molecules PI. The receptor surface is made up of four receptor subunits for acetylcholine (ACh) located at the external surface of the proteolipid. At the neighborhood, acetylcholinesterase (AChE) is found. [Copyright 1971 by the American Association for the Advancement of Science. Reproduced from E. De Robertis, *Science*, 171, 963 (1971), by permission of the author and publisher.]

the resting state, is closed in a point so as to prevent the passage of ions. Complexation of acetylcholine with a subunit of the receptor can induce a change in the macromolecule of which the receptor is an integral part and allow the passage of ions through the ionophoric part of the receptor. By occurring at a very fast rate, this reversible physicochemical change could be the primary factor responsible in synaptic transmission.

Owing to cooperative effects (Chapter 3, Section IV.D), the tetrameric structure imparts to the receptor certain advantages, probably of kinetic or energetic nature. It permits allosteric interaction between the constituting subunits. Thus, by interacting with a subunit of the receptor, acetylcholine causes its conformational change; through cooperative effects, the conformational change undergone by the first subunit is induced to another subunit, and, from this one, it propagates sequentially to the third and fourth subunits. The conformational change induced in the second subunit of the receptor enhances the interaction of its active site with the next molecule of acetylcholine, thus facilitating the action of the chemical transmitter.

REFERENCES

General

J. Cheymol, Ed., *Neuromuscular Blocking and Stimulating Agents*, 2 vols., Pergamon, Oxford, 1972.

W. F. Riker and M. Okamoto, *Annu. Rev. Pharmacol.*, 9, 173 (1969).

Symposium, "Cholinergic Mechanisms," *Ann. N.Y. Acad. Sci.*, 144, 383-935 (1967).

N. V. Khromov-Borisov and M. J. Michelson, *Pharmacol. Rev.*, 18, 1051 (1966).

R. B. Barlow, *Introduction to Chemical Pharmacology*, 2nd ed., Methuen, London, 1964.

Antimuscarinics

E. J. Ariëns, Ed., *Drug Design*, Vol. II, Academic, New York, 1971, Chap. 3 and 4.

V. G. Longo, *Pharmacol. Rev.*, 18, 995 (1966).

Ganglionic Blocking Agents

L. Bolger *et al.*, *Nature (London)*, 238, 354 (1972).

R. L. Volle, *Annu. Rev. Pharmacol.*, 9, 135 (1969).

D. A. Kharkevich, *Ganglion-Blocking and Ganglion-Stimulating Agents*, Pergamon, Oxford, 1967.

Neuromuscular Blocking Agents

P. Pauling and T. J. Pechter, *Chem.-Biol. Interactions*, 6, 355 (1973).

M. Martin-Smith, "Rational Elements in the Development of Superior Neuromuscular Blocking Agents," in E. J. Ariëns, Ed., *Drug Design*, Vol. II, Academic, New York, 1971, pp. 453-530.

E. De Robertis, *Science*, 171, 963 (1971).

A. J. Everett, *Chem. Commun.*, 1020 (1970).

A. Goldstein, L. Aronow, and S. M. Kalman, *Principles of Drug Action*, Harper & Row, New York, 1968.

D. Bovet, F. Bovet-Nitti, and G. B. Marini-Bettòlo, Eds., *Curare and Curare-like Agents*, Elsevier, Amsterdam, 1959.

W. D. M. Paton and E. J. Zaimis, *Pharmacol. Rev.*, 4, 219 (1952).

D. Bovet, *Ann. N.Y. Acad. Sci.*, 54, 407 (1951).

G. K. Moe and W. A. Freyburger, *Pharmacol. Rev.*, 2, 61 (1950).

CHAPTER 17

ADRENERGIC STIMULANTS

I. INTRODUCTION

Adrenergic stimulants, also properly called *adrenomimetics* and improperly called *sympathomimetics* and *sympathetic stimulants,* are drugs that produce effects which are similar to the responses from stimulation of adrenergic nerves. Since they have also other properties and their action varies in intensity, these drugs are used for a large number of purposes. Their main clinical applications are as vasoconstrictors, bronchodilators, central-nervous-system stimulants, mydriatics, vasopressors, anorexiants. Their use as CNS stimulants and anorexiants was already covered in preceding chapters.

Untoward effects and adverse reactions caused by adrenergic agents are varied. In vasopressors and bronchodilators, they are headache, anxiety, tremor, restlessness, and palpitation. Overdosage can cause convulsions, cardiac arrhythmias, and cerebral hemorrhage, besides minor effects such as dizziness, nausea, vomiting.

Nasal decongestants applied topically may cause stinging, burning, or dryness of the mucosa. Prolonged use can convert the congested mucosa into a swollen one either through returgescence or hypertrophy of tissue.

Ophthalmic topical preparations of adrenergic agents may cause headache, browache, irritation, blurred vision, hyperemia, tearing, and allergic conjunctivitis and dermatitis. In older patients, they can cause aqueous floaters.

II. CLASSIFICATION

Several classifications have been proposed for sympathomimetic agents, taking into account especially pharmacological effects, mode of action, and chemical structures.

A. Pharmacological Effects

On the basis of pharmacological effects, adrenergic stimulants can be divided into the following groups:

1. *Vasopressors:* amidefrine, dimetofrine, ephedrine, epinephrine, hydroxy-amphetamine, levarterenol, mephentermine, metaraminol, methamphetamine, methoxamine, phenylephrine

2. *Bronchodilators:* ephedrine, epinephrine, ethylnorepinephrine, fenoterol, hexoprenaline, isoproterenol, methoxyphenamine, orciprenaline, protokylol, pseudoephedrine, rimiterol, salbutamol, terbutaline, tetroquinol*

3. *Nasal decongestants:* cyclopentamine, ephedrine, epinephrine, fenoxazoline, hydroxyamphetamine, mephentermine, methylhexaneamine, metizoline, naphazoline, oxymetazoline, phenylephrine, phenylpropanolamine, propylhexedrine, tetrahydrozoline, tuaminoheptane, xylometazoline

4. *Mydriatics:* phenylephrine (see also Chapter 16)

5. *Ophthalmic decongestants:* ephedrine, naphazoline, phenylephrine, tetrahydrozoline.

6. *Anorexiants:* amphetamine, dextroamphetamine, methamphetamine, and several others (see Chapter 12, Section II.B.2)

7. *Antiarrhythmics:* epinephrine, isoproterenol, methoxamine

B. Mode of Action

According to the mode of action, adrenomimetics are classified in three categories: directly acting, indirectly acting, and of mixed action.

1. Directly Acting Sympathomimetics

They act through complexation with specific receptors. They are represented by epinephrine and they have a catechol nucleus, one phenolic hydroxyl at meta, one alcoholic hydroxyl at β, and one amino function at the side chain.

Sympathomimetic amines that act directly can interact either with α receptors or with β receptors or with both. These interact with α receptors: deoxyepinephrine, dopamine, m-hydroxyphenylpropylamine, nordephrine, norepinephrine, norphenylephrine, phenylephrine, synephrine. The following ones interact with β receptors: isoproterenol, isoxsuprine, metaproterenol, nylidrin, protokylol, rimiterol. Epinephrine and dihydroxyephedrine complex with both receptors. Recently, certain β_2 stimulants were developed: bamethan, bufenine, fenoterol, hexoprenaline, isoetarine, orciprenaline, quinprenaline, ritodrine, salbutamol, soterenol, terbutaline.

Salbutamol and analogous compounds are characterized by being selective β-adrenergic stimulants. Some esters of levarterenol are capable of crossing blood-brain barrier, property which is not possessed by other adrenergics.

*As bronchodilators also some drugs are used that are not sympathomimetics: decloxizine, epronizol (Eupneron), fenspiride (Viarespan), quinetalate (Phtalamaquin, Ventaire), and sodium chromoglycate (Intal, Lomudal).

2. Indirectly Acting Sympathomimetics

They act by releasing catecholamines, mainly norepinephrine, from storage granules in the sympathetic nervous terminals. They are represented by tyramine. The following ones also act indirectly: amphetamine, metamphetamine, α-methyltyramine, phenethylamine. Recent evidence indicates that amphetamines can also act directly, as α-adrenegic agents.

3. Sympathomimetics of Mixed Action

They act by both mechanisms seen above. They are represented by ephedrine

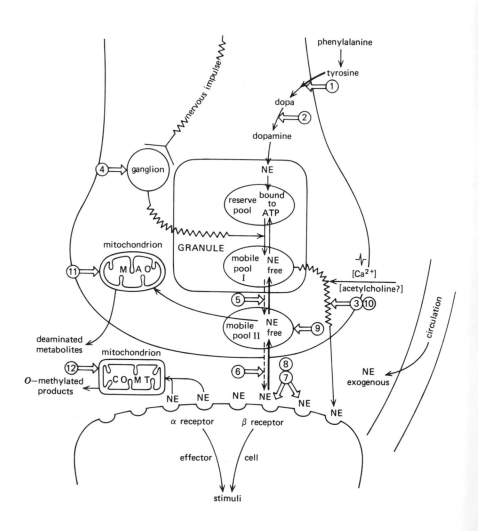

and have some of the structural characteristics of directly acting adrenomimetic amines. The following ones are also of mixed action: p-hydroxyephedrine, p-hydroxyphenylpropanolamine, metaraminol, norsynephrine, phenylpropanolamine.

It is not possible to make an absolute distinction between directly acting and indirectly acting sympathomimetics. Very likely, all these amines act, some more than the others, by both mechanisms, that is, both on receptors sites and on storage sites of catecholamines.

Adrenomimetic effects can be obtained also by the use of (a) inhibitors of monoamino oxidase, such as pargyline and tranylcypromine, which decrease the metabolism of norepinephrine and thus increase its level in the brain and other tissues; (b) ganglion blocking agents, such as trimethaphan, which decrease sympathetic tonus; (c) inhibitors of the mechanism of uptake and recapture of norepinephrine at axonal presynaptic membrane—among them, certain local anesthetics (cocaine), tricyclic antidepressives (imipramine, desipramine), some antihistamines (chlorpheniramine), the psychotropic chlorpromazine, drugs structurally related to adrenomimetics (tyramine), some adrenolytics.

The sites of action of some of the drugs discussed above are shown in Figure 17.1.

C. Chemical Structure

Stereochemistry is very important for activity of adrenergic stimulants. In a series of these drugs (isomers of epinephrine, norepinephrine, nordefrin, phenylephrine, octopamine) activity is greater in those having R configuration at β-carbon atom, the R(−)-isomers being more active than the S(+)-isomers.

Figure 17.1 Sites of action of drugs at sympathetic nervous terminal. NE = norepinephrine; MAO = monoamine oxidase; COMT = catechol-O-methyltransferase; heavy arrows = active transport; dash arrows = passive diffusion; light arrows = enzymatic biosynthesis. 1. α-methyl-L-tyrosine—it inhibits tyrosine hydroxylase, which catalyzes the rate-determining step of the catecholamine biosynthesis; 1. methyldopa—it acts as preferential substrate and thus decreases the catecholamine biosynthesis; 3. bretylium—it inhibits NE release at post-ganglionic nervous terminal, causing blockade of the activity of adrenergic nerve; 4. ganglionic blocking agents—they inhibit nervous-impulse transmission at sympathetic ganglia; 5. reserpine—it blocks the active transport from cytoplasmic mobile pool (I) to intragranular pool (II); 6. cocaine, imipramine, chlorpromazine—they block the active transport from the extracellular fluid to cytoplasmic mobile pool (I); 7. adrenolytic agents—they block α and β receptors; 8. adrenomimetic agents—they activate α and β receptors; 9. tyramine, ephedrine—they displace NE from cytoplasmic mobile pool (I), producing sympathomimetic effects; 10. guanethidine—it releases actively NE from intragranular mobile pool (II), leading eventually to depletion of reserve pool; 11. MAO inhibitors—they decrease the metabolism of free NE and thus cause accumulation of NE; 12. pyrogallol, catechol, 4-methyltropolone, 4-isopropyltropolone, 3,4-dihydroxy-α-methylpropiophenone—they inhibit COMT.

Table 17.1 Aromatic Sympathomimetic Amines

Official Name	Proprietary Name	Chemical Name	Structure
epinephrine*	Adrenalin Suprarenin Suprarenalin	R(−)-3,4-dihydroxy-α-[(methyl-amino)methyl] benzyl alcohol	
levarterenol* (norepinephrine)	Levophed	R(−)-α(aminomethyl)-3,4-dihy-droxybenzyl alcohol	
isoproterenol* (isoprenaline)	Aludrine Isuprel	3,4-dihydroxy-α-[(isopropyl-amino) methyl]-benzyl alcohol	
levonordefrin[+]	Neo-Cobefrin	(−)-α-(1-aminoethyl)-3,4-dihydr-oxybenzyl alcohol	

methoxamine*   Vasoxyl   (±)-α-(1-aminoethyl)-2,5-dimethoxy-benzyl alcohol

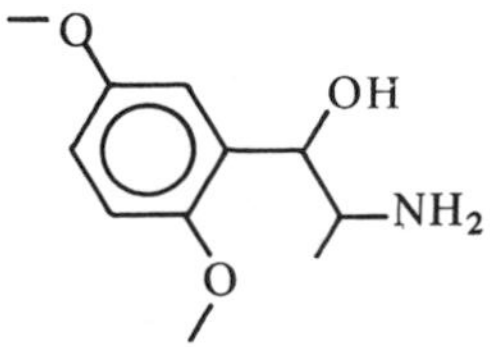

hydroxyamphetamine[+]   Paredrine   (±)-p-(2-aminopropyl)phenol

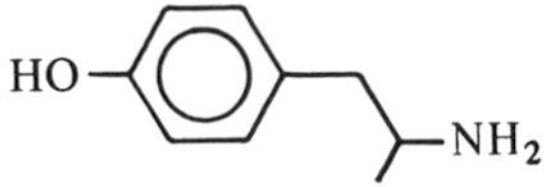

metaraminol*   Aramine   (−)-α-(1-aminoethyl)-m-hydroxy-benzyl alcohol

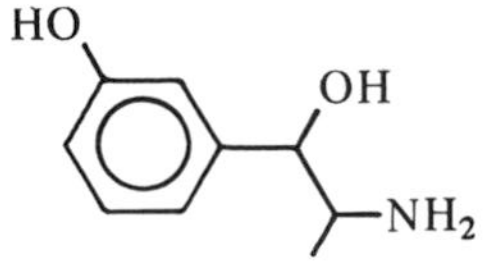

phenylephrine*   Isophrin Neo-Synephrine   R(−)-m-hydroxy-α-[(methylamino)-methyl] benzyl alcohol

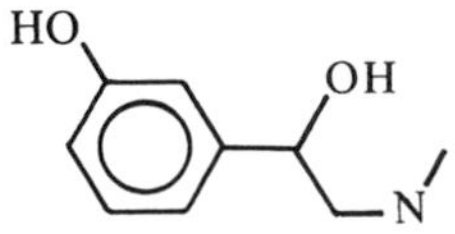

phenylpropanolamine[+]   Propadrine   (±)-1-phenyl-2-amino-1-propanol

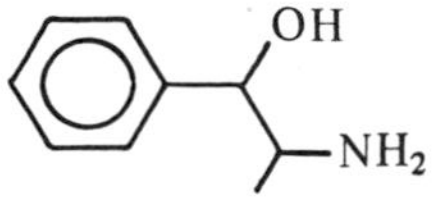

Table 17.1 Aromatic Sympathomimetic Amines (continued)

Official Name	Proprietary Name	Chemical Name	Structure
ephedrine*[+]		(−)-erythro-α-[(1-methylamino)-ethyl] benzyl alcohol	
pseudoephedrine[+]	Sudafed	(+)-threo-α-[(1-methylamino)-ethyl] benzyl alcohol	
mephentermine[+]	Wyamine	N,α,α-trimethylphenethylamine	
methoxyphenamine[+]	Orthoxine	(±)-o-(2-methylaminopropyl)anisole	

*In *USP XIX* (1975).
[+]In *NF XIV* (1975).

Table 17.2 Alicyclic Sympathomimetic Amines

Official Name	Proprietary Name	Chemical Name	Structure
cyclopentamine[+]	Clopane	N,α-dimethylcyclopentaneethyl-amine	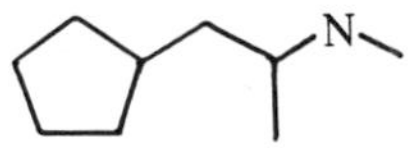
propylhexedrine[+]	Benzedrex	N,α-dimethylcyclohexaneethyl-amine	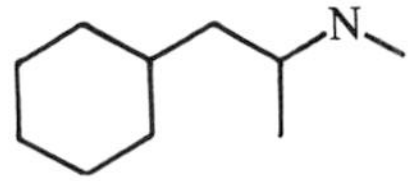

[+] In *NF XIV* (1975).

245

According to the chemical structures, adrenergic stimulants are divided into four main groups: aromatic amines, alicyclic amines, aliphatic amines and imidazoline derivatives.

1. Aromatic Amines

Those most commonly used are listed in Table 17.1.

They are used mostly as salts (hydrochloride, hydrobromide, sulfate, tartrate, bitartrate), which are water-soluble, white crystalline powders.

New compounds of this class include carbuterol, hexoprenaline, pyrbuterol, rimiterol, salbutamol, salmefamol, soterenol, trimetoquinol.

2. Alicyclic Amines

The most widely used are shown in Table 17.2. They are used as nasal decongestants. Their effects are similar to those produced by ephedrine, but they are less active as central stimulants.

3. Aliphatic Amines

Those of main application are listed in Table 17.3. They are mostly used as nasal decongestants, by topical application.

Table 17.3 Aliphatic Sympathomimetic Amines

$$R-\underset{\underset{NH_2}{|}}{CH}-CH_3$$

Official Name	Proprietary Name	Chemical Name	Structure R
tuaminoheptane[+]	Tuamine	2-aminoheptane	$CH_3-(CH_2)_4-$
methylhexaneamine	Forthane	2-amino-4-methyl-hexane	$CH_3-CH_2-\underset{\underset{CH_3}{|}}{CH}-CH_2-$

[+]In *NF XIV* (1975).

4. Imidazoline Derivatives

The most often used are seen in Table 17.4. Besides these, the following are marketed: fenoxazoline (Aturgyl), metizoline, timazoline (Pernazene), tramazoline (Dexa-Rhinaspray). These drugs are used almost exclusively as vasoconstrictors in the treatment of surface ocular irritations and local swelling of nasal mucous membranes. They are applied topically in ophthalmic preparations and as nasal decongestants.

III. MECHANISM OF ACTION

Adrenergic stimulants owe their action to complexation with specific receptors.

Table 17.4 Imidazoline Derivatives

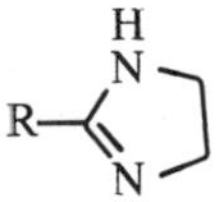

$$R-\underset{\underset{H}{N}}{\overset{N}{\diagup}}\diagdown$$

Official Name	Proprietary Name	Chemical Name	Structure R
naphazoline*	Privine	2-(1-naphthylmethyl)-2-imid-azoline	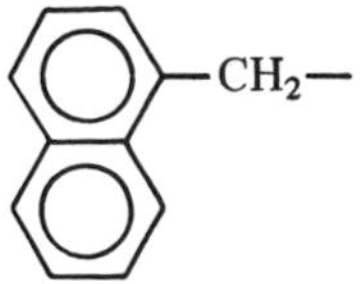 (1-naphthylmethyl group) CH₂—
tetrahydrozoline* (tetryzoline)	Tyzine Visine	2-(1,2,3,4-tetrahydro-1-naphthyl)-2-imidazoline	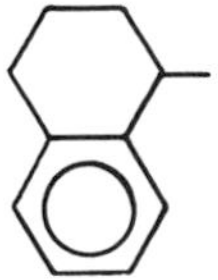 (tetrahydronaphthyl group)
xylometazoline⁺	Otrivin	2-(4-t-butyl-2,6-dimethylbenzyl)-2-imidazoline	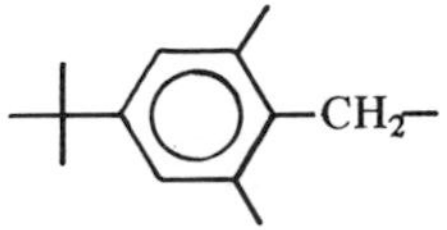 (4-t-butyl-2,6-dimethylbenzyl group) CH₂—
oxymetazoline*	Afrin	6-t-butyl-3-(2-imidazolin-2-ylmethyl)-2,4-dimethylphenol	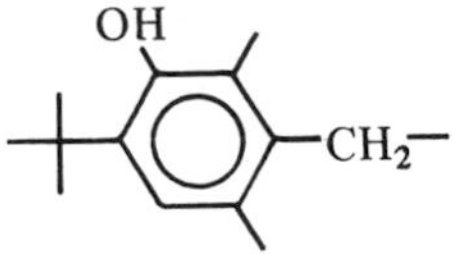 (t-butyl-dimethylphenol group) CH₂—, OH

*In *USP XIX* (1975).
⁺In *NF XIV* (1975).

It is believed that there are at least two receptors for adrenomimetics: α-adrenergic receptors and β-adrenergic receptors. Lands and coworkers postulated, however, the existence of two β receptors, β_1 and β_2. There are authors who defend the existence of other adrenergic receptors, named γ and δ.

The α-adrenergic receptors are predominantly involved in smooth-muscle excitation. The β-adrenergics are associated with inhibition of smooth-muscle tonus, including that of the intestine and the stimulation of the myocardium. The β_1 receptors are located in the heart and small intestine and the β_2 receptors, in the bronchi and vascular bed; activation of the first causes increase of lipolysis and of the second ones, of glycogenolysis.

Consequently, there are, at least, two types of adrenomimetics:

1. The α-adrenergic ones, the prototype of which is norepinephrine, although phenylephrine is the specific one

2. The β-adrenergic ones, the prototype and specific one being isoproterenol (isoprenaline)

Epinephrine, by acting on both α and β receptors, has both α- and β-adrenergic properties. This can be schematically shown as follows:

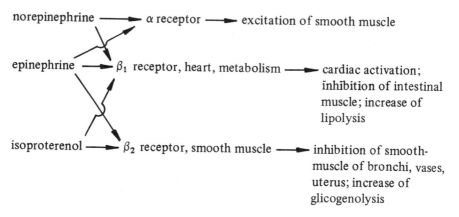

norepinephrine ⟶ α receptor ⟶ excitation of smooth muscle

epinephrine ⟶ β_1 receptor, heart, metabolism ⟶ cardiac activation; inhibition of intestinal muscle; increase of lipolysis

isoproterenol ⟶ β_2 receptor, smooth muscle ⟶ inhibition of smooth-muscle of bronchi, vases, uterus; increase of glicogenolysis

In directly acting adrenomimetics, the essential structural features are the following ones:

1. For the activation of α receptors: catechol nucleus and an amine function either unsubstituted or with a nonbulky substituent

2. For the activation of β receptors: phenolic hydroxyl function in meta at the catechol nucleus and, at the side chain, an alcoholic hydroxyl in β, and an amine with bulky group

In fact, the catechol nucleus, especially the phenolic hydroxyls, seems to be relatively more critical for activation of β receptors than of α receptors, since the nonphenolic amines (cyclopentamine, propylhexedrine, tuamino-heptane, for instance) are almost completely devoid of β activity, although α activity is maintained.

The alcoholic hydroxyl in β position plays a much more important role in the activation of the β receptor than in the activation of the α receptor.

The amine group, whose nitrogen atom in physiological pH is protonated, is of the greatest importance for the activation of α receptors, because bulky substituents diminish α agonistic activity and enhance β agonistic activity. Recent evidence points out that in the activation of the α receptor hydrogen bonding is involved between the onium group of the drug and a negatively charged group of the receptor and that activation of the β receptor results in part from dispersive forces between the alkyl group and the receptor. This alkyl group appears to be more important than the nitrogen atom.

In 1960 Belleau advanced the hypothesis that ATP is a constituent part of the adrenergic receptor and proposed that catecholamines activate adenyl cyclase to catalyze the conversion of ATP into cyclic AMP (Figure 12.1). This hypothesis was slightly modified in 1966 by Bloom and Goldman, who postulated that adrenergic receptors are substrate-enzyme complexes, that is, ATP-adenyl cyclase, their agonism being the process that facilitates the substrate utilization. In interacting with a catecholamine, the receptor is destroyed and subsequently regenerated through complexation of another ATP molecule with adenyl cyclase. Catecholamines are thus catalysts of conversion of ATP into cyclic AMP.

According to Robison and coworkers (1967), interaction of β-adrenergic stimulants with the adenyl cyclase system occurs at the allosteric site of the enzyme and activates adenyl cyclase, causing it to increase the level of cyclic AMP. Interaction of α-adrenergic stimulants decreases the effective concentration of cyclic AMP and causes an opposed effect in the cellular function.

In 1967 Belleau proposed that, at the α-receptor, norepinephrine and epinephrine act primarily as proton donors. The acceptor group of this proton is the terminal phosphate of ATP coordinated with Ca^{2+}.

As to the β receptor, Belleau advanced the hypothesis that, in its interaction with catecholamines, important roles are played by both the β-hydroxyl group and catechol nucleus. β-Hydroxyl forms a coordinate bond with the chelated atom of Mg^{2+}, now proven to be Fe^{2+} or Mn^{2+}, which assists in placing the protonated nitrogen at the neighborhood of the negative oxygen of the most internal phosphate of ATP. The catechol ring is accomodated at the lipoproteic surface of adenyl cyclase, through specific interactions, inducing it to adopt a more suitable conformation for the catalytic function. The phenolic hydroxyl groups fit sterically in the lattice of a localized water crust, thus increasing proton motility and transfering this effect to the terminal anionic site of ATP. This results in a second neutralization of charge and assists in the leaving of

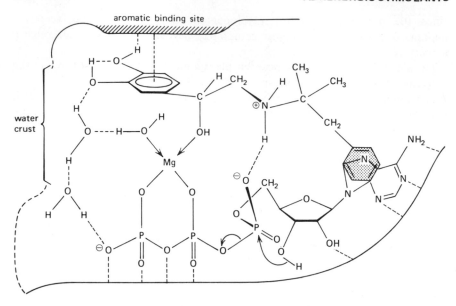

Figure 17.2 Schematic representation of the interaction of a β-sympathomimetic with adenyl cyclase. In this interaction a π bond may be formed between the aromatic ring of the drug and the adenine ring of ATP. [Reprinted from B. Belleau, *Ann. N. Y. Acad. Sci.,* 139, 580 (1967), by permission of the author and New York Academy of Sciences.]

pyrophosphate. The critical phase of the interaction with the receptor is, however, the neutralization of the positive charge of the catecholamine nitrogen by the negative charge of the oxygen of the internal phosphate group of ATP. This reaction makes possible the intramolecular nucleophilic attack of the 3-hydroxyl group from ribose to the above-mentioned phosphorus atom. At last, with the leaving of pyrophosphate, cyclic AMP is formed (Figure 17.2).

REFERENCES

General

N. Kirshner *et al., Advan. Drug Res.,* 6, 121 (1971).

G. A. Neville *et al., J. Med. Chem.,* 14, 717 (1971).

D. M. Aviado, *Sympathomimetic Amines,* Thomas, Springfield, Ill., 1970.

E. Marley, *Advan. Pharmacol.,* 3, 168 (1964).

M. Beauvallet, *Actual. Pharmacol.,* 8, 1 (1955).

Introduction

B. Pullman *et al., Int. J. Quantum Chem.,* Quantum Biology Symp. No. 1, 93 (1974).

B. Pullman *et al., J. Med. Chem.,* **15**, 17 (1972).

L. Pedersen *et al., J. Pharm. Pharmacol.,* **23**, 216 (1971).

B. Pullman *et al., Mol. Pharmacol.,* 7, 397 (1971).

J. R. Smythies *et al., Nature (London),* **231**, 185 (1971).

L. B. Kier, *J. Pharm. Pharmacol.,* **21**, 93 (1968).

L. B. Kier, *J. Pharmacol. Exp. Ther.,* **164**, 75 (1968).

Classification

H. Stormann and K. Turnheim, *Arzneim.-Forsch.,* **23**, 30 (1973).

G. A. Robison and E. W. Sutherland, *Advan. Cytopharmacol.,* 1, 263 (1971).

U. K. Terner *et al., Biochem. Pharmacol.,* **20**, 597 (1971).

G. J. Durant *et al., Progr. Med. Chem.,* 7, 124 (1970).

J. Bralet, *Prod. Probl. Pharm.,* **24**, 322 (1969).

V. A. Cullum *et al., Brit. J. Pharmacol.,* **35**, 141 (1969).

A. M. Lands *et al., Nature (London),* **214**, 597 (1967).

J. Levy and B. Tchoubar, *Actual. Pharmacol.,* **5**, 143 (1962).

D. Bovet and F. Bovet-Nitti, *Actual. Pharmacol.,* **6**, 21 (1953).

Mechanism of Action

C. Petrongolo *et al., J. Med. Chem.,* **17**, 501 (1974).

A. Korolkovas, *Rev. Bras. Clin. Ter.,* **2**, 109 (1973).

A. Arnold, *Farmaco, Ed. Sci.,* **27**, 79 (1972).

J. R. Smythies, *J. Theor. Biol.,* **35**, 93 (1972).

D. J. Triggle, *Annu. Rev. Pharmacol.,* **12**, 185 (1972).

J. M. George *et al., Mol. Pharmacol.,* 7, 328 (1971).

S. L. Pohl, *Annu. Rep. Med. Chem.,* **1970**, 233 (1971).

P. Pratesi *et al., Farmaco, Ed. Sci.,* **26**, 379 (1971).

P. N. Patil *et al., J. Pharm. Sci.,* **59**, 1205 (1970).

R. T. Brittain *et al., Advan. Drug Res.,* **5**, 197 (1970).

W. C. Bowman and M. W. Nott, *Pharmacol. Rev.,* **21**, 27 (1969).

B. Belleau, *Ann. N.Y. Acad. Sci.,* **139**, 580 (1967).

B. Belleau, *Pharmacol. Rev.,* **18**, 131 (1966).

B. Belleau, "Relationships between Agonists, Antagonists and Receptor sites," in J. R. Vane, G. E. W. Wolstenholme, and M. O'Connor, Eds., *Adrenergic Mechanisms,* Churchill, London, 1960, pp. 223-245.

L. H. Easson and E. Stedman, *Biochem. J.,* **27**, 1257 (1933).

CHAPTER 18

ADRENERGIC BLOCKING AGENTS

I. INTRODUCTION

Adrenergic blocking agents, also called *adrenolytics* and *sympatholytics*, are drugs that selectively inhibit certain responses of sympathetic stimulation or block the effects produced by sympathomimetic agents.

Adrenolytic effects, however, are obtained by drugs with different modes and sites of action (Figure 17.1), such as those that:

1. Interact with specific receptors, as in the case of α-adrenergic blocking agents and β-adrenergic blocking agents. The first ones block the effects of α receptor stimulation and the second ones, of β receptor stimulation.

2. Prevent the release of norepinephrine from its storage sites; these are adrenergic neuron blocking agents. Those mostly known are bretylium, guanethidine, betanidine, xylocholine, and debrisoquin.

3. Affect the storage of catecholamines—reserpine. It was already studied in Chapter 10, as an antipsychotic agent.

4. Inhibit the enzymes involved in the biosynthesis of norepinephrine, among them methyldopa and α-methyl-L-tyrosine, inhibitors of dopa-decarboxylase and tyrosine-hydroxylase, respectively. These and related drugs are studied in the next chapter.

Drugs studied in this chapter are mainly used as antihypertensive agents.

a-Adrenergic blocking agents have various therapeutic uses: relief of several peripheral diseases (early stages of Raynaud's syndrome, acrocyanosis, phlebitis, frostbite, phlebothrombosis), diagnosis of pheocromocytoma, and treatment of some other conditions.

β-Adrenergic blocking agents are used in certain cardiac arrhythmias, angina pectoris, essential hypertension.

Adrenergic neuron blocking agents are useful in the treatment of hypertension.

II. CLASSIFICATION

Drugs studied in this chapter can be divided into three classes: α-adrenergic blocking agents; β-adrenergic blocking agents; and adrenergic neuron blocking agents. The first class consists of substances not structurally related to norepinephrine. The second one, however, is comprised by structural analogues of isoproterenol, a β-adrenergic stimulant. The third one shows a slight similarity to sympathomimetic amines.

A. α-Adrenergic Blocking Agents

In this class are included indolethylamine alkaloids (dihydroergotamine, yohimbine), imidazole derivatives (tolazoline, phentolamine), benzodioxan derivatives (piperoxan, butamoxane), β-haloalkylamines (phenoxybenzamine), dibenzazepines (azapetine), hydrazinophthalazines (hydralazine), miscellaneous compounds (moxisylyte) (Table 18.1). Most of these drugs are used as hydrochlorides, but some as sulfate, phosphate, mesylate, which are water-solute crystalline substances.

Ergot alkaloids vary in pharmacological activity. Natural alkaloids, such as those of ergotamine group (ergotamine and ergosine) and of the ergotoxine group (ergocristine, ergocryptine and ergocornine), have moderate α-adrenergic blocking activity and marked vasoconstrictor action, for which reason they are used in the treatment of migraine, as in the case of ergotamine (Ergomar, Gynergen).

Ergonovine and semisynthetic derivatives (methylergonovine) exert no α-adrenergic blocking activity but slight vasoconstrictor and marked oxytocic activities, being used to treat migraine and as oxytocics: thus, ergonovine (Ergotrate) and methylergonovine (Methergine) are oxytocics, and methysergide (Sansert) and metergoline (Liserdol) are active in migraine.

Dihydrogenated derivatives, such as dihydroergotamine, are more active as α-adrenergic blockers than the parent compounds, being less vasoconstrictor and not oxytocic. Thus, Hydergine, which is an equiproportional mixture of dihydroergocornine, dihydroergocristine, and dihydroergocryptine mesylates, is mainly used in the treatment of peripheral spastic vasculopathy. A semisynthetic derivative recently introduced is nicergoline (Sermion).

Dihydroergotamine

It should be kept in a cool place and out of light. This drug was prepared by Stoll and Hofman in 1943 by catalytic hydrogenation of ergotamine. It is used in the treatment of migraine. The usual dose, of 1 to 2 mg, is administered by intramuscular route.

Phentolamine

Marketed as hydrochloride and mesylate; the mesylate is more stable than the hydrochloride. Phentolamine is used as diagnostic aid in pheochromocytoma

Table 18.1 α-Adrenergic Blocking Agents

Official Name	Proprietary Name	Chemical Name	Structure
ergotamine*	Gynergen		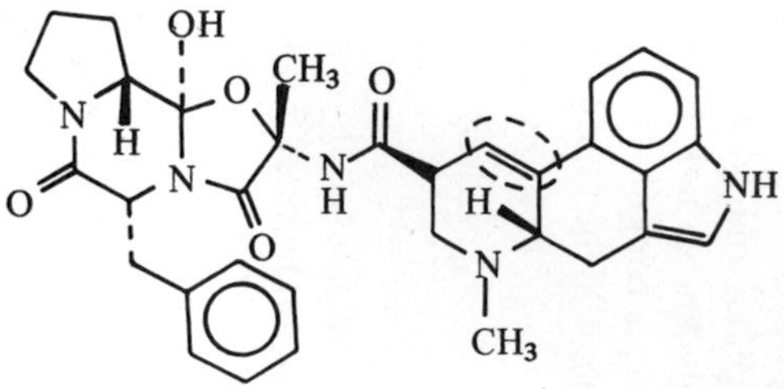
dihydroergotamine[+]	DHE 45 Dihydergot		
ergonovine[+]	Ergotrate		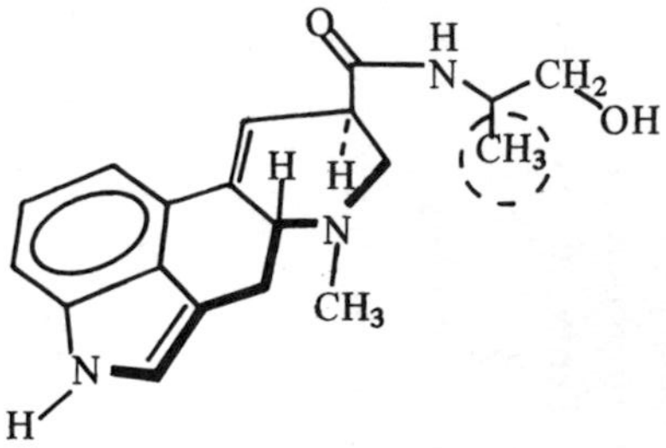

methylergonovine[+]

Methergine

N-[α-(hydroxymethyl)-propyl]-
D-lysergamide

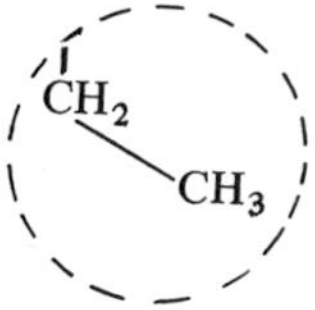

phenoxybenzamine[+]

Dibenzyline

N-(2-chloroethyl)-N-(1-methyl-2-
phenoxyethyl) benzylamine

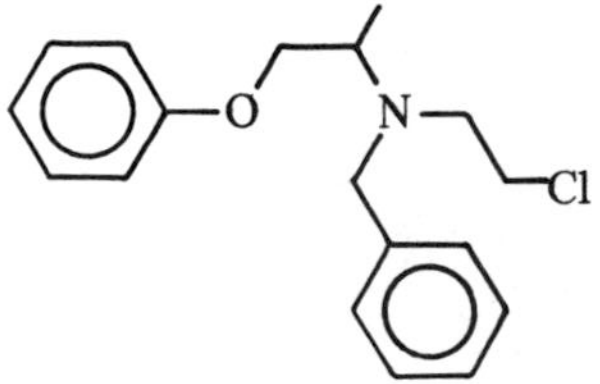

tolazoline[+]

Priscoline

2-benzyl-2-imidazoline

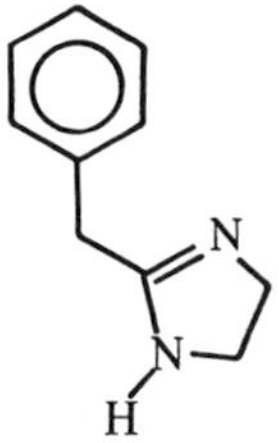

Table 18.1 α-Adrenergic Blocking Agents (continued)

Official Name	Proprietary Name	Chemical Name	Structure
phentolamine*+	Regitine	m-[N-(2-imidazolin-2-ylmethyl)-p-toluidino]phenol	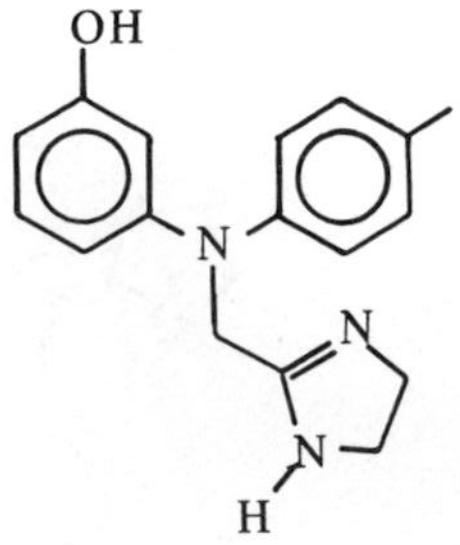
hydralazine*	Apresoline	1-hydrazinophthalazine	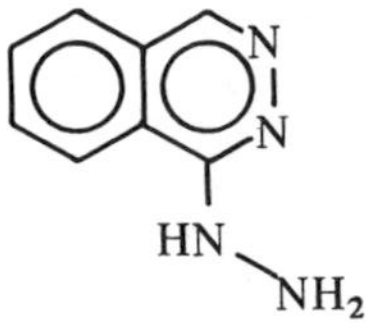

*In *USP XIX* (1975).
+In *NF XIV* (1975).

Table 18.2 β-Adrenergic Blocking Agents

Official Name	Proprietary Name	Chemical Name	Structure
dichloroisoproterenol	DCI	1-(isopropylamino)-3-(3,4-dichlorophenyl)-ethanol	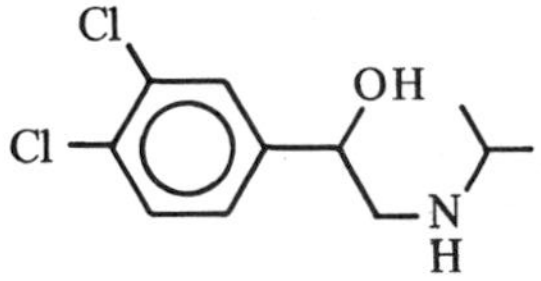
propranolol*	Inderal	1-(isopropylamino)-3-(1-naphth-yloxy)-2-propanol	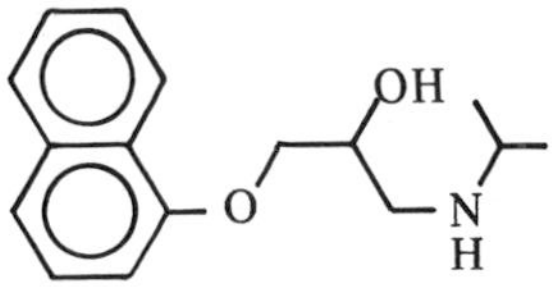
pronetalol	Alderlin	2-isopropylamino-1-(naphth-2-yl)ethanol	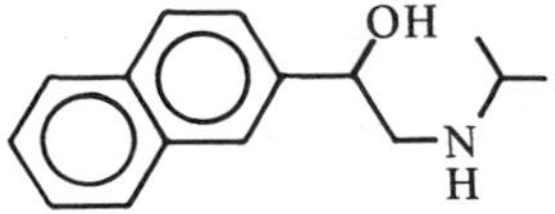

*In *USP XIX* (1975).

and as an antihypertensive agent. It is contraindicated in patients with angina pectoris.

Phenoxybenzamine

As hydrochloride, it occurs as water-insoluble, colorless crystals. It is useful in management of pheochromocytoma and peripheral vascular diseases. It is contraindicated in coronary and cerebrovascular disease. The initial dosage is 10 mg daily, which is increased by 10 mg every 4 days until the desired therapeutic effect is achieved.

B. β-Adrenergic Blocking Agents

The most widely used drugs of this class are listed in Table 18.2. Also available, besides several others, are the following ones: alprenolol (Aptime), bunitrolol, bunolol, bupranolol (Betadrenol), butaxamine, butidrine (Recatan), Doberol, metalol, nifenalol, oxprenolol (Trasicor), phenbutol, pindolol (Visken), practolol (Eraldin), sotalol (Beta-cardone), timolol (Blocarden), tiprenolol, tolamol, toliprolol.

β-Adrenergic blocking agents are arylethanolamines or aryloxypropanolamines. They may be represented by the general formula in which R may be

$$R\text{-}\underset{A}{\underbrace{\bigcirc}}\text{-}\underset{\underset{B}{OH}}{CH}\text{-}CH_2\text{-}NH\text{-}\underset{\underset{C}{CH_3}}{\overset{CH_3}{C}}\text{-}R'$$

another ring or substituents suitable to provide additional hydrophobic or other interactions with the receptor site, and R' is either H or CH_3. Therefore, the pharmacophoric moiety of β-adrenergic blocking agents consists of an aromatic (substituted) system (A) attached directly or through a methylene bridge to a α-hydroxyethylamine moiety (B) and to an alkyl residue substituted at the amino group (C).

Propranolol

It is a widely used antianginal and antiarrhythmic agent and peripheral vasodilator. In patients with angina pectoris propranolol lessens pain and increases tolerance to exercise; however, it may cause congestive heart failure. In antiarrhythmic therapy its usefulness is limited, being recommended especially in cases of digitalis intoxication.

C. Adrenergic Neuron Blocking Agents

Those mostly used are listed in Table 18.3. They are chemically related to sym-

Table 18.3   Adrenergic Neuron Blocking Agents

Official Name	Proprietary Name	Chemical Name	Structure
xylocholine bromide		choline bromide 2,6-xylyl ether	
guanethidine*	Ismelin	[2-(hexahydro-1(2H)-azocinyl)-ethyl] guanidine	
bretylium tosylate	Darenthin Hypotyl	(o-bromobenzyl)ethyldimethyl-ammonium p-toluenesulfonate	
debrisoquin	Declinax	3,4-dihydro-2- (1H)-isoquinoline-carboxamidine	
betanidine	Bethanid	1-benzyl-2,3-dimethylguanidine	

*In *USP XIX* (1975).

pathomimetic amines, but their terminal moiety is either amidine (in most of them) or quaternary nitrogen (in bretylium). Several other guanidine derivatives, besides those listed in Table 18.3, however, are marketed. The guanidine derivatives are represented by the general formula

$$R-\underset{\underset{|}{H}}{N}-C\underset{NH_2}{\overset{NH}{\diagup}}$$

Guanethidine

Used in the treatment of moderate and severe hypertension. Its untoward effects are marked orthostatic hypotension. It is contraindicated in cases of pheochromocytoma.

III. MECHANISM OF ACTION

Several mechanisms are involved in the action of sympatholytics. α-Adrenergic blocking agents can act by one of the following mechanisms:

1. Competitive antagonists of norepinephrine. They compete with this hormone for the α-adrenergic receptor. In this class are included indolethylamine alkaloids (ergot alkaloids, yohimbine), imidazoline derivatives (tolazoline, phentolamine), benzodioxane derivatives (piperoxan, butamoxane).

2. Noncompetitive antagonists of norepinephrine. This group consists of β-haloalkylamines (phenoxybenzamine).

β-Haloethylamines, fundamental structure of which is $RR'NCH_2CH_2X$, present an easily displaced group (Cl, Br, I, mesyl, tosyl) at β position in relation to an amine function of suitable basicity. They lose the X group and cyclize, forming the immonium ion. This ion can then alkylate a carboxylate or phosphate anion of different macromolecules and not necessarily of that which contain the α-adrenergic receptor, as shown in Figure 18.1. Once this alkylation was considered irreversible, but lately it was found that, with time, it reverts. The effect of phenoxybenzamine is described as chemical sympathectomy because it blocks selectively the excitatory responses of smooth muscle and myocardium.

3. Unknown mechanism. This group is comprised of dibenzazepines (azapetine) and hydrazinophthalazines (hydralazine).

β-Adrenergic blocking agents act as competitive antagonists of norepinephrine at the β receptor. According to Belleau, they owe their inhibitory effects to the bulky substituents attached to the nitrogen atom (Figure 17.2). By binding to the adenine ring of ATP, these substituents prevent proton transfer processes, very likely by displacing the adenine ring from its binding site at the

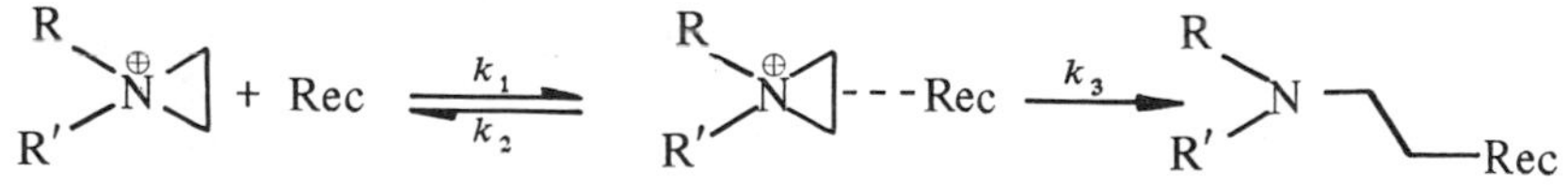

$$R = R' = -CH_2-C_6H_5 \qquad \text{dibenamine}$$

$$\left. \begin{array}{l} R = -CH_2-C_6H_5 \\ R' = -CH(CH_3)-CH_2-O-C_6H_5 \end{array} \right\} \text{phenoxybenzamine}$$

Figure 18.1 Interaction of the cyclic intermediate of β-haloethylamines with the receptor substance.

receptor surface.

These agents are structural analogues of isoproterenol and can, therefore, occupy the same β-receptor site; furthermore, the groups attached to the aromatic ring provide additional points of complexation with that receptor through van der Waals and hydrophobic interactions.

Adrenergic neuron blocking agents prevent the release of the chemical transmitter norepinephrine from its storage sites.

REFERENCES

General

A. M. Barrett, "Design of β-Blocking Drugs," in E. J. Ariëns, Ed., *Drug Design*, vol. III, Academic, New York, 1972, pp. 205-228.

A. M. Roe, *Annu. Rep. Med. Chem.*, 7, 59 (1972).

A. M. Karow *et al.*, *Prog. Drug Res.*, 15, 103 (1971).

P. Somani and A. R. Laddu, *Eur. J. Pharmacol.*, 14, 209 (1971).

O. Schier and A. Marxer, *Progr. Drug Res.*, 13, 101 (1969).

E. Schlittler, *Antihypertensive Agents*, Academic, New York, 1967.

History

D. Bovet and F. Bovet-Nitti, *Médicaments du Système Nerveau Végétatif*, Karger, Basel, 1948.

E. Fourneau and D. Bovet, *Arch. Int. Pharmacodyn. Thér.*, 46, 178 (1933).

H. H. Dale, *J. Physiol. (London)*, 34, 163 (1906).

Classification

C. T. Dollery *et al.*, *Clin. Pharmacol. Ther.*, 10, 765 (1969).

R. Ahlquist, *Annu. Rev. Pharmacol.*, 8, 259 (1968).

C. I. Furst, *Advan. Drug Res.*, 4, 133 (1967).

J. H. Biel and B. K. B. Lum, *Progr. Drug Res.*, 10, 46 (1966).

Mechanism of Action

M. Rocha e Silva and F. Fernandes, *Eur. J. Pharmacol.*, 25, 231 (1974).
A. Korolkovas, *Rev. Bras. Clin. Ter.*, 2, 333 (1973).

CHAPTER 19

INHIBITORS OF CATECHOLAMINE
BIOSYNTHESIS AND METABOLISM

I. INTRODUCTION

As it was seen in Chapter 18, adrenergic blocking effects can be achieved either by adrenergic blocking agents or by inhibitors of catecholamine biosynthesis and metabolism.

Only a few of these inhibitors are of therapeutic value. They act at various steps of catecholamine biosynthesis and metabolism by inhibiting different enzymes involved in biological processes (Figure 19.1).

II. INHIBITORS OF CATECHOLAMINE BIOSYNTHESIS

Some substances inhibit enzymes involved in catecholamine biosynthesis and, in this way, prevent their formation. Several compounds have shown to inhibit one or more biosynthetic steps, as seen in Figure 19.1. However, pharmacologically the most effective are those which inhibit the rate-limiting step, the biotransformation of L-tyrosine to DOPA.

Enzymes susceptible to inhibition of catecholamine biosynthesis are the following:

1. *Tyrosine hydroxylase.* This enzyme, for which the Union of Pure and Applied Chemistry proposed the name tyrosine 3-monooxygenase, catalyzes the rate-determining step of catecholamine biosynthesis, which is the biotransformation of L-tyrosine to DOPA.

Several compounds inhibit this enzyme:

a. Analogues of tyrosine itself. They act as competitive inhibitors of tyrosine hydroxylase. One of these compounds is α-methyl-L-tyrosine. Its methyl group at the alpha carbon retards decarboxylation and blocks transamination. It does not yet have clinical application, being used only as an experimental tool.

b. Catechols. They are not competitive inhibitors of the enzyme but act by different mechanisms. For instance, the compound of the following structure

263

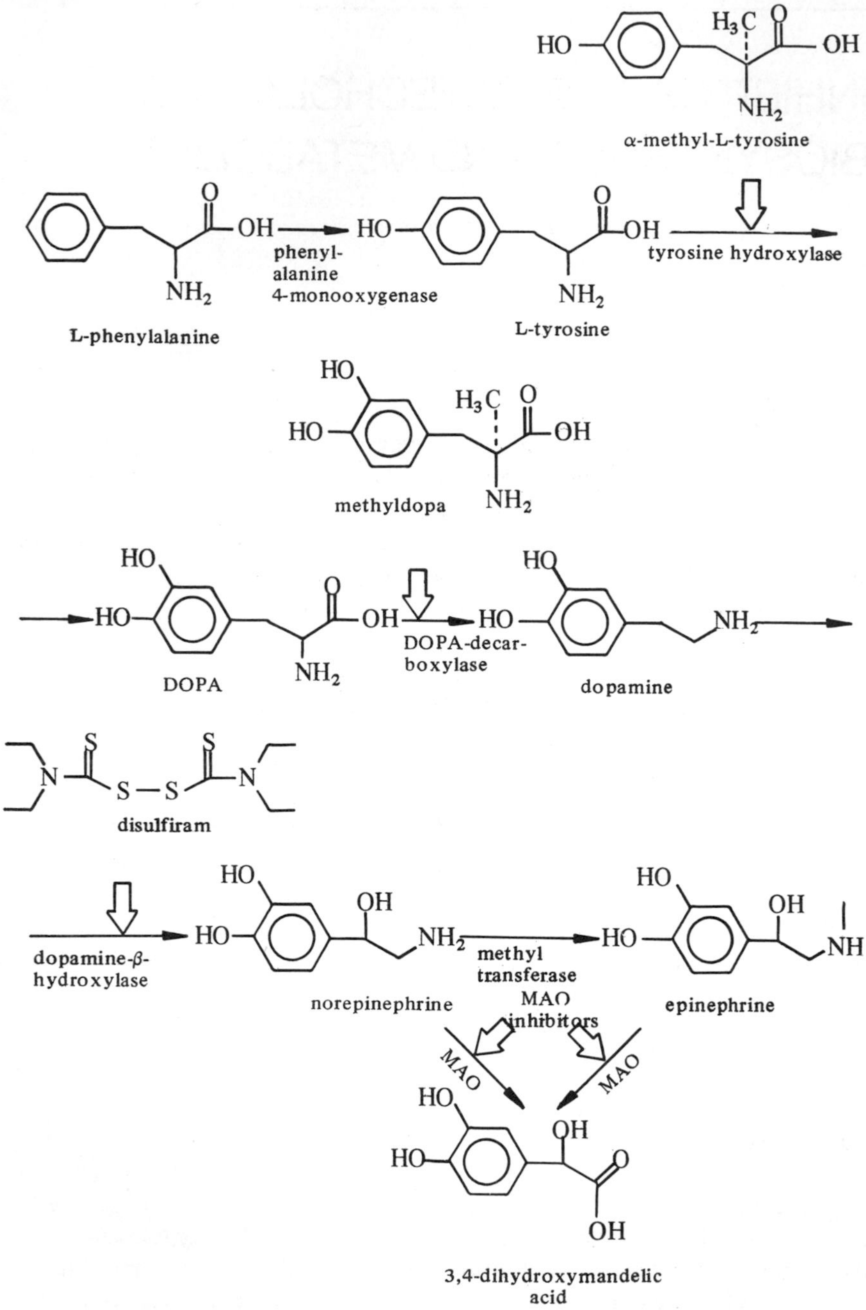

Figure 19.1 Biosynthesis and metabolism of catecholamines and their inhibitors.

acts by competing with the cofactor pteridine, thus depleting Fe^{2+}:

c. Norepinephrine itself. It acts by a feedback mechanism when the level of its concentration is high enough.

2. *DOPA-decarboxylase.* This enzyme catalyzes the biotransformation of DOPA to dopamine.

Among many other substances that inhibit this enzyme, methyldopa has found clinical application in the management of all forms of hypertension. Marketed as Aldomet, this compound, α-methyl-L-3,4-dihydroxyphenylalanine, was found to have antihypertensive properties by Sourkes in 1954. Presently it is often used to control blood pressure in patients with chronic renal insufficiency; it is usually well tolerated. Most common untoward effects are weakness, nausea, dizziness, drowsiness, nightmares, mild hepatitis. Prolonged treatment may cause alterations in the results of liver-function tests. Methyldopa acts as a competitive antagonist of DOPA. Methyldopate hydrochloride (Aldomet ester hydrochloride) has the same uses.

Methyldopa and α-methyl-L-tyrosine are also called false transmitters, because they are metabolized to α-methylnorepinephrine and metaraminol, respectively, substances that dislodge norepinephrine but have a much reduced adrenergic activity.

Two new DOPA-decarboxylase inhibitors, benserazide and carbidopa, are used as antiparkinsonism agents. A third inhibitor, brocresine, has also been tried, but it does not pass the blood-brain barrier in doses useful clinically and, therefore, it is not suitable as a therapeutic agent.

3. *Dopamine-β-hydroxylase.* This copper-dependent enzyme, also called dopamine β-monooxygenase, catalyzes the biotransformation of dopamine to norepinephrine.

This enzyme is inhibited by several substances, such as disulfiram, fusaric acid, and its analogues. Disulfiram is reduced to diethyldithiocarbamate, and this metabolite acts as a copper-chelating agent, but it has found clinical applicability since 1948 in the treatment of chronic alcoholism, when Hald and Jacobsen described its hypersensitive effects in those who had ingested alcohol. These effects with tetramethylthiuram were observed early in 1937 by Williams.

III. INHIBITORS OF CATECHOLAMINE METABOLISM

Both norepinephrine and epinephrine are deaminated by MAO to 3,4-dihydroxymandelic acid. Inhibition of this enzyme causes the concentrations of catecholamines to rise above their normal levels. Normally MAO inhibitors such as tranylcypromine (Chapter 10) produce hypertension in this way. Another MAO inhibitor, pargyline (Table 10.8) is used actually as an antihypertensive, since the action appears after several weeks of daily administration.

REFERENCES

Introduction

F. Sulser and E. Sanders-Bush, *Annu. Rev. pharmacol.*, **11**, 209 (1971).
M. Sandler and C. R. J. Ruthven, *Progr. Med. Chem.*, **6**, 200 (1969).

Inhibitors of Catecholamine Biosynthesis

A. Weissman and B. K. Koe, *Annu. Rep. Med. Chem.*, **1968**, 246 (1969).
A. Morita and L. J. Weber, *Annu. Rep. Med. Chem.*, **1967**, 252 (1968).
C. A. Stone and C. C. Porter, *Advan. Drug Res.*, **4**, 71 (1967).
S. Spector, *Pharmacol. Rev.*, **18**, 599 (1966).

CHAPTER 20

HISTAMINE AND ANTIHISTAMINIC AGENTS

I. HISTAMINE

A. Chemistry

Structurally, histamine is 4-(2-aminoethyl)-imidazole or β-imidazolylethyl-amine. It exists in two tautomeric forms, which have not been separated. At physiological pH, the amino nitrogen atom of the side chain is protonated. Histamine occurs therefore as a monovalent cation, which forms an intramolecular hydrogen bond between the amino group of the lateral chain and a nitrogen of the imidazole ring (Figure 20.1).

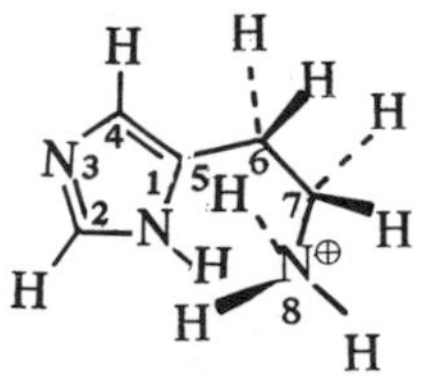

Figure 20.1 Preferred conformation of histamine.

Histamine is found in a number of animal tissues, venoms of insects (where it may reach a concentration as high as 2%), bacteria, and plants. Its concentration varies according to organs and species: in man, it is high in the skin, low in the blood; in the rabbit, it is high in the blood; in the dog, it is high in the liver.

Histamine is biosynthesized mainly at mastocytes. It is stored at heparin granules from which it can be released by several chemical compounds: antigens and simpler substances, venoms and toxins, trypsin and other proteolytic enzymes, detergents, several amines, such as the compounds 48/80 (a polymer of p-methoxy-N-methylphenylethylamine), 19/35 L (a substituted butylamine), CL-1182C (a pyridopyrimidine derivative). Histamine is formed by decarboxylation of histidine (Chapter 2, IV.E.1). It is rapidly metabolized by oxidation and N-methylation.

1. Conformation

Molecular orbital calculations by the EHT method performed by Kier have shown that histamine can exist in two preferred conformations, which are in equilibrium.

X-ray diffraction studies indicate that histamine is in its extended form, the group $\overset{\oplus}{N}H_3$ is in trans position in relation to the ring, and the plane of this ring lies perpendicular to the plane of the side chain.

The PCILO method of molecular orbital calculations applied to histamine in its cationic form—the one most likely to occur in the crystalline state, in solution and at physiological pH—has confirmed that the preferred conformation is the one shown in Figure 20.1.

Nuclear magnetic resonance studies have shown that the side chain $-CH_2-CH_2\overset{+}{N}$ of histamine has approximately equal proportions of anti (1) rotamer and both gauche (IIa and IIb) rotamers. These results confirm Kier's calculations, according to which anti and gauche forms have almost equivalent total energies:

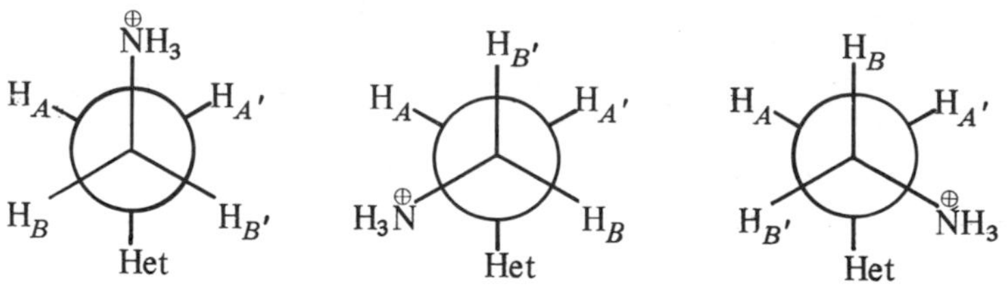

I	IIa	IIb
anti	gauche	

2. Electron Density

By the EHT method Kier observed that in the monovalent cation of histamine, although the $C-\overset{\oplus}{N}H_3$ group has a net charge of +0.811, the nitrogen, in contrast to the usual representation, bears a negative charge (−0.218) instead of a positive one.

By the PCILO method Pullman et al. found for the preferred conformation of the same ion of histamine much less marked charges, including that on the nitrogen atom of the side chain (−0.06).

3. Pharmacological Properties

Histamine produces various effects on several organs. For instance, it causes vasodilation of capillaries, and, at high doses, it can make them permeable to fluids and plasmatic proteins that, by pouring over, cause edema. It also stimulates gastric-juice secretion by the stomach (and in this way can cause ulcers), accelerates the heart beat, and inhibits contraction of the rat uterus. It is implicated in allergic phenomena and anaphylactic shock.

Histamine has no therapeutic application. As the phosphate it is used in determination of gastric acidity, as an adjunct in the differential diagnosis of peptic ulcer, gastric carcinoma, and pernicious anemia.

A histamine isomer, betazol (Histalog) or 3-(2-aminoethyl) pyrazole, stimulates gastric secretion more markedly than histamine; it is, however, less potent than histamine as to other pharmacological properties. A histamine isostere, betahistine (Serc) or 2-(β-methylaminoethyl)pyridine, inhibits histaminase; it is used in the symptomatic treatment of Ménière's syndrome.

B. Antagonists

Histamine effects can be antagonized by the use of (a) inhibitors of histamine biosynthesis, such as α-methylhistidine semicarbazide and brocresine, which are histidine decarboxylase inhibitors (Chapter 2, IV.E.1); (b) antihistaminic agents, which are competitive antagonists of histamine.

II. ANTIHISTAMINIC AGENTS

Antihistaminic agents are used primarily in the management of certain allergic disorders but are only palliative. Some of these drugs are also useful as antitussive agents (Chapter 9), antianxiety agents (Chapter 10), antipsychotic agents (Chapter 10), antiparkinsonism drugs (Chapter 11), and antiemetics, in motion sickness.

Although recommended as nasal decongestants and in rhinitis, antihistaminics are of little benefit alone in these conditions. They do not prevent or abort the common cold.

Among untoward effects, the most common are sedation, dizziness, and disturbed coordination to deep sleep. Occasionally, insomnia, tremors, irritability, and convulsions may appear, besides fatigue, excessive perspiration, headache, early menses, and other discomforts. Alcoholic beverages and other CNS depressants should not be used simultaneously with antihistaminics.

A. Classification

Antihistaminics are represented by the general formula $RR'-X-CH_2-CH_2-NR''R'''$, in which X stands for nitrogen, oxygen, or carbon. Those in which X = O have marked sedative action. Compounds with X = N are more active and more toxic. When X = C, the antihistaminics are less active but also less toxic. The terminal nitrogen must be a tertiary one; the dimethyl derivatives exert more intense activity than other homologues; the alkyl chain between X and N,

270

Table 20.1 Ethanolamines $R-O-CH_2-CH_2-N(CH_3)_2$

Official Name	Proprietary Name	Chemical Name	R
diphenhydramine*	Benadryl	2-(diphenylmethoxy)-N,N-dimethyl-ethylamine	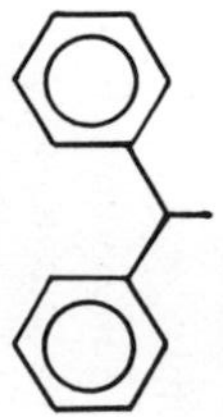
dimenhydrinate*[a]	Dramamine	diphenhydramine 8-chlorotheophylline salt	see above structure
bromodiphenhydramine[+]	Ambodryl	2-[p-bromo-α-phenylbenzyl)oxy]-N,N-dimethylethylamine	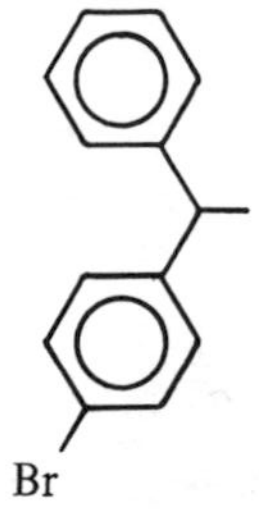

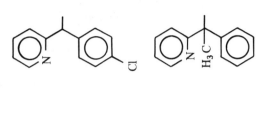

carbinoxamine[+]   Clistin   2-[p-chloro-α-(2-dimethylamino-
                                          ethoxy)benzyl]pyridine

doxylamine[+]   Decaprin   2-[α-(2-dimethylaminoethoxy)-α-
                                         methylbenzyl]pyridine

_____

*In *USP XIX* (1975).
+ In *NF XIV* (1975).
[a]Used primarily as an antiemetic.

for optimal activity, must have two carbon atoms; optimal activity is obtained when R and R' are aromatic; the introduction of groups with a -I inductive effect into the para position of the R phenyl enhances the potency. Summing up, an antihistaminic agent of this type has an ionizable amino group and a central dipole.

Chemically, antihistaminic agents belong to the following groups of structures: ethanolamines, ethylenediamines, alkylamines, piperazines, phenothiazines and miscellaneous. They are generally used as salts (hydrochloride, citrate, phosphate, succinate, maleate), which are water-soluble, white, crystalline powders.

1. Ethanolamines

Ethanolamines most widely used are shown in Table 20.1. Besides these, several others are marketed: amethobenzepine, bromazine (Ambodryl), clemastine (Tavegil), embramine (Bromadryl), medrylamine (Aphilan), moxastine (Alfadryl), phenyltoloxamine (Antin), pyroxamine (AHR-224).

Diphenhydramine

It is the prototype of this class of antihistaminics and the agent of choice for parenteral use in the treatment of anaphylactic reactions. Diphenhydramine should be used cautiously in patients with hypertension or cardiac disease. It is not recommended for topical application because of its sensitizing effects.

Carbinoxamine

Among ethanolamines, it is one that is characterized by a lower incidence of drowsiness. The usual dosage is 12 to 32 mg daily in divided doses.

2. Ethylenediamines

Ethylenediamines mostly used are listed in Table 20.2. Many others, however, are marketed, for instance chlorothen (Tagathen), tolpropamine (Pragman).

Tripelennamine

It is the prototype of the ethylenediamine class of antihistaminics. Untoward effects produced by this drug are sedation, gastrointestinal irritation, and dizziness.

3. Alkylamines

Alkylamines mostly used are in Table 20.3. Several others are marketed.

Some of these drugs are used as racemic mixtures (bromopheniramine, chlorpheniramine), although the dextrorotatory isomers (dextropheniramine, dexchlorpheniramine) are more active.

Table 20.2 Ethylenediamines

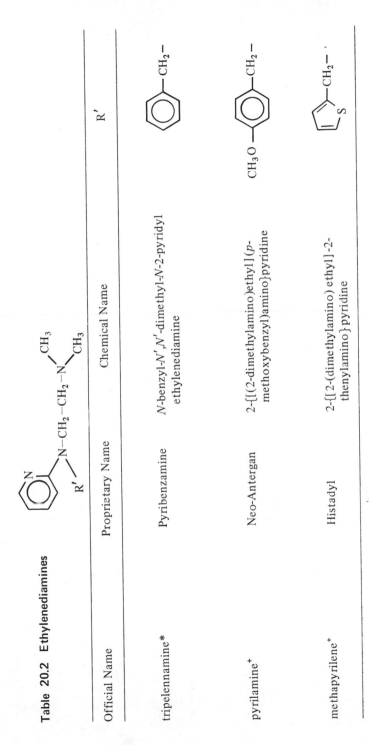

Official Name	Proprietary Name	Chemical Name	R'
tripelennamine*	Pyribenzamine	N-benzyl-N',N'-dimethyl-N-2-pyridyl ethylenediamine	
pyrilamine[+]	Neo-Antergan	2-{[(2-dimethylamino)ethyl](p-methoxybenzyl)amino}pyridine	
methapyrilene[+]	Histadyl	2-{[2-(dimethylamino) ethyl]-2-thenylamino} pyridine	

*In *USP XIX* (1975).
[+] In *NF XIV* (1975).

Table 20.3 Alkylamines

Official Name	Proprietary Name	Chemical Name	Structure
triprolidine[+]	Actidil	trans-2-[3-(1-pyrrolidinyl)-1-p-tolylpropenyl] pyridine	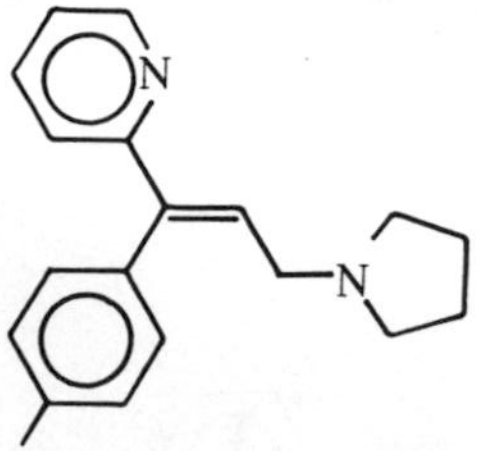
brompheniramine[+]	Dimetane	(±)-2-[p-bromo-α-[2-(dimethyl-amino)ethyl] benzyl]pyridine	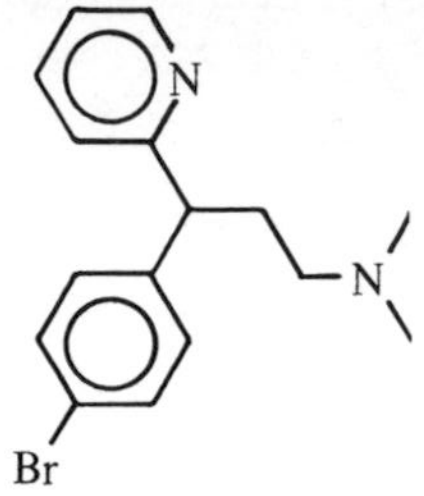
chlorpheniramine*	Chlor-Trimeton	(±)-2-{p-chloro-α-[2-(dimethyl-amino)-ethyl] benzyl} pyridine	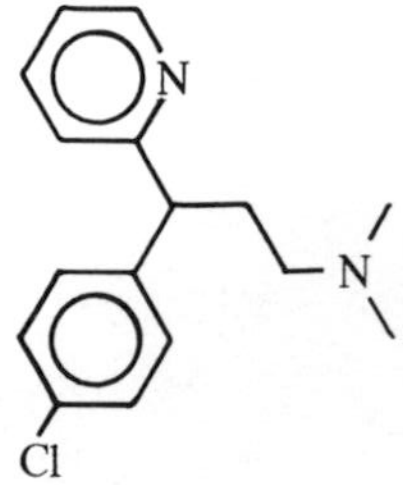

dexbrompheniramine[+]	Disomer	(+)-2-[p-bromo-α-[2-(dimethyl-amino)-ethyl]benzyl]pyridine	see brompheniramine
dexchlorpheniramine[+]	Polaramine	(+)-2-[p-choro-α-[2-(dimethyl-amino)-ethyl]benzyl]pyridine	see chlorpheniramine
dimethindene[+]	Forhistal	2-{1-[2-[2-(dimethylamino)ethyl]-inden-3-yl]ethyl}pyridine	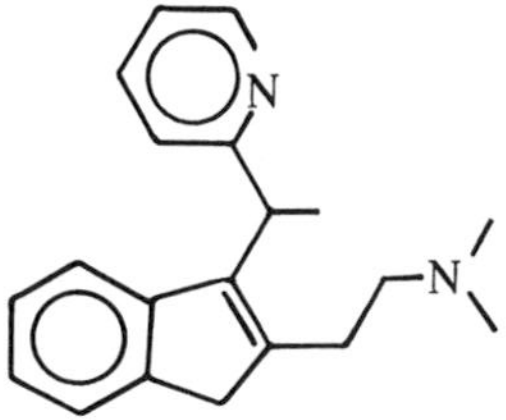

*In *USP XIX* (1975).
[+] In *NF XIV* (1975).

Chlorpheniramine

It represents isosteric replacement of O or N atoms of the ethanolamine and ethylenediamine antihistaminics. Introduction of a chlorine in the para position of the benzyl group in the older drug pheniramine gave a 20-fold increase in potency. The dextrorotatory isomer of racemic chlorpheniramine is dexchlorpheniramine. The greater potency of the latter drug shows that activity resides largely in the (+) isomer.

Triprolidine

Activity is found only in the isomer in which the pyrrolidinomethyl group is trans to the 2-pyridyl group. It has rapid onset of action, and its effects last up to 12 hours. The usual dosage is 2.5 mg three times daily.

4. Piperazines

A number of piperazine derivatives have antihistaminic activity. Besides those shown in Table 20.4, these are also marketed: buclizine (Softran), cinnarizine (Dimitron), decloxizine, homochlorcyclizine.

Some of these drugs (cyclizine and meclizine, for instance) are used in the prophylaxis and treatment of motion sickness. Others have predominantly antihistaminic action.

5. Phenothiazines

Several phenothiazines are used as antihistaminic agents, mainly those listed in Table 20.5.

Promethazine

It is a highly potent antihistaminic agent, but because of sensitizing effects it is not recommended for topical use. Other untoward effects are leukopenia, hypotension, agranulocytosis, tremors, and dystonias. The usual dosage is 12.5 to 25 mg three or four times daily.

6. Miscellaneous Antihistaminics

A number of antihistaminic agents cannot be included in one of the preceding groups. Among these agents of miscellaneous structures are those listed in Table 20.6.

Antazoline

Used topically in allergic conjunctivitis, because of its low incidence of sensitization, in contrast to other antihistaminic agents which produce marked sensitization effects.

Table 20.4 Piperazines

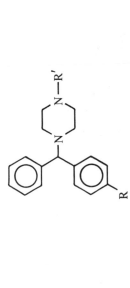

Official Name	Proprietary Name	Chemical Name	R	R'
cyclizine*[+]	Marezine	1-(diphenylmethyl)-4-methyl-piperazine	H—	—CH₃
chlorcyclizine[+]	Diparalene Perazil	1-(p-chloro-α-phenylbenzyl)-4-methylpiperazine	Cl—	—CH₃
hydroxyzine[+]	Atarax	2-[2-[4-(p-chloro-α-phenylbenzyl)-1-piperazinyl]ethoxy]ethanol	Cl—	—CH₂CH₂OCH₂CH₂OH
meclizine* (meclozine)	Bonine	1-(p-chloro-α-phenylbenzyl)-4-(m-methylbenzyl)piperazine	Cl—	

*In *USP XIX* (1975).
[+]In *NF XIV* (1975).

Table 20.5 Phenothiazines

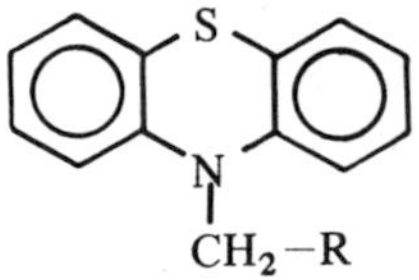

$CH_2 - R$

Official Name	Proprietary Name	Chemical Name	R
promethazine*	Phenergan	(±)-10-[(2-dimethylamino)propyl]-phenothiazine	$-CH(CH_3)N(CH_3)_2$
trimeprazine*	Temaril	(±)-10-[3-(dimethylamino)-2-methylpropyl] phenothiazine	$-CH(CH_3)CH_2N(CH_3)_2$
methdilazine	Tacaryl	(±)-10-[(1-methyl-3-pyrrolidinyl)-methyl] phenothiazine	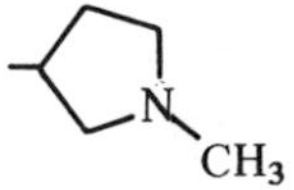

*In *USP XIX* (1975).

Cyproheptadine

It is used to relieve pruritus associated with skin disorders (urticaria, allergic dermatitis, neurodermatitis). It also has antiserotonin effects.

7. Mixtures

Several mixtures of two or more antihistaminics or antihistaminics with adrenergic agents, analgetics, adrenal corticosteroids, or antibacterial agents are marketed and used for the common cold, allergies, and other discomforts. According to the American Medical Association, these pharmaceutical preparations are irrational mixtures, because there is no sound basis for their use.

B. Mechanism of Action

Antihistaminic agents are competitive antagonists of histamine. However, those used in therapeutics do not antagonize all the effects of histamine. Certain responses induced by histamine, especially the increase of the gastric secretion in the stomach, are not antagonized by them. For this reason it is suggested that there are two receptors for histamine: the H_1 receptor and the H_2 receptor. Activation of the first results in capillary vasodilatation. Activation of the second stimulates production of gastric juice. The H_1 receptor is blocked by antihistamines, such as mepyramine. The H_2 receptor is insensitive to antihistaminic agents of therapeutic use. It is postulated that these antihistaminic agents act only on the H_1 receptor without effect on the H_2 receptor.

In 1972 Black *et al.* observed that 4-methylhistamine stimulates selectively the H_2 receptor and discovered that burimamide, which is chemically N-methyl-$N'\{4-[4(5)$-imidazolyl] butyl}thiourea, is a specific competitive antagonist of the H_2 receptor, blocking effectively gastric secretion evoked by histamine and pentagastrin (see Figure 2.6, for the structures of H_1 and H_2 histaminergics, both agonists and antagonists). Other H_2 antagonists are metiamide and cimetidine.

Barlow, Bloom and Goldman, Nauta and coworkers, Hite and Shafi'ee, and Rocha e Silva have proposed models for interaction between antihistamines and the histamine receptor.

On the basis of molecular orbital calculations, Kier advanced the hypothesis that histamine exists in two preferred conformations and can, therefore, elicit two distinct biological responses, depending on which one of the complementary receptors is present. The H_1 receptor would be complementary to the following internitrogen relationship:

$$\gtrless N \longleftarrow 4.55 \text{ Å} \longrightarrow \overset{\oplus}{N}$$

which is found in triprolidine, for instance, whereas the H_2 receptor would be

Table 20.6 Miscellaneous Antihistamines

Official Name	Proprietary Name	Chemical Name	Structure
antazoline[+]	Antistine	2-[(N-benzylanilino)methyl]-2-imidazoline	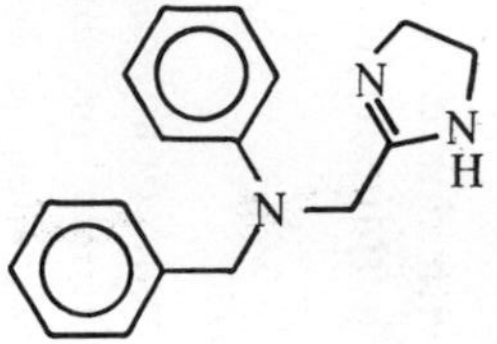
cyproheptadine[+]	Periactin	4-(5H-dibenzo[a,d]cyclohepten-5-ylidene)-1-methylpiperidine	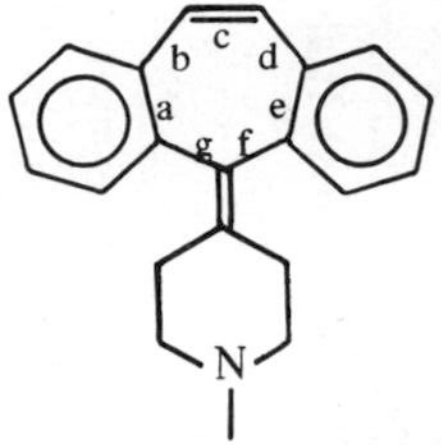

phenindamine[+]

Thephorin

2,3,4,9-tetrahydro-2-methyl-9-
phenyl-1H-indeno[2,1-c]pyridine

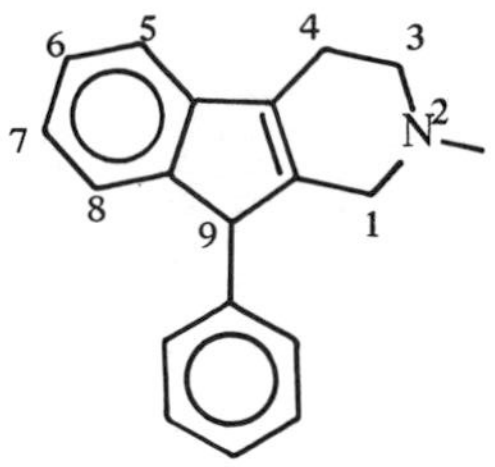

diphenylpyraline[+]

Diafen
Hispril

4-(diphenylmethoxy)1-methyl-
piperidine

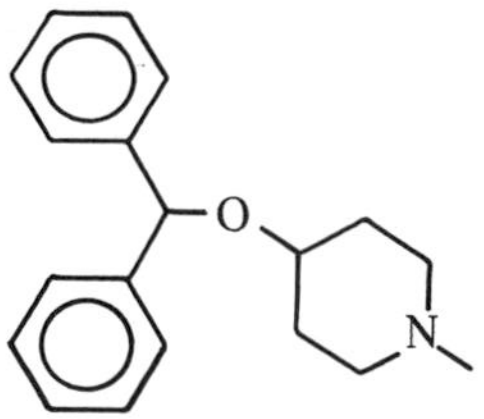

[+] In *NF XIV* (1975).

complementary to the one seen below:

$$\geqslant N \longleftarrow 3.60 \text{ Å} \longrightarrow \overset{\oplus}{N}$$

That is, in its anti conformation histamine interacts with the H_1 receptor, corresponding to the one found in guinea pig ileum, and in its gauche conformation, with the H_2 receptor, which evokes gastric secretion.

Taking Kier's findings into account, Rocha e Silva modified his original proposal (advanced in 1966) and suggested in 1969 that histamine is attracted to its specific receptor site (H_1) by (*a*) a strong electrostatic interaction between the nitrogen (N^-) of the imidazole group of the receptor site and the strongly charged protonated nitrogen (N^+) of histamonium ion; (*b*) the reciprocally inverted dipoles in the receptor peptide linkage and the carbon (C^+)–nitrogen (N^-) of the imidazole ring of the agonist (Figure 20.2).

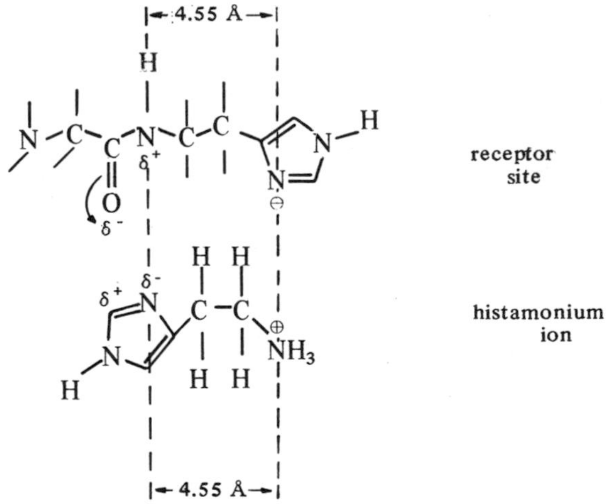

Figure 20.2   Schematic representation of the interaction of histamine with its hypothetical receptor. [After Rocha e Silva, *J. Pharm. Pharmacol.*, 21, 778 (1969)].

Antihistaminic agents, by acting as competitive antagonists of histamine, can dislodge histamine from its attachment to the specific site of the receptor. Furthermore, their bulky aromatic groups form additional bonds with the nonspecific site of the receptor through van der Waals and hydrophobic interactions. This combination between the histamine receptor and antihistaminic agents is similar to that between the acetylcholine receptor and anticholinergics represented in Figure 3.2.

REFERENCES

General

A. Korolkovas, *Cilncia Cultura,* **26**, 347 (1974).

M. Schachter, Ed., *Histamine and Antihistamines,* Vol. 1, Pergamon, Oxford, 1973.

C. D. Wood and A. Graybiel, *Clin. Pharmacol. Ther.,* **11**, 621 (1970).

G. Valette, *Prod. Probl. Pharm.,* **24**, 71 (1969).

M. Rocha e Silva, Sub-Ed., "Histamine and Antihistaminics," *Handbuch der experimentellen Pharmakologie,* Vol. XVIII/I, Springer, Berlin, 1966.

Histamine

B. Pullman and J. Port, *Mol. Pharmacol.,* **10**, 360 (1974).

C. R. Ganellin *et al., J. Med. Chem.,* **16**, 610, 616, 620 (1973).

N. S. Ham *et al., J. Med. Chem.,* **16**, 470 (1973).

W. Tautz *et al., J. Med. Chem.,* **16**, 705 (1973).

O. B. Reite, *Physiol. Rev.,* **52**, 778 (1972).

J.-L. Coubeils *et al., C.R.H. Acad. Sci.,* Ser. D, **272**, 1813 (1971).

G. Kahlson and E. Rosengren, *Physiol. Rev.,* **48**, 155 (1968).

M. Rocha e Silva, *Histamine: Its Role in Anaphylaxis and Allergy,* Thomas, Springfield, Ill., 1955.

Antihistaminic Agents

G. L. Durant *et al., J. Med. Chem.,* **18**, 830, 905 (1975).

L. Farnell *et al., J. Med. Chem.,* **18**, 662 (1975).

A. F. Harms *et al.,* "Diphenhydramine derivatives: through manipulation toward design," in E. J. Ariëns, Ed., *Drug Design,* Vol. VI, Academic, New York, 1975, pp. 1-80.

B. Pullman *et al., Mol. Pharmacol.,* **11**, 268 (1975).

A. Korolkovas and K. Tamashiro, *Rev. Paul. Med.,* **83**, 295 (1974).

M. Rocha e Silva *et al., Eur. J. Pharmacol.,* **17**, 333 (1972).

J. W. Black *et al., Nature (London),* **236**, 385 (1972).

M. Rocha e Silva *et al., Eur. J. Physiol.,* **17**, 333 (1972).

M. Rocha e Silva, *Physiol. Chem. Phys.,* **2**, 505 (1970).

W. T. Nauta *et al.,* "Diarylcarbinol Ethers: Structure Activity Relationships, A Physico-Chemical Approach," in E. J. Ariëns, Ed., *Physico-Chemical Aspects of Drug Action,* Pergamon, Oxford, 1968, pp. 305-325.

B. Idson, *Chem. Rev.,* **47**, 307 (1950).

CHAPTER 21

LOCAL ANESTHETICS

I. INTRODUCTION

Local anesthetics are agents that reversibly block the generation and the conduction of impulses along a nerve fiber. They are used to abolish the pain sensation in restricted areas of the body. Their action results from their ability to depress impulses from afferent nerves of the skin, surfaces of the mucosa, and muscles to the central nervous system.

These agents are widely used especially in surgery, dentistry, and ophthalmology to achieve either a partial or a complete, but necessarily reversible, block of the transmission of impulses in peripheral nerves or nerve endings.

Several techniques are used to induce local anesthesia with drugs. The most common are:

1. Surface anesthesia, which is accomplished by applying drugs, topically to skin and mucous membranes, as creams, ointments, aerosols, solutions, jellies, or suppositories. Many drugs are suitable for this anesthesia, such as benzocaine, butacaine, cyclomethycaine, diperodon, dyclonine, hexylcaine, phenacaine, piperocaine, pramoxine, proparacaine, tetracaine.

2. Spinal anesthesia, in which local anesthetics are injected into the spinal subarachinoid space. The drug of choice is tetracaine. Other drugs used are dibucaine, lidocaine, mepivacaine, piperocaine, and procaine.

3. Infiltration and block anesthesia, which results from the injection of a solution of a local anesthetic into the desired area. Several drugs are used for this purpose, such as chloroprocaine, hexylcaine, lidocaine, meprylcaine, metabutethamine, prilocaine, procaine, propoxycaine, pyrrocaine, tetracaine.

4. Epidural and caudal anesthesia, which is accomplished by injecting the appropriate drug into the epidural space. The drugs of choice are lidocaine, prilocaine, and mepivacaine. Others also used are chloroprocaine, piperocaine, procaine, and tetracaine.

5. Intravenous anesthesia, which is seldom used, owing to the low margin of safety and short duration of action. In this technique the preferred drugs are lidocaine and procaine.

In order to prolong and intensify the effect produced by local anesthetics, vasoconstrictors, most commonly epinephrine, are usually added to solutions of those agents. Other vasoconstrictors recommended are levarterenol, nordefrin, and phenylephrine.

It was once customary to add hyaluronidase to solutions of local anesthetics with the purpose of facilitating their diffusion in infiltration and block anesthesia. This practice is no longer recommended, because of the incidence of systemic reactions.

Overdosage of local anesthetics and rapid systemic absorption cause adverse systemic reactions, involving mainly (a) the central nervous system, with the following symptoms: vomiting, nausea, euphoria, dizziness, and eventually convulsions, coma, respiratory and heart failure, and death; (b) the cardiovascular system, with bradycardia, hypotension, and a shocklike state. These conditions are relieved by the administration of ultrashort-acting or short-acting barbiturates or skeletal-muscle relaxants.

Some local reactions, primarily of allergic or cytotoxic nature, can also occur, such as pain, edema, skin discoloration, neuritis, and eczematoid dermatitis.

II. CLASSIFICATION

A. General Structure

Most local anesthetics are structurally related to cocaine (Table 21.1) and can be represented by the general formula, which is the *local anesthesiophoric* moiety:

$$\text{Ar}-\overset{\displaystyle O}{\overset{\displaystyle \|}{C}}-O(CH_2)_n-\overset{\displaystyle R_1}{\underset{\displaystyle |}{N}}-R_2$$

lipophilic center	intermediate chain	hydrophilic center

Some of these three constituent groups can be replaced by isosteric moieties. It is essential for anesthetic activity that a balance exists between the lipophilic and the hydrophilic parts of the molecule. Furthermore, in all local anesthetics of the ester and amide types, the carbonyl group is activated by the partially positive charge on the carbon atom. This is made possible by conjugated double bonds, which allow the π framework over the aromatic ring to delocalize toward the carbonyl oxygen.

As to duration of effect, it depends upon the rate of hydrolysis by nonspecific enzymes and upon the hydrophobicity of the compounds. Thus, in the series of local anesthetics, the duration of effect increases progressively in the sequence: procaine < lidocaine < prilocaine < mepivacaine.

Table 21.1 Ester Derivatives

Official Name	Proprietary Name	Chemical Name	Structure
Benzoic acid esters			
cocaine*+		2β-carbomethoxy-3β-benzoxy-tropane	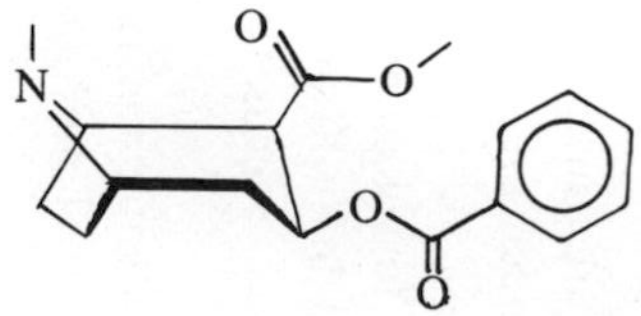
hexylcaine+	Cyclaine	1-(cyclohexylamino)-2-propyl benzoate	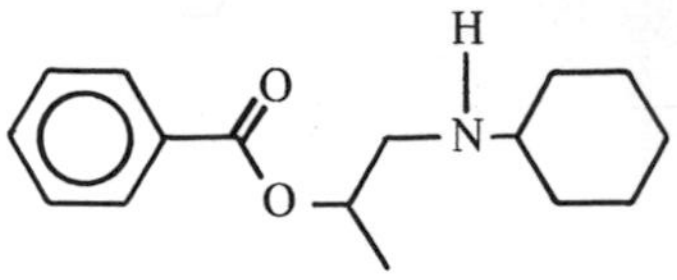
isobucaine+	Kincaine	2-(1-isobutylamino)-2-methyl-1-propyl benzoate	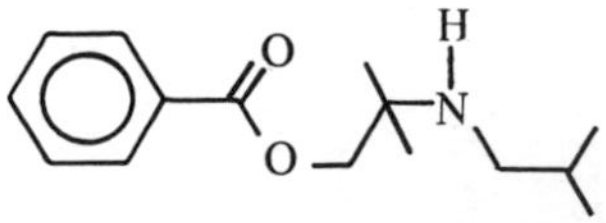

meprylcaine[+]

Oracaine

2-methyl-2-(propylamino)-1-propyl benzoate

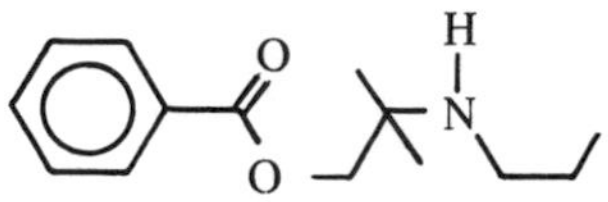

p-*Aminobenzoic acid esters*

procaine*[+]

Novocaine

2-(diethylamino)ethyl *p*-amino-benzoate

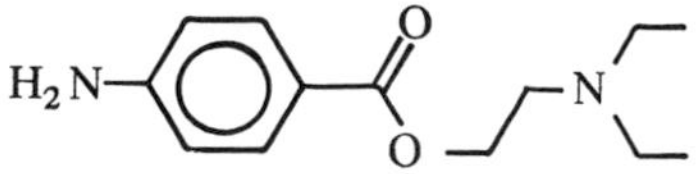

chloroprocaine[+]

Nesacaine

2-(diethylamino)ethyl 4-amino-2-chlorobenzoate

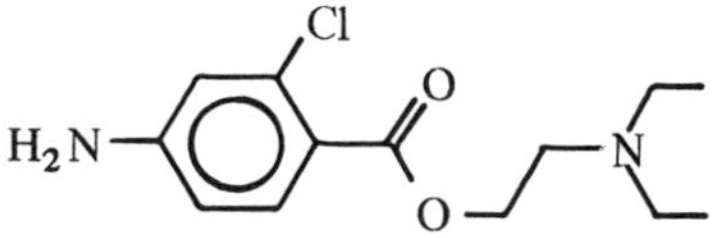

butacaine[+]

Butyn

3-di-*n*-butylaminopropyl *p*-amino-benzoate

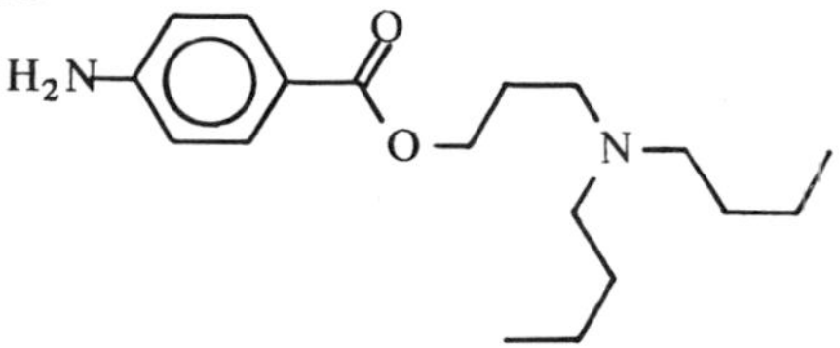

Table 21.1 Ester Derivatives (continued)

Official Name	Proprietary Name	Chemical Name	Structure
propoxycaine[+]	Blockaine Ravocaine	2-(diethylamino)ethyl 4-amino-2- propoxybenzoate	
benoxinate[+]	Dorsacaine	2-(diethylamino)ethyl 4-amino- 3-*n*-butoxybenzoate	
benzocaine[+]	Anesthesin	ethyl *p*-aminobenzoate	

butamben[+]

Butesin

butyl p-aminobenzoate

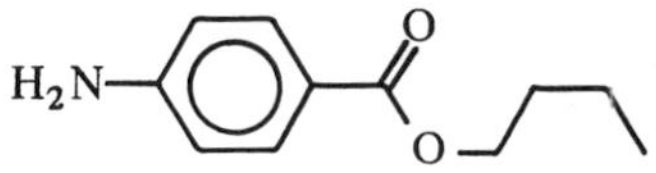

tetracaine*[+]

Pontocaine

2-(dimethylamino)ethyl p-(butyl-amino)benzoate

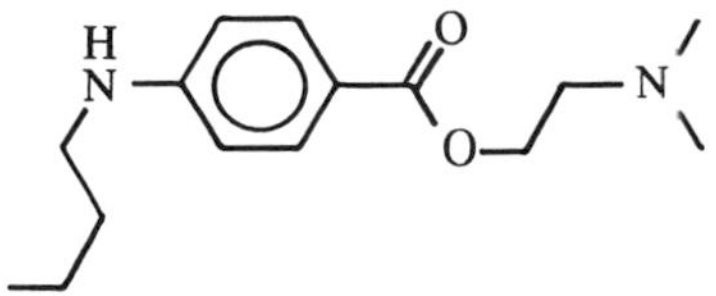

m-Aminobenzoic acid esters

proparacaine*
(proxymetacaine)

Ophthaine

2-(diethylamino)ethyl 3-amino-4-propoxybenzoate

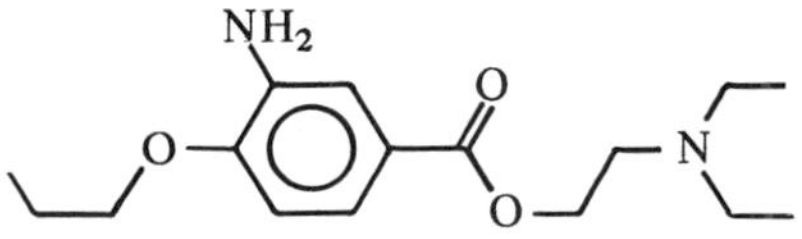

Table 21.1 Ester Derivatives (continued)

Official Name	Proprietary Name	Chemical Name	Structure
p-Alkoxybenzoic acid esters			
cyclomethycaine	Surfacaine	3-(2-methylpiperidino)propyl *p*-cyclohexyloxybenzoate	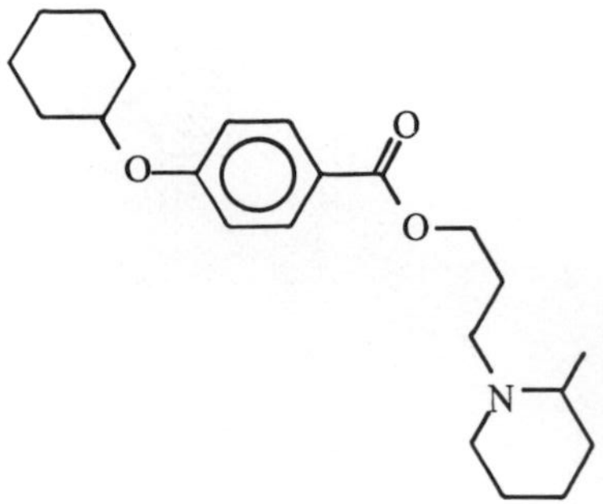
parethoxycaine	Intracaine	2-diethylaminoethyl *p*-ethoxy-benzoate	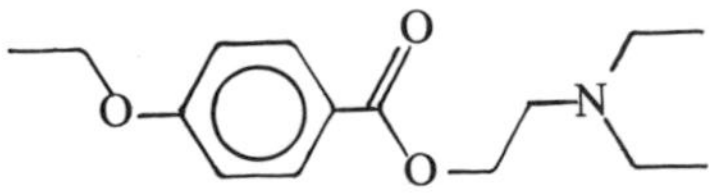

*In *USP XIX* (1975).
+ In *NF XIV* (1975).

B. Classes of Local Anesthetics

Local anesthetics presently used in therapeutics can be grouped in the following three classes: ester derivatives, amide derivatives, and miscellaneous local anesthetics. Practically all in the free-base form are liquids. For this reason, most of these drugs are used as salts (hydrochloride, sulfate, picrate, nitrate, and borate) which occur generally as water-soluble, odorless, crystalline solids.

1. *Ester Derivatives*

The most widely used are listed in Table 21.1. Besides these, many others are marketed, for instance, butethamide (Unacaine), piperocaine (Metycaine).

As can be seen in Table 21.1, these substances are esters of one of the following acids: benzoic, *p*-aminobenzoic, *m*-aminobenzoic, *p*-alkoxybenzoic. However, Apothesine, the first local anesthetic developed in the United States, is an ester of cinnamic acid; piridocaine is an ester of anthranilic acid. As esters, these drugs are easily hydrolyzed both *in vitro* and *in vivo* with loss of activity.

Cocaine

It is extracted from the leaves of *Erythroxylon coca* Lamarck and other species of the same genus, or it may be synthesized from ecgonine. Cocaine is a levorotatory substance, occurring as colorless crystals or as a white, crystalline powder, sparingly soluble in water. It is a tertiary amine; it forms readily crystallizable salts, such as hydrochloride, hydrobromide, citrate, borate, benzoate, and salicylate.

Cocaine is used topically in the eye, nose, ear, throat, vagina, and rectum, but it should not be injected or ingested. With the introduction of safer and more potent drugs, its use has declined.

Procaine

Used as hydrochloride, borate, and nitrate. Pharmaceutical preparations are marketed which contain it with other local anesthetics (propoxycaine, tetracaine) or adrenomimetics (phenylephrine, levarterenol, levonordefrin).

Procaine is the prototype of local anesthetics of parenteral use. It is also used intravenously in the treatment of cardiac arrhythmias. It is contraindicated for patients treated with digitalis, anticholinesterases, or succinylcholine. Since it is hydrolyzed to *p*-aminobenzoic acid, it should not be used with sulfa drugs of which it is a competitive antagonist (see Chapter 31). The dosage varies according to the type of anesthesia for which it is used.

2. *Amide Derivatives*

Those listed in Table 21.2 have wide therapeutic application.

This class of drugs is comprised of three groups: (*a*) basic amides, such as

Table 21.2 Amide Derivatives

Official Name	Proprietary Name	Chemical Name	Structure
lidocaine*	Xylocaine	2-diethylamino-2′,6′-acetoxylidide	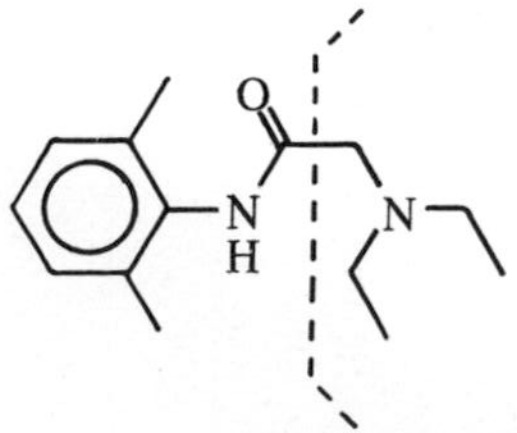
pyrrocaine[+]	Endocaine	1-pyrrolidinoaceto-2′,6′-xylidide	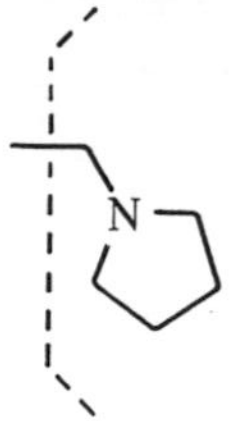
mepivacaine*[a]	Carbocaine	(±)-1-methyl-2′,6′-pipecoloxylidide	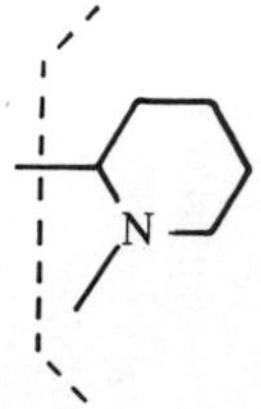

prilocaine[*][+]

Citanest

2-(propylamino)-o-propionotoluidide

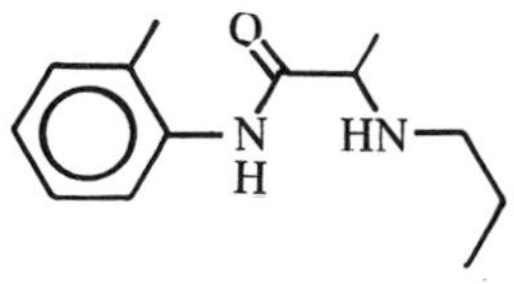

dibucaine[+]
(cincocaine)

Nupercaine

2-butoxy-N-[(2-diethylamino)-
ethyl] cinchoninamide

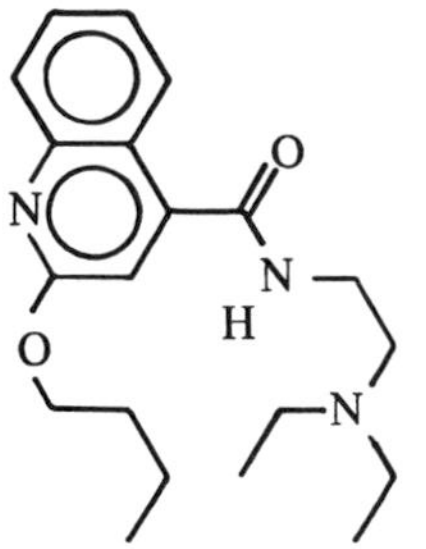

---

[*] In *USP XIX* (1975).
[+] In *NF XIV* (1975).
[a] Also hydrochloride with levonordefrin as injection, in *NF XIV* (1975).

293

Table 21.3 Miscellaneous Local Anesthetics

Official Name	Proprietary Name	Chemical Name	Structure
dyclonine[+]	Dyclone	4'-butoxy-β-piperidinopropio-phenone	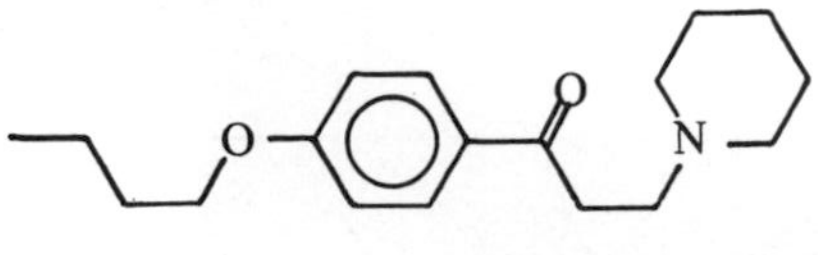
phenacaine[+]	Holocaine	*N,N'-bis*(*p*-ethoxyphenyl)-acetamidine	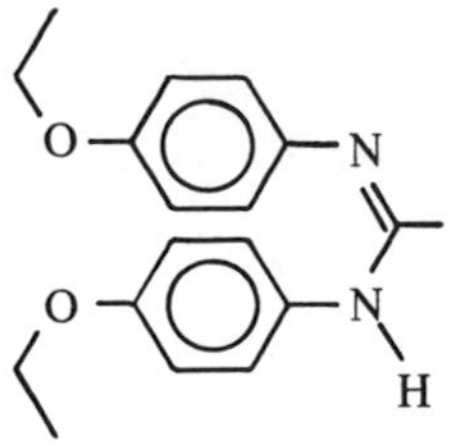
pramoxine[+] (pramocaine)	Tronothane	4-[3-(*p*-butoxyphenoxy)propyl]-morpholine	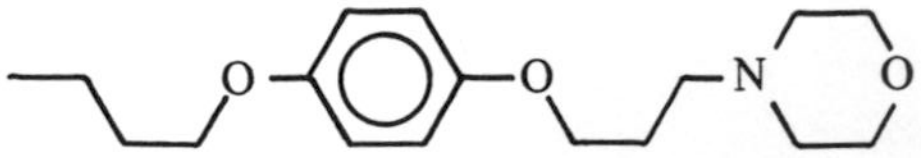

dimethisoquin[+]
(quinisocaine)

Quotane

3-butyl-1-[(2-dimethylamino)-
ethoxy]isoquinoline

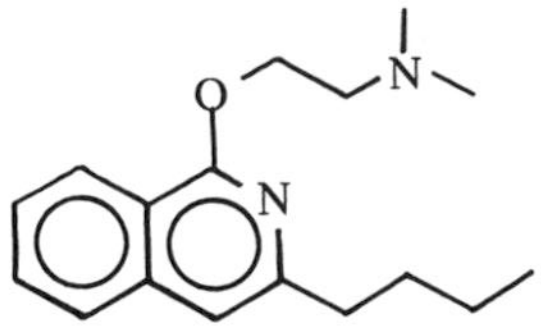

diperodon[+]

Diothane

3-(1-piperidyl)-1,2-propanediol
di(phenylurethan)

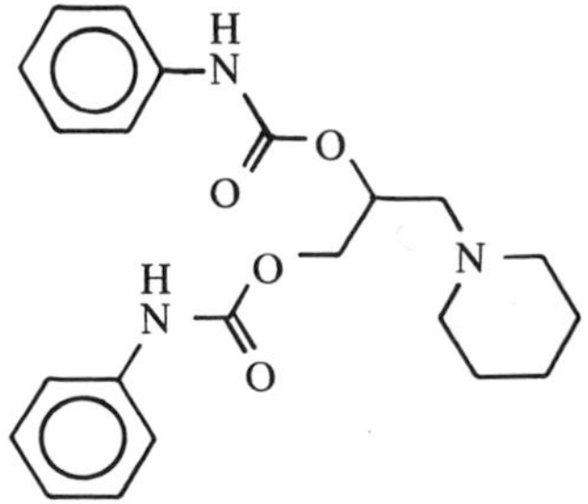

[+] In *NF XIV* (1975).

dibucaine; (b) anilides, toluidides, and 2,6-xylidides, whose prototype is lidocaine; (c) tertiary amides, represented by oxetacaine.

The first two groups can be considered as resulting from the replacement of the oxygen ester atom of ester derivatives by the isosteric NH group. These substances are therefore more stable and more resistant to hydrolysis than the parent esters.

Lidocaine

Both the free base and the hydrochloride are crystalline powders with a characteristic odor. It is the most stable of all local anesthetics known, being extremely resistant to hydrolysis. Besides its use as a local anesthetic, it is also used intravenously in the treatment of arrhythmias. Lidocaine has a more rapid onset than procaine, and its action is longer-lasting and twice as potent. In order to prolong its action as a local anesthetic, it is administered with epinephrine. The usual dosage for injection is 300 mg without epinephrine and 500 mg with epinephrine.

3. Miscellaneous Types

Besides those shown in Table 21.3, others are used, such as amolanone (Amethone), fomocaine (Panacain), propicaine (Falicain).

Dyclonine

Structurally, it is a ketone. It is used primarily for surface anesthesia. Since it is a tissue irritant, it should not be injected or infiltrated into tissues.

III. MECHANISM OF ACTION

Local anesthetics cross nerve sheaths in the nonionized form, but they interact with acceptors, located at neural membrane, in the ionized form, and thus they stabilize the potential of that membrane by blocking nervous conduction (Figure 21.1). They prevent depolarization of the neural membrane and thus block both the generation and the propagation of the nerve impulse. This depolarization results from interference with the flux of Na^+ and K^+ ions across the membrane. In consequence, the negative electrical potential necessary for the propagated discharge is not developed:

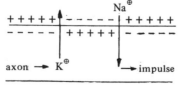

propagation of nerve impulse

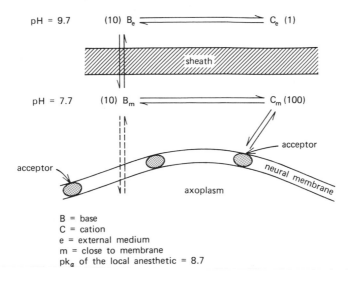

pH = 9.7   (10) B_e ⇌ C_e (1)

sheath

pH = 7.7   (10) B_m ⇌ C_m (100)

acceptor

acceptor

neural membrane

axoplasm

B = base
C = cation
e = external medium
m = close to membrane
pk_a of the local anesthetic = 8.7

Figure 21.1 Mechanism of action of local anesthetics at the molecular level. [Adapted from J. M. Ritchie and P. Greengard, *Annu. Rev. Pharmacol.*, 6, 405 (1966).]

Effects of some local anesthetics on the nerves can be antagonized by calcium; this indicates that both local anesthetics and calcium compete for the same binding sites.

Although most authors admit the existence of a *receptor* for local anesthetics, some authors prefer to apply the term *acceptor* to the site of interaction of these drugs with an organic macromolecule. *Acceptor* differs from *receptor* in terms of specificity; that is, *acceptor* usually presents less selectivity of binding, and it was not genetically designed to interact with unnatural substrates.

According to Feinstein, the acceptor of local anesthetics is an acidic phospholipid. Recent researches, in which phospholipid model membranes were used, seem to confirm that the site of action of local anesthetics in biological membranes is acidic phospholipid and that their action is inhibition of the ability of these phospholipids to bind to Ca^{2+}. Furthermore, the same investigations suggest that acidic phospholipids may be directly involved in the mechanism of excitation of biological membranes, although the results obtained do not prove necessarily that phospholipids are closely involved with the mechanism of generation of action potential.

Recently Büchi and Perlia proposed that, after penetrating into nerve membrane, the local anesthetic molecule binds to the reactive groups of the phosphatide structure as it is shown in Figure 21.2. Procaine-acceptor interaction is through (*a*) ion-dipole attraction between their ionic and polarized groups and (*b*) hydrophobic interactions between their lipophilic moieties. This

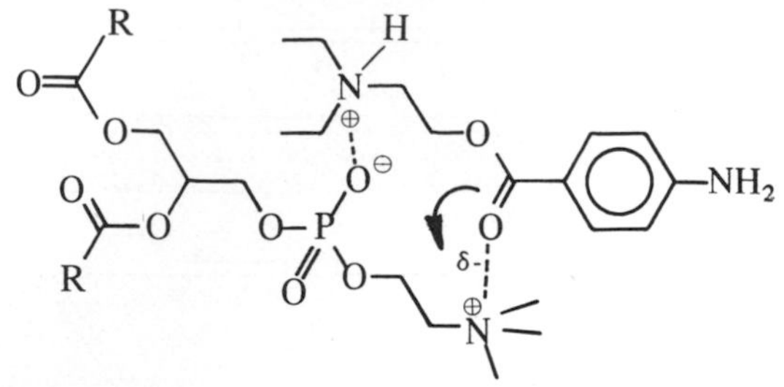

Figure 21.2 Procaine-acceptor interaction. [After Büchi and Perlia.]

interaction displaces the calcium ions and causes structural and functional alterations of the nerve membrane by obstruction and dehydration. The net result is interruption of nerve conduction and, consequently, local anesthesia.

REFERENCES

General

Symposium on Local Anaesthesia, *Br. J. Anaesth.,* **47**, Supplementary Edition, 1975.

P. Lechat, Ed., *Local Anesthetics,* Vol. I, Pergamon, Oxford, 1971.

J. Adriani, *The Pharmacology of Anesthetic Drugs,* 5th ed., Thomas, Springfield, Ill., 1970.

J. Büchi, *Grundlagen der Arzneimittelforschung und der synthetischen Arzneimittel,* Birkhauser, Basel, 1963.

Classification

F. G. Rudo and J. C. Krantz Jr., *Br. J. Anaesth.,* **46**, 181 (1974).

F. P. Luduena, *Annu. Rev. Pharmacol.,* **9**, 503 (1969).

J. Büchi *et al., Arzneim. Forsch.,* **18**, 610, 791 (1968).

J. Büchi, *Pharm. Acta Helv.,* **42**, 534 (1967).

Mechanism of Action

B. R. Fink, Ed., *Molecular Mechanisms of Anesthesia,* Raven Press, New York, 1975.

A. Korolkovas, *Rev. Bras. Clin. Ter.,* **2**, 61 (1973).

J. L. Coubeils and B. Pullman, *Mol. Pharmacol.,* **8**, 278 (1972).

W. R. Glave and C. Hansch, *J. Pharm. Sci.,* **61**, 589 (1972).

D. Papahadjopoulous, *Biochim. Biophys. Acta,* **263**, 169 (1972).

G. M. Rosen and S. Ehrenpreis, *Trans. N.Y. Acad. Sci.,* **34**, 255 (1972).

J. Büchi and K. Perlia, "The Design of Local Anesthetics," in E. J. Ariëns, Ed., *Drug Design,* Vol. III, Academic, New York, 1972, pp. 243-391.

J. M. Ritchie and P. Greengard, *Annu. Rev. Pharmacol.,* **6**, 405 (1966).

M. B. Feinstein, *J. Gen. Physiol.,* **48**, 357 (1964).

PART 4

DRUGS ACTING ON THE CARDIOVASCULAR, HEMATOPOIETIC, AND RENAL SYSTEMS

In this part, drugs that affect cardiovascular function, as well as those that act on blood vessels and the renal system, are presented. Cardiovascular agents include cardiotonic, antiarrhythmic, antihypertensive, vasodilator, and hypocholesteremic agents. They are considered in Chapter 22. Drugs acting on the blood are called hematological agents. They are comprised of antianemics, coagulants, anticoagulants, and plasma expanders. These drugs are considered in Chapter 23. Agents that affect the renal system are called diuretics. They are dealt with in Chapter 24.

REFERENCES

C. K. Friedberg, Ed., *Current Status of Drugs in Cardiovascular Disease,* Grune & Stratton, New York, 1969.

G. G. Rowe, *Annu. Rev. Pharmacol.,* 8, 95 (1968).

A. N. Brest and J. H. Moyer, Eds., *Cardiovascular Drug Therapy,* Grune & Stratton, New York, 1965.

H. J. Sanders, *Chem. Eng. News,* 43(10), 130, and 43(12), 74 (1965).

G. Fawaz, *Annu. Rev. Pharmacol.,* 3, 57 (1963).

CHAPTER 22

MISCELLANEOUS CARDIOVASCULAR AGENTS

I. INTRODUCTION

Cardiovascular diseases occupy first place as *causa mortis* in civilized countries; they are responsible for about 54% of all deaths in the United States. This group of diseases is comprised of heart diseases and diseases of blood vessels and lymphatics.

In the treatment of heart diseases, the drugs chiefly used are cardiotonics, antiarrhythmics, and antihypertensives, besides other types of drugs, such as diuretics.

For treatment of some diseases of blood vessels and lymphatics, the following measures are taken: surgery and administration of vasodilators, antihypertensive agents, and drugs for atherosclerosis, besides anticoagulants and other agents.

II. CARDIOTONICS

Cardiotonics are drugs that increase the contractile force of the heart and exert important actions on cardiac excitability, automaticity, conduction velocity, and refractory periods. They are indicated mainly for congestive heart failure, atrial fibrillation, atrial flutter, and paroxysmal atrial tachycardia.

In the treatment of tachyarrhythmia or acute ventricular failure, when prompt action is necessary, the drugs of choice are ouabain and deslanoside, because they have a rapid onset of action and can be administered intravenously. For less acute, chronic, or stabilized cardiac failure, the drugs usually prescribed are digitalis leaf or digitoxin which, given orally, exert longer action. Digoxin, by oral or intravenous routes, shows an intermediate action between these two groups. It is moderately rapid-acting and has a relatively brief duration of action. The therapeutic index of all cardiotonic drugs is roughly the same. The margin of safety is narrow: the therapeutic dose is 50 to 60% of the toxic dose.

Accumulated overdosage or prolonged use of digitalis glycosides results in digitalis intoxication, whose early symptoms are anorexia, salivation, vomiting, nausea, and diarrhea, which are not serious. However, these drugs may cause also ventricular extrasystole; this condition usually disappears by interrupting therapy, but sometimes it is necessary to administer chronotropic drugs such as epinephrine or isoproterenol. A common side effect in chronic congestive heart failure treated with digitalis and a diuretic is hypokalemia, which is corrected by oral or intravenous administration of potassium chloride and temporary suspension of digitalis therapy.

A. Classification

Cardiotonics most often used are listed in Table 22.1. Besides these, there are the following less frequently used drugs: digitalis, digitalis powdered leaf, lanatoside A (Adigal), α-acetyldigoxin (Cedigocine), β-acetyldigoxin (Novodigal), strophanthin K (Kombetin, Strofosid), proscillaridine (Proscillid), convallotoxin (Convallopan), peruvoside (Encordin), thevetin (Thevanil), gitoformate (Formiloxine), medigoxine, pengitoxin (Cordoval).

These drugs are extracted from certain species of the following families of plants: *Scrofulariaceae, Liliaceae, Apocynaceae,* and *Ranunculaceae.*

Digitalis glycosides are naturally occurring compounds consisting usually of a steroidal genin attached to a sugar moiety, most commonly D-galactose, D-glucose and L-rhamnose. Chemically, these steroids can be divided into two main classes—cardenolides and bufadienolides—represented by the prototypes digitoxigenin and bufogenin, respectively.

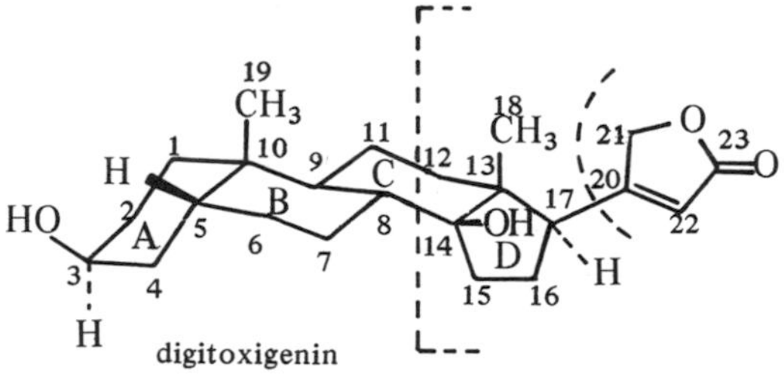

digitoxigenin

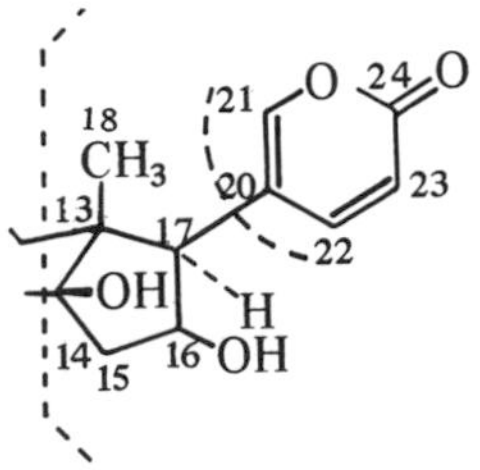

bufogenin

For cardiotonic activity three features are considered essential: (a) an unsaturated lactone ring at the 17β position: (b) a β-oxygen function at the C-14 position; (c) cis-configuration between rings A and B and between C and D. Furthermore, the sugar moiety, once considered important for transport of these drugs to the site of action, is not essential for activity.

B. Mechanism of Action

The digitalis glycosides act at the molecular level as inhibitors of membrane ATPase, and in this way they inhibit Na^+ and K^+ transport across the myocardial cellular membrane. However since Ca^{2+}, like cardiac glycosides, potentiates heart contractility and also inhibits the membrane ATPase, it is hypothesized that Ca^{2+} and digitalis act synergically on the myocardial membrane. According to Repke, cardiotonic drugs have a specific receptor located on "transport ATPase" system.

III. ANTIARRHYTHMIC DRUGS

Antiarrhythmic agents are drugs used in the treatment of disturbances of the heart rate and rhythm. An arrhythmia is any abnormality of the initiation or propagation of the heart-beat stimulus.

For treatment of atrial tachyarrhythmia the drugs of first choice are deslanoside, acetyldigitoxin, digitalis, digoxin. digitoxin, lanatoside C, and ouabain. In atrial fibrillation and flutter, digitalics are not always enough. In this disorder other drugs should usually be administered, either after or associated with digitalis glycosides, in order to restore the sinus to its normal rhythm. These drugs are quinidine, procainamide, and lidocaine.

Other valuable drugs used as antiarrhythmic agents are certain adrenergic agents (epinephrine, isoproterenol, methoxamine), beta-adrenergic blockers (propranolol), and anticonvulsants (phenytoin). This last drug is used in digitalis overdosage.

Untoward effects of antiarrhythmic agents are usually a consequence of drug overdosage. Tachyarrhythmias caused by digitalis glycosides are treated with oral or intravenous administration of potassium chloride. Excessive and sometimes even normal doses of quinidine cause headache, vertigo, nausea, diarrhea, palpitation, syncope, and other symptoms of cinchonism. Procainamide produces nausea, bitter taste, weakness, mental depression, agranulocytosis, and several other side effects, including hypersensitivity reactions. Propranolol can cause hypotension and bradycardia, which are treated with intravenous administration of isoproterenol.

Table 22.1 Cardiac Glycosides (Cardiotonics)

Official Name	Proprietary Name	Structure
digoxin*	Davoxin Lanoxin	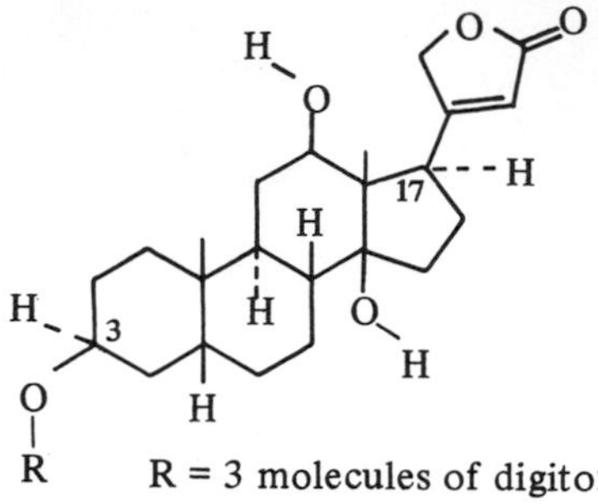 R = 3 molecules of digitoxose
lanatoside C	Cedilanid	R = 3 molecules of digitoxose + acetyl + glucose
deslanoside⁺	Cedilanid-D	R = 3 molecules of digitoxose + 1 molecule of glucose
digitoxin*	Crystodigin Digitaline Nativelle Myodigin Purodigin	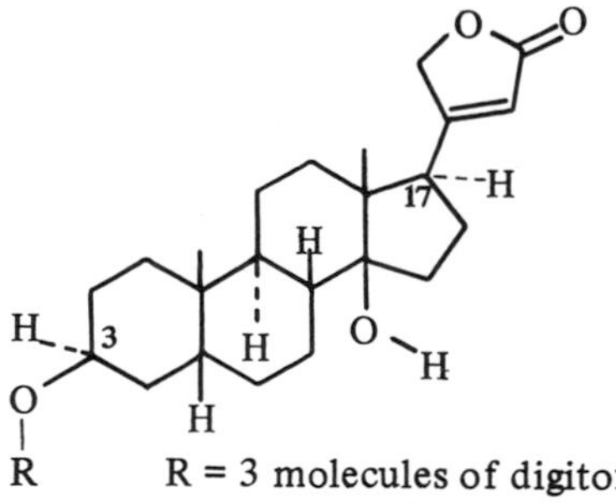 R = 3 molecules of digitoxose
acetyldigitoxin⁺	Acylanid	R = 3 molecules of digitoxose + acetyl
ouabain*		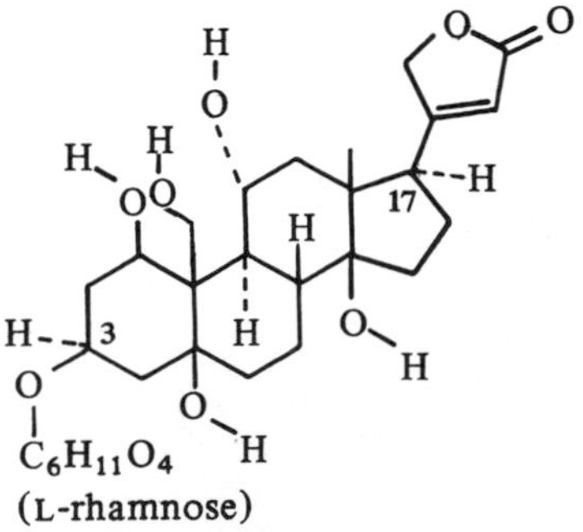

Table 22.1 Cardiac Glycosides (Cardiotonics) (continued)

Official Name	Proprietary Name	Structure
gitalin	Gitaligin	mixture of amorphous glycosides prepared from *Digitalis purpurea*

*In *USP XIX* (1975).
+In *NF XIV* (1975).

A. Classification

Drugs having antiarrhythmic activity belong to several different chemical classes. Those most commonly used, mainly as hydrochlorides, are listed in Table 22.2.

Szekeres and Papp proposed the following pharmacological classification of antiarrhythmic drugs: (a) specific antiarrhythmic agents—those acting indirectly, through the regulation of the autonomic activity of the heart; (b) nonspecific antiarrhythmic agents—those acting directly on the myocardium by altering some basic properties of the cardiac function, such as automaticity, excitability, conductivity, and refractoriness; quinidine is the prototype of these agents, and most antiarrhythmic drugs used in therapeutics belong to this class.

Structure-activity relationship studies have shown that the essential feature in almost all antiarrhythmic drugs seems to be a tertiary amine group. In fact, even diethylaminoethanol, the product of hydrolysis of procaine, has antiarrhythmic properties.

B. Mechanism of Action

The mechanism of action of cardiac glycosides was studied in the preceding section.

The antiarrhythmic action of quinidine results from its binding to myocardial cellular membrane, through its quinoline ring, and to lipids or lipoproteins; the quinuclidine moiety, with its protonated nitrogen atom, causes one or more of the following effects: (a) repulsion of cations, including sodium; (b) thickening of the membrane by hydration; (c) chelation of calcium ions. These interactions correct the pathological movement of ions at the myocardium and produce antiarrhythmic effects.

Sympathomimetic drugs, such as epinephrine and isoproterenol, owe their cardiac effects to stimulation of adrenergic receptors: α receptor stimulation raises the blood pressure and reflexly increases vagus tone and terminates a

Table 22.2 Antiarrhythmic Drugs

Official Name	Proprietary Name	Chemical Name	Structure
quinidine*	Cardioquin Quinaglute Quinidex Quinora	6-methoxy-α-(5-vinyl-2-quinuclidinyl)-4-quinolinemethanol	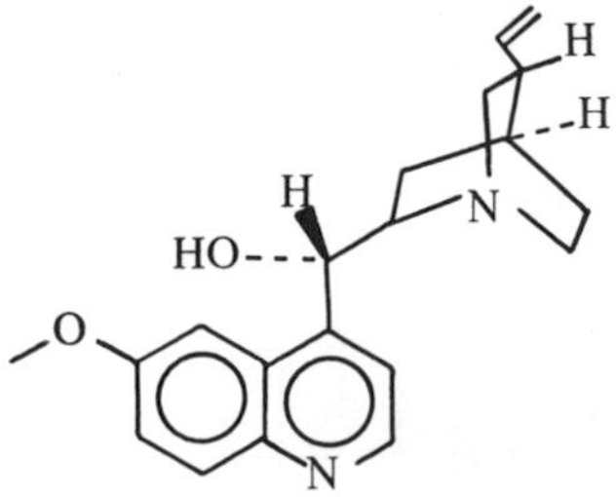
procainamide*	Pronestyl	p-amino-N-[2-(diethylamino)-ethyl] benzamide	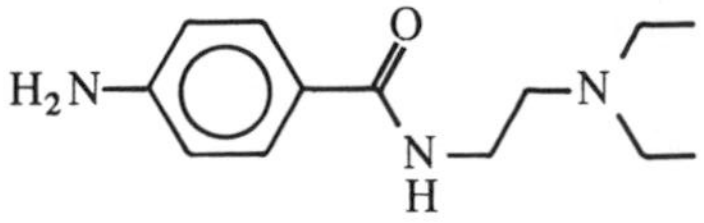
chloroprocainamide	Isoritmon	p-amino-o-chloro-N-[2-(diethyl-amino)ethyl] benzamide	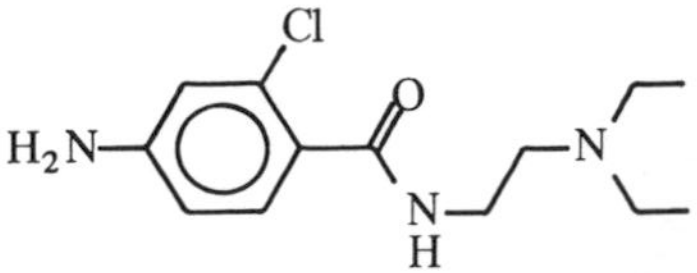

*In *USP XIX* (1975).

paroxysm of atrial tachycardia; β receptor stimulation increases automaticity and improves conduction.

The antiarrhythmic effect produced by propranolol and other β-adrenergic blocking agents results from their interaction with the cellular transport of Ca^{2+}, and not through β-adrenergic blockade.

IV. ANTIHYPERTENSIVE AGENTS

Antihypertensive agents are drugs used in the treatment of hypertension—a condition in which systolic pressure exceeds 160 mm Hg or diastolic pressure exceeds 95 mm Hg. There are two main types of hypertension: *essential* or *primary* hypertension and *secondary* hypertension.

Essential hypertension affects 10% of the population and constitutes 80% of the total cases of hypertension. It affects women more often than men. It is characterized by diastolic hypertension whose cause is unknown in about 85% of cases. There is a strong evidence that it may be a hereditary disorder. This form of hypertension may take a benign (or gradual) or malignant (or accelerated) course.

Secondary hypertension results from known causes. It may be subdivided into four different forms: renal, neurogenic, endocrine, and cardiovascular.

Renal hypertension is the most common cause of secondary hypertension. It may be related to essential hypertension. In fact, it was discovered that renin, a kidney proteolytic enzyme, after its release from storage sites, acts on a blood globulin, angiotensinogen, which eventually yields angiotensin, a potent vasopressor polypeptide, according to the following sequence:

$$\text{angiotensinogen} \xrightarrow{\text{renin}} \text{angiotensin I} \xrightarrow{\text{converting enzyme}} \text{angiotensin II}$$

(renin substrate in blood)	(a decapeptide with no pressor activity)	(an octapeptide with potent pressor activity)

Angiotensin II produces other effects which are also implicated in hypertension: (a) release of catecholamines from the adrenal medulla; (b) stimulation of secretion of aldosterone, which is involved in sodium and water retention; (c) a positive inotropic effect on the myocardium. However, there is no direct evidence that the renin-angiotensin system is the causative factor of essential hypertension.

Neurogenic hypotension is caused by lesions of vasomotor centers and those leading to an increase in cerebrospinal-fluid pressure.

Endocrine hypertension results from endocrine disorders, such as pheochromocytoma, Cushing's syndrome, and primary aldosteronism.

Cardiovascular hypertension is a consequence of constriction of the aorta and usually is treated by surgery.

Several types of drugs are presently available for treatment of hypertension, especially of the essential type, with the purpose of lowering blood pressure to normal levels, if possible, or to the lowest level that the patient will tolerate. Those mostly used in chronic hypertension, in order of preference, are methyldopa, reserpine, and hydralazine. In malignant hypertension, oral diuretics and guanethidine or methyldopa are the most useful drugs. In hypertensive crises, several drugs can be used, some by intravenous or parenteral routes, others by oral route.

Antihypertensive drugs may cause untoward effects, including among them lethargy, weakness, and orthostatic hypotension.

A. Classification

Considering that antihypertensive agents are comprised of widely varying chemical structures, they are better classified according to their mechanism of action. Thus, there are drugs acting on central or reflex mechanisms, drugs acting on the autonomic nervous system, drugs acting directly on peripheral blood vessels, and drugs acting by other mechanisms.

Drugs may be used alone, but, frequently, it is necessary to use a combination of drugs in order to maintain an effective reduction of hypertension.

1. Drugs Acting on Central or Reflex Mechanisms

This class of drugs includes sedatives, such as barbiturates, and antianxiety agents, such as meprobamate. The main drugs of this class used as antihypertensive agents are mebutamate and clonidine.

Veratrum alkaloids, such as alkavervir (Veriloid), cryptenamine (Unitensen), protoveratrines A (Protalba) and B (Provell), once widely used and acting by a central mechanism, are now considered obsolete.

Clonidine

Marketed as Catapres, this new drug is chemically 2-(2,6-dichloroanilino)-2-imidazoline:

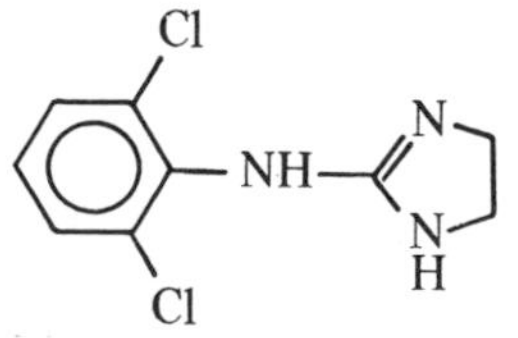

Used as hydrochloride, a water-soluble, white, crystalline powder, clonidine is useful in serious hypertensive forms and in patients who fail to respond to other therapy. Adverse reactions are marked sedation and dry mouth, which can be alleviated by taking it in conjunction with chlorthalidone.

Some analogues of clonidine having a similar mechanism of action are under study: oxazoline, xylazine.

2. Drugs Acting on the Autonomic Nervous System

This class can be subdivided into three groups: (*a*) adrenergic blocking agents; (*b*) antiadrenergic agents; (*c*) ganglionic blockers.

In the first group are included α-adrenergic blocking agents (phenoxybenzamine and phentolamine), β-adrenergic blocking agents (propranolol), and adrenergic neuron blocking agents (guanidine derivatives, including those recently introduced: guancydine and guanadrel). These drugs were studied in Chapter 18. With the exception of guanidine derivatives, they are not widely used as antihypertensive agents.

Antiadrenergic agents, the second group, are widely used. Among them are found (*a*) methyldopa and methyldopate (already studied in Chapter 19); (*b*) *Rauwolfia* alkaloids, such as *Rauwolfia serpentina* root (Raudixin, Rauval), reserpine, alseroxylon (Rauwiloid), deserpidine, rescinnamine, syrosingopine, which were studied in Chapter 10, Section II, as antipsychotic drugs; (*c*) pargyline, a MAO inhibitor, already studied in Chapter 10, Section IV, as an antidepressant drug.

A combination of dihydroergocristine-clopamide-reserpine, marketed as Brinerdine, is more active than any of its components or a combination of two of them.

In the third group, ganglionic blockers, are found several drugs, studied in Chapter 16, Section III, but only a few are used, and even so rarely, as antihypertensive agents: mecamylamine, trimethaphan, chlorisondamine, pentolinium, trimethidinium.

3. Drugs Acting Directly on Peripheral Blood Vessels

In this class are included thiazide diuretics (which are studied in Chapter 24), diazoxide, hydrazinophthalazine derivatives (hydralazine and dihydralazine), minoxidil, and sodium nitroprusside (Nipride).

Diazoxide (USP)

Marketed under the name of Hyperstat (7-chloro-3-methyl-2*H*-1,2,4-benzothiadiazine-1,1-dioxide):

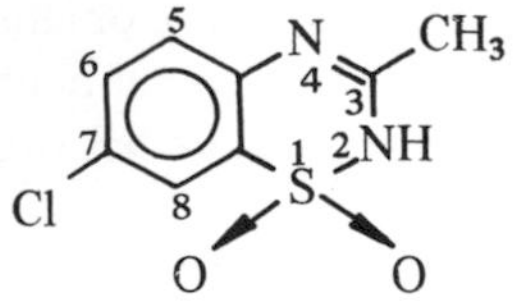

diazoxide 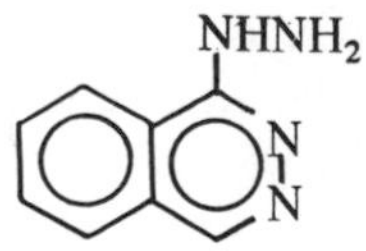   hydralazine

It is a nondiuretic thiazide, being effective in the management of most hypertensive crises and particularly indicated in cases which require a prompt reduction in blood pressure. Diazoxide is administered by intravenous route.

Hydralazine (USP)

Chemically it is 1-hydrazinophthalazine. It is given as the hydrochloride with other drugs in control of chronic hypertension.

Sodium Nitroprusside

Its structure is $Na_2Fe(CN)_5NO \cdot 2H_2O$, and it occurs as a crystalline red substance which is easily soluble in water. This inorganic compound is used when other agents have shown ineffectiveness. It has rapid action and produces short-lasting effects.

4. Drugs Acting by Unknown Mechanisms

Several drugs of miscellaneous structure and mechanism of action are here included. Among others are the following ones: (a) magnesium salts, especially thiosulfate and nicotinate; (b) bufeniode (Proclival), a derivative of bufenine; (c) tocopherylquinone (Eutrophyl, Ipotensil, Métorène), a derivative of benzoquinone; (d) amiquisin, a quinoline derivative; (e) indoramin, a derivative of indole; (f) oxylidine, a derivative of 3-quinuclidinol developed in the Soviet Union; (g) prazosin and trimazosin, quinazoline derivatives; (h) molsidomine (Morial), a sydnone derivative; (i) methylapogalantamine, used in USSR.

In the treatment of severe hypertension, combined therapy (chlorothiazide, methyldopa, and propranolol) is often used. Another combination is chlorothiazide, propranolol, and hydralazine. For emergencies in hospitals, the following drugs are often injected: trimethaphan, reserpine, and sodium nitroprusside.

B. Mechanism of Action

The several sites of action of antihypertensive agents are shown in Figure 22.1. These drugs act, therefore, by different mechanisms.

Sedatives and antianxiety drugs, such as mebutamate, owe their antihypertensive effects, if any, to general sedation they produce. As to clonidine, it

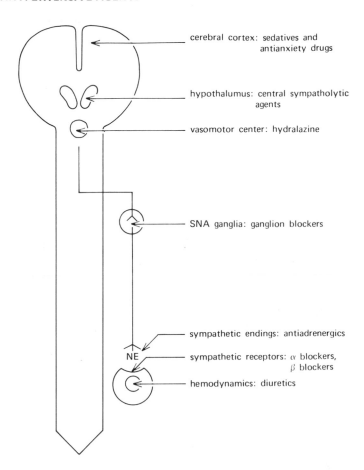

cerebral cortex: sedatives and
                antianxiety drugs

hypothalumus: central sympatholytic
                agents

vasomotor center: hydralazine

SNA ganglia: ganglion blockers

sympathetic endings: antiadrenergics

sympathetic receptors: α blockers,
                β blockers

hemodynamics: diuretics

NE

Figure 22.1 Sites of action of antihypertensive agents.

stimulates central α receptors, leading to an inhibition of sympathetic tone.

Antiadrenergic agents owe their hypotensive effects to depletion of catechol-amines (*Rauwolfia* alkaloids and *Veratrum* alkaloids), prevention of their release (bretylium), interference with their biosynthesis (methyldopa), or inhibition of metabolism (pargyline).

Adrenergic blocking agents act as hypertensive drugs owing to competitive antagonism with catecholamines for specific α- and β-adrenergic receptors or owing to adrenergic neuron blocking (guanethidine).

Ganglion blockers produce antihypertensive effects by interrupting autonomic-nervous-system transmission.

Diuretics cause hypertensive effects by decreasing plasmatic volume. Hydralazine produces peripheral vasodilation by a direct action on smooth muscle of the
arteriolar bed.

V. VASODILATOR DRUGS

Vasodilator drugs are agents that increase blood flow. They are divided into
coronary vasodilators and peripheral vasodilators. The first ones are used for
relief of pain in angina pectoris attacks. The second ones have limited clinical
value, being useful in peripheral vasoconstrictive conditions.

Coronary vasodilators are more properly called *antianginal agents,* because
their beneficial effects do not derive exclusively from vasodilation: thus,
dipyridamole (Persantine), although being a potent coronary vasodilator, is not
effective in angina pectoris. Some antianginal drugs are used in the treatment of
acute attacks, whereas others are used for long-term prophylaxis.

Peripheral vasodilators belong mainly to classes of drugs already studied:
adrenergic stimulants and adrenergic blockers. They lack selectivity; that is, none
of them can produce vasodilation in specific ischemic areas. Furthermore, when
used in large doses, they can cause postural hypotension, being contraindicated
in patients with angina pectoris, coronary thrombosis, and cerebrovascular
disease.

A. Classification

Vasodilators can be divided into two classes: (*a*) coronary vasodilators; (*b*)
peripheral vasodilators.

1. *Coronary Vasodilators*

The drugs most widely used are listed in Table 22.3. Several others, however, are
marketed. Some of them, such as sodium nitrate, papaverine, aminophylline, and
dipyridamole (Persantine), have been used for some time, but now the American
Medical Association (AMA) does not recommend them for angina pectoris,
because their effectiveness in this disease is unproved. As to dipyridamole, the
AMA further states that "carefully controlled double-blind studies have shown
that it is of no value in preventing or relieving acute attacks of angina pectoris."
Actually, in very severe ischemia, high doses of dipyridamole induce anginal
attacks.

Amyl Nitrite

It is a volatile yellowish liquid with a pungent taste and ethereal odor. Amyl

Table 22.3 Antianginal Drugs Most Widely Used

Official Name	Proprietary Name	Chemical Name	Structure
amyl nitrite[+]		isopentyl nitrite	$(CH_3)_2CHCH_2CH_2ONO$
pentaerythritol tetranitrate[+]	Pentritol Peritrate	2,2-bis(hydroxymethyl)-1,3-propanediol tetranitrate	$(CH_2ONO_2)_4\overline{C}$
nitroglycerin*	Nitroglyn Nitrospan Vasoglyn	glyceryl trinitrate	CH_2ONO_2 $-CHONO_2$ $-CH_2ONO_2$
erythrityl tetranitrate[+]	Cardilate	erythritol tetranitrate	CH_2ONO_2 $-CHONO_2$ $-CHONO_2$ $-CH_2ONO_2$
propranolol*	Inderal	1-(isopropylamino)-3-(1-naphthyloxy)-2-propanol	see Table 18.2
isosorbide dinitrate*	Isordil Sorbitrate	1,4:3,6-dianhydrosorbitol 2,5-dinitrate	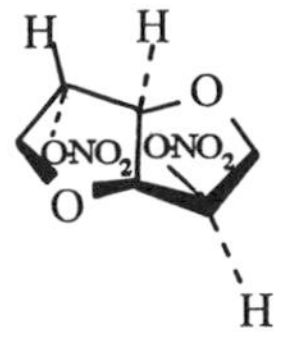

*In *USP XIX* (1975).

[+] In *NF XIV* (1975).

313

nitrite is taken only by inhalation, being used for rapid treatment of acute attack of angina pectoris.

Nitroglycerin

It is the oldest and the most effective antianginal drug. It occurs as a colorless oil with a sweet, burning taste. Nitroglycerin is taken sublingually, being the preferred drug for treatment of acute attacks and prevention of attacks of angina pectoris; by oral route, it has no reliable action. The usual dosage is 0.2 to 0.6 mg, providing complete relief in 1 to 3 minutes. Untoward effects are headache and hypotension. It is also extensively used as an explosive in dynamite.

Isosorbide Dinitrate

It occurs as a poorly water-soluble, colorless, crystalline powder. It is also a latent form of nitrate and is, therefore, long acting. Configuration is important for activity: only the *exo-endo* isomer is active: the *endo-endo* isomer (isomannide dinitrate) is practically inactive, whereas the *exo-exo* isomer (isoidide dinitrate) is more active but has not yet been introduced into therapeutics. Taken sublingually, it relieves angina pectoris attacks in 2 or 3 minutes and its action lasts 1 to 2 hours. Given by oral route, action starts in 30 minutes and lasts 4 to 6 hours. Hence, it is useful as prophylatic drug but may cause hypotensive reactions and other disturbances.

Propranolol

This drug was already studied as a β-adrenergic blocking agent. By decreasing cardiac work, its main effect in angina pectoris is reduction in the amount of oxygen required by the heart. It has several contraindications.

2. Peripheral Vasodilators

In this group are included the following types of drugs:

1. α-Adrenergic blocking agents. The most commonly used are phentolamine, phenoxybenzamine, azapetine, and tolazoline (see Chapter 18).
2. β-Adrenergic stimulants. Among them, mainly isoxsuprine and nylidrin (Table 22.4).
3. Miscellaneous agents. The principal ones are cyclandelate and papaverine (Table 22.4). Once niacin and nicotinyl alcohol were also advocated as peripheral vasodilators, but they act primarily on dermal vessels; "alleged improvement in peripheral vascular disease is probably psychological" (AMA).

B. Mechanism of Action

Vasodilators produce their effect by exerting a direct musculotrophic action. In the case of antianginal drugs such as nitrates, this results in an increase of myo-

Table 22.4 Peripheral Vasodilators

Official Name	Proprietary Name	Chemical Name	Structure
isoxsuprine[+]	Vasodilan	p-hydroxy-α-(1-[(1-methyl-2-phenoxyethyl)amino]ethyl)-benzyl alcohol	
nylidrin[+]	Arlidin	p-hydroxy-α-(1-[(1-methyl-3-phenylpropyl)amino]ethyl)-benzyl alcohol	
cyclandelate	Cyclospasmol	3,5,5-trimethylcyclohexyl mandelate	
papaverine[+] and derivatives			see Table 16.2

[+] In *NF XIV* (1975).

cardial perfusion and reduction of oxygen utilization.

Recent evidence suggests that nitroglycerin acts by increasing oxygen tension in the endocardium, due to a redistribution of blood from the epicardium. Other coronary vasodilators, such as amiodarone, dipyridamole, lidoflazine, papaverine, prenylamine, and verapamil, exert a selective dilator action on small coronary blood vessels. This action might be a consequence of their inhibition of adenosine diffusion across cellular membranes where adenosine, which is one mediator of autoregulatory control of myocardial blood flow, is deaminated.

Propranolol and similar agents act by blocking β-adrenergic receptors in the heart (where almost all adrenergic receptors are of the β type), thus preventing excessive oxygen consumption and reducing cardiac work, which allows the heart to function efficiently at a lower rate.

As to peripheral vasodilators, it has been suggested that they act by complexing with specific receptors, which were discussed in preceding chapters. However, it is doubtful that this mechanism of action may explain peripheral vasodilation activity of them all.

Cyclandelate and papaverine and derivatives act by a direct relaxant effect on smooth muscle.

VI. HYPOCHOLESTEROLEMIC AGENTS

Hypocholesterolemic agents, also called *hypolipemic, antilipemic* and *hypolipidemic* agents are drugs used in the treatment of atherosclerosis, a very common disease characterized by lipid deposit in the intima of arteries followed by calcification.

Atherosclerosis is a cause of coronary heart disease, the most prevalent form of heart diseases, which occupy the first place as *causa mortis* in the affluent Western countries, such as the United States, Finland, Canada, and England, where the death rates per 100,000 inhabitants are, respectively, 450, 390, 360, and 250.

Several factors have been implicated in causing atherosclerosis, the main ones being hypercholesterolemia, hypertension, and cigarette smoking. Other risk factors are heredity, low vital capacity, diet, lack of physical activity, obesity, stress, gout, diabetes, environmental factors. However, atherosclerosis is usually related to lipids and lipoproteins, and especially cholesterol.

Five different types of primary hyperlipoproteinemia are distinguished. Recommended treatment is shown in Table 22.5.

Drugs most commonly used as hypolipidemic agents are oxandrolone, clofibrate, dextrothyroxine, and cholestyramine resin. Several others, such as estrogens, triparanol (Mer-29), niacin, sitosterols (Cytellin), neomycin, and heparin, have been used but they are no longer recommended, owing to the untoward effects, some of which are very serious.

Table 22.5 Therapy of Hyperlipoproteinemia

Type	Diet	Drug of Choice
I	1. Restriction of fat to about 25 g per day 2. Supplementation with medium chain-length triglycerides	None effective at present
II	1. Low-cholesterol diet (less than 300 mg per day) 2. Increased intake of polyunsaturated fats	1. cholestyramine, 16 to 32 g/day 2. dextrothyroxine 3. nicotinic acid
III.	1. Balanced diet (40% of calories fat, 40% carbohydrates) 2. Reduction to ideal body weight 3. Low-cholesterol diet (less than 300 mg per day)	1. clofibrate, 2 g/day 2. oxandrolone, 5 to 7.5 mg/day 3. dextrothyroxine 4. nicotinic acid
IV	1. Reduction to ideal body weight 2. Increased intake of polyunsaturated fats 3. Modest restriction of carbohydrates	1. oxandrolone, 5 to 7.5 mg/day 2. clofibrate, 2 g/day 3. nicotinic acid, 3 to 6 g/day
V	1. Reduction to ideal body weight 2. Increased intake of protein 3. Reduction of fat to less than 70 g per day 4. Restriction of carbohydrates when possible	1. oxandrolone, 5 to 7.5 mg/day 2. nicotinic acid, 3 to 6 g/day 3. clofibrate

Source. After Levy and Fredrickson, *Postgr. Med.*, **47**, 130 (1970), modified.

Table 22.6 Hypocholesterolemic Agents

Official Name	Proprietary Name	Chemical Name	Structure
oxandrolone[+]	Anavar Lipidex	17β-hydroxy-17α-methyl-2-oxa- 5α-androstan-3-one	
clofibrate*	Atromid-S	ethyl 2-(p-chlorophenoxy)-2- methylpropionate	
dextrothyroxine[+]	Choloxin	D-3-[4-(4-hydroxy-3,5-diiodo- phenoxy)-3,5-diiodophenyl]- alanine	
cholestyramine resin*	Cuemid Questran		

*In *USP XIX* (1975).
[+] In *NF XIV* (1975).

For treatment of hypercholesterolemia it is also recommended that vegetable oils of sunflower, safflower, soya, corn, and cottonseed, which are rich in mono-unsaturated and polyunsaturated fatty acids, should be used in the diet instead of oils containing saturated fatty acids. This change of diet reduces the elevated concentrations of low-density proteins present in hyperlipoproteinemic patients.

A. Classification

Substances most commonly used as hypocholesteremic agents belong to several classes of chemical structures, as shown in Table 22.6. Several other chemical compounds are, however, either already marketed or under investigation.

B. Mechanism of Action

Theoretically, hypocholesterolemic drugs might act, as pointed out by Bach, by one of the following mechanisms:

1. Inhibition of the biosynthesis of cholesterol or its precursors. Some drugs are known to block this synthesis at certain steps. Thus, triparanol blocks it at the desmosterol level, but this results in the undesirable accumulation of desmos-terol at the aortic tissues. The most attractive step to be blocked is the one between mevalonic acid and squalene, because precursors of squalene can be disposed of by the liver.

2. Lowering of triglyceride levels and inhibition of lipid mobilization. This can be accomplished in three different ways: (*a*) inhibition of triglyceride lipase activity, which results in a decrease of the rate of triglyceride hydrolysis; (*b*) blockade of the action of free-fatty-acid-releasing hormones; (*c*) inhibition of free-fatty-acids (FFA) binding to albumin (Figure 22.2).

3. Lowering of β- and pre-β-lipoprotein levels.

4. Plaque reversal.

5. Acceleration of lipid excretion and inhibition of cholesterol absorption.

A simpler classification of hypocholesterolemic agents as to their mechanism of action is that of Braun and Rabinowitz. By bringing it up to date, we can place these agents in three main classes:

1. Agents preventing absorption of exogenous cholesterol: boxidine, linolex-amide, neomycin, polyene macrolides, β-sitosterol

2. Agents increasing excretion or degradation of cholesterol: cholestyramine, colestipol, dextrothyroxine, estrogens, heparin, pectin

3. Agents inhibiting endogenous cholesterol synthesis: azacholesterols, clofi-brate and analogues, nicotinic acid, pyridinolcarbamate, trifluperidol

As to drugs most commonly used in therapy in the United States (Table 22.6), the mechanism of action at the molecular level is not established. It is known,

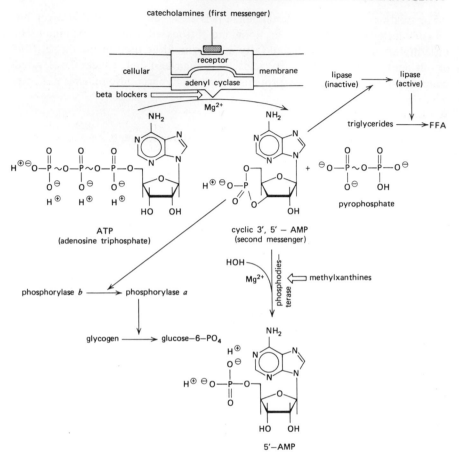

Figure 22.2 Biosynthesis of triglycerides and the site of action of beta blockers.

however, that cholestyramine binds bile acids and prevents their reabsorption; this results in an increased biotransformation of cholesterol to bile acids and a decrease of reabsorption of cholesterol.

The mechanism of action of clofibrate is not clear. It has been suggested that it can act by one or more of the following ways: inhibition of cholesterol biosynthesis, reduced synthesis of lipoproteins, inhibition of hepatic release of lipoproteins, increase of cholesterol turnover.

As to oxandrolone, it is believed that it inhibits excessive biosynthesis of triglycerides.

Dextrothyroxine appears to act by increasing the rate of oxidative elimination of cholesterol.

REFERENCES

Introduction

D. T. Mason, Ed., *Congestive Heart Failure: Mechanisms, Evaluation and Treatment,* Yorke Medical Books, New York, 1975.

K. L. Melmon, Ed., *Cardiovascular Drug Therapy,* Davis, Philadelphia, 1974.

M. Sokolow and E. Jawetz, "Heart and Great Vessels," in M. A. Krupp and M. J. Chatton, Eds., *Current Diagnosis and Treatment,* Lange Medical Publications, Los Altos, Calif., 1972, pp. 146-226.

L. Werko, *Ann. Intern. Med.,* 74, 278 (1971).

J. R. Parratt, *Progr. Med. Chem.,* 6, 11 (1969).

Cardiotonics

J. C. Allen and A. Schwartz, *Ann. N. Y. Acad. Sci.,* 242, 646 (1974).

P. Harris and L. Opie, Eds., *Calcium and the Heart,* Academic, New York, 1971.

T. Akera *et al., J. Pharmacol. Exp. Ther.,* 173, 145 (1970).

R. W. Jelliffe *et al., Ann. Intern. Med.,* 72, 453 (1970).

H. J. Nord, *Ann. Intern. Med.,* 72, 649 (1970).

C. Fisch and B. Surawicz, Eds., *Digitalis,* Grune and Stratton, New York, 1969.

F. Radt, *Steroids (VI): Cardanolides, Bufanolides, Homosteroids,* Springer, Berlin, 1969.

R. H. Thorp and L. B. Cobbin, *Cardiac Stimulant Substances,* Academic, New York, 1967.

T. C. West and N. Toda, *Annu. Rev. Pharmacol.,* 7, 145 (1967).

K. Repke, *Internist,* 7, 418 (1966).

I. M. Glynn, *Pharmacol. Rev.,* 16, 381 (1964).

Antiarrhythmic Agents

E. Gibson, *Drugs,* 7, 8 (1974).

B. N. Singh and D. E. Jewitt, *Drugs,* 7, 426 (1974).

T. Bigger, Jr., *Advan. Intern. Med.,* 18, 251 1972).

A. L. Bassett and B. F. Hoffman, *Annu. Rev. Pharmacol.,* 11, 143 (1971).

T. Lawrence, *Top. Med. Chem.,* 3, 333 (1970).

D. T. Mason *et al., Clin. Pharmacol. Ther.,* 11, 460 (1970).

L. Szekeres and J. G. Papp, *Progr. Drug Res.,* 12, 292 (1968).

Antihypertensive Agents

D. W. DuCharme, *Annu. Rep. Med. Chem.,* 9, 50 (1974).

E. D. Freis, *Clin. Pharmacol. Ther.,* 13, 627 (1972).

K. Okamato, Ed., *Spontaneous Hypertension,* Springer, Berlin, 1972.

A. J. Wohl, *Mol. Pharmacol.,* 6, 189, 195 (1970).

A. Grollman, *Clin. Pharmacol. Ther.,* 10, 755 (1969).

O. Schier and A. Marxer, *Progr. Drug Res.,* 13, 101 (1969).

A. D. Bender, *Top. Med. Chem.,* **1**, 177 (1967).

E. Schittler, Ed., *Antihypertensive Agents,* Academic, New York, 1967.

C. A. Stone and C. C. Porter, *Advan. Drug Res.,* **4**, 71 (1967).

F. Gross, Ed., *Antihypertensive Therapy: Principles and Practice,* Springer, Berlin, 1966.

W. S. Peart, *Pharmacol. Rev.,* **17**, 143 (1965).

Vasodilator Drugs

B. N. C. Prichard, *Drugs,* **7**, 55 (1974).

R. Charlier, "Antianginal Drugs," *Handbuch der experimentellen Pharmakologie,* Vol. XXXI, Springer, Berlin, 1971.

A. Bollinger, *Dent. Med. Wochenschr.,* **94**, 2347 (1969).

C. T. Dollery *et al., Clin. Pharmacol. Ther.,* **10**, 765 (1969).

J. D. Fitzgerald, *Clin. Pharmacol. Ther.,* **10**, 292 (1969).

Hypocholesterolemic Agents

D. Kritchevsky, Ed., "Hypolipidemic Agents," *Handbuch der experimentellen Pharmakologie,* Vol. XLI, Springer, Berlin, 1975.

G. F. Holland and J. N. Pereira, *Annu. Rep. Med. Chem.,* **9**, 172 (1974).

E. Ginter, *Science,* **179**, 702 (1973).

P. A. Lehmann F., *J. Med. Chem.,* **15**, 404 (1972).

G. J. Nelson, *Blood Lipids and Lipoproteins,* Wiley-Interscience, New York, 1972.

H. R. Casdorph, *Treatment of the Hyperlipidemic States,* Thomas, Springfield, Ill., 1971.

C. J. Glueck, *Metabolism,* **20**, 691 (1971).

Nutrition Society Symposium, "Evolution of Present Concepts Concerning the Action of Lipotropic Agents," *Federation Proc.,* **30**, 130-176 (1971).

Pharmacology Society Symposium, "Pharmacology of Hypolipidemic Drugs," *Federation Proc.,* **30**, 827-856 1971).

W. L. Holmes and W. M. Bortz, Eds., "Biochemistry and Pharmacology of Free Fatty Acids," *Progr. Biochem. Pharmacol.,* **6**, 1-395 (1971).

F. L. Bach, "Antilipemic Agents," in A. Burger, Ed., *Medicinal Chemistry,* 3rd ed., Wiley-Interscience, New York, 1970, pp. 1123-1171.

G. A. Braun and J. L. Rabinowitz, *Top. Med. Chem.,* **3**, 91 (1970).

R. I. Levy and D. S. Fredrickson, *Postgrad. Med.,* **47**, 130 (1970).

W. L. Bencze *et al., Progr. Drug. Res.,* **13**, 217 (1969).

R. J. Havel, *Advan. Intern. Med.,* **15**, 117 (1969).

W. L. Holmes *et al., Drugs Affecting Lipid Metabolism,* Plenum Press, New York, 1969.

F. G. Schettler and G. S. Boyd, Eds., *Atherosclerosis: Pathology, Physiology, Aetiology, Diagnosis and Clinical Management,* Elsevier, Amsterdam, 1969.

T. Shimamoto, *Acta Path. Jap.,* **19**(1), 15 (1969).

T. Shimamoto and F. Numano, *Atherogenesis,* Excerpta Medical Foundation, Amsterdam, 1969.

F. D. Gunstone, *An Introduction to the Chemistry and Biochemistry of Fatty Acids and Their Glycerides,* Chapman and Hall, London, 1969.

C. J. Miras *et al.*, Eds., "Recent Advances in Atherosclerosis," *Progr. Biochem. Pharmacol.*, 4, 1-626 (1968).

R. Paoletti, *Actual. Pharmacol.*, 21, 157 (1968).

D. S. Fredrickson *et al.*, *New Engl. J. Med.*, 276, 34, 148, 215, 273 (1967).

D. Kritchevsky *et al.*, Eds., "Drugs Affecting Lipid Metabolism," *Progr. Biochem. Pharmacol.*, 2, 1-499; 3, 1-511 (1967).

R. Paoletti, Ed., *Lipid Pharmacology*, Academic, New York, 1964.

D. Kritchevsky, *Cholesterol*, Wiley, New York, 1958.

CHAPTER 23

HEMATOLOGICAL AGENTS

I. INTRODUCTION

Hematological agents are substances that act on blood. According to the effect produced or sought, they are named antianemics, hemostatics, anticoagulants, and plasma substitutes.

II. ANTIANEMIC AGENTS

Antianemic agents are drugs used in the treatment of deficiency anemias. These anemias are caused by inadequate levels in the body of specific chemical substances, especially iron, vitamin B_{12} or folic acid, that are essential for normal maturation of erythrocytes. Antianemic agents are, therefore, an example of replacement therapy.

There are three main types of anemias: (a) normocytic anemia, caused by loss, destruction, or defective formation of blood; (b) hypochromic microcytic anemia, resulting from deficiency of iron; (c) macrocytic or megaloblastic anemia, a result of deficiency of vitamin B_{12} or folic acid. Hematinic agents are useful for the last two types of anemias.

Iron is a normal constituent of the human body, where it is widely distributed, both in inorganic and organic forms, in the total amount of about 3.5 to 4.5 g; 70% of this amount is functional or essential iron, thus called because it plays physiological roles and occurs in hemoglobin, myoglobin, and intracellular iron-containing enzymes, and 30% is nonessential or storage iron, which occurs in ferratin and hemosiderin. Iron is best absorbed in its ferrous state, but it is complexed to protein or heme as ferric iron.

Heme is the iron-containing porphyrin unit of hemoglobin. Considering that the contents of iron in hemoglobin is 0.33%, loss of 100 ml of blood means that 50 mg of iron is lost. This loss may occur in bleeding and also in cases of infestations by worms. Major causes of iron-deficiency anemias are, however, inadequate intake, excessive loss, or, although seldom, inadequate absorption of iron. They are treated with iron-containing preparations.

Megaloblastic anemias are caused predominantly by deficiency of vitamin

B_{12} or folates, which results from several factors, such as inadequate ingestion or absorption of these substances. These anemias are treated with either vitamin B_{12} or folates, depending on which substance is causing anemia.

Before treatment, an accurate diagnosis should be made in order to ascertain deficiency of which substance is causing the anemic state. Antianemic agents are not interchangeable. Thus deficiency of vitamin B_{12} should not be treated with folic acid and vice versa.

Iron preparations are contraindicated in cases of hemochromatosis or hemosiderosis. Oral preparations can cause gastrointestinal disturbances, especially in pregnant women. Parenteral preparations can cause fever, nausea and vomiting, urticaria, arthralgias, and other untoward reactions, including fatal anaphylaxis. Overdoses of iron are harmful and can produce severe reactions, even death.

Vitamin B_{12} preparations are clinically useful in the treatment of pernicious anemia, nutritional macrocytic anemia, and some cases of tropical or nontropical sprue, that is, in states of vitamin B_{12} deficiency. Although some liver preparations, which are a source of vitamin B_{12}, are marketed—for instance, liver injection (Pernaemon)—they are "outmoded and should no longer be used," according to the American Medical Association. Vitamin B_{12} preparations are preferably administered by intramuscular or deep subcutaneous injection. Oral preparations during prolonged period of time in pernicious anemia with neurologic complications may cause permanent damage. Irreversible neurologic damage develops if a patient does not receive vitamin B_{12} at regular intervals for the rest of his life.

Folates are indicated in cases of folic acid deficiency. Folates should never be used to treat pernicious anemia, because they mask diagnosis and produce irreversible neurologic damage. They are rapidly and preferentially absorbed from the small intestine; for this reason the oral route is the one of choice except when the folate deficiency is caused by poor absorption. No toxic reactions have been reported.

A. Classification

Antianemic agents may be divided into two main classes: (*a*) hematinic agents; (*b*) agents for megaloblastic anemias.

1. Hematinic Agents

Two groups of hematinic agents are available:

1. Oral preparations. They are ferrous salts of inorganic and organic acids: ferric ammonium citrate (now considered obsolete), ferrocholinate (Ferrolip), ferroglycine sulfate (Kelferon), ferrous ascorbate (Ascofer), ferrous fumarate, ferrous gluconate, ferrous succinate (Ferromyn), ferrous sulfate, sodium ferredetate (Ferrostrene, Sytron).

2. Parenteral preparations. They are complex salts and ferrous and ferric chelates: dextriferron, ferrotseron, iron dextran, iron-polysaccharide complex (Feraplex), iron sorbitex, polyferose (Jefron), sodium ferric complex of pantoic acid (Ferronascin), saccharated iron oxide (Proferrin).

Hematinic agents most widely used are listed in Table 23.1.

Table 23.1 Hematinic Agents

Official Name	Proprietary Name	Chemical Name	Structure
ferrous sulfate*+	Feosol	hepthydrated ferrous sulfate	$FeSO_4 \cdot 7H_2O$
ferrous fumarate*	Ircon		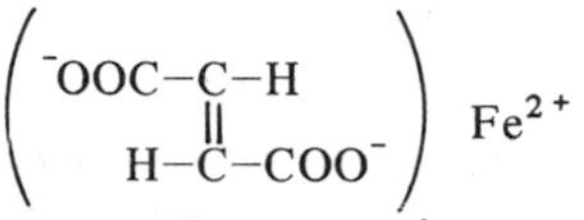 $\left(\begin{array}{c} {}^-OOC-\overset{\text{C}}{\underset{\parallel}{}}-H \\ H-C-COO^- \end{array}\right) Fe^{2+}$
ferrous gluconate+	Fergon	dihydrated ferrous gluconate	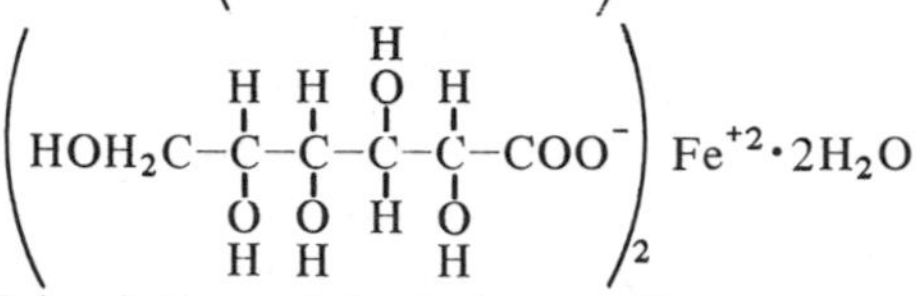 $\left(HOH_2C-\overset{H}{\underset{O}{C}}-\overset{H}{\underset{O}{C}}-\overset{O}{\underset{H}{C}}-\overset{H}{\underset{O}{C}}-COO^-\right)_2 Fe^{+2} \cdot 2H_2O$
iron dextran injection*	Imferon	sterile colloidal solution of ferric hydroxide complexed with partially hydrolyzed dextran	
iron sorbitex injection+	Jectofer	complex of iron, sorbitol, and citric acid	
dextriferron injection	Astrafer	sterile colloidal solution of ferric hydroxide complexed with partially hydrolyzed dextrin	

*In *USP XIX* (1975).
+In *NF XIV* (1975).

2. Agents for Megaloblastic Anemias

For treatment of megaloblastic anemias two groups of agents are used:

1. Vitamin B_{12} and derivatives: cyanocobalamin (Berubigen, Redisol, Rubramin, Sytobex), hydroxocobalamin (AlphaRedisol) (Figure 23.1)
2. Folates: folic acid (Folvite), calcium folinate (leucoverin, Leucoverin)

Cyanocobalamin (USP)

Chemically it is (5,6-dimethylbenzimidazolyl)cyanocobamide (Fig. 23.1), in accordance with a nomenclature especially developed for this very complex molecule. It is a neutral cobalt coordination complex containing the inorganic ion tightly bound to six octahedrally coordinated ligands. It occurs as a highly hygroscopic, red, odorless, tasteless, crystalline powder. This cobalt-containing

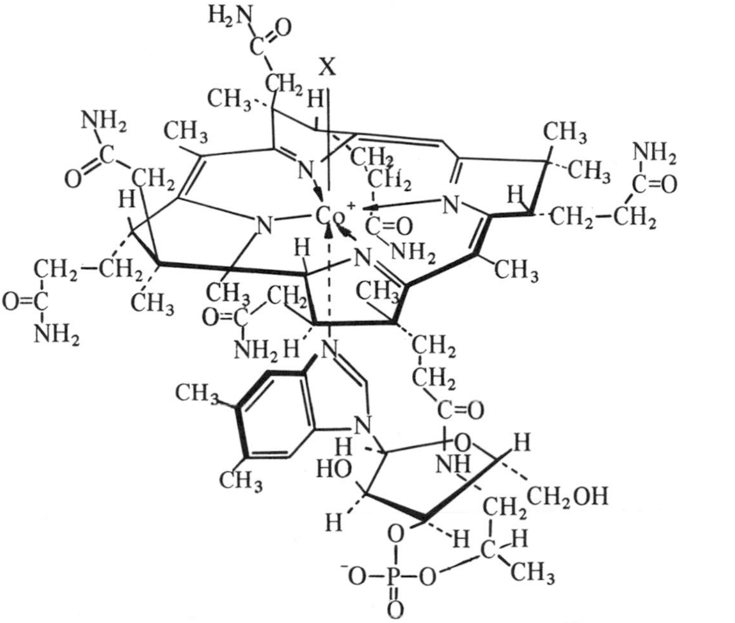

X	Name
CN	cyanocobalamin (vitamin B_{12})
OH	hydroxocobalamin (vitamin B_{12a})
H_2O	aquocobalamin (vitamin B_{12b})
NO_2	nitritocobalamin (vitamin B_{12c})
5'-deoxyadenosyl	cobamamide (coenzyme B_{12})

Figure 23.1 Chemical structure of cobalamins

vitamin B_{12} is either isolated from liver or produced by fermentation induced by certain microorganisms, especially of the genus *Streptomyces* (*griseus, olivaceus* and *aureofaciens*). Structurally, it is a substituted corrin (corrin is the name given to the macroring system of cyanocobalamin devoid of peripheral substituents). Because cobalt shares in the resonance of the corrin ring system, this atom is held very firmly. Although it is a weak polybasic acid, it may be considered an essentially neutral internal salt.

The only established clinical use of cyanocobalamin is in the treatment of confirmed vitamin B_{12} deficiency states, in which it is the drug of choice. It has no therapeutic value in infectious hepatitis, multiple sclerosis, neuropathies, poor appetite, malnutrition, allergies, thyrotoxicosis, aging, sterility, and other conditions for which has been incorrectly prescribed. No adverse reactions have been reported. It is administered by several routes.

Folic acid (USP)

It corresponds chemically to pteroylmonoglutamic acid, or *N*-[*p*([2-amino-4-

hydroxy-6-pteridinyl)methyl] amino)-benzoyl] glutamic acid. It occurs as a
slightly water-soluble, odorless, tasteless, yellow or yellowish-orange powder.
The folate sodium (Folvite solution) is water-soluble. Aqueous solutions of the
free acid as well as of the sodium salt are stable to air and can be sterilized
by autoclaving. Folic acid is decomposed readily by sunlight or ultraviolet light.

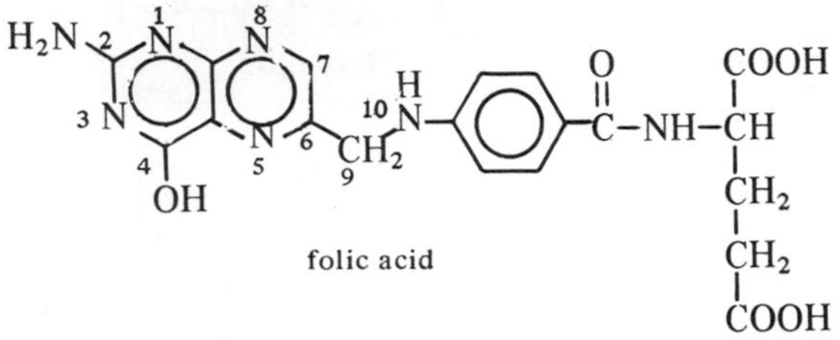

folic acid

Folic acid and its sodium salts are used specifically for control of folic acid
deficiency states. The usual dosage by intramuscular, oral, or subcutaneous
route is 5 to 15 mg daily.

3. Mixtures

Several antianemic mixtures are marketed, containing two or more of the
following drugs or groups of drugs: iron salts, all vitamins including cyanocobalamin and folic acid, all the trace minerals, liver extract, corticotropin, and
intrinsic factor. According to the American Medical Association, "their use
for treating anemias should be discouraged. They are an added expense to the
patient when a completely adequate and less expensive regimen is available."

B. Mechanism of Action

Hematinic agents are drugs which replace specific chemical substances that
erythrocytes require for their normal maturation. They act, therefore, as those
substances.

Iron preparations supply the iron needed for the normal physiological processes
of the body. It is incorporated into hemoglobin and myoglobin, which are
involved in oxygen transport and cellular respiration, respectively, and other
macromolecules which require iron.

Cyanocobalamin and hydroxocobalamin replace natural cobalamins, which
are essential for normal growth and nutrition and for performance of several
physiological processes, such as protein and DNA synthesis.

Folates replace folic acid in its metabolic functions. *In vivo* folic acid is
enzymatically reduced to derivatives of tetrahydrofolic acid, the coenzyme
forms that act as carriers of one-carbon units (formyl, formimino, hydroxy-

methyl, methyl) in several methylations, especially in the biosynthesis of methionine, choline, serine, histidine, purines, and pyrimidines, and, therefore, in DNA and RNA synthesis.

III. HEMOSTATICS

Hemostatics are substances used in hemorrhagic diseases to stop abnormal bleeding by causing hemostasis. Because coagulation or clot formation is the most complex and probably the most important phase of hemostasis, hemostatics are also called *coagulants*.

Several theories were advanced to explain the mechanism of hemostasis. One of the most accepted is the *enzymatic cascade* hypothesis proposed by Macfarlane. According to it, 13 factors are involved in hemostasis. Their intervention is schematically represented in Figure 23.2. It is seen that thrombin, the enzyme

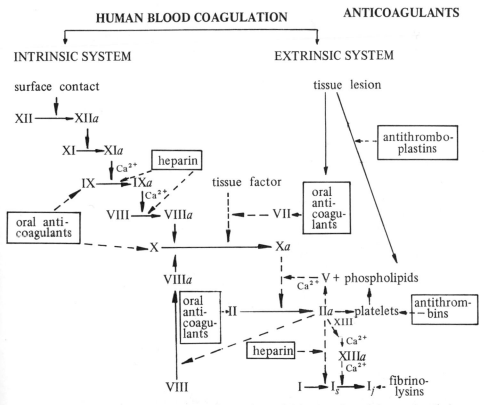

Figure 23.2 Schematic representation of the sequential intervention of factors involved in blood coagulation and of the site of action of anticoagulants. The letter *a* after the number of a factor indicates that it is the activated form. The symbol II*a* corresponds to thrombin, the I_s refers to soluble fibrinogen, and I_j means insoluble fibrinogen. The thick continuous lines indicate transformations, those discontinuous refer to interactions.

involved in clotting process, can be formed from prothrombin by two processes: intrinsic and extrinsic processes. In the intrinsic or intravascular mechanism, or spontaneous process, which is free of tissue or extravascular thromboplastin, thrombin is synthesized by a sequence of enzymatic reactions involving factors XII, XI, IX, VIII, X, V, and II, phospholipids and calcium ions. In the extrinsic or extravascular mechanism, factors VII, X, and V and calcium ions participate. Thrombin thus formed catalyzes the conversion of fibrinogen into fibrin. In this reaction each molecule of fibrinogen yields 4 fibrinopeptides and fibrin, which polymerizes in orderly disposed fibers, forming a soft clot. By the catalytic action again of thrombin, factor XIII is activated, yielding fibrinoligase (FSF*), a transamidating enzyme which cross-links fibrin, thereby forming the hard clot of fibrin (Figure 23.3).

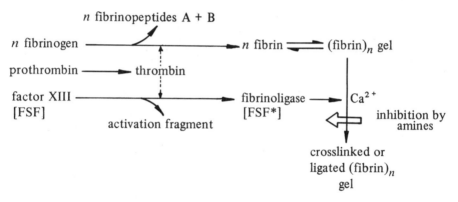

Figure 23.3 Schematic representation of the clotting mechanism in vertebrates. Thrombin, a hydrolytic enzyme, exerts dual catalytic action. It is involved in the transformation of fibrinogen to fibrin, as well as in the conversion of fibrin-stabilizing factor to fibrinoligase, a transamidating enzyme which cross-links fibrin, yielding the hard clot. [After L. Lorand, *Ann. N. Y. Acad. Sci.*, 202, 6 (1972).]

Deficiency of one or other clotting factors prevents hemostasis and causes certain diseases. These diseases are treated by systemic hemostatics. Surface bleeding and capillary oozing are controlled by topical hemostatics.

Adverse reactions to hemostatics vary greatly, according to the agent used.

A. Classification

Hemostatics most widely used (Table 23.2) are divided in two main classes:

1. Systemic hemostatics: blood components, heparin antagonists, vitamins K and analogues, and miscellaneous agents

2. Topical hemostatics: absorbable gelatin film (Gelfilm), absorbable gelatin

Table 23.2 Systemic Hemostatics

Official Name	Proprietary Name	Chemical Name	Structure
phytonadione* (phytomenadione) menadione[+]	Konakion Mephyton	2-methyl-3-phytyl-1, 4-naphthoquinone 2-methyl-1,4-naphthoquinone	see text
menadione sodium bisulfite[+]	Hykinone	2-methyl-1,4-naphthoquinone sodium bisullite	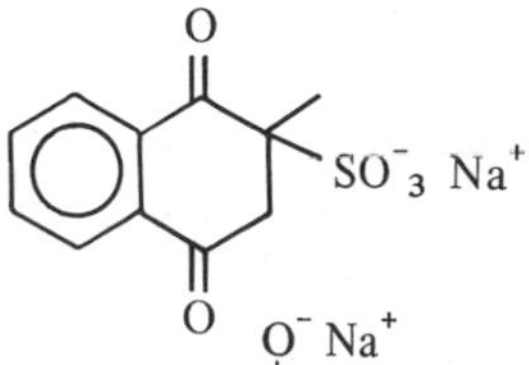
menadiol sodium diphosphate[+]	Kappadion Synkayvite	tetrasodium 2-methylnaphthohydroquinone diphosphate	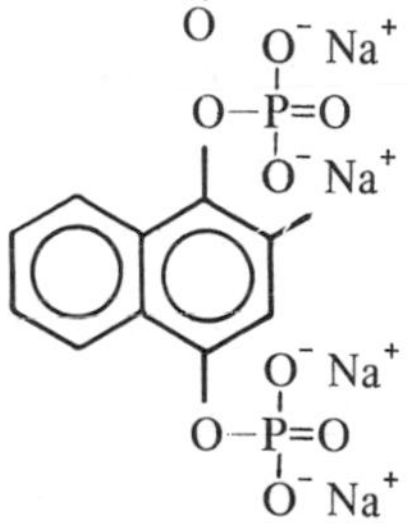
vitamin K_5	Synkamin	2-methyl-4-amino--1-naphthol	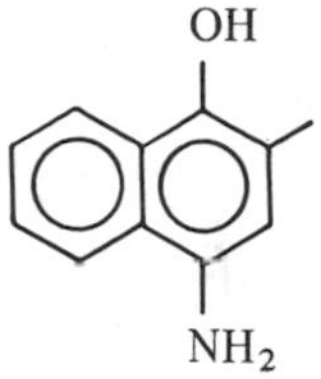
aminocaproic acid[+]	Amicar	6-aminohexanoic acid	$H_2N(CH_2)_5COOH$

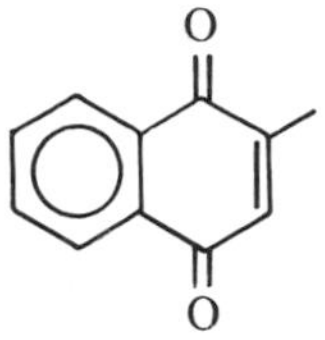

*In *USP XIX* (1975).
[+]In *NF XIV* (1975).

sponge (Gelfoam), ferric chloride, fibrin foam, oxidized cellulose (Oxycel, Surgicel), thrombin, thromboplastin

In some countries vitamin P, as well as its derivatives and analogues, are marketed as capillary protecting agents.

1. *Systemic Hemostatics*

Systemic hemostatics are administered systemically. By taking into account the source of origin, mechanism of action, or chemical structure, the following groups can be distinguished:

1. Blood components: antihemophilic factor (Hemofil, Humafac), cryoprecipitated antihemophilic factor, factor IX complex (Proplex), fibrinogen (Parenogen)
2. Heparin antagonists: protamine sulfate
3. Vitamins K and analogues: acetomenaphtone (Davitamon K), menadiol sodium diphosphate, menadiol sodium sulfate, menadione, menadione sodium bisulfite, menadoxime, naftazone, phytonadione, vitamin K_5
4. Miscellaneous agents: aminocaproic acid, p-(aminomethyl)benzoic acid (Gumbix), calcium alginate (Calgitex), conjugated estrogens (Premarin), carbazochrome salicylate (Adrenosem), etamsvlate (Dicynene), naphthazone (Maemostop), cis-pentabenzyloxy quercetine (Parietrope), tranexamic acid (Transamin), and some snake venoms

a. *Blood Components*

Blood components are prepared from fresh human plasma and are used instead of whole blood. The most common adverse reaction is acute viral hepatitis.

b. *Heparin Antagonists*

Heparin antagonists are strongly basic and, as positively charged substances, form complexes with the negatively charged heparin or heparinoids, thus neutralizing them. The only one still used is protamine sulfate, a basic protein isolated from sperm of some fishes (species of the genera *Salmo, Trutta, Onchorhynchus*). Years ago hexadimetrine bromide and tolonium chloride were also used but now are obsolete and abandoned.

c. *Vitamins K and Analogues*

Vitamins K and analogues are structurally modified fat-soluble naphthoquinones. The absolute configuration of vitamin K_1 or phytomenadione, called phytonadione in the United States, was determined by Mayer and associates in 1964; it has two asymmetric centers, $7'$ and $11'$ carbons.

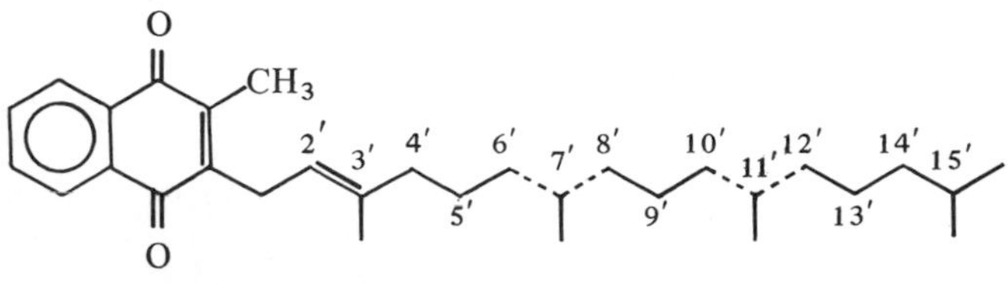

This group of hemostatics is used in the treatment of hypoprothrombinemia, which occurs rarely, because vitamin K is largely distributed in foods, besides being produced by bacteria of the colon. However, poor absorption and some other conditions may result in an inadequate supply of vitamin K to the liver.

Adverse reactions are rare. In infants, however, vitamins K and analogues can cause severe hyperbilirubinemia and kernicterus, hemoglobinuria, and hemolytic anemia. Preparations of this group, with the exception of phytonadione and vitamin K_5, cause hemolysis of red cells of certain patients, especially of infants, with deficiency of glucose-6-phosphate dehydrogenase (G6PD).

Although recent synthetic analogues of vitamin K are more economical, phytonadione is still the most widely used, because it produces a more rapid and more prolonged physiological action.

d. Miscellaneous Hemostatics

Tranexamic Acid

It occurs as a water-soluble, white, crystalline powder. Of the two geometric isomers, only the trans is active. The antifibrinolytic activity of this drug, marketed in Japan and Europe, is several times higher than that of aminocaproic acid. Its side effects are thrombotic and nonthrombotic complications and even intravascular clotting.

tranexamic acid

2. Topical Hemostatics

Topical or local hemostatics are substances applied to the oozing surface. They control oozing only from minute vessels but not bleeding from arteries or veins.

B. Mechanism of Action

The hemostatic action of fibrinogen is due to its transformation to insoluble fibrin when added to a solution containing thrombin.

Protamine sulfate acts as an antidote for heparin by complexation with heparin or heparinoids as a result of its strongly basic nature.

As to vitamins K and analogues, although the exact mechanism of action of these hemostatics is still not known, it is certain that they act on one of the final steps in the production of clotting factor subsequent to polypeptide chain synthesis on the ribosomal level. At this genetic level they induce the synthesis of prothrombin and related clotting factors, apparently by stimulating mRNA formation.

Aminocaproic acid acts as a potent competitive inhibitor of profibrinolysin activator and a weak plasmin inhibitor. Tranexamic acid and p-(aminomethyl)-benzoic acid act by a similar mechanism: their antifibrinolytic action results from their potent inhibition of plasmin.

Carbazochrome, etamsylate, naphthazone, and sodium naphthyonate act by still unknown mechanisms. However, it is very likely that they are capillary protectors rather than hemostatics.

Local hemostatics act *in situ* by various mechanisms. Thrombin clots whole blood, plasma, or a solution of fibrinogen as result of its catalytic action shown in Figure 23.3. Oxidized cellulose forms an artificial clot without entering into normal physiological coagulation process. Absorbable gelatin acts as either a sponge or a film in order to prevent bleeding. Ferric chloride is a strong astringent, and this explains its antihemorrhagic action in small bleeding.

IV. ANTICOAGULANTS

Anticoagulants are agents that prolong the coagulation time of blood. They are used in several cardiovascular disorders, such as rheumatic heart disease, myocardial infarction, pulmonary embolism, cerebral vascular disease, peripheral vascular disease, venous thrombosis. They are also useful in special cases in ophthalmology, otolaryngology, obstetrics, and orthopedic surgery. The anticoagulants presently used act by inhibiting either the action or the formation of one or more clotting factors studied in the previous section of this chapter.

In this section are included antithrombic agents, which are substances that have the property of lysing or dissolving preformed fibrin or preventing the formation of thrombus and, therefore, used in the treatment of thrombosis.

Adverse reactions are caused by anticoagulants but rarely. However, these substances are absolutely contraindicated in some special diseases, such as hemophilia, hemorrhagic blood dyscrasias, active ulcerative disease of the gastrointestinal tract, subacute bacterial endocarditis, severe kidney or liver disease, open ulcerative wounds, severe hypertension. Since they cross the placental barrier, they should be used carefully during pregnancy. Caution should be taken when given to lactating mothers.

A. Classification

Anticoagulants are classified according to various criteria: mode of action, mode of administration, and chemical structure. Two large groups are usually distinguished: (*a*) the direct-acting, heparin and heparinoids, administered parenterally; (*b*) the indirect-acting, or oral anticoagulants. In the first class are also included sodium edetate, citrates, and heparin, which are added to whole blood in order to store it. The second class is comprised of coumarin and indandione derivates and miscellaneous agents.

By taking chemical structure as the criterion, anticoagulants may be divided

Table 23.3 Anticoagulants

Official Name	Proprietary Name	Chemical Name	Structure
heparin*	•••	•••	see text
dicumarol*	Dicumarol	3,3'-methylenebis(4-hydroxycoumarin)	see text (II, R = H)
ethyl biscoumacetate	Tromexan	ethyl bis(4-hydroxy-2-oxo-2H-1-benzo-pyran-3-yl)acetate	see text (II, R = CO$_2$Et)
warfarin*+	Warcoumin	3-(α-acetonylbenzyl)-4-hydroxycoumarin	see text (I)
acenocoumarol+	Sintrom	3-(α-acetonyl-4-nitro-benzyl)-4-hydroxy-coumarin	see text (I, 4-nitrobenzyl replaces benzyl)
~~phenprocoumon~~+	Liquamar	3-(α-ethylbenzyl)-4-hydroxycoumarin	see text (I, ethyl replaces acetonyl)
phenindione+	Hedulin Danilone	2-phenyl-1,3-indandione	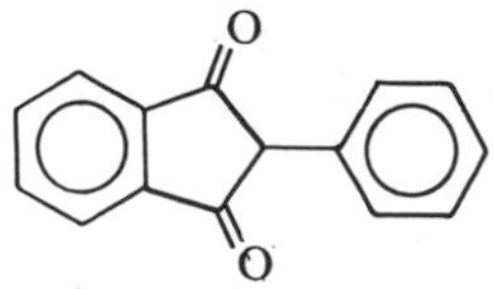
diphenadione+	Dipaxin	2-(diphenyl-acetyl)1,3-indandione	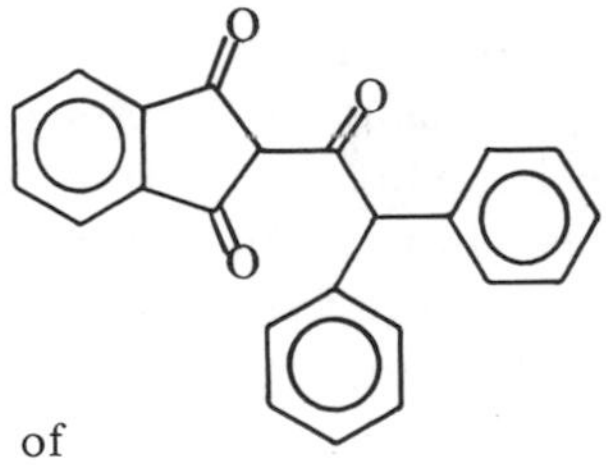
sodium citrate solution*	•••	sterile solution of sodium citrate in water for injection	
citrate dextrose solution*	•••	sterile solution of citric acid, sodium citrate, and dextrose in water for injection	
citrate phosphate dextrose solution*	•••	sterile solution of citric acid, sodium citrate, sodium diphosphate, and dextrose in water for injection	

*In *USP XIX* (1975).
+In *NF XIV* (1975).

in the following five classes:

1. Heparin and heparinoids
2. Coumarin derivatives
3. Indandione derivatives
4. Miscellaneous agents: magnesium citrate, salts of the rare earth metals (especially of cerium and neodymium: Ceramil, Isothrodym, and Phlogodyn), snake venoms
5. Anticoagulants for storage of whole blood: citrate dextrose, citrate phosphate dextrose, sodium citrate, sodium edetate, heparin, polyanethole sulfate (Liquiod)

Anticoagulants most widely used are listed in Table 23.3.

1. Heparin and Heparinoids

Heparin is a normal component of the body. It acts both *in vitro* and *in vivo*. It is formed in mast cells but is found especially in liver and lungs, although in small amounts: 10 tons of bovine liver yield only 1 kg of heparin. Chemically, it is characterized by the sulfaminic linkage, which is otherwise unknown in nature. By hydrolysis it yields several fractions, some of which have also anticoagulant action.

Heparin can be isolated from different sources, but it varies accordingly. Thus, pork heparin is not identical to dog heparin and beef heparin. Furthermore, two fractions, α-heparin and β-heparin, have been eletrophoretically separated from pure heparin.

Heparinoids are obtained from heparin through suitable chemical reactions, such as oxidation, mild hydrolysis, condensation with appropriate reagents. Another way to obtain heparinoids consists in exhaustively esterifying either some naturally occurring mucopolysaccharides or synthetic polymers with chlorosulfonic acid. Other heparinoids are thus called, owing to their structural resemblance with heparin.

Two structural features are associated with anticoagulant activity of this class of drugs: (*a*) the degree of dissociation of all the ionizable groups; (*b*) the molecular size and shape.

Owing to their strongly acidic sulfate groups, heparin and heparinoids are readily neutralized by basic compounds, such as protamine sulfate, losing thereby their anticoagulant activity.

Heparin

Commercially isolated from lung and liver of mammals, it is a sulfated mucopolysaccharide, consisting of alternating D-glucuronic acid and 2-amino-2-deoxy-D-glucose moieties, forming disaccharide units:

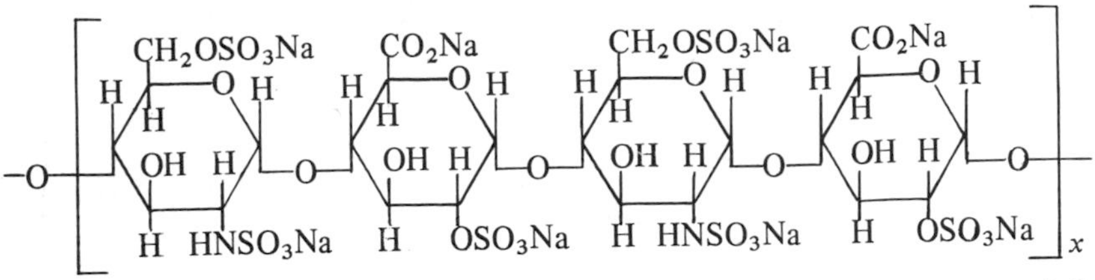

Heparin is marketed as both sodium salt and solution; the latter is used for storing blood. Heparin has a direct and immediate action. Therefore, when a rapid anticoagulant effect is needed, heparin is the drug of choice. It is given by parenteral or intravenous routes; orally or sublingually, it is inactive.

Protamine sulfate (USP) possesses an anticoagulant action but, paradoxically, it counteracts the action of heparin and thus is used as an antidote for heparin. It is injected intravenously.

2. *Coumarin Derivatives*

This class consists of 4-hydroxycoumarins. Their keto isomers are closely related to vitamin K. Apparently, for anticoagulant activity the essential requirements are a 4-hydroxycoumarin residue and a hydrogen or a hydrocarbon substituent in the 3 position. They act only *in vivo*, not *in vitro*. Some are monocoumarins (phenprocoumon, warfarin, acenocoumarol, Diocoumine), whereas others are dicoumarins (bishydroxycoumarin, coumetarol, cyclocumarol, ethyl biscoumacetate, Thioperan).

They are administered by oral route, with the exception of sodium warfarin, which can also be given intravenously or intramuscularly.

As to the duration of action, they are (*a*) intermediate acting: acenocoumarol, warfarin, and bishydroxycoumarin; (*b*) long acting: phenprocoumon. All of them, however, are almost devoid of adverse reactions. Their anticoagulant effect is overcome by phytonadione.

Nuclear magnetic resonance and infrared spectra indicate that coumarins and dicoumarols form intramolecular hydrogen bonds. Their structures are, therefore, as shown:

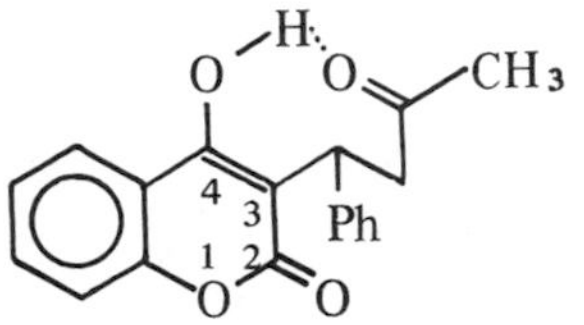

I 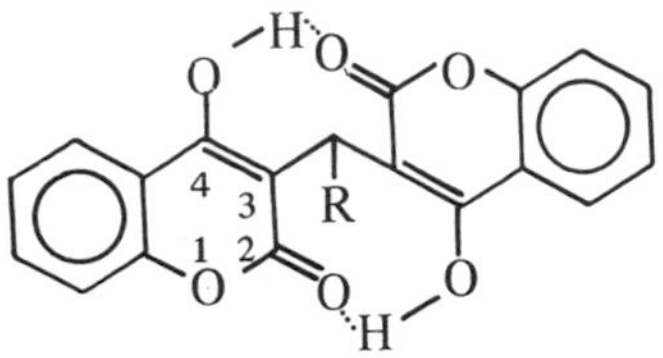   II

Untoward effects are rare and not serious: alopecia, dermatitis, urticaria, gastrointestinal irritation, leukopenia. They may cause hemorrhagic complications; in this case phytonadione should be given.

3. *Indandione Derivatives*

Their pharmacological and therapeutic action is very similar to that of coumarin derivatives. They are administered orally. Among those mostly used, phenindione is short acting, being quickly absorbed and eliminated, whereas diphenadione and anisindione are long acting, because they are slowly excreted.

These anticoagulants act only *in vivo*, not *in vitro*.

Indandione derivatives can cause severe untoward effects, such as leukopenia, leukocytosis, agranulocytosis, hepatitis, jaundice, severe dermatitis, massive edema, and albuminuria. These reactions, some of them fatal, have been observed especially with phenindione. Owing to these serious side effects, this class of anticoagulants should be reserved for those patients who cannot tolerate coumarin derivatives. Their anticoagulant effect is antagonized by phytonadione.

4. *Miscellaneous Agents*

Salts of the rare earth metals are used clinically, although seldom, because they produce hemoglobinuria. They are administered intravenously and produce their action immediately, which persists up to about 24 hours. Their coagulant effect is antagonized by phytonadione.

Magnesium citrate is used to prevent postoperative thrombus formation and as prophylactic agent.

Snake venoms with anticoagulant activity are extracted from several families, especially Hydropheidae and Elapidae. Their use in therapeutics is not established.

5. *Anticoagulants for Storage of Whole Blood*

These substances are used only for storage of blood and not as therapeutic agents.

B. Mechanism of Action

The mechanism of action of anticoagulants varies according to the class they belong to.

Heparin acts primarily by inactivating thrombin. It exerts also catalytic effect on antithrombin activation, presumably by a two-step mechanism: it first replaces the inhibitor of antithrombin by forming a complex with the latter, and then it is replaced by thrombin.

Heparinoids may have a different mechanism of action: they act by mobilizing plasma protein-bound heparin.

As to coumarin derivatives, it has been suggested that they act by competitive antagonism with vitamin K. However, evidence does not favor this hypothesis. Rather, it indicates that coumarins inhibit irreversibly the transport of vitamin K to its intracellular site of action in the liver. Much larger doses of vitamins K can surmount this inhibition, probably because vitamins K can enter the cell by an alternate mechanism which is not susceptible to inhibition by coumarins.

Indandiones act at the same site and by the same mechanism as coumarins. In short, coumarins and indandione derivatives have an essentially identical mechanism of action. They inhibit the synthesis of factors II (prothrombin), VII (proconvertin), IX (Christmas), and X (Stuart-Prower) in the liver. This explains why their therapeutic action develops slowly.

Salts of the rare earth metals interfere with the activity of factor VII, thus depressing thromboplastin formation, presumably by forming ionic complexes with it.

Snake venoms act possibly by two different mechanisms: (a) primarily by enzymatic destruction of thromboplastin; (b) secondarily, by dissolution of fibrin, owing to a specific protease.

Anticoagulants for storage of whole blood act by complexing with Ca^{2+}, preventing in this way clot formation.

C. Antithrombotic Agents

Antithrombotic agents are substances that either prevent the formation of a thrombus or dissolve it. There are two different types of thrombi: venous thrombus and arterial thrombus. The first is the *red thrombus,* which is very similar to the one formed *in vitro;* it consists of a fibrin network entrapping the formed elements of the blood. The second one is the *white thrombus,* composed mainly of platelets. All forms of venous thrombosis and embolism result from red thrombus. Thomboembolism is probably the most prevalent disease entity in middle-age population of most civilized countries, such as the United States.

Antithrombotic agents may be divided into three main classes: inhibitors of platelet aggregation, anticoagulants, and fibrinolytic agents.

1. Inhibitors of Platelet Aggregation

Several substances act by this mechanism. For instance, aspirin, dipyridamole, ditazole (Ageroplas), flurbiprofen, glyceryl guaiacolate, indomethacin, prostaglandin analogues, sudoxicam, sulfinpyrazone. Aspirin, dipyridamole, ditazole, and a few other ones are already being used clinically.

2. Anticoagulants

Some oral anticoagulants, such as warfarin, have shown effectiveness in the prevention of postoperative venous thrombosis and thromboembolic complications after myocardial infarction. They exert no action, however, on arterial thrombi, which are essentially an aggregation of platelets and not blood clots.

Anticoagulants are not effective in the prevention of myocardial infarctions and similar disease states caused by arterial thrombosis.

3. Fibrinolytic Agents

In the fibrinolytic process four main components are involved: profibrinolysin (plasminogen), fibrinolysin (plasmin), activators, and inhibitors. In its natural state profibrinolysin, which is a beta-globulin, is inactive, but activators convert it to fibrinolysin, a proteolytic enzyme which has the property of digesting fibrin to soluble polypeptides:

Fibrinolytic agents have the property of dissolving preformed fibrin. They are active only on recently formed thrombi; lysis is not achieved in thrombi older than 72 hours. Several substances are known to act by this mechanism. Among many others, are the following enzymes: arvin, brinase, erythrokinase, fibrinolysin, reptilase, streptokinase-streptodornase, urokinase. Synthetic fibrinolytic agents include Ateroid, azapropazone, bencyclane, bisobrin, flufenamic acid, indomethacin, mefenamic acid, niflumic acid, phenylbutazone, and solbutamol.

Some of these substances act in venous thrombosis. Others are active in arterial thrombosis. Most of them, however, exert action on both venous and arterial thromboses.

Serious hemorrhage caused by fibrinolytic agents is treated by fibrinolytic inhibitors, especially aminocaproic acid (Amicar). Fibrinolytic activity is enhanced by certain compounds, especially anabolic steroids (ethyl estrenol, stanozolol) and hypoglycemic agents: gliclazide, phenformin, tolbutamide.

V. PLASMA SUBSTITUTES

Plasma substitutes are substances used to replace blood or its components to restore or maintain circulating blood volume. These substances are not expected to perform a variety of blood functions. Blood is a very complex fluid, containing several components and performing many functions: transport of oxygen and metabolic substrates to tissues, removal of carbon dioxide and metabolic products, and maintenance of the adequate concentration of ions and other solutes in extracellular fluids.

Under normal conditions the circulating blood volume is kept constant. But hemorrhages, burns, diarrheas, vomiting, and other pathological conditions cause decrease of blood volume. The dehydration thus produced is overcome by several means. In the past, for deep hemorrhage most physicians preferred transfusion of whole blood. However, since sterile blood components "can now be conveniently prepared, indiscriminate use of whole blood is a practice to be discouraged. Component transfusion therapy is superior," according to the American Medical Association. These components are normal human serum albumin (Albumisol), packed human blood cells, antihemophilic human plasma, and plasma protein fraction (Plasmanate). For temporary maintenance of blood volume, plasma substitutes are enough.

Transfusion of whole blood or one of its components has become routine operation in most hospitals. However, this practice has many disadvantages: (1) blood and blood components are very expensive to collect, store, and administer; (2) there is always the risk of transfering pathogenic microorganisms, such as cytomegalic inclusion virus, those causing hepatitis (in 12 to 17.5% of cases), toxoplasmosis, syphilis, malaria, Chagas' disease; (3) febrile reactions can occur, due to pyrogens or to leukocyte or platelet agglutinins which may be present; (4) allergic responses, such as generalized itching, urticarial rashes, and bronchospasms, are possible; (5) the danger of hemolysis, which is sometimes lethal, as a consequence of mislabeling of specimens and other causes; (6) the blood or plasma may be contaminated with bacteria, and this may result in systemic bacteremia; (7) circulatory overload is a hazard, especially in the aged patients, in the very young, and in those with pulmonary or cardiac disease; (8) massive transfusion of stored blood increases the risk of cardiac arrest, hyperkalemia, thrombocytopenia, and other complications.

Considering all these disadvantages of blood and components of blood, some physicians are resorting more and more to plasma substitutes, also called, although incorrectly, *blood substitutes, plasma volume expanders,* and *plasma expanders.*

Certain properties are required from a substance used as blood substitute. Among them are the following ones: an adequate colloid-osmotic effect, viscosity suitable for intravenous administration, lack of toxicity, antigenicity and pyrogenicity, easy sterilization, retention of 50% in the body from 6 to 12 hours, long stability, absence of the hazard of causing damage to plasma components or body organs. No ideal plasma substitute is presently available. The plasma substitutes replace only some of blood plasma components (such as water, salt, glucose, amino acids, proteins, etc.), but by exerting colloidal osmotic pressure they are effective in restoring and maintaining circulatory blood volume.

A. Classification

Several substances have been proposed and tested as plasma substitutes. They may be divided into the following classes:

1. Polysaccharides: dextran, hydrodextran (Hydrox), Levan, galacturonic acid polymer, starch and its derivatives (hydroxyethyl starch, carboxymethyl starch)

2. Synthetic polymers: polyvidone

3. Proteins: gelatin (Haemaccel), Isinglass, polyhydroxygelatin (Gelifundol), Plasmagel

The three most widely used are dextran, gelatin, and polyvidone.

As initial supportive therapy, besides other preparations (dextrose, Ringer's solution, sodium DL-lactate, glucose and sucrose, and amino acid or protein hydrolysates), the following ones are used: sodium chloride injection, balanced electrolyte (Normosol-R, Plasma-Lyte 148), and dextrose in lactated Ringer's solutions. The last preparation has been used successfully, even for open-heart surgery.

Dextran

It is a biosynthetic polysaccharide obtained by fermentation of sucrose by *Leuconostoc mesenteroides* and *Leuconostoc dextranicum,* followed by partial hydrolysis of the high-molecular-weight product thus obtained and fractionation of the resulting products. It consists of a chain of glucose molecules joined mainly by glycosidic linkages, that is, of α-1,6-linked glucopyranose units with 1,4- and 1,3-linkages at the points of branching.

From the above-mentioned fermentation several products having different molecular weights can result. Those mostly used in the United States are (*a*) dextran 40, marketed as Gentran 40, LMD 10%, and Rheomacrodex, with molecular weight 40,000; (*b*) dextran 75, marketed as Gentran 75, Infukoll, Macrodex, and Plasmodex, with molecular weight 75,000. In other countries are used also dextran 45, dextran 110 (especially in England, and marketed under the names of Dextraven 110 and Intradex), and dextran 150.

Storage may cause precipitation of dextran from solution. Heating in water bath or autoclave at 100°C for short period of time redissolves it.

Dextran is used as plasma expander in emergency treatment of shock. Untoward effects are usually mild, mainly of hypersensitivity type of reactions. It is contraindicated in patients with known hypersensitivity, severe congestive failure, renal failure, and some other conditions.

Gelatin

Of the several modified gelatins that have been prepared, the one marketed as Haemaccel is obtained from bovine bone gelatin of molecular weight around 100,000 by thermal degradation to polypeptides having molecular weight of 12,000 to 15,000, which are cross-linked with diisocyanate. Haemaccel is used in 3.5% aqueous solution containing Na^+, K^+, Ca^{2+}, and Cl^- and having an osmotic pressure of 350 to 390 mm H_2O. Its half-life is about 4 hours. It is not stored in the body, being eliminated by the kidneys, largely unchanged.

Polyvidone

Also called polyvinylpyrrolidone, marketed as Kollidon, Periston, Plasdone, and known for short as PVP, it is a synthetic polymer.

Polyvidone is a water-soluble, hygroscopic, white powder that can be stored indefinitely. Owing to the presence of the amido group, it shows proteinlike properties in binding water and adsorbing physiological and nonphysiological products. It is now rarely used as blood substitute, because of its tendency to be stored in the tissue instead of being entirely eliminated. This property is used advantageously: several preparations, especially penicillin, adrenal corticol hormones, insulin, morphine, and antiseptics, incorporate a 25% solution of PVP of molecular weight 25,000 to 40,000 as a retarding agent. For instance, povidone-iodine (Betadine), a PVP-iodine complex which releases iodine slowly, is used as a topical antiseptic. Furthermore, PVP finds extensive application in the cosmetic industry.

B. Mechanism of Action

Dextran, polyvidone, and gelatin act as plasma substitutes by compensating for loss of blood. They are, therefore, only plasma expanders, to wit, they increase circulating blood volume and, in cases of shock and hemorrhage, in which this volume decreases, they increase arterial pressure. However, they are not useful in treatment of anemia or hypoproteinemia. Actually, none of them is able to perform blood hemodynamic functions, such as oxygen transport. Furthermore, they lack a clotting system, do not contain globulins and leukocytes that are involved in body's immune defense mechanism, and are deprived of the buffering ability and enzymatic activity normally provided by erythrocytes.

REFERENCES

Antianemics

B. M. Babior, Ed., *Cobalamin: Biochemistry and Pathophysiology,* Wiley Biomedical, New York, 1975.

R. B. Woodward, *Pure Appl. Chem.,* **34**, 145 (1973).

J. M. Pratt, *Inorganic Chemistry of Vitamin B_{12},* Academic, London, 1972.

H. Rembold and W. L. Gyure, *Angew. Chem. Int. Ed. Engl.,* **11**, 1061 (1972).

J. R. Derrick and M. M. Guest, *Dextrans: Current Concepts of Basic Actions and Clinical Applications,* Thomas, Springfield, Ill., 1971.

J. C. Dreyfus, Ed., *Hematopoietic Agents,* Pergamon, Oxford, 1971.

L. Hallberg *et al.,* Eds., Colloquia Geigy, *Iron Deficiency,* Academic, London, 1970.

R. L. Blakley, *The Biochemistry of Folic Acid and Related Pteridines,* Wiley, New York, 1969.

I. Chanarin, *The Megaloblastic Anemias,* Blackwell, Oxford, 1969.

Hemostatics

O. D. Ratnoff and B. Bennett, *Science,* 179, 1291 (1973).

L. Lorand, *Ann. N.Y. Acad. Sci.,* 202, 6 (1972).

N. Heimburger and H. Trobisch, *Angew. Chem. Int. Ed. Engl.,* 10, 85 (1971).

R. E. Olson, *Acta Vitaminol. Enzymol.,* 25, 122 (1971).

W. H. Seegers, *Annu. Rev. Physiol.,* 31, 269 (1969).

Anticoagulants

L. M. Aledort, Ed., "Recent Advances in Hemophilia," *Ann. N. Y. Acad. Sci.,* 240, 1-426 (1975).

R. Labbe-Bois *et al., J. Med. Chem.,* 18, 85 (1975).

J. Ehrlich and S. S. Stivala, *J. Pharm. Sci.,* 62, 517 (1973).

V. V. Kakkar and A. J. Jouhar, Eds., *Thromboembolism: Diagnosis and Treatment,* Churchill Livingstone, London, 1972.

K. Laki, Ed., "The Biological Role of the Clot-Stabilizing Enzymes: Transglutaminase and Factor XIII," *Ann. N. Y. Acad. Sci.,* 202, 1-348 (1972).

J. Hampton, *Amer. J. Cardiol.,* 27, 659 (1971).

N. Heimburger and H. Trobisch, *Angew. Chem. Int. Ed. Engl.,* 10, 85 (1971).

F. Markwardt, Ed., "Anticoagulantien," *Handbuch der experimentellen Pharmakologie,* Vol. XXVII, Springer, Berlin, 1971.

D. Ogston and A. S. Douglas, *Drugs,* 1, 228, 303 (1971).

W. W. Coon and P. W. Willis III, *Clin. Pharmacol. Ther.,* 11, 312 (1970).

R. A. O'Reilly and P. M. Aggeler, *Pharmacol. Rev.,* 22, 35 (1970).

J. M. Schor, Ed., *Chemical Control of Fibrinolysis-Thrombolysis: Theory and Clinical Applications,* Wiley-Interscience, New York, 1970.

L. B. Jaques, *Anticoagulant Therapy,* Thomas, Springfield, Ill., 1965.

Plasma Substitutes

B. A. Myhre, *Quality Control in Blood Banking,* Wiley, New York, 1974.

R. W. Beal, *Drugs,* 6, 127 (1973).

J. H. Kay, *J. Amer. Med. Assoc.,* 226, 1230 (1973).

T. H. Maugh II, *Science,* 179, 669 (1973).

L. C. Clark, Jr., Ed., "Symposium on Inert Organic Liquids for Biological Oxygen Transport, *Federation Proc.,* 29, 1695-1820 (1970).

G. W. Simmons, Jr. *et al., J. Amer. Med. Assoc.,* 213, 1032 (1970).

U. F. Gruber, *Blutersatz,* Springer, Heidelberg, 1968.

CHAPTER 24

DIURETICS

I. INTRODUCTION

A. Definition

Defined often improperly as substances that increase urinary volume, diuretics are drugs that act primarily by promoting the excretion of Na^+ or Cl^- or HCO_3^- ions which constitute the main electrolytes of the extracellular fluid. They also act by decreasing tubular reabsorption, a process which involves the active transport of electrolytes and other solutes from tubular urine to the tubular cells and then to the extracellular fluid. They are used in the relief of edema and as adjuvants in the management of hypertension.

Untoward effects of diuretics vary according to the class to which they belong. Some effects are minor, whereas others may be very serious, including death.

B. Renal Physiology

The functional unit of the kidney is the nephron, which consists of a glomerulus, proximal and distal convoluted tubules, the loop of Henle, and a collecting duct (Figure 24.1).

Urine is first formed at the glomerulus, where by ultrafiltration the urinary fluid is freed from its nonfilterable cellular components, such as red and white blood cells and the plasma proteins. During its passage along the lumen of the nephron, this practically proteinfree fluid undergoes many alterations in its composition: some of its components are reabsorbed almost completely into the renal blood vessels (Table 24.1); for instance, 98 to 99% of the water together with various electrolytes (Na^+, K^+, Cl^-, and HCO_3^-), glucose, and urea filtered at the glomerulus is reabsorbed in the tubule. The remainder, about 1.5 liters/day, is excreted as urine.

It has been recently suggested that Na^+-K^+ ATPase is involved in energizing renal Na^+ transport and consequently that some diuretics might act by inhibiting this enzyme.

Diuretics can affect three main physiological processes: (a) glomerular filtration, (b) tubular reabsorption, (c) tubular secretion.

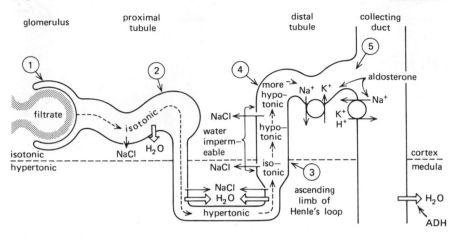

Figure 24.1 Schematic representation of the anatomic and functional subdivisions of the nephron, showing the sites of action of various diuretics. (Reproduced from P. J. Cannon and M. K. Kilcoyne, "Ethacrynic Acid and Furosemide: Renal Pharmacology and Clinical Use," in C. K. Friedberg, Ed., *Current Status of Drugs in Cardiovascular Disease*, Grune and Stratton, New York, 1969, pp. 243-262.)

Table 24.1 Quantities Filtered, Excreted, and Reabsorbed by Kidneys of Man (Mean Normal Values)

	Plasma concn, meq/l	meq/24 hour			Percentage reabsorbed
		Filtered	Excreted	Reabsorbed	
Sodium	140	23,900	171	23,729	99.3
Chloride	103	19,500	171	19,329	99.1
Bicarbonate	27	5,100	2	5,098	99.9
Potassium	4	684	51	633	80.6
		ml/hour			
Water	94%	169,000	1,500	167,570	99.1
Glucose	...	...	...	...	100.0

Source. After R. F. Pitts, *The Physiological Basis of Diuretic Therapy*, Thomas, Springfield, Ill., 1959.

II. CLASSIFICATION

Several classifications have been proposed for diuretics. By taking into account the effect produced, three types of diuretics can be distinguished:

1. Diuretics properly said: those that increase only the excretion of water

and not of electrolytes
2. Natriuretics: those that increase the excretion of sodium
3. Saluretics: those that increase the excretion of sodium and chloride

According to their chemical constitution, diuretics used clinically and under investigation may be divided into the following classes:

1. Xanthines
2. Organomercurials
3. Pyrimidines (no longer commercially available)
4. Pyrazines
5. s-Triazines
6. Acylphenoxyacetic acids
7. Sulfonamides
8. Thiazides and related drugs
9. Pteridines
10. Steroidal spirolactones
11. Miscellaneous: ammonium chloride, ammonium nitrate, glucose, L-lysine monohydrochloride, mannitol, sucrose, trifluocin, urea

Diuretics most widely used have been divided into the following six classes: xanthines (Table 24.2), osmotic diuretics (Table 24.2), mercurial compounds (Table 24.3), carbonic anhydrase inhibitors (Table 24.4), thiazides and related sulfonamide compounds (Table 24.5), and miscellaneous diuretics (Table 24.6).

Some mixtures are also used: Diazide (combination of triamterene with hydrochlorothiazide) and Aldactazide (combination of spironolactone with hydrochlorothiazide).

A. Xanthines

Theophylline and aminophylline, its ethylenediamine salt, are sometimes used as diuretics as such or in conjunction with organometallic diuretics. The principal use of aminophylline, however, is as a bronchodilator and antiasthmatic.

B. Osmotic Diuretics

The most commonly used are mannitol and urea.

Mannitol

D-Mannitol is used mainly as an adjunct in treatment of oliguria and anuria. Excessive doses can cause death. It should be injected slowly and carefully in order to avoid extravasation. It is contraindicated in patients with congestive heart failure.

Table 24.2 Xanthine and Osmotic Diuretics

Official Name	Proprietary Name	Chemical Name	Structure
aminophylline*	Carena Inophyllin	theophylline ethylene- diamine	see Table 12.2
theophylline*⁺	Theocin		see Table 12.2
mannitol*	Osmitrol	D-mannitol	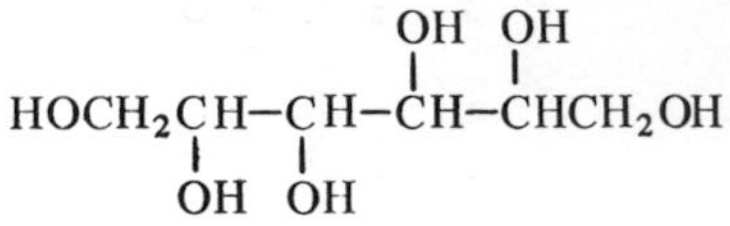 $\begin{array}{c} \quad\quad\quad OH\ \ OH \\ \quad\quad\quad \vert\ \ \ \vert \\ HOCH_2CH-CH-CH-CHCH_2OH \\ \quad\quad \vert\ \ \ \vert \\ \quad\quad OH\ OH \end{array}$
urea*	Ureaphil	carbamide	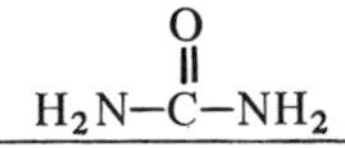 $\begin{array}{c} O \\ \parallel \\ H_2N-C-NH_2 \end{array}$

*In *USP XIX* (1975).
⁺In *NF XIV* (1975).

C. Mercurial Compounds

Organomercurial diuretics present the general structure $RCH_2\underset{\underset{\displaystyle OY}{|}}{C}HCH_2HgX$,

that is, a chain of at least 3 carbon atoms, one atom of mercury at one end of the chain and a hydrophilic group separated from Hg at the other end.

On the R, Y, and X substituents depend the potency and the side effects. Of greater influence is the R group, which can be aromatic, heterocyclic, or alicyclic, usually attached to the propyl chain through a carbamoyl group. In general Y is methyl and X, theophylline, which has diuretic activity by itself, although weak.

These compounds are saluretics; that is, they inhibit the tubular reabsorption of sodium, chloride, and water. Their main use is in the treatment of congestive heart failure. Those mostly used (meralluride and mercaptomerin) should be administered by intramuscular or subcutaneous routes; intravenous injection should be avoided, because severe systemic reactions and even sudden death may result. Other untoward reactions are gastric distress, vertigo, and stomatitis.

Besides those listed in Table 24.3, the following ones are also used: mercurophylline (Mercuzanthin), merethoxyllin procaine (Dicurin Procaine), and mersalyl (Salyrgan).

Meralluride

Used as sodium salt, it is administered by parenteral route, producing prompt diuresis.

Mercaptomerin

In the form of sodium salt, it is one of the most widely used organomercurial diuretics. It occurs as either hygroscopic white powder or amorphous solid, which is easily soluble in water. It contains 39% of mercury and should be kept in well-closed containers. It is administered by intramuscular and subcutaneous routes.

D. Carbonic Anhydrase Inhibitors

Of those official in the United States (Table 24.4), the most widely used is acetazolamide, both in the free form and as the sodium salt.

Acetazolamide

This diuretic is a slightly water-soluble, white to faintly yellowish-white, crystalline powder. Acetazolamide was used as an oral diuretic, but now thiazides are the preferred drugs by this route. It is still used, however, in the treatment of narrow-angle glaucoma. The usual dosage by oral or intravenous routes for production of diuresis in congestive heart failure is 250 to 375 mg once daily in the morning.

Table 24.3 Organomercurial Diuretics

$$R-CH_2-CH-CH_2-Hg-X$$
$$\underset{O-R'}{|}$$

Official Name	Proprietary Name	Chemical Name	R	R'	X
chlormerodrin*	Neohydrin	1-[3-(chloromercuri)-2-methoxy-propyl]-urea	$H_2N-CO-NH$	$-CH_3$	$-Cl$
meralluride⁺	Mercuhydrin	N-(2-methoxy-3-[(1,2,3,6-tetra-hydro-1,3-dimethyl-2,6-dioxo-purin-7-yl)mercuri]-propyl) carbamoyl succinamic acid	$HOOCCH_2CH_2$ $OC-NH-CO$ $NH-$	$-CH_3$	theophylline
mercaptomerin sodium*	Thiomerin	disodium N-[3-(carboxymethyl-thiomercuri)-2-methoxypropyl]-α-camphoramate	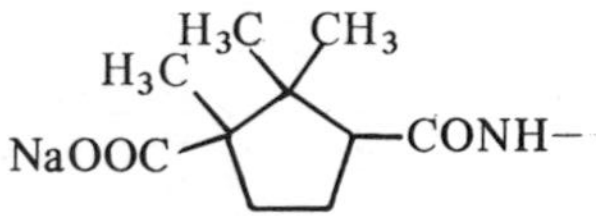 (see structure)	$-CH_3$	$-SCH_2COONa$

*In *USP XIX* (1975).
⁺In *NF XIV* (1975).

Table 24.4 Carbonic Anhydrase Inhibitors

Official Name	Proprietary Name	Chemical Name	Structure
acetazolamide*	Diamox	5-acetamido-1,3,4-thiadia-zole-2-sulfonamide	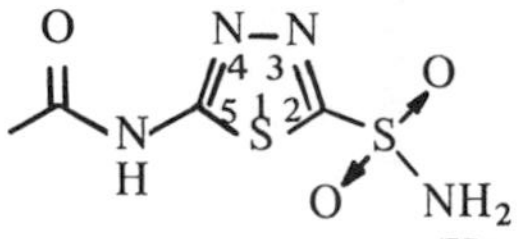
methazolamide*	Neptazane	4-methyl-2-sulfamoyl-Δ^2-1,3,4-thiadiazolin-5-ylidene)acetamide	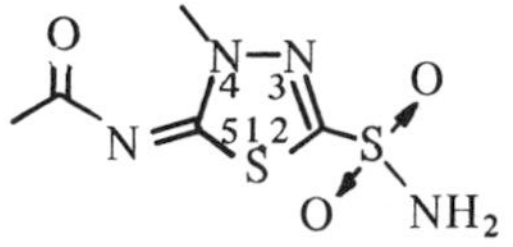
dichlorphenamide*	Daranide	4,5-dichloro-1,3-benzene-disulfonamide	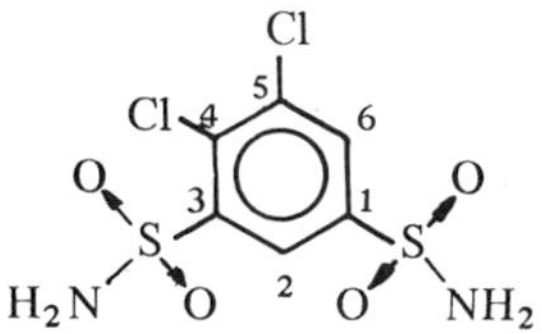
ethoxzolamide*	Cardrase	6-ethoxybenzothiazole-2-sulfonamide	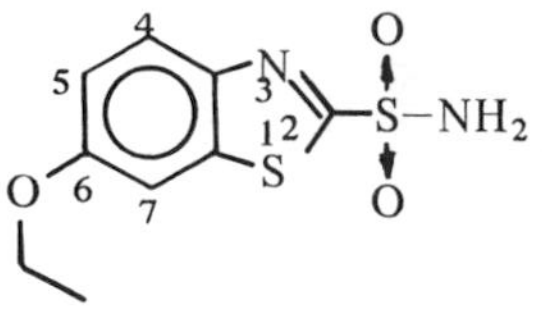

*In *USP XIX* (1975).

E. Thiazides and Related Sulfonamide Compounds

Thiazides and related drugs (Table 24.5) are saluretics, as are organomercurial compounds, depressing the reabsorption of sodium, chloride, and water. They also increase the urinary excretion of potassium and bicarbonate ions. All thiazides and related sulfonamide agents are almost identical in their action, differing only in duration of action and dosages. Thus, the duration of action of chlorothiazide is 6 to 12 hours; hydrochlorothiazide and benzthiazide, 12 to 18; hydroflumethiazide, cyclothiazide, and quinethazone, 18 to 24; bendro-flumethiazide, more than 18; methyclothiazide and trichlormethiazide, more than 24; polythiazide, 24 to 48; chlorthalidone, 24 to 72. Doses also vary widely: chlorothiazide, 500 to 1000 mg; cyclothiazide, 1 to 2 mg.

These drugs are used in all types of cardiac decompensation, being used instead of other diuretic agents. They may cause hypokalemia and other electrolyte imbalances. These conditions can be corrected by ingestion of foods rich in potassium, such as bananas and orange juice, or oral solution of potassium chloride. Minor untoward effects, which can be corrected by re-ducing the dose, are fatigue, dizziness, and gastrointestinal disorders. Since thiazides decrease the renal excretion of uric acid, they tend to raise the levels of this acid and may cause attacks of acute gouty arthritis.

Chorothiazide

It is the prototype of the thiazides and occurs as a slightly water-soluble, white, crystalline powder. Chlorothiazide is useful in the treatment of edema.

Chlorthalidone

This drug, an aromatic sulfonamide, occurs as a practically water-insoluble, white or yellow-white, almost odorless, insipid, crystalline powder. Although being a phthalimide derivative with a sulfonamide moiety and not a thiazide, its actions, uses, and untoward effects are similar to those of chlorothiazide.

Furosemide

It occurs as a white or almost white, crystalline powder, which is practically insoluble in water. Being light-sensitive, it should be kept in dark glass closed containers. Furosemide is a very potent saluretic agent, causing greater diuretic effect than thiazides or acetazolamide. By oral route, action starts within 1 hour and lasts 6 hours; by intramuscular or intravenous routes, action is immediate and lasts 1 hour. It is used in the treatment of edema, particularly in the one associated with severe renal or hepatic insufficiency in patients who cannot tolerate, or do not respond to, other diuretics. It should be used, therefore, only when less potent but safer diuretics are ineffective.

Table 24.5 Thiazides and Related Sulfonamide Diuretics

Official Name	Proprietary Name	Chemical Name 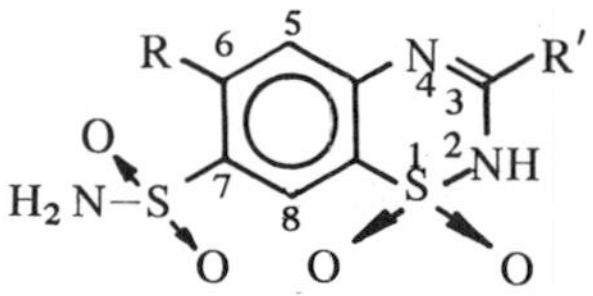	R	R'	R''
Thiazides					
chlorothiazide*+	Diuril	6-chloro-2H-1,2,4-benzothiadiazine-7-sulfonamide 1,1-dioxide	—Cl	—H	
benzthiazide+	Exna	3-[(benzylthio)methyl]-6-chloro-2H-1,2,4-benzothiadiazine-7-sulfonamide 1,1-dioxide	—Cl	—CH$_2$SCH$_2$— (phenyl)	
Hydrothiazides					
hydrochlorothiazide*	Esidrex	6-chloro-3,4-dihydro-2H-1,2,4-benzothiadiazine-7-sulfonamide 1,1-dioxide	—Cl	—H	—H
hydroflumethiazide+	Saluron	3,4-dihydro-6-(trifluoromethyl)-2H-1,2,4-benzothiadiazine-7-sulfonamide 1,1-dioxide	—CF$_3$	—H	—H
bendroflumethiazide+	Benuran Naturetin	3-benzyl-3,4-dihydro-6-(trifluoromethyl)-2H-1,2,4-benzothiadiazine-7-sulfonamide 1,1-dioxide	—CF$_3$	—CH$_2$— 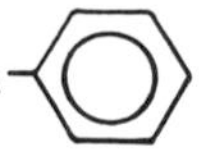 (phenyl)	—H

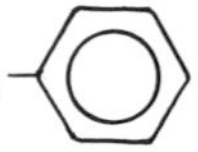

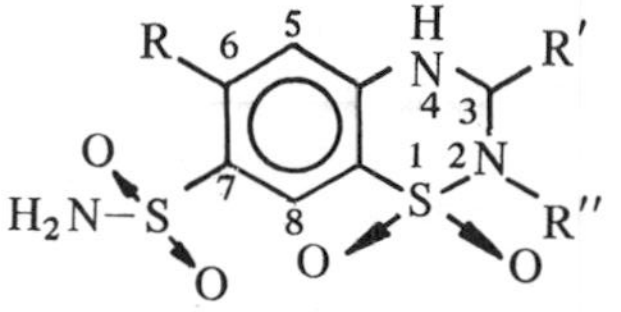

Table 24.5 Thiazides and Related Sulfonamide Diuretics (continued)

Official Name	Proprietary Name	Chemical Name			
polythiazide[+]	Renese	6-chloro-3,4-dihydro-2-methyl-3-[(2,2,2-trifluoroethyl)thio]methyl-2H-1,2,4-benzothiadiazine-7-sulfonamide 1,1-dioxide	$-Cl$	$-CH_2SCH_2CF_3$	$-CH_3$
methylclothiazide[+]	Enduron	6-chloro-3-chloromethyl-3,4-dihydro-2-methyl-2H-1,2,4-benzothiadiazine-7-sulfonamide 1,1-dioxide	$-Cl$	$-CH_2Cl$	$-CH_3$
trichlormethiazide[+]	Metahydrin Naqua	6-chloro-3-dichloromethyl-3,4-dihydro-2H-1,2,4-benzothiadiazine-7-sulfonamide 1,1-dioxide	$-Cl$	$-CHCl_2$	$-H$
cyclothiazide[+]	Anhydron	6-chloro-3,4-dihydro-3-(5-norbornen-2-yl)-2H-1,2,4-benzothiadiazine-7-sulfonamide 1,1-dioxide	$-Cl$		$-H$

Related sulfonamides

chlorthalidone*	Hygroton	2-chloro-5-(1-hydroxy-3-oxo-1-isoindolinyl)benzenesulfonamide	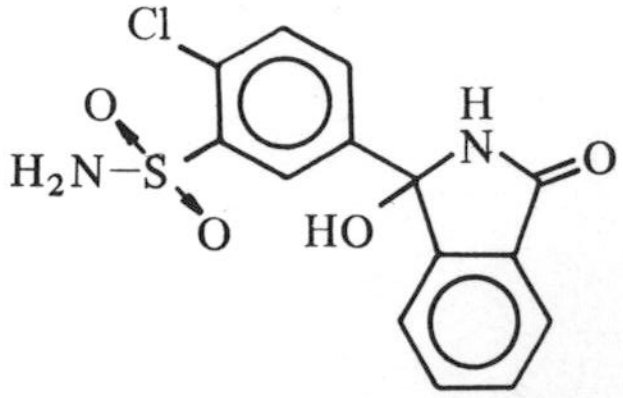

quinethazone[†]	Hydromox	7-chloro-2-ethyl-1,2,3,4-tetrahydro-4-oxo-6-quinazolinesulfonamide 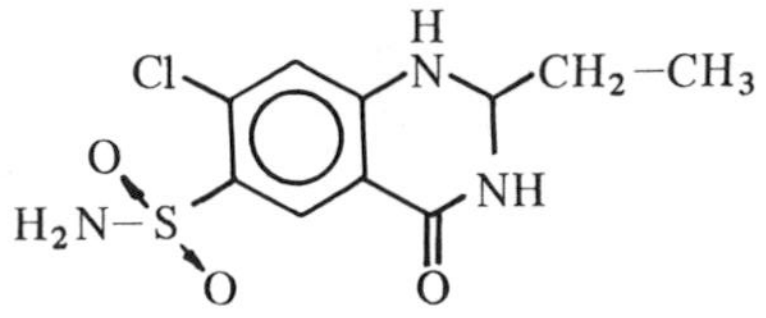
furosemide[*]	Lasix	4-chloro-N-furfuryl-5-sulfamoylanthra-nilic acid 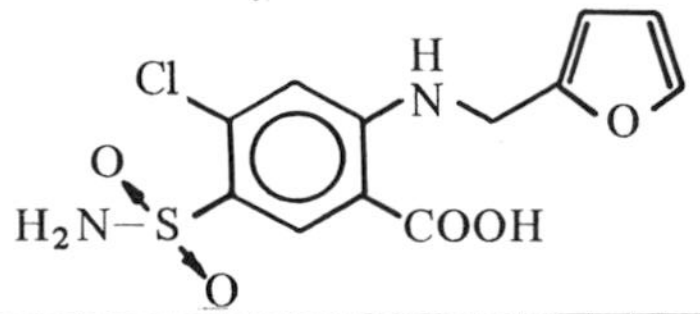

*In *USP XIX* (1975).
†In *NF XIV* (1975).

F. Miscellaneous Diuretics

This class is comprised of diuretics of varied chemical constitution (Table 24.6).

Ethacrynic Acid

It is a white, crystalline powder, which is slightly soluble in water. Ethacrynic acid is a saluretic agent, used both as free acid and as sodium salt. It increases the excretion of sodium and chloride and, to a lesser extent, that of potassium. By oral route it is used in the control of edema associated with congestive heart failure and idiopatic edema. Since it has a rapid onset of action, it may be effective in acute pulmonary edema and as an adjunct in control of hypertensive crises. Doses should be determined for each patient individually. Untoward effects are similar to those of thiazides. It should be used only when the less potent but safer drugs are ineffective.

Triamterene

This pteridine derivative occurs as a poorly water-soluble, yellow crystalline tabular substance. It is also a saluretic agent: it promotes excretion of water, sodium, and chloride but retention of potassium. It is used in the treatment of edema associated with congestive heart failure, hepatic cirrhosis, and the nephrotic syndrome. For better results, it should be used with a thiazide. Triamterene is a suitable adjunct to thiazides, because it prevents the hypokalemia caused by those drugs.

Spironolactone

This synthetic steroid occurs as a yellowish crystalline powder, with slight mercaptanlike odor. It is stable to air and practically insoluble in water. Spironolactone promotes excretion of sodium and chloride, being therefore a saluretic agent. It is most effective when administered concomitantly with a thiazide or organomercurial compound. Spironolatone is used in disorders caused by an excess of production of aldosterone. Although almost devoid of untoward effects, it is contraindicated in acute renal insufficiency, hyperkalemia, and some other conditions. The initial dosage, by oral route, is 100 mg daily in four divided doses for 5 days.

III. MECHANISM OF ACTION

Little is known about the mechanism of action of diuretics at the molecular level. For this reason, these agents are usually divided into three main classes: (a) osmotic diuretics: urea, mannitol, sorbitol, sucrose; (b) acid-forming salts:

Table 24.6 Miscellaneous Diuretics

Official Name	Proprietary Name	Chemical Name	Structure
ammonium chloride[+]	Darammon	ammonium chloride	NH_4Cl
ethacrynic acid*	Edecrin	[2,3-dichloro-4-(2-methyl-enebutyryl)-phenoxy] acetic acid	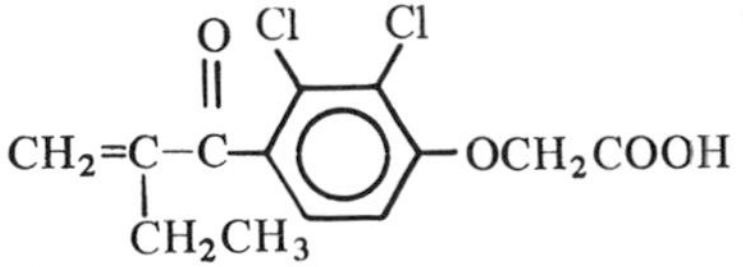 $CH_2=C-C-$ ... CH_2CH_3 ... OCH_2COOH, Cl, Cl
triamterene*	Dyrenium	2,4,7-triamino-6-phenyl-pteridine	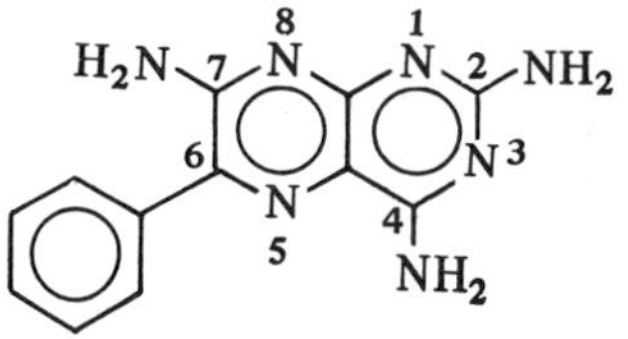 H_2N, NH_2, NH_2 (pteridine ring, positions 1,2,3,4,5,6,7,8)
spironolactone*	Aldactone	17-hydroxy-7α-mercapto-3-oxo-17α-pregn-4-ene-21-carboxylic acid γ-lactone 7-acetate	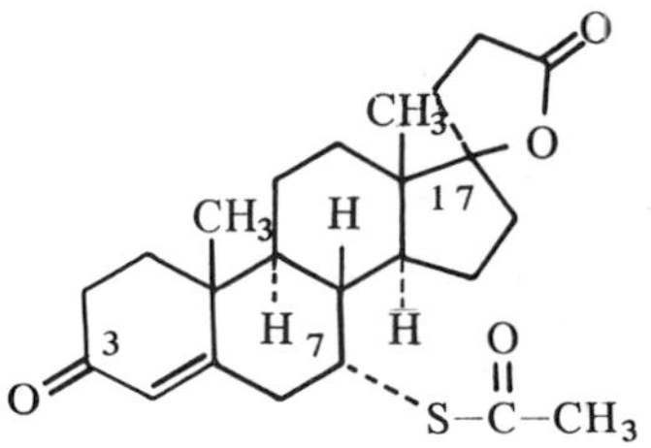 $S-C-CH_3$, steroid structure (positions 3, 7, 8, 17)

*In *USP XIX* (1975).
[+]In *NF XIV* (1975).

ammonium chloride, ammonium nitrate, calcium chloride, potassium salts; (c) inhibitors of renal tubular transport: xanthines, organomercurials, carbonic anhydrase inhibitors, aromatic sulfonamides, benzothiadiazines and related drugs, aldosterone antagonists, acylphenoxyacetic acids, pteridines, and pyrazines.

A. Osmotic Diuretics

These agents are poorly absorbed by the renal tubules. Thus, by osmotic effect, they prevent the reabsorption of water, which is excreted in urine.

B. Acid-Forming Salts

They act by the following mechanism: the cations of these salts are either metabolized or poorly absorbed, leaving an excess of the anion (e.g., chloride), which is excreted, together with an equivalent amount of sodium and the corresponding volume of water.

C. Inhibitors of Renal Tubular Transport

They act by several different mechanisms, which are related to the chemical constitution of each class of diuretics.

1. Xanthines

The diuretic activity of xanthines results from a decrease in tubular electrolyte reabsorption and an increase of glomerular filtration rate. Xanthines are competitive inhibitors of nucleotide phosphodiesterase, an enzyme that regulates degradation of cyclic-3',5'-AMP, as shown in Figure 12.1. This inhibition results in cardiac stimulation and consequent increase of renal blood flow and of rate of glomerular filtration.

2. Organomercurial Diuretics

Organomercurial compounds owe their diuretic action to inhibition of ATPase of the tubular membrane which is responsible for the active reabsorption of sodium through the nephron. They interact with at least one sulfhydryl group of ATPase through their Hg atom. The active species is the mercuric ion, which attaches specifically to two receptor sites, one being a sulfhydryl and the second another sulfhydryl or a phenolic hydroxy group, amino group, carboxyl group, or imidazole ring, as shown in Figure 24.2.

3. Carbonic Anhydrase Inhibitors

Carbonic anhydrase, a metalloenzyme containing one zinc atom per protein molecule of molecular weight 30,000, catalyzes the reversible hydration of carbon dioxide. The product of this reaction is carbonic acid, which dissociates into H^+ and HCO_3^-:

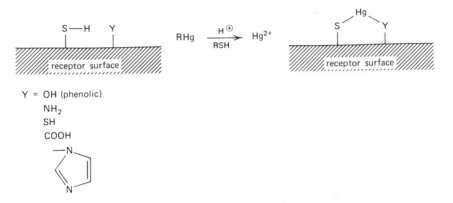

Figure 24.2 Reaction of organomercurial diuretics with the thiol groups of ATPase. (After Miller and Farah.)

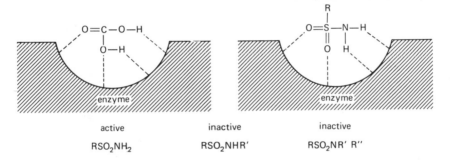

Figure 24.3 Complexation of activator and inhibitors of carbonic anhydrase with its active site through hydrogen bonding. (After R. O. Roblin, unpublished.)

$$H_2O + CO_2 \rightleftharpoons H_2CO_3 \rightleftharpoons H^+ + HCO_3^-$$

Inhibition of this enzyme reduces the concentration of H^+ ion, with the consequent delay in the exchange Na^+/H^+, at the tubular urine, promotes the exchange Na^+/K^+, and prevents the conversion of HCO_3^- to carbonic acid that is necessary for the reabsorption of bicarbonate. As a result, excretion of Na^+, K^+, and HCO_3^- ions through the kidneys is increased and urinary pH also increases.

Owing to the structural similarity of the sulfonamide group to carbonic acid, diuretic sulfonamides inhibit carbonic anhydrase, as shown in Figure 24.3, although the antagonism is not of the competitive type. The sulfonamide group should not contain substituents; otherwise diuretics acting by this mechanism lose, either partially or completely, their activity, because attachment to the receptor is weakened.

This decrease in activity might be explained also by results obtained in molecular orbital calculations performed by Yonezawa and coworkers. Based on their data, they suggested that for carbonic anhydrase activity it seems indispensable that there be a high nucleophilicity by the nitrogen atom of the sulfonamide group. This nitrogen atom would inhibit carbonic anhydrase by reacting with it the following way:

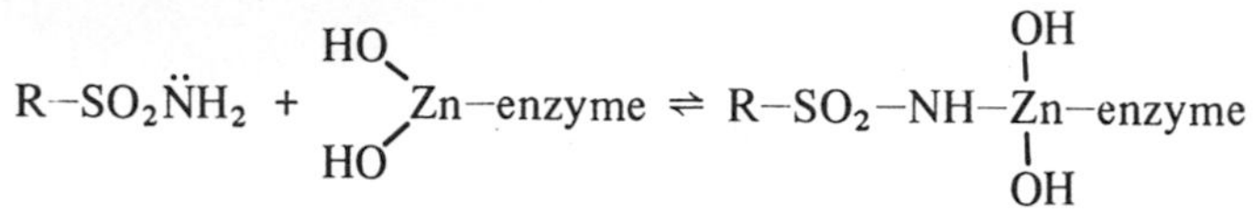

$$R\text{--}SO_2\ddot{N}H_2 \; + \; \underset{HO}{\overset{HO}{\diagdown}}Zn\text{--enzyme} \; \rightleftharpoons \; R\text{--}SO_2\text{--}NH\text{--}\underset{OH}{\overset{OH}{\underset{|}{\overset{|}{Zn}}}}\text{--enzyme}$$

A third explanation has been given by Taylor and coworkers, who represented the inhibition of carbonic anhydrase by sulfonamide as shown in Figure 24.4.

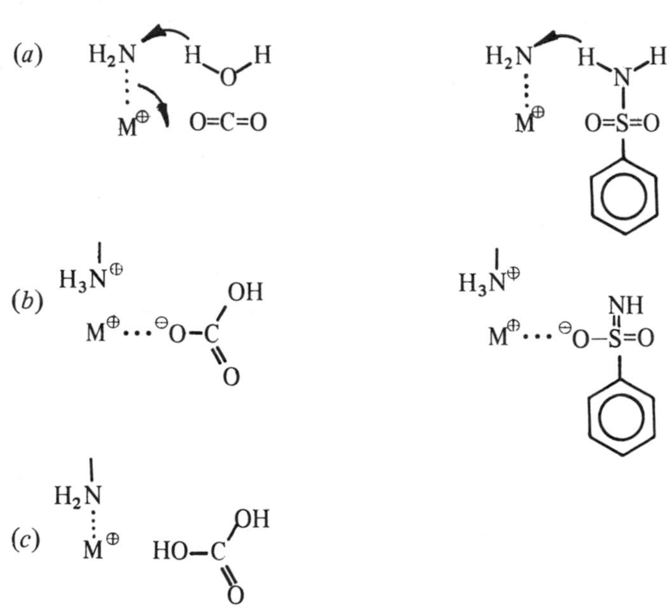

Figure 24.4 Mechanism of action of sulfonamide diuretics as inhibitors of carbonic anhydrase. [After Taylor *et al., Biochemistry*, **9, 2638, 3894** (1970).]

In 1967, by X-ray diffraction methods, Fridborg and coworkers determined the three-dimensional structure of the complex formed between human carbonic anhydrase C, now known as carbonate hydrolase EC 4.2.1.1 and whose tridimensional structure was recently determined at 2.0 Å resolution, and acetoxymercurisulfonamide, which is a modified inhibitor of this enzyme.

They observed that the inhibitor inserts itself into a narrow slit in the enzyme's cavity and, through the sulfonamide group, is bound to the zinc present in carbonic anhydrase. Another molecule of the same inhibitor binds to the only sulfhydryl group of the enzyme through the mercury atom.

4. Aromatic Sulfonamides

The main two representatives are furosemide and chlorthalidone. They are poor inhibitors of carbonic anhydrase, and, therefore, their mechanism of action is not ascribed to this effect.

Furosemide blocks Na^+ reabsorption in the ascending limb, affects osmolal concentration gradients in the renal medulla, and has a direct cellular action on the loop of Henle and on the proximal tubule. It is very likely that it acts through inhibition of membrane permeability and inhibition of the sodium pump.

Kakeya and coworkers found a close relationship between diuretic activity and electronic characteristics of the sulfonamide group: this activity increases with the increase of the dissociation constant.

5. Benzothiadiazines

Owing to the presence of the sulfonamide group, it was once thought that benzothiadiazines act by inhibiting carbonic anhydrase. Today, however, it is admitted that they exert a direct effect on renal tubular transport of sodium and chloride, but this effect does not derive necessarily from inhibition of carbonic anhydrase, although they also act, but weakly, by this mechanism.

6. Other Diuretics

Owing to its structural resemblance with aldosterone, spironolactone competes with this adrenocortical steroid for the receptor sites which are responsible for Na^+ reabsorption in the distal renal tubule.

Ethacrynic acid was designed to act the same way as organomercurial diuretics, that is, by selective reaction with thiol groups of renal cellular proteins, thus causing an inhibition of sulfhydryl-catalyzed enzyme systems. However, it is now known that it does not react with thiol groups of ATPase, but it acts directly on energy-generating systems. It inhibits the transport of Na^+ out of the ascending limb of the loop of Henle and also Na^+ transport in both the proximal and distal parts of the nephron.

Triamterene acts directly on the renal transport system for Na^+ at the renal distal tubule causing a moderate increase in the excretion of Na^+ and, to a lesser extent, of Cl^- and HCO_3^-. This drug blocks cation transport by binding to receptors at essential hydrophilic atoms (N-1 or N-8) and reinforcing hydrophobic sites.

REFERENCES

General

W. N. Suki and G. Eknoyan, Eds., *The Kidney in Systemic Disease,* Wiley Biomedical, New York, 1975.

K. D. G. Edwards, Ed., "Drugs Affecting Kidney Function and Metabolism" *Progr. Biochem. Pharmacol.,* 7, 1-538 (1972).

H. Herkan, Ed., *Diuretica,* Springer, Berlin, 1969.

E. F. Cafruny, *Annu. Rev. Pharmacol.,* 8, 131 (1968).

J. M. Sprague, *Top. Med. Chem.,* 2, 1 (1968).

J. E. Baer and K. H. Beyer, *Annu. Rev. Pharmacol.,* 6, 261 (1966).

G. deStevens, *Diuretics: Chemistry and Pharmacology,* Academic, New York, 1963.

Introduction

B. R. Rennick, *Annu. Rev. Pharmacol.,* 12, 141 (1972).

J. Orloff and M. Burg, *Annu. Rev. Physiol.,* 33, 83 (1971).

D. W. Seldin, Ed., "Symposium: The Physiology of Diuretic Agents," *Ann. N.Y. Acad. Sci.,* 139, 273 (1966).

Classification

L. L. Sakaletzky *et al., J. Med. Chem.,* 12, 977 (1969).

J.-J. Godfroid, *Chim. Ther.,* 3, 376 (1968).

H. J. Laragh, *Ann. Intern. Med.,* 67, 606 (1967).

A. F. Lant *et al., Clin. Pharmacol. Ther.,* 7, 196 (1966).

Mechanism of Action

Y. Shinagawa and Y. Shinagawa, *Int. J. Quantum Chem.,* Quantum Biology Symp. no. 1, 169 (1974).

G. R. Zins, *Annu. Rep. Med. Chem.,* 8, 83 (1973).

A. Liljas *et al., Nature New Biol.,* 235, 131 (1972).

R. H. Prince and P. R. Wooley, *Angew. Chem. Int. Engl. Ed.,* 11, 408 (1972).

E. J. Landon and L. R. Forte, *Annu. Rev. Pharmacol.,* 11, 171 (1971).

P. W. Taylor *et al., Biochemistry,* 9, 2638, 3894 (1970).

D. D. Fanestil, *Annu. Rev. Med.,* 20, 223 (1969).

N. Kakeya *et al., Chem. Pharm. Bull.,* 17, 1010, 2000, 2558 (1969).

T. Yonezawa *et al., Mol. Pharmacol.,* 5, 446 (1969).

E. J. Cafruny, *Pharmacol. Rev.,* 20, 89 (1968).

K. Fridborg *et al., J. Mol. Biol.,* 25, 505 (1967).

T. B. Miller and A. E. Farah, *J. Pharmacol. Exp. Ther.,* 136, 10 (1962).

PART 5

CHEMOTHERAPEUTIC AGENTS

Chemotherapeutic agents are drugs used in the treatment of infectious diseases. These diseases are caused by certain species of metazoa, protozoa, fungi, bacteria, rickettsia, and viruses. Drugs active on these pathogenic agents are studied in the following 11 chapters (25 to 35).

The term *chemotherapy,* which literally means "chemical therapy" or "chemical treatment," was coined in 1913 by Paul Ehrlich, the father of modern chemotherapy. He uttered the famous phrase: *Corpora non agunt nisi fixata* (Drugs do not act unless they bind). Ehrlich's effort was directed toward the discovery of the "magic bullet," that is, a chemical substance with selective action on the parasite but devoid of toxic effects to the host.

Ehrlich defined chemotherapy as the use of drugs to injure an invading organism without causing injury to the host. According to Ehrlich, therefore, chemotherapeutic agents are chemical substances with high parasitotropism and low or no organotropism. In other words, they have *selective toxicity,* being harmful as much as possible to the invading organism but innocuous to the host.

Although some authors impart to chemotherapy a broader meaning, such as treatment of any disease by chemicals, including infectious and noninfectious diseases, such as psychic disorders, most authors adhere to Ehrlich's definition, as is done in this textbook.

Some differences may be pointed out between chemotherapy and pharmacodynamics; among them are the following:

1. Chemotherapeutic agents are used in the treatment and cure of infectious disease; pharmacodynamic agents are used for relief and correction of abnormal functions.

2. Chemotherapeutic agents usually exert an irreversible action, by attaching strongly, sometimes through a covalent bond, to special moieties of macromolecules of the invading organism; pharmacodynamic agents should preferably produce reversible results, by forming weak bonds with pharmacological receptors.

3. In chemotherapeutic agents the all-or-nothing effect is unobjectionable, whereas pharmacodynamic agents are expected to give a graded response, according to the doses administered.

4. Potential chemotherapeutic agents are often easily screened, because in many cases it is very simple to isolate the invading organism and study it separately; pharmacodynamic agents are more difficult to test, because it is not possible as yet to isolate receptor molecules.

REFERENCES

C.-G. Heden and T. Illeni, Eds., *New Approaches to the Identification of Microorganisms,* Wiley Biomedical, New York, 1975.

A. Albert, *Selective Toxicity,* 5th ed., Methuen, London, 1973.

W. B. Pratt, *Fundamentals of Chemotherapy,* Oxford University Press, London, 1973.

P. D. Hoeprich, Ed., *Infectious Diseases,* Harper & Row, New York, 1972.

C. E. G. Smith, Ed., "Research in Diseases of the Tropics," *Brit. Med. Bull.,* **28,** 1-95 (1972).

Z. M. Bacq, Ed., *Fundamentals of Biochemical Pharmacology,* Pergamon, Oxford, 1971, pp. 417-570.

T. J. Franklin and G. A. Snow, *Biochemistry of Antimicrobial Action,* Academic, New York, 1971.

H. Seneca, *Biological Basis of Chemotherapy of Infections and Infestations,* Davis, Philadelphia, 1971.

A. Burger, Ed., *Medicinal Chemistry,* 3rd ed., Wiley-Interscience, New York, 1970, Part I.

H. F. Conn, Ed., "Efficacy of Antimicrobial and Antifungal Agents," *Med. Clin. N. Amer.,* **54,** 1075-1350 (1970).

R. J. Schnitzer and F. Hawking, Eds., *Experimental Chemotherapy,* 5 vols., Academic, New York, 1963-1967.

B. A. Newton and P. E. Reynolds, Eds., *Biochemical Studies of Antimicrobial Drugs,* University Press, Cambridge, 1966.

T. S. Work and E. Work, *The Basis of Chemotherapy,* Interscience, New York, 1948.

CHAPTER 25

INTRODUCTION TO
CHEMOTHERAPEUTIC AGENTS

I. BASIC CONSIDERATIONS

Chemotherapeutic agents are chemicals that have selective action on parasites. They are used for the treatment or control of diseases caused by pathogenic invading organisms or cells. According to the parasites on which they act, they are classified as anthelmintic, antiprotozoal, antifungal, antibacterial, antiviral, and antineoplastic agents.

Some chemotherapeutic agents may have one or both of the following effects: (*a*) *static,* when they inhibit further growth or multiplication of the invading organism or cell; (*b*) *cidal,* when they kill or destroy it. For instance, we have both bacterio*static* and bacteri*cidal* agents. Static or cidal effects depend on several factors, such as concentration of the drug, pH, temperature, duration of action, metabolic phase of the invader, presence of interfering substance. Thus, drugs with static effects may exert cidal effects if the doses are increased.

The ideal chemotherapeutic agent is one with selective toxicity to the parasite and innocuity to the host. Since no such an agent exists and very likely never will be developed or found, the relative efficiency and safety of chemotherapeutic agents is indicated by the so-called *chemotherapeutic index,* which is expressed by the relationship

$$\frac{\text{maximal tolerated dose by the host}}{\text{minimal curative dose}}$$

The greater this index, the better the chemotherapeutic agent because of its greater safety to the patient.

In chemotherapy the close interrelationship among three different entities should always be considered: drug, parasite, and host. Each one acts on the remaining two. Thus, the drug acts predominantly on the parasite but also, although less intensively, on the host. The parasite can interfere with both drug and some normal physiological or metabolic processes of the host. In a similar way, host can act on both drug, either activating or inactivating it through enzymatic pathways, and parasite, by phagocytosis, for instance.

II. STRATEGY OF CHEMOTHERAPY

As it was pointed out in Chapter 3, Section V.C., the strategy of chemotherapy consists in exploring the morphological and biochemical differences between the invading cell or parasite and the host. These differences are not many, and those that exist are slight, because cells in all living organisms are morphologically and biochemically very similar. Some of these differences, which allow a selective action by chemotherapeutic agents on parasites or invading cells, have already been found and are being explored, as it will be seen in the following chapters.

III. HISTORICAL DEVELOPMENT

Most chemotherapeutic agents were introduced during the last 40 years. A few were known at the beginning of the century. Surprisingly, some drugs still useful against microbes and other parasites date back many centuries before our time.

Thus, to ancient civilizations, such as those of the Chinese, Hindus, Mayas, and Incas, several efficient chemotherapeutic preparations were known. Chinese Emperor Shen Nung (2735 B.C.) compiled a book on therapeutic herbs; one of them is *Ch'ang Shang,* which has been shown in modern times to contain antimalarial alkaloids. The *Papyrus Ebers,* of the Egyptians (about 1500 B.C.), recommends drugs from plant, animal, and mineral kingdoms, and some of these drugs have been found active as chemotherapeutic agents. Chaulmoogra fruit was used by ancient Indians for the treatment of leprosy. Brazilian Indians and peoples of the Far East used ipecacuanha root, from which emetine is extracted, for diarrhea and dysentery, which is now known to be caused by amebiasis. Incas of Peru used cinchona bark, which contains quinine, for the treatment of malaria. Extracts from *Chenopodium anthelminticum,* which contains ascaridol, were used by old Hebrews, Romans, Mayas, and Aztecs, for the treatment of infestations by worms.

In the fourth century B.C., Hippocrates (about 460-370 B.C.), "the father of medicine," recommended metallic salts for some diseases which now are known to be caused by pathogenic organisms. Another Greek, Dioscorides (fl. c. A.D. 50), in his book *De Materia Medica,* for the same purpose prescribed natural products. The naturalist Galenus (A.D. 131-201), born in Pergamon but settled in Rome, advocated herbal mixtures for all types of ailments. The teachings of these three ancient authorities, although advancing some fields of medicine, actually retarded its progress during many centuries.

The Arabs, who conquered a great part of the ancient civilized world in the 7th and 8th centuries, spread their culture, including their medical practice, to European countries. Rhazes and Avicenna, both born in Persia, are the most outstanding figures of Arabian medicine, from which Europeans learned some useful techniques and new drugs, such as mercurial ointment; incidentally, mercury was indicated for alleviation of the symptoms of syphilis by Cumanus in 1495.

In the Middle Ages, the central medical figure was Theophrastus Bombastus von Hohenheim, called Paracelsus (1493-1541), the father of iatrochemistry (*iatro*, in Greek, means physician). He contended that the body was a chemical laboratory, and, therefore, diseases could be cured by administration of chemicals. He introduced tartar emetic, still used for treatment of schistosomiasis.

From Paracelsus's time on, chemotherapy in Europe made slight progress: ipecacuanha and cinchona bark were brought from South America (Brazil and Peru, respectively) in the 17th century. Progress in microbiology and parasitology further advanced chemotherapy, with certain discoveries shown in Table 25.1. The great impetus was given by Paul Ehrlich (1854-1915), called "the father of modern chemotherapy."

IV. DRUG RESISTANCE

A. Causes

The widespread use of chemotherapeutic agents has created a great medical problem: the rapid rise of forms or strains of parasites resistant to several presently available drugs. The emergence of these resistant forms or strains has caused serious clinical consequences, because it has limited the usefulness of many drugs.

Microbial drug resistance was observed long ago. Paul Ehrlich described trypanosomes resistant to arsenicals as early as 1907. Bacterial resistance to sulfonamides and antibiotics appeared soon after the introduction of these agents into medical and veterinary practice. Resistance to other chemotherapeutic agents is known and has been reported. This phenomenon is most common because living organisms—and parasites are living organisms—are naturally endowed with the ability to adapt themselves to the toxic hazards of chemotherapeutic agents.

B. Origin of Drug-Resistant Forms

Drug-resistant forms of parasites originate in one of the following two ways:

Table 25.1 Introduction of Chemotherapeutic Agents

Drug	Synthesized or Isolated by (Year)	Evaluated by (Year)	Used for Treatment of
quinine	Pelletier and Caventou (1820)		Malaria
phenol	Hunt (1849)	Lister (1865)	Topical bacterial infections
benzyl benzoate	Canizzaro (1854)	Sachs (1900)	Scabies
atoxyl	Béchamp (1863)	Thomas and Breinl (1906)	Trypanosomiasis
arsphenamine	Ehrlich and Bertheim (1909)	Ehrlich (1910)	Syphillis
emetine	Pelletier and Magendie (1817)	Vedder (1911)	Amebiasis
tartar emetic		Gaspar Vianna (1912)	Leishmaniasis
		Christopherson (1918)	Schistosomiasis
tryparsamide	Jacobs and Heildelberger (1919)	Brown and Pearce (1919)	Trypanosomiasis
suramin	Fourneau et al. (1916-1923)	Handel and Joettar (1920)	Trypanosomiasis
		Mayer and Zeiss (1920)	Trypanosomiasis
stibophen	Schmidt (1920)	Uhlenhuth et al. (1924)	Schistosomiasis
quinacrine (mepacrine)	Mauss and Mietzch (1932)	Kikuth (1932)	Malaria
prontosil	Mietzsch and Klarer (1932)	Domagk (1935)	Systemic infections by Gram+ bacteria
sulfanilamide	Gelmo (1908)	Tréfouel et al. (1935)	Systemic infections by Gram+ bacteria
benzylpenicillin	Fleming (1929) Chain et al. (1939)	Chain et al. (1939)	Systemic infections by Gram+ bacteria
tyrothricin	Dubos (1939)	Dubos (1939)	Topical bacterial infections
sulfadiazine	Roblin et al. (1940)	Fenistone et al. (1940)	Systemic infections by Gram+ bacteria

Drug	Reference	Reference	Disease
streptomycin	Waksman et al. (1943)	Waksman et al. (1943)	Tuberculosis
aminosalicyclic acid	Seidel and Bittner (1901)	Lehmann (1944)	Tuberculosis
chloroquine	Andersag et al. (1939)		Malaria
lucanthone	Mauss (1940)	Kikuth (1945)	Schistosomiasis
amodiaquine	Burckhalter et al. (1946)	Loeb et al. (1946)	Malaria
diethylcarbamazine	Kushner et al. (1946)	Kikuth et al. (1946)	Filariasis
chlortetracycline	Duggar et al. (1947)	Porter (1946)	Several infectious diseases
chloramphenicol	Burkholder (1947)	Hewitt et al. (1947)	Several infectious diseases
	Long and Troutman (1949)	Duggar et al. (1947)	
		Ehrlich et al. (1947)	
dapsone	Fromm and Wittmann (1908)	Cochrane (1949)	Leprosy
piperazine	Cloez (1853)	Boismaré (1950)	Ascariasis
mercaptopurine	Elion et al. (1951)	Elion et al. (1951)	Leukemia
erythromycin	McGuire et al. (1952)	McGuire et al. (1952)	Systemic infections by Gram+ bacteria
isoniazid	Mayer and Mally (1912)	Fox (1952)	Tuberculosis
nitrofurantoin	Hayes (1952)	Mintzer et al. (1953)	Systemic bacterial infections
tetracycline	Boothe et al. (1953)	English et al. (1953)	Several infectious diseases
	English et al. (1953)	Minieri et al. (1953)	
	Minieri et al. (1953)		
bialamicol	Burckhalter et al. (1946)	Thompson et al. (1955)	Amebiasis
amphotericin B	Gold et al. (1955)	Gold et al. (1955)	Fungal infections
amopyroquine	Burckhalter and Nobles (1952)	Thompson et al. (1958)	Malaria
kanamycin	Umezawa et al. (1958)	Umezawa et al. (1958)	Systemic infections by Gram− bacteria
metronidazole	Anon (1957)	Cosar and Julou (1959)	Trichomoniasis
idoxuridine	Prusof (1959)	Hermann (1961)	Herpes simplex

Table 25.1 Introduction of Chemotherapeutic Agents (continued)

Drug	Synthesized or Isolated by (Year)	Evaluated by (Year)	Used for Treatment of
thiabendazole	Brown et al. (1961)	Brown et al. (1961)	Helminthiasis
nalidixic acid	Lesher et al. (1962)	Lesher et al. (1962)	Systemic infections by Gram⁻ bacteria
amantadine	Stetter et al. (1960)	Davies (1964)	Influenza A_2
hycanthone	Rosi et al. (1956)	Rosi et al. (1965)	Schistosomiasis
tetramisole	Janssen (1966)	Janssen (1966)	Helminthiasis
pyrantel	Austin (1966)	Austin (1966)	Helminthiasis
mebendazole	Janssen (1970)	Janssen (1970)	Helminthiasis

(1) mutations; (2) genetic transfer.

Mutations free from drug resistance probably are not usually induced by drugs. They occur spontaneously, even in the absence of drug, although with low frequency, that is, one gene mutation per 10^5 to 10^{10} cells per cell division, the same order of magnitude as other mutations. Chemotherapeutic agents do not induce mutations; they act merely as screening agents in selecting resistant forms by destroying the sensitive population without injuring the resistant mutants. Spontaneous mutations are, however, minor contributors to the global problem of drug resistance.

Genetic transfer is accomplished by three different mechanisms. In the sequence of increasing importance they are genetic transformation, phage transduction, and conjugation.

In the mechanism of genetic transformation, which does not contribute substantially to the medical problem of drug resistance, the property of drug resistance is transferred to sensitive cells by treating them with naked DNA extracted from drug-resistant mutants.

In the process of phage transduction, which occurs in both Gram-positive and Gram-negative bacteria, genetic information is transferred from the donor to the recipient cell through DNA carried inside a phage. Phage transduction is responsible, for instance, for the resistant forms of *Staphylococcus aureus,* in which drug resistance is imparted by extrachromosomal genetic elements called plasmids.

The mechanism of conjugation is involved in the transfer of drug-resistance genes among various groups of Enterobacteriaceae. First observed in Japan in 1959-1960, now it is known to occur worldwide. The genetic materials associated with this mechanism of drug resistance are bacterial cytoplasmic elements called R factors, which consist of double-stranded DNA existing in circular form. Recent studies indicate that certain R factors are composed of two elements, resistance plasmids and transfer factors, both being independent replicons because they can duplicate themselves. R factors must have existed before the era of modern chemotherapy, because the conjugation process is not directly related to chemotherapeutic agents. The conjugation process occurs by a sort of contact or mating between cells bearing an R factor (R^+) in which the R factor is transferred from one cell to another. R factors are transferred not only to cells of the same species but also to those of related species. For instance, various bacteria, such as *E. coli, Aerobacter, Salmonella, Shigella, Proteus, Serratia,* and *V. cholerae,* can act as hosts for R factors. Furthermore, R factors can be transferred from a pathogenic microorganism, such as *Salmonella,* to a pathogenic bacterium, such as *E. coli,* and back again to other pathogenic bacteria. The problem of drug resistance by transfer of R factors has been aggravated by the widespread use and especially abuse of modern antimicrobial drugs in animal feeds, a practice that selects strains of resistant microorganisms which are subsequently transmitted to humans who eat raw meats or meat products.

C. Biochemical Mechanisms

Many possible mechanisms can explain the resistance of microorganisms to chemotherapeutic agents. According to Bernard Davis, they are:

1. Conversion of an active drug to an inactive derivative by enzyme(s) produced by resistant forms of the microorganism. Examples: inactivation of penicillins by beta-lactamase and inactivation of chloramphenicol by acetylation.

2. Modification of the target site of the chemotherapeutic agent. Examples: alteration of DNA-dependent RNA polymerase in mutants resistant to rifamycins, loss of ribosomes sensitive to streptomycin in resistant forms of *Mycobacterium tuberculosis.*

3. Loss of cell permeability to a drug. Examples: diminished permeability of bacterial cells to tetracyclines, loss of transport function of neoplastic cells resistant to antimetabolites.

4. Increased production of the enzyme inhibited by the chemotherapeutic agent. Example: excess of anthranilate synthetase by bacterial mutants resistant to 5-methyltryptophan.

5. Increased concentration of a metabolite that antagonizes the inhibitor. Examples: certain mutants resistant to sulfonamides.

6. Enhancement of an alternative metabolic route bypassing the inhibited pathway. Example: pathways of purine and pyrimidine nucleotide metabolism.

7. Decreased requirement of a product of the inhibited metabolic system.
Example: disappearance of the need of formylglycinamidine ribonucleotide, the product of amidation of the ribonucleotide of formylglycinamide in the purine nucleotide pathway, when an alternative route is used in this biosynthesis.

D. Control

Drug resistance is usually controlled by the following measures:

1. Prevention of the emergence of drug-resistant mutants during treatment. In some cases this may be accomplished by administration of high "loading" doses of chemotherapeutic agent. In other cases, combined therapy is recommended. For instance, to prevent the outgrowth of streptomycin-resistant mutants of *M. tuberculosis* combined therapy is used with two or more structurally unrelated drugs, such as streptomycin and isoniazid or streptomycin and *p*-aminosalicylic acid. Another example of combined therapy is found in the attempt of preventing the emergence of resistant cells in neoplasias during treatment with chemotherapeutic agents.

2. Prevention of the spreading of resistant mutants. The spread of drug-resistant strains of microorganisms resulted from the careless use of chemo-

therapeutic agents. To prevent further spreading, it is necessary to exert a much stricter control over the use of these agents, limiting it as much as possible.

3. Elimination of resistant strains after their emergence during treatment. This is accomplished by switching to drugs to which the infectious microorganism is sensitive. In therapy by penicillins, resistant mutants can be treated by beta-lactamase-resistant penicillins, such as cloxacillin or methicillin.

E. Superinfection

The use of wide-spectrum chemotherapeutic agents may result in the disturbance of the equilibrium existing among normal flora in the intestinal, vaginal, and respiratory tracts. This situation is advantageous to opportunistic microorganisms, such as *Candida albicans, Staphylococcus, Proteus,* and *Pseudomonas,* to overgrow and thus cause superinfection. This phenomenon may occur with all antibiotics but it is less common or unknown with other chemotherapeutic agents.

F. Microbial Persistence

Some microbes have the ability of surviving during drug exposure in the tissues despite susceptibility to the drug *in vitro.* Thus some microorganisms may exist temporarily in a spheroplast or L-form state, in which they are not sensitive to penicillins, antibiotics that act on cell wall.

Microbial persistence is due either to the protection afforded to the microbes by their surroundings, such as the interior of a cell, or to the refractoriness to chemotherapeutic agents because the pathogens are at a peculiar stage—for instance, the stationary phase of growth—of their life cycle.

V. PRESENT STATUS

Nowadays chemotherapeutic agents are available for use against most parasitic diseases. However, agents are still needed for some of them, such as Chagas' disease, for which no effective drug exists, although nifurtimox has shown some usefulness in the acute form of the disease. Better drugs are needed for the treatment of cancer and viral infections, as well as several other parasitoses: schistosomiasis, malaria, leishmaniasis, other types of helminthiasis, and leprosy, to cite just a few.

REFERENCES

Basic Considerations

S. B. Pessoa and A. V. Martins, *Parasitologia Médica,* 9th ed., Guanabara-Koogan, Rio de Janeiro, 1974.

L. Rey, *Parasitologia,* Guanabara-Koogan, Rio de Janeiro, 1973.

G. Piekarski, *Medizinische Parasitologie,* Springer, Berlin, 1972.

S. B. Pessoa, *Parasitologia Medica,* 8th ed., Guanabara-Koogan, Rio de Janeiro, 1972.

H. W. Brown, *Basic Clinical Parasitology,* Appleton-Century-Crofts, New York, 1969.

A. H. Rose, *Chemical Microbiology,* 2nd ed., Butterworth, London, 1968.

B. D. Davis *et al., Microbiology,* Harper & Row, New York, 1967.

A. W. Jones, *Introduction to Parasitology,* Addison-Wesley, Mass., Reading, 1967.

W. D. Foster, *A History of Parasitology,* Livingstone, Edinburgh, 1965.

L. G. Goodwin and R. H. Nimmo-Smith, Eds., *Drugs, Parasites and Hosts,* Little, Brown, Boston, 1962.

Strategy of Chemotherapy

T. von Brand, *Biochemistry of Parasites,* 2nd ed., Academic, New York, 1973.

A. Korolkovas, *Rev. Bras. Clin. Ter.,* **1**, 769 (1972).

R. M. Hochster and J. M. Quastel, Eds., *Metabolic Inhibitors: A Comprehensive Treatise,* 5 vols., Academic, New York, 1963-1972.

M. Florkin and B. T. Scheer, Eds., *Chemical Zoology,* 9 vols., Academic, New York, 1967-1974.

W. M. Hryniuk and J. R. Bertino, *Advan. Inter. Med.,* **15**, 267 (1969).

B. R. Baker, *Design of Active-Site-Directed Irreversible Enzyme Inhibitors,* Wiley, New York, 1967.

W. G. van der Kloot, *Federation Proc.,* **26**, 975 (1967).

J. L. Webb, *Enzyme and Metabolic Inhibitors,* 3 vols., Academic, New York, 1963-1966.

N. O. Kaplan and M. Friedkin, *Advan. Chemother.,* **1**, 499 (1964).

D. R. Lincicome, *Ann. N. Y. Acad. Sci.,* **133**, 360 (1963).

Eighth Symposium of the Society for General Microbiology, *The Strategy of Chemotherapy,* University Press, Cambridge, 1958.

Historical Development

R. Benviste and J. Davies, *Annu. Rev. Biochem.,* **42**, 471 (1973).

W. F. Bynum, *Bull. Hist. Med.,* **44**, 518 (1970).

G. Sonnedecker, Revisor, *Kremers and Urdang's History of Pharmacy,* 3rd ed., Lippincott, Philadelphia, 1963.

C. D. Leake, *The Old Egyptian Medical Papyri,* University of Kansas Press, Lawrence, 1952.

A. Castiglioni, *A History of Medicine,* Knopf, New York, 1947.

C. E. K. Mees and J. R. Baker, *The Path of Science,* Wiley, New York, 1946.

P. de Kruif, *Microbe Hunters,* Harcourt, Brace, New York, 1932.

Drug Resistance

V. Krčméry, L. Rosival, and T. Watanabe, Eds., *Bacterial Plasmids and Antibiotic Resistance,* Springer, Berlin, 1972.

E. Dulaney and A. I. Laskin, Eds., "The Problem of Drug-resistant Pathogenic Bacteria," *Ann. N.Y. Acad. Sci.,* **182,** 1-415 (1971).

S. Mitsuhashi, Ed., *Transferable Drug Resistance Factor R,* University Park Press, Baltimore, 1971.

W. V. Shaw, *Advan. Pharmacol. Chemother.,* **9,** 131 (1971).

S. Goto and S. Kuwahara, *Progr. Antimicrob. Anticancer Chemother.,* **1,** 419 (1970).

M. G. Sevag and S. J. De Couray, Jr., *Top. Med. Chem.,* **3,** 107 (1970).

J. S. Kiser *et al., Advan. Appl. Microbiol.,* **11,** 77 (1969).

L. H. Schmidt, *Annu. Rev. Microbiol.,* **23,** 427 (1969).

E. S. Anderson, *Annu. Rev. Microbiol.,* **22,** 131 (1968).

E. Meynell *et al., Bacteriol. Rev.,* **32,** 55 (1968).

Present Status

A. J. Gordon and S. G. Gilgore, *Progr. Drug Res.,* **16,** 194 (1972).

R. G. Denkewalter and M. Tishler, *Progr. Drug Res.,* **10,** 11 (1966).

CHAPTER 26

ORGANOMETALLIC COMPOUNDS

I. INTRODUCTION

Organometallic compounds were among the first chemotherapeutic agents. Most of them, however, have been superseded by safer and more potent drugs, especially antibiotics. They present, nevertheless, great historical interest.

As derivatives of heavy metals, these drugs are usually extremely toxic and intensively irritant. Notwithstanding, some of them are used to a certain extent, in special cases, even today.

II. CLASSIFICATION

Organometallic compounds of chemotherapeutic interest are classified as arsenicals, antimonials, and bismuthals. Less important are mercurials and silver salts, both used as antiseptics and studied in Chapter 30.

In all organometallic compounds seen in this chapter, the metal is bound to C atom either directly (C—M) or through another atom (C—X—M). As to comparative stability and ease of preparation, they may be placed in the following sequence: As > Sb > Bi, in C—M, and Bi > Sb > As, in C—X—M. The chemotherapeutic specificity follows the sequence: As > Sb > Bi.

A. Arsenicals

Derivatives of arsenic used in medicine have arsenic atom linked to carbon atom directly. Arsenicals had widespread use for the treatment of syphilis, trypanosomiasis, and several other parasitic diseases. Since the antibiotic era, their use in syphilis was abandoned. But they continue to be used in the treatment of amebic dysentery and of the late neurological stage of African trypanosomiasis.

The therapeutically useful arsenicals can be either pentavalent or trivalent, the latter being the more active and more toxic. Arsenicals belong to one of the following structural types: (a) arsonic acids, $ArAsO(OH)_2$; (b) arsenoso compounds, $ArAs=O$; (c) arseno compounds, $ArAs=AsAr$; (d) thioarsenites, $ArAs(SR)_2$. Contrary to the way they are usually represented (Table 26.1),

376

arsenoso and arseno compounds actually are associated and do not contain either arsenic-oxygen or arsenic-arsenic double bonds:

oxophenarsine

arsphenamine

Compounds of arsenic most widely used are listed in Table 26.1. Besides these, there are several others which are either obsolete or less commonly used: arsenamide, arsphenamine (Salvarsan), arsthinol (Balarsen), butarsen, dichlorphenarsine (Chlorarsen), diphetarsone, melarsonyl (Trimelarsen), neoarsphenamine (Neosalvarsan), sodium arsanilate (Atoxyl), sulfarsphenamine (Myosalvarsan), thiocarbarsone, treparsol.

Many arsenical compounds are active in human African trypanosomiasis. Some are useful in intestinal amebiasis: acetarsone, arsthinol, carbarsone, diphetarsone, glycobiarsol, thiocarbarsone. Still others are or were used in the treatment of syphilis: arsphenamine, neoarsphenamine, oxophenarsine, sulfarsphenamine. Carbarsone and glycobiarsol have also trichomonacidal action. Acetarsone and carbarsone are useful in the treatment of balantidiasis, although tetracyclines are the drugs of choice. Arsenamide has shown activity in filariasis.

Melarsoprol

It is a water-insoluble yellowish powder. Melarsoprol is the drug of choice in the treatment of the late stage of human African trypanosomiasis, being active against most strains of *Trypanosoma rhodesiense* and *T. gambiense,* even against some tryparsamide-resistant ones. Untoward effects, which are common, include *reactive encephalopathy,* the most serious one and often lethal. Others are hemolytic reactions in those patients with G6PD deficiency. Melarsoprol is contraindicated during epidemics of influenza. It is administered intravenously.

B. Antimonials

Although new and less toxic drugs are available, antimonials still have their place in therapy. Two types of antimony compounds are in clinical use: (*a*) organoantimony compounds containing a carbon-antimony bond; (*b*) compounds in which antimony is linked to the organic part of the molecule through oxygen or sulfur.

To the first type belong aromatic stibonic acids, generally represented as $ArSbO(OH)_2$, although their accurate structure is uncertain. They are obtained by reacting antimony(III) oxide and antimony(III) chloride with diazonium salts.

Table 26.1 Arsenic Compounds

Official Name	Proprietary Name	Chemical Name	Structure
carbarsone[+]	Ameban Amebarsone Amibiarson Leucarsone	N-carbamoylarsanilic acid	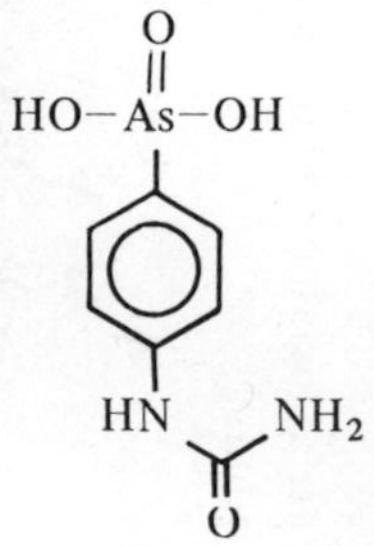
acetarsone	Stovarsol	3-acetamido-4-hydroxy-benzenearsonic acid	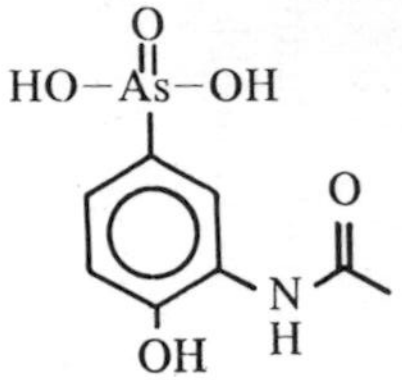
tryparsamide	Tryparsone Trypotane	monosodium N-(carbamoyl-methyl)arsanilate	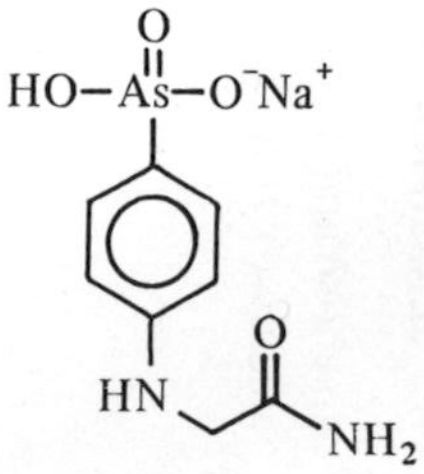

oxophenarsine	Mapharsen	2-amino-4-arsenosophenol 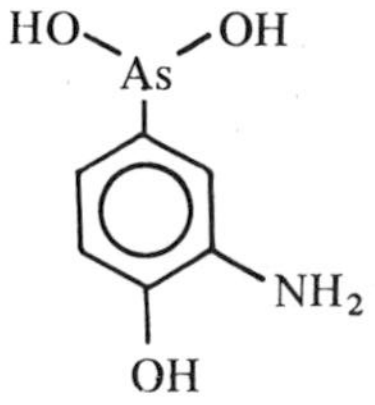
glycobiarsol[+]	Milibis	bismuthyl N-glycoloyl-arsanilate 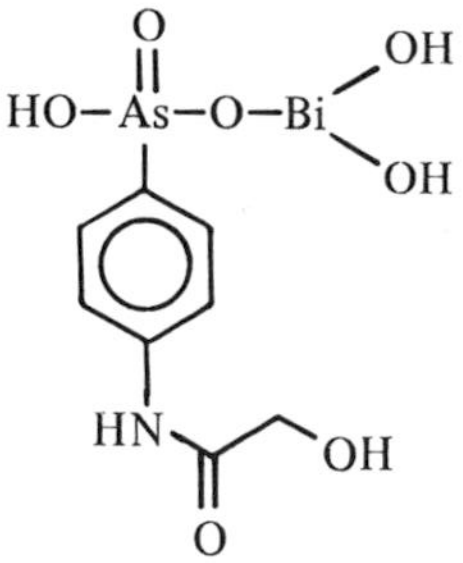
melarsoprol	Arsobal	2-[p-(4,6-diamino-s-triazin-2-ylamino)-phenyl]-1,3,2-dithiar-solane-4-methanol 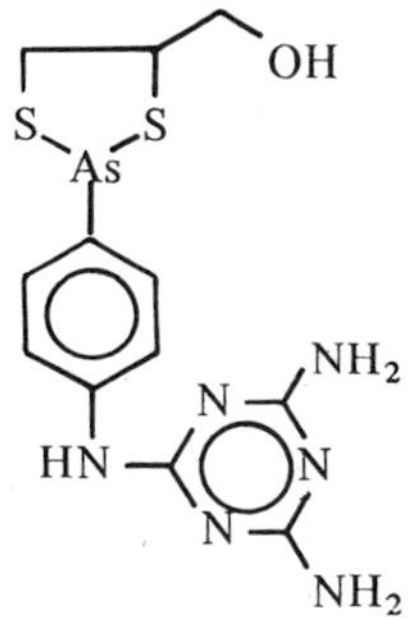

[+]In *NF XIV* (1975).

Table 26.2 Antimony Compounds

Official Name	Proprietary Name	Chemical Name	Structure
antimony potassium tartrate (tartar emetic)*	···	antimonyl potassium tartrate	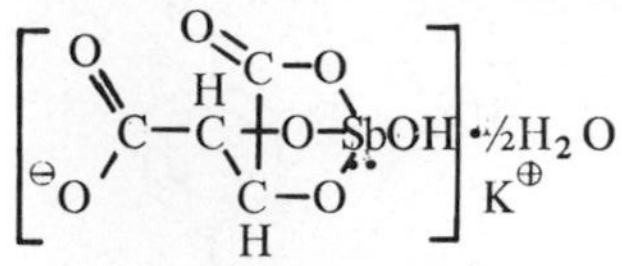
stibophen[+]	Fuadin	pentasodium antimony-bis[catechol-2,4-disulfonate]	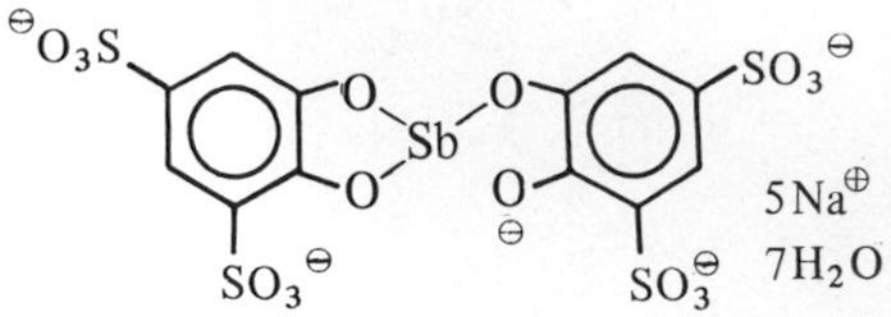
sodium stibocaptate	Astiban	hexasodium salt of the S,S-diester of the cyclic thioantimonate (III) of 2,3-dimercaptosuccinic acid	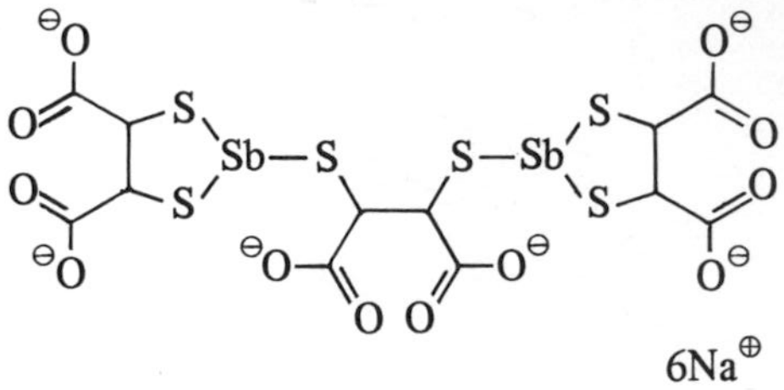

*In *USP XIX* (1975).
[+]In *NF XIV* (1975).

Few compounds of this type have therapeutic value.

In the second type are found most of the antimony compounds used in medicine. They are prepared from antimony(III) oxide, but in the synthesis of a few of them antimony(V) oxide is the starting material.

Antimonials most often used are listed in Table 26.2. Several others, however, are marketed: anthiomaline, antimony sodium thioglycollate, captostibone, ethylstibamine, glucantime, MSb, solusfibosan (Stibanose), ureastibamine.

Compounds of antimony are used in the treatment of (a) leishmaniasis: antimony potassium tartrate, stibophen, anthiomaline, glucantime, solustibosan, stibocaptate, sodium stibogluconate; (b) schistosomiasis: stibophen, antimony potassium tartrate, stibocaptate.

Antimony Potassium Tartrate

Also called tartar emetic, this compound, when anhydrous, is a white powder; when containing water of crystallization, it occurs as colorless, odorless, transparent crystals. It is the most active but also the most toxic of trivalent antimonials. It is the drug of choice in the treatment of infestations by *Schistosoma japonicum,* curing 90% of patients. Untoward effects are nausea, vomiting, diarrhea, cough, colic, bradycardia, syncope, hypotension, and other disorders. It is administered by intravenous route very slowly in freshly prepared 0.5% solution on alternate days 2 hours after a light meal. Tartar emetic is prepared by refluxing potassium hydrogen tartrate with freshly precipitated antimony(III) oxide in aqueous solution.

C. Bismuth Compounds

Bismuth compounds are now considered obsolete. But before the advent of antibiotics, they were used in the treatment of syphilis and several other spirochetal diseases. Among other bismuthals, the following are or were marketed: bismuth soluble, bismuth potassium tartrate, bismuth sodium tartrate, bismuth sodium thioglycollate (Thio-Bismol), bismuth sodium triglycollamate (Bistrimate).

The only organic bismuth compound still used in chemotherapy is glycobiarsol; it is active in amebic dysentery (Tables 26.1 and 29.1). Milk of bismuth (bismuth subnitrate) is listed in *NF XIV* (1975) as an antacid and astringent. In general, bismuth compounds are either dangerous or useless.

III. MECHANISM OF ACTION

Arsenicals, antimonials, and bismuthals act fundamentally by a similar mechanism. They combine with sulfhydryl groups present in essential enzyme systems of the parasite by forming with them a covalent bond and thus cause their toxic

Table 26.3 Antidotes

Official Name	Proprietary Name	Chemical Name	Structure
dimercaprol*	BAL	2,3-dimercapto-1-propanol	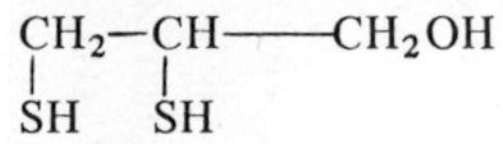 $CH_2-CH-CH_2OH$; SH SH
penicillamine*	Cuprimine	D(−)-3-mercaptovalline	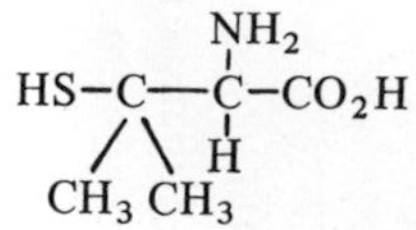 $HS-C-C-CO_2H$ with NH_2, H, CH_3 CH_3
edetate disodium*	Endrate	disodium ethylenediamine-tetraacetate	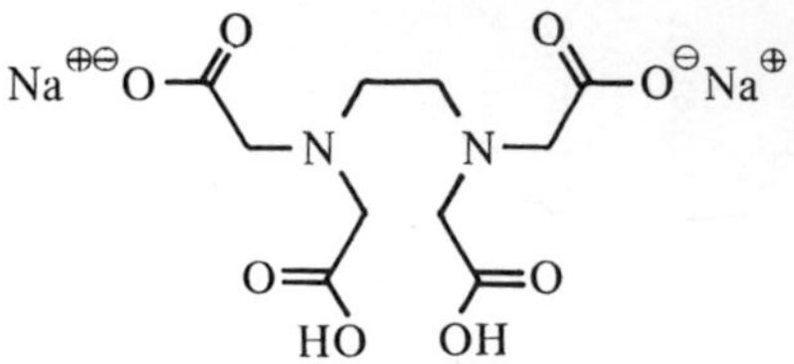 disodium EDTA structure
edetate calcium disodium*	Calcium disodium Versenate	calcium disodium ethylene-diaminetetraacetate	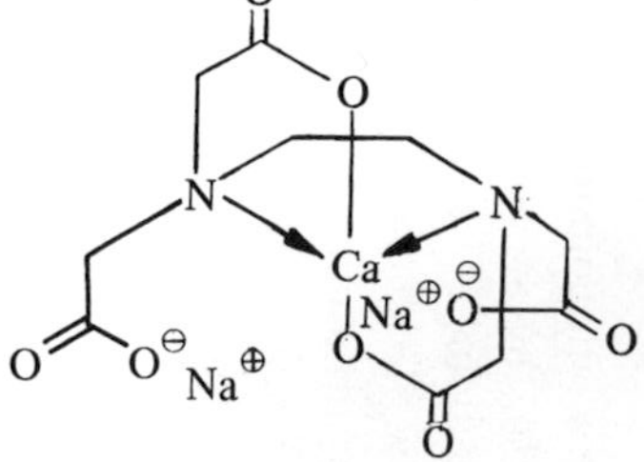 calcium disodium EDTA structure

deferoxamine mesylate*	Desferal	N-[5-(3-[(5-aminopentyl)hydroxy-carbamoyl] propionamido)pentyl]-3-([5-(N-hydroxyacetamido)pentyl]-carbamoyl)propionohydroxamic acid mesylate 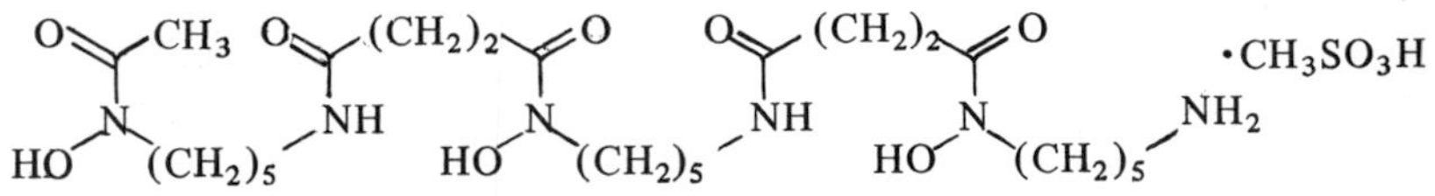

oxine		8-hydroxyquinoline 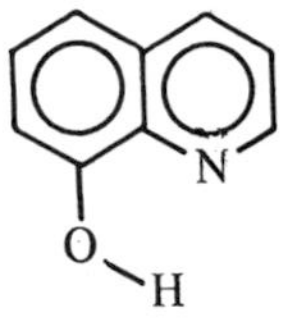

---

*In *USP XIX* (1975).

effect. These compounds, however, act only when in trivalent form. For this reason, arsonic acids should be reduced and arseno compounds have to be oxidized to arsenoso compounds, which are the active forms:

$$R{-}As{=}O \; + \; 2R'{-}SH \; \xrightarrow[-H_2O]{} \; R{-}As\!\begin{smallmatrix} S{-}R' \\ S{-}R' \end{smallmatrix}$$

In the case of antimonials, the reaction is similarly with trivalent compounds. Support for this mechanism is provided by the fact that dimercaprol is an antidote for poisoning, by arsenic, antimony, and bismuth:

$$
\begin{array}{ccl}
CH_2{-}SH & & CH_2{-}S \\
| & & | \quad\;\; X{-}R \\
CH\;{-}SH \;\; + \;\; X{-}R \rightarrow & CH\;{-}S & \qquad X = As, \; Sb, \; or \; Bi \\
| & & | \\
CH_2{-}OH & & CH_2{-}OH
\end{array}
$$

In the case of antimonial schistosomicides, such as antimony potassium tartrate, stibophen, and stibocaptate, for instance, they combine with sulfhydryl groups on phosphofructokinase of both schistosomes and hosts. But mammalian phosphofructokinase is 80 times less sensitive to those drugs than the enzyme of schistosomes. Therefore, antimonials exert a selective toxic action on the phosphofructokinase of schistosomes. This enzyme catalyzes the conversion of fructose-6-phosphate in fructose-1,6-diphosphate in the glycolytic pathways of glycogen and glucose:

$$
\begin{array}{lcl}
glycogen + P_i & antimonials & lactic \; acid \\
\;\; | & \downarrow & \uparrow \\
glucose{\rightarrow}glucose{-}6{-}phosphate \;\; ATP & \;\;\;\; ADP & \\
\;\;\;\; \Updownarrow & & \\
fructose{-}6{-}phosphate & \longrightarrow & fructose{-}1,6{-}diphosphate
\end{array}
$$

IV. ANTIDOTES

To overcome the toxic effects of arsenicals, antimonials, and bismuthals, as well as other poisons, some powerful antidotes have been either discovered or rationally designed. Those most widely used are shown in Table 26.3. Several others are marketed: CDTA (Chel CD), AET (Antiradon), diethylenetriamine-pentaacetic acid (DTPA). Antidotes are reactive agents; several act by the mechanism represented in Section III of this chapter.

Dimercaprol

It was designed in 1940 by Peters and coworkers to be a specific antidote for the toxic war gas Lewisite. It occurs as a clear, colorless liquid that should be kept in a cool place in well-closed containers. Sulfhydryl groups of this agent compete with those of cellular enzymes for heavy metals, such as As, Hg, and Au, to form stable mercaptides that are excreted in the urine. Dimercaprol has also some action in poisoning by Cu, but penicillamine is the drug of choice in this intoxication. The value of dimercaprol in poisoning by Sb and Bi is uncertain. It should not be used in Fe, Cd, or Se poisoning, because the complexes of dimercaprol with these metals are more toxic than the metal alone. Untoward effects are mild and temporary, and some of them can be relieved by antihistamines. It is administered intramuscularly.

Penicillamine

It is a water-soluble, white crystalline powder obtained by degradation of penicillin. The active form is D. Penicillamine was prepared by Abraham and associates in 1943 and evaluated by Walshe in 1956. It chelates Cu, Fe, Hg, Pb, and heavy metals, forming soluble complexes that are excreted in the urine. Untoward effects are hypersensitivity reactions and hematological disorders. The usual dosage is 250 mg four times daily.

Disodium Edetate

Introduced by Sidbury and coworkers in 1953, it is a water-soluble, white, crystalline powder used in the treatment of intoxication by Pb, but it is also valuable in poisoning by Cu, Ni, Zn, Cd, Cr, Mn, and Ca. It is not useful in intoxication by Hg, As, or Au. It may cause thrombophlebitis, chills, fever, muscle cramps, urinary urgency, and other conditions.

Deferoxamine Mesylate

It was isolated in 1960 from *Streptomyces pilosus.* This chelating agent occurs as a white crystalline powder, which is soluble in water. It removes iron from the tissues but it does not displace it from essential proteins (hemoglobin) involved in the iron transport mechanism. It has great affinity for Fe(III) but low for Fe(II). It is administered by injection.

Oxine

It is a water-insoluble, white, crystalline powder. Besides being a powerful antidote for Cu, Zn, Pb, Sr, Ca, Mg, Ni, and Fe, it is active in bacterial and other parasitic infestations—amebiasis and trichomoniasis, for instance. In the form of salts, as citrate, tartrate, benzoate, and sulfate, it is used as an antiseptic.

REFERENCES

General

S. C. Harvey, "Heavy Metals," in L. S. Goodman and A. Gilman, Eds., *The Pharmacological Basis of Therapeutics,* 5th ed., Macmillan, New York, 1975, pp. 924-945.

G. O. Doak and L. D. Freedman, "Arsenicals, Antimonials, and Bismuthals," in A. Burger, Ed., *Medicinal Chemistry,* 3rd ed., Wiley-Interscience, New York, 1970, pp. 610-626.

History

E. A. H. Friedheim *et al., Amer. J. Trop. Med. Hyg.,* 3, 714 (1954).

M. Marquardt, *Paul Ehrlich,* Schuman, New York, 1951.

Classification

D. V. Frost, *Federation Proc.,* 26, 194 (1967).

D. Grdenic and B. Kamenar, *Acta Cryst.,* 19, 196 (1965).

K. Hedberg *et al., Acta Cryst.,* 14, 369 (1961).

S. E. Rasmussen and J. Danielsen, *Acta Chem. Scand.,* 14, 1862 (1960).

F. F. Blicke and F. D. Smith, *J. Amer. Chem. Soc.,* 52, 2946 (1930).

Mechanism of Action

E. Bueding, *Biochem. Pharmacol.,* 18, 1541 (1969).

A. Korolkovas, *Rev. Fac. Farm. Bioquim. S. Paulo,* 5, 5 (1967).

E. Bueding and J. Fisher, *Biochem. Pharmacol.,* 15, 1197 (1966).

H. J. Saz and E. Bueding, *Pharmacol. Rev.,* 18, 871 (1966).

T. E. Mansour, *Advan. Pharmacol.,* 3, 129 (1964).

W. O. Foye, *J. Pharm. Sci.,* 50, 93 (1961).

H. Eagle and G. O. Doak, *Pharmacol. Rev.,* 3, 107 (1951).

C. Voegtlin and H. W. Smith, *J. Pharmacol. Exp. Ther.,* 15, 475 (1920).

Antidotes

W. G. Levin, "Heavy-Metal Antagonists," in L. S. Goodman and A. Gilman, Eds., *The Pharmacological Basis of Therapeutics,* 5th ed., Macmillan, New York, 1975, pp. 912-923.

M. B. Chenoweth, *Clin. Pharmacol. Ther.,* 9, 365 (1968).

F. P. Dwyer and D. P. Mellor, *Chelating Agents and Metal Chelates,* Academic, New York, 1964.

J. F. Frederick, Ed., "Chelation Phenomenon," *Ann. N.Y. Acad. Sci.,* 88, 281-531 (1960).

S. Chaberek and A. E. Martell, *Organic Sequestering Agents,* Wiley, New York, 1959.

M. B. Chenoweth, *Pharmacol. Rev.,* 8, 57 (1956).

A. E. Martell and M. Calvin, *Chemistry of the Metal Chelate Compounds,* Prentice-Hall, New York, 1952.

CHAPTER 27

ANTHELMINTIC AGENTS

I. INTRODUCTION

Anthelmintic agents are drugs used to combat any type of helminthiasis, a disease caused by worms. They act either by destroying the helminths or by expelling them from infested patients. Helminthiasis is the most widespread and common parasitic disease in the world, and it shows a tendency to increase in importance. Although some infestations may go unnoticed, others are debilitating and deadly; such is the case with hookworm disease and schistosomiasis. Heavy infestations with several other worms may be fatal or at least cause serious anemia.

Helminths of interest in medicinal chemistry comprise two phyla of metazoan animals; (1) *Nemathelminthes,* whose most important class is *Nematoda;* (2) *Platyhelminthes,* with many classes, such as *Cestoidea* and *Digenea* (formerly *Trematoda,* which embraced also minor groups), that are the most important. A third phylum is *Acanthocephala,* but species of this phylum infest man only occasionally, because they are animal parasites.

By taking into account the way helminths are lodged in the intestinal tract, they can be divided into the following groups: (*a*) worms attached to the intestinal wall: *Taenia solium, Taenia saginata, Trichuris trichiura, Trichinella spiralis;* (*b*) worms totally embedded in the mucosa: *Strongyloides stercoralis;* (*c*) worms unattached in the intestinal tract: *Ascaris lumbricoides, Enterobius vermicularis.*

There are more than 25 different species of worms that infest mankind. According to a survey published in 1947, helminths infest over 1,100,000,000 members of the human population, including about 40,000,000 Americans. Now this number has increased. It is not unusual that multiple infestations occur in some individuals living in underdeveloped countries. The estimated worldwide number of patients infested with the main types of helminthiasis are (*a*) hookworm disease: 700,000,000; (*b*) ascariasis: 860,000,000, of whom 3,000,000 are in the United States; (*c*) trichuriasis: 475,000,000; (*d*) enterobiasis: 285,000,000, of whom 24,000,000 are in the United States and Canada; (*e*) strongyloidiasis: 34,900,000, of whom 400,000 are in the United States; (*f*) onchocerciasis: 50,000,000; (*g*) trichinosis: 27,800,000, of whom 21,100,000

387

are in the United States and Canada; (h) filariasis: 190,000,000; (i) teniasis: 70,000,000; (j) schistosomiasis: 200,000,000, of whom 8,000,000 are in Brazil. For this reason and because helminths, especially flukes and roundworms, infest also animals that are useful to man, the expression *this wormy world,* used in 1947 by Stoll to describe our planet, is still appropriate. It is also understandable why so much research is aimed at the introduction of new anthelmintic agents. For instance, more than 250,000 compounds have already been screened for activity in schistosomiasis.

In a rational approach to the treatment of infestations by helminths several factors should be considered: (a) the nature of the helminth; (b) the life cycle of the parasite; (c) the reservoir hosts; (d) the intermediate animal host; (e) the site of the human infestation; (f) the definitive human host; (g) the drug to be used in the therapy.

Sometimes comprehensive measures, and not chemotherapy alone, should be resorted to in order to achieve the expected results. For instance, in the case of schistosomiasis, combat against this parasitosis includes blockade of the life cycle of the parasite in one or more phases of its life cycle: (a) destruction of schistosoma ova by special means; (b) prevention of snail-harboring waters from contamination by infestant ova; (c) destruction of miracidia; (d) extermination of snails that are intermediate hosts by molluscicides; (e) prevention of penetration of human skin by cercariae; (f) treatment of the infested patient.

In regard to schistosomiasis, the monumental irrigation schemes devised by the governments—as in Egypt and Brazil, for instance—to improve the living standard of their peoples have resulted ironically in the spread of parasitosis, owing to the increase in the network of waterways which harbor snails as intermediate hosts of the worm. To control the breeding of snails, chemicals are used as molluscicides: copper sulfate, sodium pentachlorophenoxide, niclosamide, N-tritylmorpholine, isobutyltriphenylbutylamine. An alternative method of exterminating snails is through certain fish that eat them.

Screening of potential anthelmintic agents is performed *in vivo* because this method supplies more reliable information than *in vitro* tests. For the assessment of the degree of parasiticidal or antiparasitic activity the following criteria are used: (a) disappearance or decrease of number of ova in animal feces; (b) death of helminths; (c) elimination of helminths from blood; (d) migration of worms within the host to an organ in which they are destroyed by phagocytosis.

Presently efficient drugs are available for most types of helminthiasis. Some of these drugs have broad-spectrum action, and for this reason they are used in mixed infestations. However, for *T. trichiura* no unequivocally effective drug is marketed.

Adverse reactions common to all anthelmintic agents are nausea and vomiting. Other untoward effects depend on the drug used, but all anthelmintics are potentially toxic, especially to patients with severe renal, hepatic, or cardiac disease and to children under 1 year of age. Some helminths, such as schistosomes and hookworms, cause severe anemia. In the case of schistosomiasis, antianemic drugs should be administered before schistosomicidal drugs. In infestations by hookworms, iron should be given before, or concomitantly with, specific therapy.

II. CLASSIFICATION

Some anthelmintics have a broad-spectrum activity. However, most drugs are specific for one or two parasitic infestations. Therefore, the existing anthelmintic drugs can be divided into the following three classes: (*a*) drugs active on nematodes; (*b*) drugs active on cestodes; (*c*) drugs active on digeneae (trematodes).

No longer recommended because of their toxicity and now considered obsolete are the following drugs: ascaridol, carbon tetrachloride, difenan (Bulotan), dithiazanine iodide (Delvex), Egressin, β-naphthol, oil of chenopodium, pelletierine, and santonin.

Some new drugs have shown promising antiparasitic activity and are being submitted to further tests; among many other are the following ones:

1. Human helminthiasis: anthelmycin, carbendazole, cephamycin, dehydroemetine, furapromidium, haloxon, parbendazole, paromomycin, rafoxanide

2. Animal helminthiasis: buquinolate, dimetridazole, levamisole, methyl benzoquate, metichlorpindol, monensin, morantel

Drugs which show activity in animal helminthiasis are subsequently tested in human infestations, often with equivalent results. Sometimes, however, active drugs are too toxic, which precludes their use in humans.

Drugs of choice and alternate drugs for the various helminth infestations are shown in Table 27.1.

A. Drugs Active on Nematodes

Anthelmintic agents most widely used mainly in infestations by nematodes are listed in Table 27.2. However, several other drugs are used in:

1. Human helminthiasis: acetarsol, 1-bromo-2-naphthol, dichlorvos (Atgard, Dichlorman, Equigard, Vapona), diphezyl (considered by the Russians as the drug of choice in trichuriasis), gentian violet, dymanthine, Mantomide, morantel, stilbazium iodide (Monopar), tetramisole (Ascaridil)

2. Animal helminthiasis: buquinolate, butonate, cambendazole, dichlorophen

Table 27.1 Drugs of Choice and Secondary Drugs Used in the Treatment of Infestations by Helminths

Parasitic Disease	Helminth	Drug of Choice	Alternate Drugs
Nematoda (Roundworms)			
Ancylostomiasis	*Ancylostoma duodenale* (hookworm)	bephenium hydroxy-naphthoate	tetrachloroethylene, thiabendazole, mebendazole
	Necator americanus (hookworm)	tetrachloroethylene	bephenium hydroxynaphthoate, mebendazole
Ascariasis	*Ascaris lumbricoides* (roundworm)	thiabendazole piperazine	pyrantel pamoate, thiabendazole, bephenium hydroxynaphthoate, hexylresorcinol, mebendazole. tetramisole
Enterobiasis	*Enterobius vermicularis* (pinworm)	piperazine	pyrvinium pamoate, pyrantel pamoate, thiabendazole, mebendazole, stilbazium iodide
Filariasis	*Wuchereria bancrofti* *Brugia malayi* *Loa loa*	diethylcarbamazine	
Onchocerciasis	*Onchocerca volvulus*	diethylcarbamazine	suramin
Larva migrans (creeping eruption)	*Ancylostoma braziliense* *Ancylostoma caninum*	thiabendazole	diethylcarbamazine
Strongyloidiasis	*Strongyloides stercoralis*	thiabendazole	pyrvinium pamoate, corticotropin
Trichinosis	*Trichinella spiralis*	thiabendazole	
Trichuriasis	*Trichuris trichiura* (whipworm)		bephenium hydroxynaphthoate, hexylresorcinol, mebendazole, dichlorvos, thiabendazole
Dracontiasis	*Dracunculus medinensis* (guinea worm, serpent worm)	niridazole	

Class / Disease	Organism	Drug of choice	Alternative drugs
Cestoidea *(Tapeworms)*			
Intestinal teniasis	*Diphyllobothrium latum* (fish tapeworm) *Hymenolepis nana* (dwarf tapeworm) *Hymenolepis diminuta* *Dipylidium caninum* *Taenia solium* (pork tapeworm) *Taenia saginata* (beef tapeworm)	quinacrine (mepacrine)	niclosamide aspidium oleoresin hexylresorcinol dichlorophen
Cysticercosis	*Echinococcus granulosus* *Taenia solium*	niclosamide	
Digenea *(Trematoidea)* *(Flukes)*			
Liver flukes	*Fasciola hepatica* *Dicrocoelium dendriticum*	bithionol	chloroquine emetine
	Clonorchis sinensis *Opisthorchis felineus* *Fasciolopsis buski*	hexylresorcinol	chloroquine gentian violet tetrachloroethylene, aspidium oleoresin, stilbazium iodide
Intestinal flukes	*Heterophyes heterophyes* *Metagonimus yokogawai*	hexylresorcinol	tetrachloroethylene, aspidium oleoresin, piperazine, bephenium hydroxynaphthoate
	Gastrodiscoides hominis *Watsonius watsoni*	tetrachloroethylene bithionol	
Lung flukes	*Paragonimus westermani*		tartar emetic, emetine, chloroquine, sulfonamides

Table 27.1 Drugs of Choice and Secondary Drugs Used in the Treatment of Infestations by Helminths (continued)

Parasitic Disease	Helminth	Drug of Choice	Alternate Drugs
Schistosomiasis	Schistosoma haematobium	stibophen	stibocaptate, niridazole, hycanth-one, lucanthone, trichlorfon
	Schistosoma japonicum	antimony potassium tartrate	
	Schistosoma mansoni	stibophen	niridazole, hycanthone, lucan-thone, tartar emetic, oxamniquine

(Anthiphen), dichlorvos, disophenol, haloxon, methyridine (Promintic), naphtha-lophos, ruelene, thenium closylate (Bancaris), trichlorfon

Drugs used for the treatment of filariasis only reduce the number of micro-filariae in the peripheral circulation or kill the adult forms of the nematodes, but have little or no effect on the symptoms.

An important group of drugs active on nematodes is cyanine dyes. They have an amidinium ion system, which is characterized by a quaternary nitrogen atom linked to a tertiary nitrogen atom by a conjugated carbon chain of alternate double and single bonds:

$$-\overset{+}{N}=C(-C=C)_n-N= \quad\longleftrightarrow\quad =N-C(=C-C)_n=\overset{+}{N}-$$

This amidinium ion system with resonance structures is apparently essential for anthelmintic activity, because it may participate in charge-transfer interactions. The most useful drugs of this group are pyrvinium pamoate and stilbazium iodide. They are almost insoluble in water. For this reason they are poorly absorbed from the intestinal tract of the patient, remaining for a long time in contact with the intestinal parasites and injuring them.

Pyrvinium

Used as the pamoate, a red salt of cyanine dye, stable to air, light, and heat. This drug is very effective in the treatment of infestations by *E. vermicularis* and *S. stercoralis;* in most cases, one or two doses cure about 90% of infestations. Untoward effects are gastrointestinal disturbances. Pyrvinium stains stool bright red and stains clothing if vomited.

Piperazine

It is marketed either as the free base or more frequently in the form of salts, all forming hexahydrate in solution: citrate, tartrate, adipate, phosphate, calcium edetate. The free base occurs as water-soluble volatile crystals. Piperazine produces some transient untoward effects, such as vomiting, nausea, headache, diarrhea, vertigo, abdominal cramps.

Thiabendazole

It occurs as a slightly water-soluble, white, crystalline substance. It has a broad-spectrum anthelmintic activity, being considered the drug of choice in strongy-loidiasis and larva migrans. It causes transient dizziness, nausea, vomiting, diarrhea, anorexia. The usual dosage is 25 mg per kg of body weight twice daily after meals.

Table 27.2 Drugs Used Mainly in Infestations by Nematodes

Official Name	Proprietary Name	Chemical Name	Structure
tetrachloroethylene*		tetrachlorethene	$Cl_2C=CCl_2$
hexylresorcinol[+]	Crystoids	4-hexylresorcinol	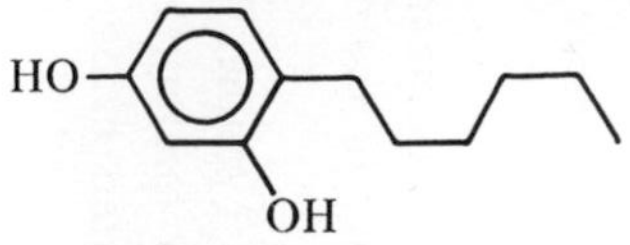
piperazine*[+]	Antepar Arthriticin Perin Piperat		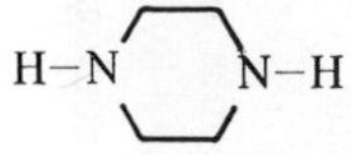
diethylcarbamazine*	Hetrazan	N,N-diethyl-4-methyl- -1-piperazinecarboxamide	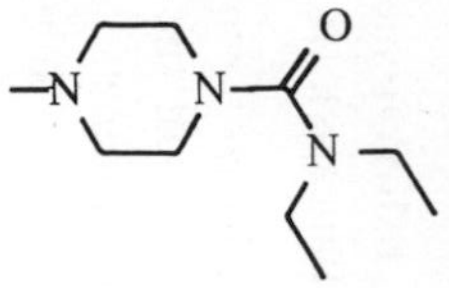
pyrvinium pamoate*	Pamovin Povan Vanquil Vanquin	6-(dimethylamino)-2-[2-(2,5- dimethyl-1-phenylpyrrol-3- yl)-vinyl]-1-methylquino- linium pamoate	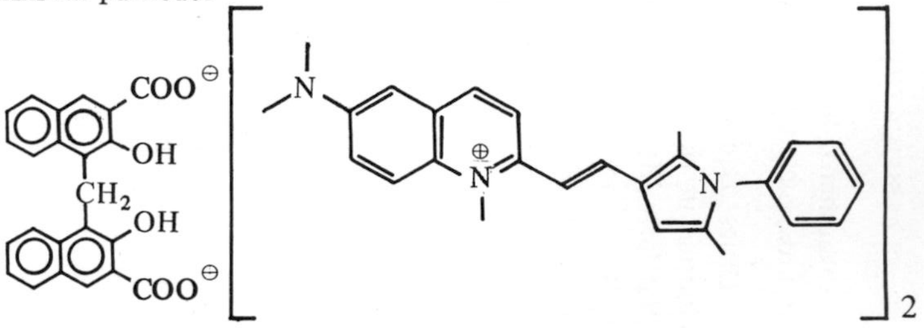

bephenium hydroxynaphtoate*	benzyldimethyl(2-phen-oxyethyl)ammonium 3-hydroxy-2-naphthoate	Alcopar Befeniol Nemex
pyrantel pamoate*	1,4,5,6-tetrahydro-1-methyl-2-[trans-2-(2-thienyl)vinyl]pyrimidine pamoate	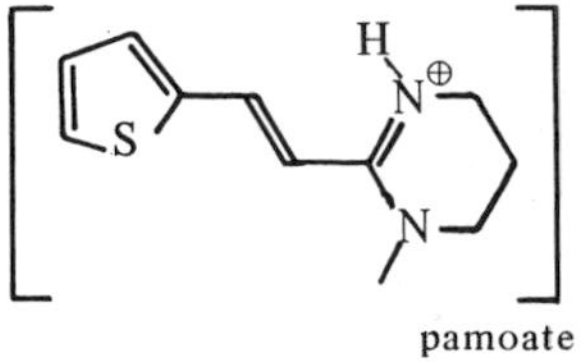 Piranver
thiabendazole*	2-(4-thiazolyl)benzimidazole	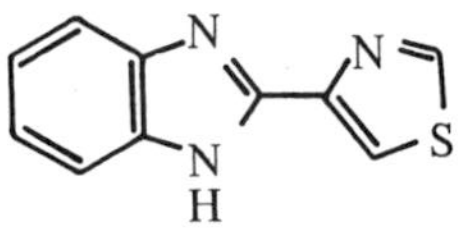 Mintezol
mebendazole	methyl(5-benzoyl-2-benzimidazole)carbamate	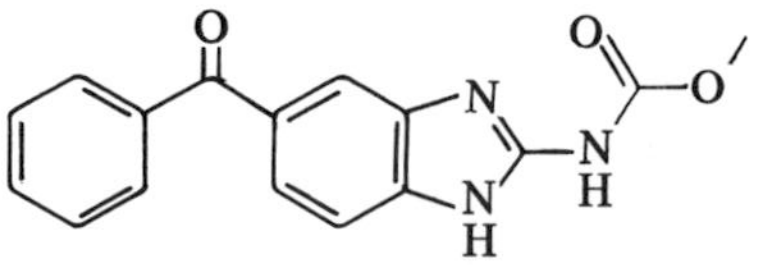 Pantelmin
bitoscanate	p-phenylene-bis-(isothiocyanate)	 Jonit

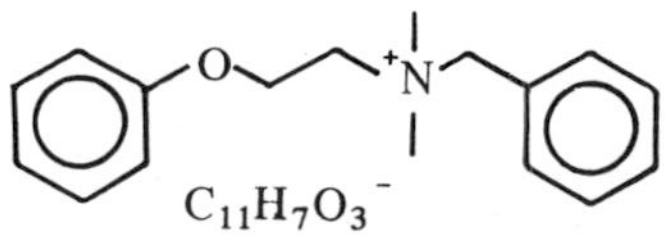

$C_{11}H_7O_3^-$

pamoate

*In *USP XIX* (1975).
†In *NF XIV* (1975).

395

Bephenium Hydroxynaphthoate

It is a water-soluble, yellow, odorless, crystalline powder, with a bitter taste. It is active on both main species of hookworm, being the drug of choice in the treatment of infestations by *A. duodenale,* performing 80 to 98% of cures. It has also some activity in infestations by *A. lumbricoides* and *T. trichiura.* The usual dosage is 5 g twice daily in 1 day for *A. duodenale* and 5 g daily for 3 days for *N. americanus.*

B. Drugs Active Against Cestodes

Table 27.3 lists the most commonly used agents in infestations by cestodes. Other drugs, however, are used. Among them are the following: arecoline, dichlorophen, paromomycin. In veterinary medicine, butamidine and resorantel (Terenol) have shown activity.

Treatment of cestode infestations include the following cares: (*a*) removal of the scolex of the worm, otherwise within 6 to 12 months new proglottids will be formed; (*b*) a highly nutritious diet to the patient, rich in calories, iron, protein, and vitamins, especially to children; (*c*) pre- and postsaline purgation, to enhance the action of the drug, lessen its toxicity, and remove the paralyzed helminth; (*d*) low-residue diet the night before the administration of the drug.

Quinacrine Hydrochloride

It is a sparingly water-soluble crystalline powder with a bitter taste. Solutions are unstable and should not be stored. It is the drug of choice for tapeworm infestations. After treatment, a saline purge is required to expel worms, which are stained bright yellow, and the scolex is looked for; if not found, a patient is considered cured when stools are free of worm eggs and segments for 3 to 6 months. Special preparation of the patient is required before administration of the drug. Care should be taken to prevent vomiting in cases of infestations by *T. solium* because of the hazard of causing cysticercosis. Quinacrine has other useful activity as an anticonvulsant, antimalarial, and antineoplastic agent.

C. Drugs Active on Digeneae (Trematodes)

Search for drugs active on trematodes, now placed in the class *Digenea,* has resulted in several useful agents, some of which are in Table 27.4. However, other drugs are also used, among them the following ones: antimony derivatives (such as antimony sodium tartrate, stibocaptate, and stibophen, studied in Chapter 26), amphotalide (Schistosomide), bithionol sulfoxide (Bitin S), chloroquine, dehydroemetine, emetine, gentian violet, metrifonate, oxyclozanide, tetrachloroethylene, trichlorfon (Dipterex), tris(*p*-aminophenyl)carbonium (TAC). In veterinary medicine, brotianide (Dirian), elioxanide, and rafoxanide have shown activity in infestations by *F. hepatica.*

Table 27.3 Drugs Used in Infestations by Cestodes

Official Name	Proprietary Name	Chemical Name	Structure
aspidium oleoresin			see text
quinacrine hydrochloride* (mepacrine)	Atabrine	6-chloro-9-([4-diethylamino)-1-methylbutyl]amino)-2-methoxyacridine dihydrochloride	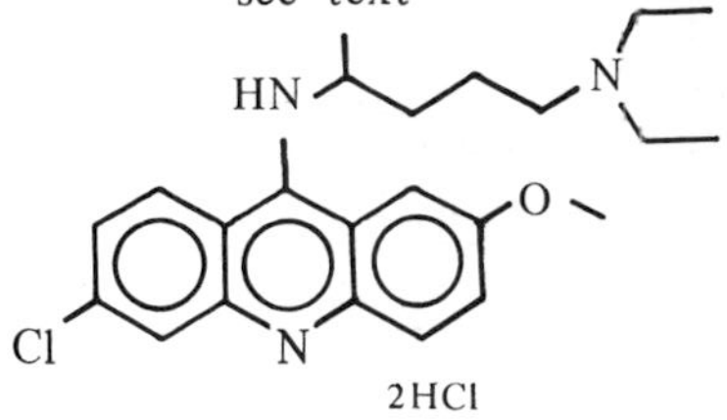 (structure) 2HCl
niclosamide	Cestocide Yomesan	2',5-dichloro-4'-nitrosalicylanilide	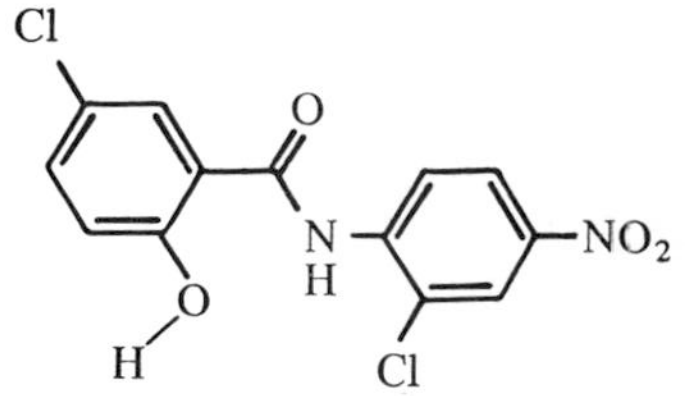 (structure)

*In *USP XIX* (1975).

397

Table 27.4 Drugs Active Mainly against Infestations by Trematodes

Official Name	Proprietary Name	Chemical Name	Structure
bithionol	Bithin	2,2'-thiobis(4,6-dichloro-phenol)	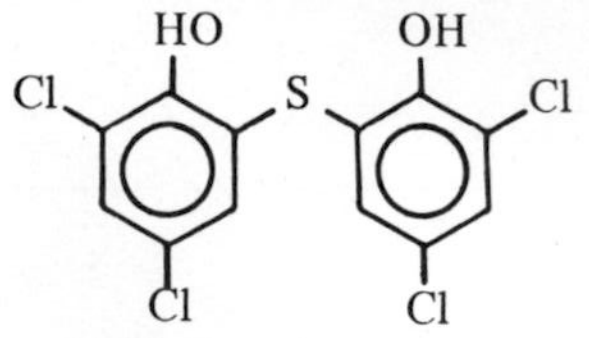
niridazole	Ambilhar	1-(5-nitro-2-thiazolyl)-2-imidazolidinone	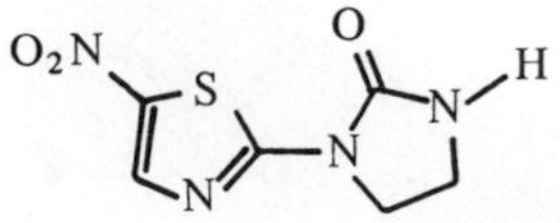
lucanthone	Miracil D Nilodin	1-([2-(diethylamino)ethyl]-amino)-4-methylthioxan-then-9-one	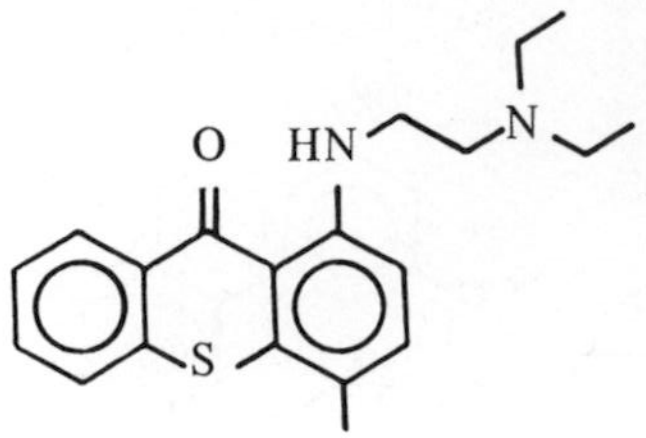
hycanthone	Etrenol	1-([2-diethylamino)ethyl]-amino)-4-(hydroxymethyl)-thioxanthen-9-one	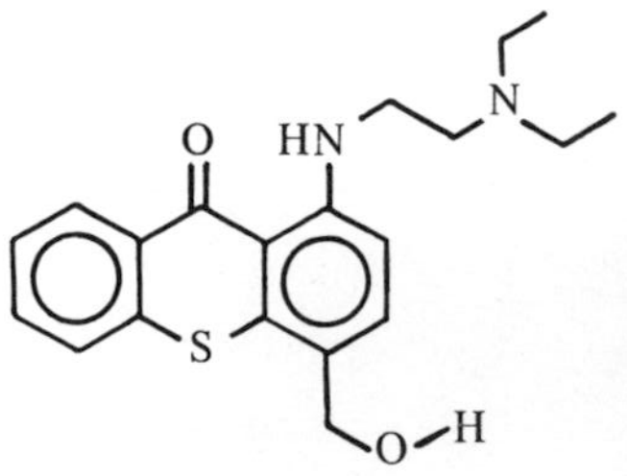

oxamniquine   Mansil   1,2,3,4-tetrahydro-2-
-[(isopropylamino)-
methyl]-7-nitro-6-
-quinolinomethanol

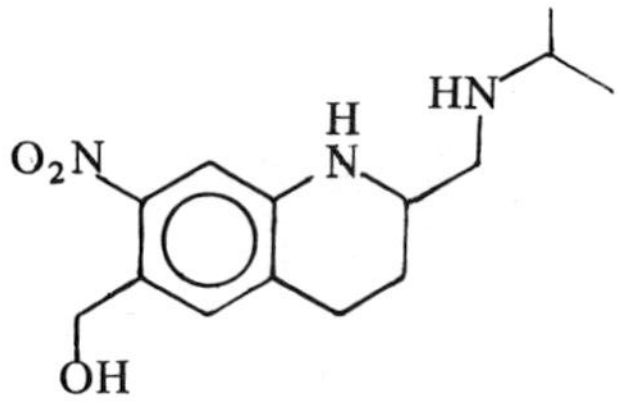

III. MECHANISM OF ACTION

Anthelmintic agents may act by one or more of the following five mechanisms:

1. Direct action, causing narcosis, paralysis, or death of the helminth, with its subsequent elimination. Piperazine acts this way. It paralyzes the musculature of *A. lumbricoides,* either by a neuromuscular blockade or by a hyperpolarization mechanism. But it also inhibits succinic dehydrogenase of the worm. Similarly tetrachloroethylene exerts a paralyzing action on the worm, and purgation eliminates it; however, this drug may act by another mechanism: interference in the process of lysosomal intracellular digestion of nutrients by nematodes.

2. Irritant action, burning the worm tissues. This mechanism is operative in hexylresorcinol and related compounds, which are vermicidal on hookworms, *A. lumbricoides,* and *T. trichiura.*

3. Mechanical action, causing disturbance to the worm, thus forcing it to migrate and subsequently be destroyed by phagocytosis. Diethylcarbamazine acts by this mechanism in filariasis. It quickly kills the microfilariae and renders them vulnerable to phagocytosis or sterilizes adult females of the helminths that cause the infestation.

4. Enzymatic action, digesting the worm by a proteolytic agent. Niclosamide and dichlorophen act as uncouplers of oxidative phosphorylation. After this initial attack of these drugs, helminths become highly vulnerable to the proteolytic enzymes of the host intestine and undergo partial digestion.

5. Antimetabolic action, interfering through different mechanisms on the metabolism of the helminth. This last mechanism is the most common. Several anthelmintic agents act by inhibiting specific enzymes of the worms. For instance, pyrantel (which acts also as a depolarizing block agent), tris(*p*-aminophenyl)carbonium, dichlorvos, haloxon, trichlorfon, and some other agents inhibit the worm's acetylcholinesterase. Thiabendazole and tetramisole inhibit succinate dehydrogenase (fumarate reductase), the latter one possibly by a hydrolysis product; this inhibition cuts off the worm's energy supply and results in its paralysis and elimination from the gut. Antimonial schistosomicides inhibit phosphofructokinase, as was seen in Chapter 26. Niridazole may affect nitroreductase, but its schistosomicidal action is due to inhibition of schistosome phosphorylase inactivation; this results in the availability of more active phosphorylase, which catalyzes the degradation of glycogen, and this causes a decrease in the glycogen stores.

Bephenium inhibits glucose transport and aerobic muscle glycolysis in parasite.

Cyanine dyes (pyrvinium and stilbazium) interfere with respiratory enzyme systems and also with the absorption of exogenous glucose in intestinal helminths.

Mebendazole inhibits metabolic reactions related to glucose uptake in the worm. This inhibition causes glycogen depletion and a decrease in ATP production. However, apparently it does not interfere with similar reactions in mammals.

Lucanthone, hycanthone, chloroquine, and quinacrine complex with worm's DNA by intercalation, as seen in Figures 3.1, 28.4, and 28.5. This complexation disturbs the worm's normal DNA structure and function, which results in anthelmintic action.

REFERENCES

General

R. Cavier, Ed., *Chemotherapy of Helminthiasis,* Vol. 1, Pergamon, Oxford, 1973.

T. W. M. Cameron, *Advan. Parasitol.,* **2**, 1 (1964).

J. M. Watson, *Medical Helminthology,* Baillière, Tindall and Cox, London, 1960.

Introduction

D. L. Lee, *Advan. Parasitol.,* **10**, 347 (1972).

H.-K. Lim and D. Heyneman, *Advan. Parasitol.,* **10**, 192 (1972).

Z. Pawlowski and M. G. Schultz, *Advan. Parasitol.,* **10**, 269 (1972).

A. D. Berrie, *Advan. Parasitol.,* **8**, 43 (1970).

A. S. da Cunha, organizador, *Esquistossomose mansoni,* Sarvier and Universidade de São Paulo, São Paulo, 1970.

B. Dawes and D. L. Hughes, *Advan. Parasitol.,* **8**, 259 (1970).

P. Jordan and G. Webbe, *Human Schistosomiasis,* Thomas, Springfield, Ill., 1970.

S. B. Kendall, *Advan. Parasitol.,* **8**, 251 (1970).

G. S. Nelson, *Advan. Parasitol.,* **8**, 173 (1970).

B. A. Southgate, *J. Trop. Med. Hyg.,* **73**, 235 (1970).

H. W. Brown, *Clin. Pharmacol. Ther.,* **10**, 5 (1969).

J. Pellegrino and N. Katz, *Advan. Parasitol.,* **6**, 233 (1968).

W. P. Rogers and R. I. Sommerville, *Advan. Parasitol.,* **6**, 327 (1968).

"Schistosomiasis: A Symposium," *Bull. N. Y. Acad. Med.,* **44**, 230-374 (1968).

A. Korolkovas, *Rev. Fac. Farm. Bioquim. S. Paulo,* **5**, 5 (1967).

J. Pellegrino, *Parasitol. Rev.,* **21**, 112 (1967).

O. D. Standen, *Trans. Roy. Soc. Trop. Med. Hyg.,* **61**, 563 (1967).

O. Felsenfeld, *The Epidemiology of Tropical Diseases,* Thomas, Springfield, Ill., 1966.

J. F. Maldonado, *Helmintiasis del hombre en América,* Cientifico-Médico, Barcelona, 1965.

S. Yamaguti, *Systema Helminthum,* 5 vols., Interscience, New York, 1958-1963.

Classification

P. B. Hulbert *et al., Science,* **186**, 647 (1974).

K. S. Warren, *Schistosomiasis: The Evolution of a Medical Literature,* MIT Press, Cambridge, 1974.

A. Davis, *Drug Treatment in Intestinal Helminthiases,* World Health Organization, Geneva, 1973.

S. Archer and A. Yarinsky, *Progr. Drug Res.,* **16**, 11 (1972).

J. W. McFarland, *Progr. Drug Res.,* **16**, 157 (1972).

R. S. Desowitz, *Annu. Rev. Pharmacol.,* **11**, 351 (1971).

P. E. Hartman *et al., Science,* **172**, 1058 (1971).

S. H. Rogers and E. Bueding, *Science,* **172**, 1057 (1971).

D. K. Hass, *Top. Med. Chem.,* **3**, 171 (1970).

L. M. Werbel, *Top. Med. Chem.,* **3**, 125 (1970).

T. E. Gibson, *Advan. Parasitol.,* **7**, 349 (1969).

F. C. Goble, Ed., "The Pharmacological and Chemotherapeutic Properties of Niridazole and Other Antischistosomal Compounds," *Ann. N.Y. Acad. Sci.,* **160**, 423-946 (1969).

J. R. Douglas and N. F. Baker, *Annu. Rev. Pharmacol.,* **8**, 213 (1968).

J. E. D. Keeling, *Advan. Chemother.,* **3**, 109 (1968).

A. Korolkovas, *Rev. Fac. Farm. Bioquim. S. Paulo,* **6**, 115, 147, 153 (1968).

G. Lämmler, *Advan. Chemother.,* **3**, 153 (1968).

B. H. Kean, *Ann. Intern. Med.,* **67**, 461 (1967).

P. Thompson, *Annu. Rev. Pharmacol.,* **7**, 77 (1967).

Mechanism of Action

J. P. Brugmans *et al., J. Amer. Med. Assoc.,* **217**, 313 (1971).

F. F. Soprunov, *Tr. Vses. Inst. Gel'mintol.,* **17**, 189 (1971).

I. E. Weinstein and E. Hirschberg, *Progr. Mol. Subcell. Biol.,* **2**, 232 (1971).

M. L. Aubry *et al., Brit. J. Pharmacol.,* **38**, 332 (1970).

C. Bryant, *Advan. Parasitol.,* **8**, 139 (1970).

P. Eyre, *J. Pharm. Pharmacol.,* **22**, 26 (1970).

R. K. Prichard, *Nature (London),* **228**, 684 (1970).

H. Van den Bossche and P. A. J. Janssen, *Biochem. Pharmacol.,* **18**, 35 (1969).

E. Bueding, *Biochem. Pharmacol.,* **18**, 1541 (1969).

R. A. Carchman *et al., Biochim. Biophys. Acta,* **179**, 158 (1969).

A. Allison, *Advan. Chemother.,* **3**, 253 (1968).

H. J. Saz and E. Bueding, *Pharmacol. Rev.,* **18**, 871 (1966).

CHAPTER 28

ANTIMALARIAL AGENTS

I. INTRODUCTION

Antimalarial agents are drugs used for the treatment or prophylaxis of malaria. The name *malaria* derives from Italian and literally means *bad air,* for it was thought that the disease resulted from effluvia from the marshes. An alternative term, *paludism,* is of French origin and has the same meaning.

This parasitic disease has been known to man for centuries. It is probably the most disseminated endemia. It once covered an area where 1935 million people lived. At the end of 1974, 789 million, corresponding to 41% of the figure, lived in areas from where malaria has been eradicated or eradication was under way. However, 1,136 million (59% of the originally infected areas) of people reside where malaria is still endemic or epidemic. Therefore, despite the great efforts and achievements made by the program of the World Health Organization for the eradication of malaria since 1950, this disease continues to be a major problem in many regions of the world. It is estimated that even now it infects 150 million persons each year, killing at least 1 million of its victims and disabling the rest for long periods.

A. Biology of the Infection

Malaria is caused by protozoa of the genus *Plasmodium* which is introduced through the bite of the female *anopheles* mosquito. A species of plasmodium which causes the disease in birds will not infect man and vice versa. However, human malaria has been transmitted to the monkey. The four species of human malaria are malignant tertian caused by *P. falciparum,* benign tertian caused by *P. vivax,* benign quartan caused by *P. malariae,* and a benign tertian infection caused by *P. ovale.* Tertian and quartan refer to the reproductive cycle of the parasite or the time interval between a patient's fever peaks, which are 48 hours for tertian and 72 hours for quartan. *Falciparum* and *vivax* malaria are by far the most prevalent of the four types. Attacks of the former may be fatal; however, the symptoms, if properly treated with quinine or better still with one of the modern drugs, are unlikely to recur. Attacks of *vivax* malaria, on the other hand, are seldom fatal, but the symptoms

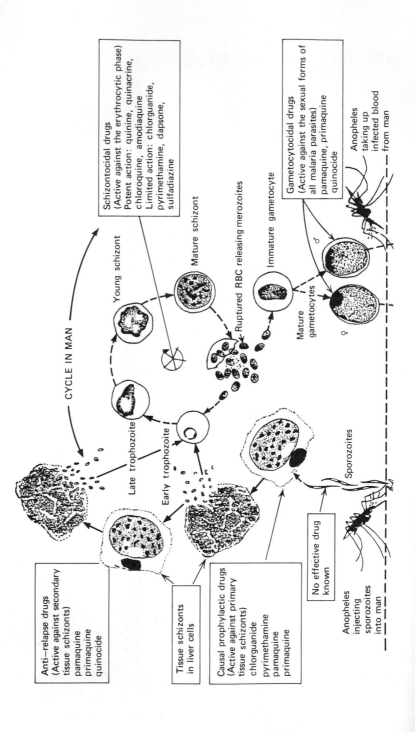

Schizontocidal drugs
(Active against the erythrocytic phase)
Potent action: quinine, quinacrine,
chloroquine, amodiaquine
Limited action: chlorguanide,
pyrimethamine, dapsone,
sulfadiazine

Gametocytocidal drugs
(Active against the sexual forms of
all malaria parasites)
pamaquine, primaquine
quinocide

Anopheles
taking up
infected blood
from man

Young schizont

Mature schizont

Ruptured RBC releasing merozoites

Immature gametocyte

♂

Mature
gametocytes

♀

CYCLE IN MAN

Late trophozoite

Early trophozoite

Sporozoites

No effective drug
known

Anopheles
injecting
sporozoites
into man

Anti–relapse drugs
(Active against secondary
tissue schizonts)
pamaquine
primaquine
quinocide

Tissue schizonts
in liver cells

Causal prophylactic drugs
(Active against primary
tissue schizonts)
chlorguanide
pyrimethamine
pamaquine
primaquine

404

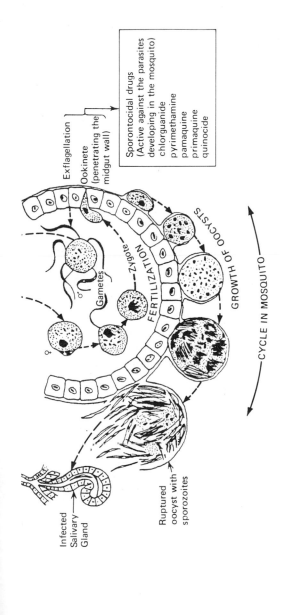

Exflagellation

Ookinete
(penetrating the
midgut wall)

Sporontocidal drugs
(Active against the parasites
developing in the mosquito)
chlorguanide
pyrimethamine
pamaquine
primaquine
quinocide

GROWTH OF OOCYSTS

FERTILIZATION

Zygote

Gametes

♂

♀

CYCLE IN MOSQUITO

Infected
Salivary
Gland

Ruptured
oocyst with
sporozoites

Figure 28.1 Classification of antimalarial drugs in relation to the different stages of the life cycle of the parasite. [After L. J. Bruce-Chwatt, *Bull. Wld. Hlth. Org.,* 27, 287 (1962), slightly modified.]

often recur periodically, even after long lapses of time.

One may conclude from a study of the biology of the infection that to rid the earth of the human malarial parasite either the mosquito or man must be completely destroyed. Toward the former objective, great progress outside the field of chemotherapy has been made with the widespread used of (*a*) insecticides: chlorophenothane (DDT), chlordane, aldrin, dieldrin, benzene hexachloride (BHC), and several others; and (*b*) insect-repellents: diethyltoluamide (Detamide), ethohexadiol, butopyronoxyl (Indalone), dimethyl phthalate (Mipax), oil of citronelia, besides other substances.

B. Types of Action of Antimalarial Agents

According to the stage of the life cycle of the parasite at which they act (Figure 28.1), antimalarial agents are classified as indicated in Table 28.1, to wit:

1. *Blood schizontocide* or *suppressive agent:* a drug which acts on the asexual parasites in the blood.

2. *Tissue schizontocide* or *casual prophylactic agent:* a drug which acts on asexual parasites in the tissues. This may be (*a*) a primary-tissue schizontocide: a drug which acts on preerythrocytic forms; (*b*) a secondary-tissue schizontocide: a drug which acts on secondary exoerythrocytic forms.

3. *Gametocyticide:* a drug which acts on the sexual forms of the parasites.

4. *Sporontocide:* a drug which acts on the sporogenic forms in the mosquito.

The effects produced by antimalarial drugs may be:

1. *Suppression* or *clinical prophylaxis:* the prevention of clinical symptoms by action on asexual parasites in the blood. It may be temporary or permanent.

2. *Clinical cure:* relief of the immediate symptoms of an attack and apparent recovery of the patient without necessarily complete elimination of the infection.

3. *Radical cure:* complete elimination of the parasite, both blood and tissue phases, from the body so that relapse cannot occur.

C. Resistance to Antimalarial Agents

Drug-resistant strains of malarial parasites in experimental animals were first observed in 1945. In 1961 resistance of human plasmodia to chloroquine and other 4-aminoquinolines was described. This discovery was soon followed by reports of the increasing appearance, in many parts of the world, of strains of drug-resistant malarial parasites, not only to 4-aminoquinolines but to other antimalarial drugs as well. Now drug-resistant strains of plasmodia exist in Cambodia, Thailand, Malaya, Brazil, Colombia, Venezuela, and parts of the Phillipines and Vietnam.

Several theories have been proposed to explain the emergence of drug-resistant malarial parasites. The actual mechanism responsible for this phenomenon is essentially the same that was discussed in Chapter 25, Section IV, to wit: (a) inactivation of the drug by the *Plasmodium;* (b) decreased penetration of the drug—chloroquine, for instance—into the protozoan cell; (c) alteration of the *Plasmodium* metabolism so as to eliminate the step or steps previously susceptible to the drug.

Chlorguanide, pyrimethamine, and related drugs that inhibit plasmodial synthesis of folic acid provoke drug resistance readily, whereas quinine, pamaquine, quinacrine, chloroquine, amodiaquine, and primaquine, that act on plasmodial synthesis of nucleic acids, give rise to drug-resistant strains only with difficulty.

The drugs of choice in resistant strains are (a) in acute malaria: pyrimethamine in combination with sulfonamides or sulfones; (b) for suppression of malaria: primaquine.

II. CLASSIFICATION

Antimalarial agents most widely used are listed in Table 28.2. Besides these, several other drugs are marketed and used: (a) acridines: acriquine, aminoacrichin, azacrin (the latter one designed by conjunction of features of 4-aminoquinolines and 8-aminoquinolines); (b) 8-aminoquinolines: quinocide; (c) 4-aminoquinolines: santoquine, oxychloroquine, cycloquine, amopyroquine; (d) amidino ureas: nitroguanil; (e) sulfonamides: sulfadiazine, sulfalene, sulfadoxine, sulfadimethoxine, sulfamethoxypyridazine; (f) sulfones: dapsone, acedapsone. For treatment of drug-resistant strains, several combinations of antimalarial agents are used. Some of these combinations, such as Camoprim (amodiaquine plus primaquine) and chloroquine plus primaquine, have a rational basis and are marketed.

The very useful synthetic antimalarial drugs evolved through molecular modification of methylene blue, quinine, and sulfanilamide, as seen in Figure 2.7 in the case of chlorguanide and Figure 28.2 for other antimalarials.

Drugs used in malaria belong to the following chemical classes: cinchona alkaloids, acridines, 8-aminoquinolines, 4-aminoquinolines, pyrimidines, biguanides, dihydrotriazines, sulfonamides, sulfones. The first four groups present the following general structural features:

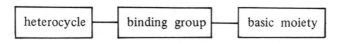

The heterocycle is either quinoline or acridine. The binding group is either an

Table 28.1 Antimalarial Agents Classified According to the Stages of the Life Cycle of the Plasmodia

Stage	Type of Therapy	Effective Drugs	Drug Action
Sporozoites	True causal prophylaxis	None	Destruction of sporozoites or prevention of red-cell invasion by sporozoites
Exoerythrocytic (primary)	Causal prophylaxis (primary-tissue schizontocides)	primaquine chlorguanide pyrimethamine	Destruction of primary-tissue schizonts
Exoerythrocytic (secondary)	Radical cure	primaquine	Complete elimination of schizonts from body so that relapses cannot occur
Erythrocytic	Suppressive (blood schizontocidal)	*Potent action:* quinine quinacrine chloroquine hydroxychloroquine amodiaquine amopyroquine *Limited action:* chlorguanide chlorproguanil pyrimethamine pyrimethamine } + sulfones and sulfonamides chlorguanide	*Prophylactic use* (clinical prophylaxis): prevention of symptoms by destruction of schizonts as they are released from tissue into blood stream *Therapeutic use:* termination of acute attacks and complete elimination of schizonts from blood stream
Erythrocytic	Clinical	Blood schizontocides	Relief of symptoms without necessarily achieving complete elimination of infection

			Destruction of gametocytes
Sexual	Gametocidal	primaquine quinocide	
Mosquito	Sporontocidal	primaquine pyrimethamine chlorguanide	Rendering gametocytes noninfective, thus preventing sporogenous development

Source. Adapted from J. R. diPalma, in J. R. diPalma, Ed., *Drill's Pharmacology in Medicine,* 4th ed., McGraw-Hill, New York, 1971, p. 1773.

Table 28.2 Antimalarial Agents Most Widely Used

Official Name	Proprietary Name	Chemical Name	Structure
quinine*			see Figure 28.3
primaquine*		8-[(4-amino-1-methylbutyl)amino]-6-methoxyquinoline	see Figure 28.2
quinacrine* (mepacrine)	Atabrine Atebrin	6-chloro-9-{[4-(diethylamino)-1-methyl-butyl]amino}-2-methoxyacridine	see Figure 28.2
chloroquine*	Aralen	7-chloro-4-({[4-(diethylamino)-1-methyl-butyl]amino}quinoline	see Figure 28.2
hydroxychloroquine*	Plaquenil	7-chloro-4-{4-[ethyl(2-hydroxyethyl)-amino]-1-methylbutylamino}quinoline	
amodiaquine*	Camoquin	4-[(7-chloro-4-quinolyl)amino]-α-(diethyl-amino)-o-cresol	see Figure 28.2
pyrimethamine*	Daraprim	2,4-diamino-5-(p-chlorophenyl)-6-ethyl-pyrimidine	see Figure 28.2
chlorguanide (proguanil)	Guanatol Paludrine	1-(p-chlorophenyl)-5-isopropylbiguanide	see Figure 28.2
cycloguanil	Camolar	4,6-diamino-1-(p-chlorophenyl)-1,2-dihy-dro-2,2-dimethyl-s-triazine	see Figure 28.2

*In *USP XIX* (1975).

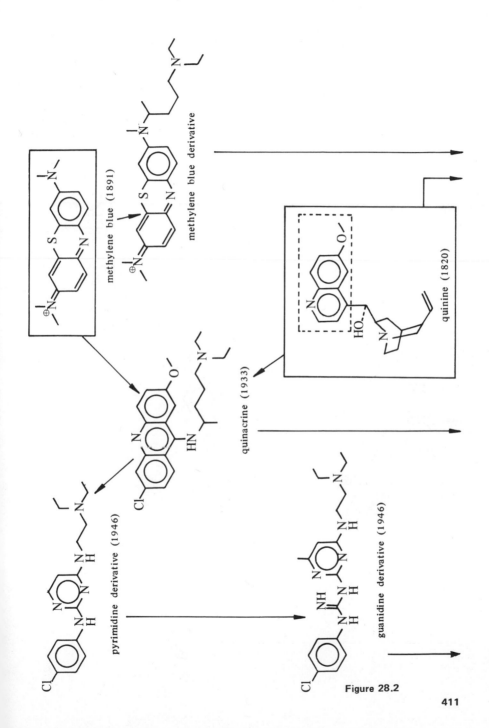

methylene blue (1891)

methylene blue derivative

quinine (1820)

quinacrine (1933)

pyrimidine derivative (1946)

guanidine derivative (1946)

Figure 28.2

411

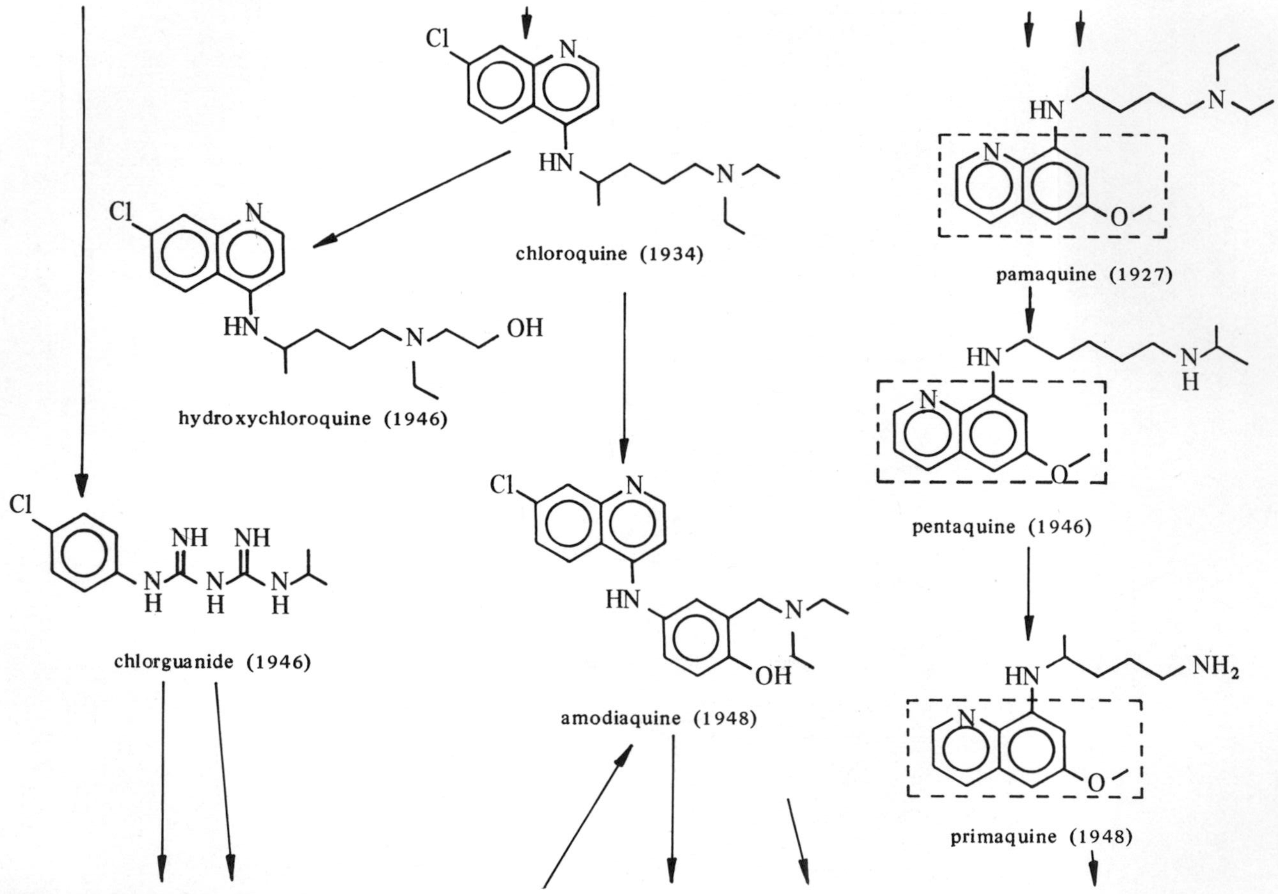

pamaquine (1927)

pentaquine (1946)

primaquine (1948)

chloroquine (1934)

amodiaquine (1948)

hydroxychloroquine (1946)

chlorguanide (1946)

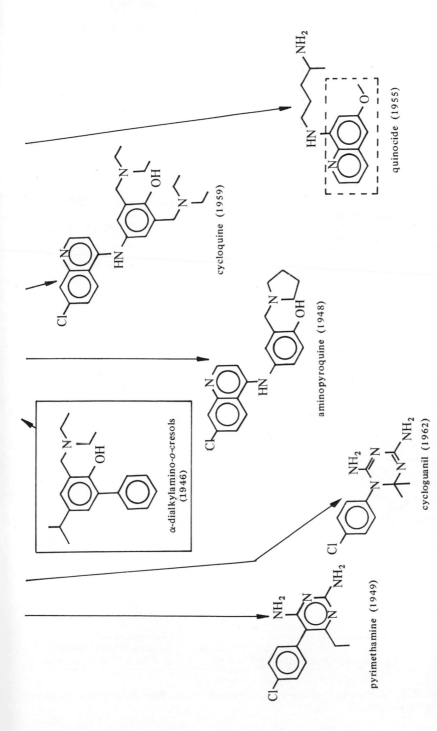

Figure 28.2 Genesis of antimalarial drugs through molecular modification of methylene blue and incorporation of moieties of quinine and α-dialkylamino-*o*-cresols.

413

alcohol (in cinchona alkaloids) or an amine residue. The basic moiety is quinuclidine (in cinchona alkaloids) or a basic chain.

Sulfonamides and sulfones are used in the free form. Other antimalarial agents are used either as free bases or, preferably, as salts (hydrochloride, sulfate, phosphate), which occur as water-soluble, odorless, white or nearly white powders (in general), having a bitter taste.

A. Cinchona Alkaloids

Cinchona alkaloids are extracted from cinchona bark. Several *Cinchona* species, of *Rubiaceae* family, produce these alkaloids. Cinchona is a naturally occurring South American plant. Now it is cultivated extensively in the Congo and especially in Indonesia.

The most important alkaloids found in this plant are two pairs of isomeric compounds: (*a*) quinine and quinidine; (*b*) cinchonine and cinchonidine. Of the four, quinine is the one with greatest antimalarial activity. These alkaloids exist also in the *epi* forms, which are inactive.

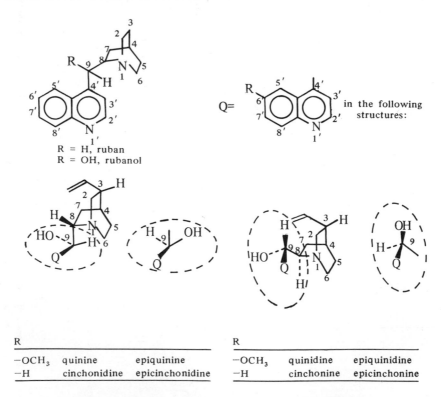

R		
$-OCH_3$	quinine	epiquinine
$-H$	cinchonidine	epicinchonidine

R		
$-OCH_3$	quinidine	epiquinidine
$-H$	cinchonine	epicinchonine

Figure 28.3 Absolute configurations of cinchona alkaloids. Note differences in configuration at C-8 and C-9.

Cinchona alkaloids are derived from ruban, the name Rabe gave to the common structural skeleton. The corresponding alcohol is called rubanol. The structures of ruban, rubanol, and cinchona alkaloids appear in Figure 28.3. Quinine is used as an antimalarial agent, whereas quinidine is an antiarrhythmic drug. Other cinchona alkaloids have no therapeutic use.

Quinine

As a free base, it occurs as a levorotatory, odorless, white, crystalline powder, which is only slightly soluble in water and has an intense bitter taste. It crystallizes with 3 molecules of water of crystallization. Since quinine is a diacidic base, it forms two types of salts: the neutral salts and the acid bisalts. The first involves only the tertiary nitrogen in the quinuclidine nucleus, because this nitrogen is more basic than the quinoline nitrogen, whereas the acid salts involve both basic nitrogens, yielding acidic compounds.

Several salts of quinine are, or have been, used: sulfate, bisulfate, hydrochloride, dihydrochloride, hydrobromide, salicylate, phosphate, ethylcarbamate. Sulfate salt is administered by oral route, and the dihydrochloride is used parenterally in preventing or terminating an acute attack by 4-aminoquinoline-resistant strains of *P. falciparum*. Quinine is ineffective in infections by other species of *Plasmodium*. Furthermore, it is not always effective in achieving radical cure of all infections by *P. falciparum*. To reduce recrudescence of malaria attacks, quinine is used in combination with pyrimethamine.

A common untoward effect of quinine, even with therapeutic doses, is cinchonism, which is characterized by tinnitus, altered auditory acuity, and nausea. Higher doses may cause other undesirable effects, including irreversible auditory or visual damage.

The usual dosage for treatment of acute attacks in nonimmune patients is 650 mg of sulfate or dihydrochloride salts every 8 hours. For treatment of infections caused by *P. falciparum*, 650 mg of sulfate combined with 50 mg of pyrimethamine daily for 3 days, or a combination of quinine sulfate, pyrimethamine, and dapsone given orally.

Quinine is also used as an antipyretic analgetic and in diagnosis of myasthenia gravis. Its older uses, as local anesthetic, stomachic, sclerosing agent, or to induce labor, are dangerous and obsolete.

B. 8-Aminoquinolines

Antimalarial agents derived from 8-aminoquinolines are used mainly on exoerythrocytic forms and on gametocytes. The most active and less toxic is primaquine, the only one now widely used. In the past, pamaquine and pentaquine were used.

Primaquine Phosphate

It occurs as an orange-red, crystalline substance. Its use is in prevention of relapses caused by *P. vivax, P. malariae,* or *P. ovale.* In combination with chloroquine, it is used in the prophylaxis of malaria; with amodiaquine, it is active in all stages of malarial plasmodia. Primaquine causes several untoward effects: abdominal discomfort, headache, pruritus, methemoglobinemia, leukopenia, agranulocytosis. The most serious one, however, is acute hemolytic anemia in patients with deficiency of glucose-6-phosphate dehydrogenase (G6PD), the enzyme that catalyzes the initial oxidative step of the pentose phosphate pathway of glucose metabolism. This deficiency is inherited and occurs prevalently in people of dark-skinned races from the Mediterranean-African area.

C. Acridines

Several acridines have antibacterial action. The one once used in malaria therapy is quinacrine (Atabrine), also called mepacrine. Now, however, less toxic drugs are available, and quinacrine is seldom used as an antimalarial agent. It is, however, the drug of choice in some tapeworm infections (Chapter 27) and in symptomatic giardiasis (Chapter 29).

D. 4-Aminoquinolines

Antimalarial agents derived from 4-aminoquinolines are the drugs of choice for treatment of acute attacks of plasmodia. They achieve a radical cure in non-resistant strains of *P. falciparum* and a clinical cure in infections by other species of plasmodia. However, resistant strains against these agents have been reported from several areas.

Chloroquine

As a free base, it occurs as an odorless, white or yellowish crystalline powder. It is used also as phosphate and hydrochloride, but the latter salt does not crystallize well. In combination with primaquine, it is used in prophylaxis of malaria in endemic areas. Among other adverse reactions, it produces gastrointestinal disorders, central nervous stimulation, and reversible interference with visual accomodation. Overdosage can cause serious reactions, including death.

Amodiaquine

Used as dihydrochloride, a yellow substance, in the prevention or termination of acute attacks. In combination with primaquine it is active against all stages of malaria. It is also useful in the treatment of giardiasis and extraintestinal amebiasis. Untoward effects may be vomiting, nausea, and diarrhea.

E. Pyrimidines

The only antimalarial agents derived from pyrimidine are pyrimethamine and trimethoprim (Syraprim).

Pyrimethamine

Used as a free base, a water-insoluble, odorless, tasteless, crystalline white powder. It is a prophylactic agent. However, plasmodia strains resistant to it are common. For the treatment of acute attacks caused by these strains, pyrimethamine is used in combination with dapsone, sulfisoxazole, and quinine. Pyrimethamine is also useful for the treatment of toxoplasmosis.

F. Biguanides

Biguanides act as primary-tissues schizonts, blood schizonts, and gametocytes, usually as prophylactic agents. Resistant strains against biguanides, with cross-resistance to pyrimethamine, have been found in some areas. Biguanides used as antimalarial agents are chlorguanide (proguanil) and chlorproguanil (Lapudrin).

Chlorguanide

Internationally called proguanil, it is stable in air but slowly darkens on exposure to light. *In vivo* it is metabolized to a cyclic compound, which is the active substance and was introduced in therapy under the name of cycloguanil. Chlorguanide presents the advantage of not causing great incidence of untoward effects.

G. Dihydrotriazines

The only dihydrotriazine used as an antimalarial agent is cycloguanil, a metabolite of chlorguanide.

Cycloguanil Pamoate

It is a repository form of the free base. It provides a protection against infections by *P. falciparum* for several months, being administered by intramuscular route.

H. Sulfonamides

These agents are discussed in detail in Chapter 31, because their main use is as antibacterial agents. Those used as antimalarial agents are methachloridine, sulfadiazine, sulfadoxine, sulfalene, sulfamethoxypyridazine, sulfamonomethoxine.

I. Sulfones

Sulfones are studied in Chapter 32, for their main application is as antilepral agents.

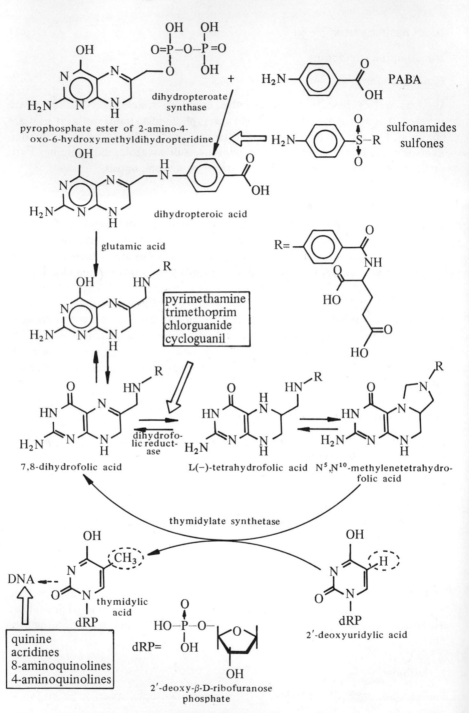

Figure 28.4 Sites of action of antimalarial agents.

III. MECHANISM OF ACTION

Several mechanisms have been proposed for the action of antimalarials. The one advanced by Schönhöfer and widely accepted in the past postulated that for antimalarial activity it was necessary to have a quinoline structure susceptible to oxidation to a quinoid form. This hypothesis is of historical interest, because the evidence now available does not prove the conclusion that 5,6-quinones are the active form of the antimalarial aminoquinolines.

Nevertheless, antimalarial agents act by various mechanisms. At the molecular level they act primarily by either inhibiting enzymes involved in the biosynthesis of precursors of DNA or forming molecular complexes with the DNA itself, thus blocking the synthesis of DNA and RNA of the plasmodia by inhibiting DNA and RNA polymerases. Suggested sites of action of these drugs are shown in Figure 28.4.

Quinine, 8-aminoquinolines and 4-aminoquinolines, owing to their flat ring system, intercalate *in vitro* between base pairs in double-helical DNA and, by their side chains, are electrostatically attached to phosphate groups, as shown in Figures 3.1 and 28.5. Molecular orbital calculations indicate that the flat aromatic ring of these antimalarials, especially in the protonated form which is the structure present during interaction with the malaria parasite, has a low (between 0 and -0.5β) LEMO value, whereas the guanine-cytosine base pair has a high ($+0.487\beta$) HOMO value (Table 3.1). This allows a charge transfer complexation between both entities.

Quinine was considered for a long time a general protoplasmic poison, devoid of appreciable selective action on plasmodia. Now, however, it appears that its antimalarial action is a direct consequence of its attachment to DNA, in the following way: (*a*) the quinoline ring intercalates between the base pairs of double-helical DNA, forming a charge transfer complex; (*b*) the alcoholic hydroxyl group forms a hydrogen bond with one of the DNA bases; (*c*) the quinuclidine moiety projects into one of the grooves of DNA, and its tertiary aliphatic amino group, which is protonated, forms an ionic bond with the negatively charged phosphate group of the deoxyribose-phosphate backbone of the double helix of DNA. This complexation by three portions of the molecule results in a decrease in the effectiveness of the parasite DNA to act as a template.

In the case of aminoquinolines, complexation with DNA is thought to occur (*a*) through both ionic attraction between the protonated nitrogen atoms of the side chain of these drugs and the negatively charged phosphate groups of the complementary strands of double-helical DNA across the minor groove and (*b*) a more specific charge transfer or hydrophobic interaction involving the aromatic ring parts of aminoquinolines and guanine-cytosine bases of DNA. In chloroquine, the 7-chlorine atom is electrostatically attracted to the guanine

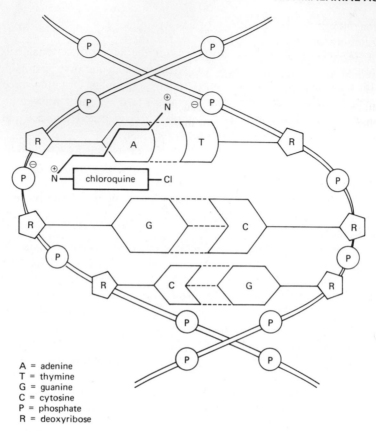

A = adenine
T = thymine
G = guanine
C = cytosine
P = phosphate
R = deoxyribose

Figure 28.5 Mechanism of action of chloroquine at the molecular level.

2-amino group; this ionic bond is thought to be responsible for the guanine specificity of complexation between chloroquine and DNA. Bonding of chloroquine to DNA results in inhibition of DNA to act as a template for DNA and RNA polymerases, because it prevents the required separation of complementary strands of the parent double-helical DNA (Figure 28.5).

Amodiaquine acts in a similar way. However, recent evidence from Burckhalter's laboratories suggests that the aromatic side chain of this drug might be involved in the process of intercalation along with the quinoline ring. (See dust jacket of this textbook.)

Pyrimethamine, trimethoprim, chlorguanide, and cycloguanil owe their antimalarial activity to selective inhibition of dihydrofolate reductase—which converts dihydrofolic acid to tetrahydrofolic acid—of the parasite. This inhibition interferes with plasmodial biosynthesis leading to the formation of purine and pyrimidine bases and eventually of DNA. Although these agents have no selective action on the enzyme of the parasite, they bind to dihydrofolate reductase of plasmodia much more tightly than to the same enzyme of the host. Furthermore, the host is not dangerously affected by this blocked reaction, because the folinic acid needed is supplied by food.

Sulfonamides and sulfones act as antimalarials by inhibiting dihydropteroate synthase, the enzyme that catalyzes the condensation of pyrophosphate ester of 2-amino-4-oxo-6-hydroxymethyldihydropteridine with p-aminobenzoic acid. This interference prevents the incorporation of p-aminobenzoic acid into dihydropteroic acid and results in the parasite's death (Figure 28.4).

REFERENCES

General

R. M. Pinder, *Malaria,* Scientechnica, Bristol, 1973.

P. E. Thompson and L. M. Werbel, *Antimalarial Agents: Chemistry and Pharmacology,* Academic, New York, 1972.

E. A. Steck, *The Chemotherapy of Protozoan Diseases,* Walter Reed Army Institute of Research, Washington, D. C., 1972.

R. M. Pinder, *Progr. Med. Chem.,* 8, 231 (1971).

World Health Organization, *Tech. Rep. Ser.,* no. 467, 1971.

W. Peters, *Chemotherapy and Drug Resistance in Malaria,* Academic, London, 1970.

Introduction

A. Fletcher and B. Maegraith, *Advan. Parasitol.,* 10, 31 (1972).

B. Maegraith and A. Fletcher, *Advan. Parasitol.,* 10, 49 (1972).

F. A. Neva *et al., Ann. Intern. Med.,* 73, 295 (1970).

I. N. Brown, *Advan. Immunol.,* 11, 268 (1969).

R. D. Powell and W. D. Tigertt, *Annu. Rev. Med.,* 19, 81 (1968).

History

J. H. Burckhalter, *Trans. Kansas Acad. Sci.,* 53, 433 (1950).

J. H. Burckhalter *et al., J. Amer. Chem. Soc.,* 70, 1363 (1948).

F. H. S. Curd and F. L. Rose, *J. Chem. Soc.,* 729 (1946).

R. C. Elderfield *et al., J. Amer. Chem. Soc.,* 68, 1516, 1524 (1946).

F. Y. Wiselogle, Ed., *A Survey of Antimalarial Drugs, 1941-1945,* 2 vols., Edwards, Ann Arbor, Mich., 1946.

R. B. Woodward and W. F. Doering, *J. Amer. Chem. Soc.,* 67, 860 (1945).

Classification

R. S. Rozman, *Annu. Rev. Pharmacol.,* **13**, 127 (1973).

H. L. Ammon and L. A. Plastas, *J. Med. Chem.,* **16**, 169 (1973).

W. H. G. Richards, *Advan. Pharmacol. Chemother.,* **8**, 121 (1970).

M. H. Brooks *et al., Clin. Pharmacol. Ther.,* **10**, 85 (1969).

E. F. Elslager, *Progr. Drug Res.,* **13**, 170 (1969).

L. G. Hunsicker, *Arch. Intern. Med.,* **123**, 645 (1969).

P. E. Thompson, *Annu. Rev. Pharmacol.,* **7**, 77 (1967).

Mechanism of Action

J. W. Corcoran and F. E. Hahn, Eds., "Mechanism of Action of Antimicrobial and Anti-tumor Agents," *Antibiotics,* **3**, 58, 203, 274, 304, 516 (1975).

A. Korolkovas, *Rev. Bras. Clin. Ter.,* **4**, 183 (1975).

V. E. Marquez, J. W. Cranston, R. W. Ruddon, and J. H. Burckhalter, *J. Med. Chem.,* **17**, 856 (1974).

C. D. Fitch, *Antimicrob. Agents Chemother.,* **3**, 545 (1973).

L. J. Bruce-Chwatt, *Amer. J. Trop. Med. Hyg.,* **21**, 731 (1972).

C. H. Lantz and K. Van Dyke, *Exp. Parasitol.,* **31**, 255 (1972).

C. H. Lantz and K. Van Dyke, *Biochem. Pharmacol.,* **21**, 891 (1972).

J. H. Burckhalter and V. E. Marquez *et al., J. Med. Chem.,* **15**, 36 (1972).

G. E. Bass *et al., J. Med. Chem.,* **14**, 275 (1971).

W. E. Gutteridge and P. I. Trigg, *Trans. Roy. Soc. Trop. Med. Hyg.,* **64**, 12 (1970).

F. E. Hahn, *Progr. Antimicrob. Anticancer Chemother.,* **2**, 416 (1970).

C. R. Morris *et al., Mol. Pharmacol.,* **6**, 240 (1970).

K. Van Dyke *et al., Science,* **169**, 492 (1970).

R. Ferone *et al., Mol. Pharmacol.,* **5**, 49 (1969).

H. Sternglanz *et al., Mol. Pharmacol.,* **5**, 376 (1969).

J. A. Singer and W. P. Purcell, *J. Med. Chem.,* **10**, 754 (1967).

F. E. Hahn *et al., Mil. Med.,* Suppl. 9, **131**, 1071 (1966).

G. H. Hitchings and J. J. Burchall, *Advan. Enzymol.,* **27**, 417 (1965).

L. S. Lerman, *J. Mol. Biol.,* **10**, 367 (1964).

P. E. Thompson *et al., Amer. J. Trop. Med. Hyg.,* **12**, 481 (1963).

F. Schönhöfer, *Hoppe-Seyler's Z. Physiol. Chem.,* **274**, 1 (1942).

CHAPTER 29

ANTIPROTOZOAL AGENTS

I. INTRODUCTION

Antiprotozoal agents are drugs used in prophylaxis or treatment of parasitic diseases caused by protozoa. Many species of protozoan parasites invade man or animals of interest to man.

These species belong to three different subphyla of the phylum *Protozoa*, as shown:

I. Subphylum *Sarcomastigophora*
    A. Superclass *Mastigophora*
        1. Class *Zoomastigophora*
            a. Order *Kinetoplastida: Trypanosoma gambiense, T. rhodesiense, T. cruzi, Leishmania donovani, L. tropica, L. braziliense*
            b. Order *Retortamonadida: Chilomastix mesnili*
            c. Order *Diplomonadida: Giardia lamblia*
            d. Order *Trichomonadida: Pentatrichomonas hominis, Trichomonas tenax, T. vaginalis*
    B. Superclass *Sarcodina*
        1. Class *Rhizopodea*
            a. Order *Amoebida: Entamoeba histolytica, E. coli, E. hartmanni, E. gingivalis, Endolimax nana, Dientamoeba fragilis, Iodamoeba butschlii*
II. Subphylum *Sporozoa*
        1. Class *Teleosporea*
            a. Order *Cocciddida: Isospora hominis, I. belli*
            b. Order *Haemosporidiida: Plasmodium vivax, P. falciparum, P. malariae, P. ovale*
        2. Class *Toxoplasmea: Toxoplasma gondii*
III. Subphylum *Ciliophora*
            a. Order *Trichostomatida: Balantidium coli*

In the preceding chapter, drugs used in infections caused by species of *Plasmodium* were discussed. In this chapter the agents used in infections caused by other parasitic protozoa are studied. For didactic reasons, these

423

drugs are divided into the following groups: antiamebic agents, antitrypanosomal agents, antileishmanial drugs, antitrichomonal agents, agents active in other protozoan infections, drugs used in veterinary protozoan infections (Table 29.1).

II. ANTIAMEBIC AGENTS

Amebiasis is one of the major causes of illness and death in many countries, especially in Africa, where the morbidity by this parasitic infection is 83.3 per 100,000 inhabitants. However, even in the United States about 5% of the population is infected by *Entamoeba histolytica*, in areas with poor sanitation.

This parasite, as well as other species of the same order, undergoes five successive stages, all in the human intestine, in its life cycle: trophozoite, precyst, cyst, metacyst, and metacystic trophozoite. The infection begins with the ingestion of cyst, which measures 10 to 15 μ. At the small intestine each cyst releases 8 metacysts, which grow into trophozoites, which are 25 μ long. Trophozoite is the mature form of ameba. At this stage it multiplies, invades tissues, and produces lesions. The trophozoite phagocytes bacteria and other food particles and can multiply indefinitely in the intestine. It is the predominant stage in a diarrheic or dysenteric stool. At a given moment, a trophozoite rounds up to form a precyst, which secretes a thin, tough membrane in which it wraps itself, giving origin to a cyst. Maturation of this stage, which occurs in the large bowel or the expelled stool, is reached after two successive divisions of the nucleus. With the formation of the quadrinucleate cyst, *E. histolytica's* life cycle is completed.

Two main types of amebiases are described: (*a*) intestinal amebiasis; (*b*) hepatic amebiasis. For both types metronidazole is presently the drug of choice. For treatment of intestinal amebiasis, the so-called *luminal* or *contact amebicides* are used. For treatment of hepatic amebiasis, drugs called *tissue amebicides* are principally used.

All antiamebic drugs may cause gastrointestinal discomforts, such as nausea and vomiting, diarrhea or constipation, and anorexia.

A. Classification

Drugs used as antiamebic agents belong to the following chemical classes:

1. Active principles from plants: emetine, dehydroemetine, conessine, berberine
2. Antibiotics: tetracycline, chlortetracycline, oxytetracycline, paromomycin, erythromycin, fumagillin

Table 29.1 Drugs of Choice and Secondary Drugs Used in the Treatment of Infections by Protozoal Parasites

Parasitic Disease	Protozoan	Drug of Choice	Alternate Drugs
Amebic infections			
Amebiasis	*Entamoeba histolytica*		
Intestinal amebiasis		metronidazole	carbarsone
			diiodohydroxyquin
			iodochlorhydroxyquin
			paromomycin
			tetracycline
			oxytetracycline
			chlortetracycline
			erythromycin
			niridazole
Asymptomatic intestin-al amebiasis		diiodohydroxyquin	metronidazole
		iodochlorhydroxyquin	
		paromomycin	
Chronic nondysenteric intestinal amebiasis		metronidazole	tetracycline + chloroquine +
			emetine or dehy-droemetine
Hepatic amebiasis		metronidazole	dehydroemetine or emetine + chloro-quine + tetracycline
Amebic diarrhea	*Entamoeba coli*	carbarsone	chlorbetamide
			paromomycin
Amebic diarrhea	*Endolimax nana*	carbarsone	glycobiarsol
Dientamoeba diarrhea	*Dientamoeba fragilis*	diiodohydroxyquin	carbarsone
	Iodamoeba butschlii	diiodohydroxyquin	glycobiarsol
			carbarsone + erythromycin

Table 29.1 Drugs of Choice and Secondary Drugs Used in the Treatment of Infections by Protozoal Parasites (continued)

Parasitic Disease	Protozoan	Drug of Choice	Alternate Drugs
Flagellate infections			
Cutaneous leishmaniasis	*Leishmania tropica*	sodium stibogluconate	ethylstibamine cycloguanil pamoate amphotericin B quinacrine dehydroemetine
Mucocutaneous leishmaniasis	*Leishmania braziliensis*	meglumine antimoniate	pentamidine isethion-ate ethylstibamine amphotericin B
Visceral leishmaniasis	*Leishmania donovani*	hydroxystilbamidine	pentamidine amphotericin B sodium stibogluconate
Chagas' disease	*Trypanosoma cruzi*		
Acute		nifurtimox	
Chronic			
Gambian sleeping sickness	*T. gambiense*		
Early (hemolymphatic) stage		suramin sodium	pentamidine melarsonyl
Late (meningoencephalitic) stage		melarsoprol	melarsonyl nitrofurazone tryparsamide
Rhodesian sleeping sickness	*T. rhodesiense*		
Early febril stage		suramin sodium	melarsonyl
Late stage		melarsoprol	melarsonyl nitrofurazone

Disease	Organism		
Sporozoan infections			
Toxoplasmosis	*Toxoplasma gondii*	pyrimethamine + trisulfa combination	spiramycin
Pneumocystosis	*Pneumocystis carinii*	pentamidine isethionate	
Ciliate infection			
Balantidiasis	*Balantidium coli*	oxytetracycline	carbarsone diiodohydroxyquin
Chilomastix diarrhea	*Chilomastix mesnili*	carbarsone	
Giardiasis	*Giardia lamblia*	metronidazole	quinacrine furazolidone chloroquine amodiaquine nifuratel
Urogenital trichomoniasis	*Trichomonas vaginalis*	metronidazole	furazolidone diiodohydroxyquin iodochlorhydroxyquin povidone-iodine tinidazole
Intestinal trichomoniasis	*Trichomonas hominis*	acetarsone	carbarsone

Table 29.2 Main Antiamebic Drugs

Official Name	Proprietary Name	Chemical Name	Structure
metronidazole*	Flagyl	2-methyl-5-nitroimidazole-1-ethanol	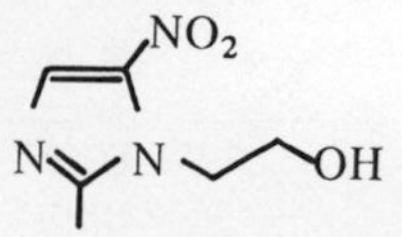
diiodohydroxyquin*	Diodoquin Yodoxin	5,7-diiodo-8-quinolinol	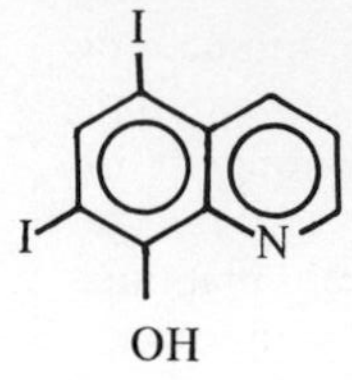
iodochlorhydroxy-quin⁺	Vioform	5-chloro-7-iodo-8-quinolinol	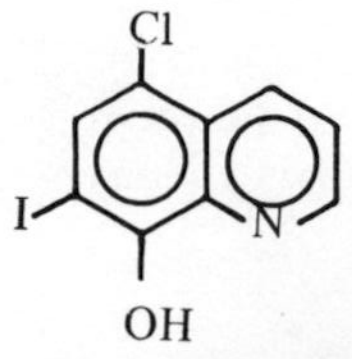

emetine*

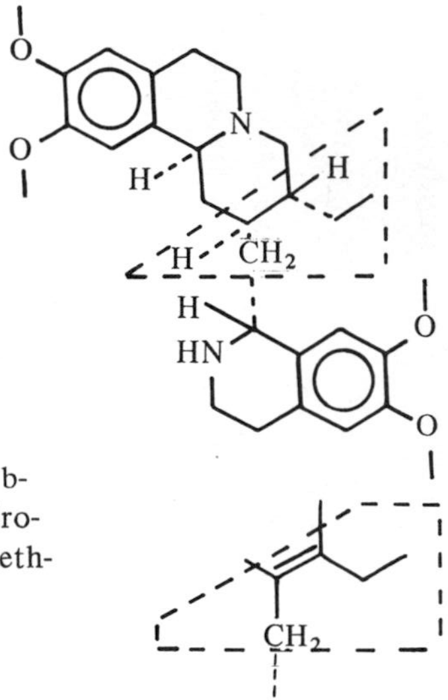

see Table 26.1
see Table 26.1

dehydroemetine   Mebadin   3-ethyl-9,10-dimethoxy-1,6,7,11b-tetrahydro-2-[(1,2,3,4-tetrahydro-6,7-dimethoxy-1-isoquinolyl)methyl]-4H-benzo[a]quinolizine

carbarsone[+]   see Chapter 26   p-ureidobenzenearsonic acid
glycobiarsol[+]   Milibis   bismuthyl N-glycoloylarsanilate
chloroquine*[#]   see Chapter 28
tetracycline*[+]   see Chapter 33
paromomycin[+]   see Chapter 33   Humatin

*In USP XIX (1975).
[+]In NF XIV (1975).

429

3. Arsenicals: carbarsone, glycobiarsol, acetarsol, arsthinol

4. 8-Quinolinol derivatives: diiodohydroxyquin, iodochlorhydroxyquin, chiniofon (Yatren), clamoxyquine (Clamoxyl)

5. Quinoline derivatives: chloroquine, hydroxychloroquine, quinacrine

6. α-Amino-*o*-cresols: bialamicol (Camoform)

7. Haloacetamides: chlorbetamide (Mantomide), clefamide (Mebinol), diloxanide (Entamide), diloxanide furoate (Furamide), teclozan (Falmonox), etofamide (Kitnos)

8. Quinones: phanquone (Entobex)

9. Nitroheterocyclic compounds: metronidazole, niridazole, tinidazole (Fosigyn), panidazole, nifuratel (Macmiror), nimorazole (Naxogin)

Several drugs listed above were already studied in the preceding chapters. Amebicides most widely used are listed in Table 29.2.

Metronidazole

It is the drug of choice for the treatment of both types of amebiases, being used alone. It is also an antitrichomonal agent. Metronidazole is contraindicated in cases of blood dyscrasia and central-nervous-system disease. Its use during the first three months of pregnancy is not recommended. This drug also has been advocated for treatment of chronic alcoholism, because it produces a disulfiramlike effect by causing an accumulation of acetaldehyde, an intermediate in the oxidation of ethanol to carbon dioxide and water.

Diiodohydroxyquin

It is useful mainly in intestinal amebiasis, but it acts also in asymptomatic intestinal amebiasis. This drug, as well as iodochlorhydroxyquin, is contraindicated in cases of hypersensitivity to iodine, or liver disease. Owing to the presence of iodine atoms, both drugs for several months interfere with results of thyroid-function tests. Diiodohydroxyquin is prepared through reaction of an alkaline solution of iodine (NaIO) with an alkaline solution of 8-hydroxyquinoline.

Emetine

This alkaloid is levorotatory and contains two basic nitrogens; therefore, it forms salts, such as hydrochloride, quite easily. Although having emetic action, its main use is in the treatment of both types of amebiasis. Its most serious adverse reactions occur in the cardiovascular system. During treatment patients should be kept in a hospital and remain in bed.

B. Mechanism of Action

Amebicides act by several mechanisms. Some act directly, killing the etiological agent. Others act indirectly, against supporting intestinal bacteria necessary for survival of the amebae. A third class acts both ways, directly and indirectly.

To the first class, called *luminal* or *contact amebicides,* belong carbarsone, diiodohydroxyquin, iodochlorhydroxyquin, and paromomycin. It was once thought that in diiodohydroxyquin and iodochlorhydroxyquin the iodine atom was necessary for amebicidal action. However, compounds structurally similar to these two but without iodine or other halogens prepared by Burckhalter and coworkers have shown antiamebic activity. Therefore, now it is believed that these two compounds and relatives—clamoxyquine, for instance—owe their amebicidal action to chelating properties of the 8-quinolinol moiety. Carbarsone is first reduced to a trivalent arsenical and then it exerts its amebicidal action by combining with thiol groups of the parasite. Paromomycin inhibits synthesis of proteins of *E. histolytica* by interfering in the phase of translation.

Drugs of the second class, comprised of tetracycline, chlortetracycline, oxytetracycline, and erythromycin, act indirectly in the intestinal wall and lumen. They modify intestinal flora required by amebae for their survival. They interfere in the translation process of the synthesis of proteins. Tetracyclines and emetine and its derivative dehydroemetine prevent the binding of aminoacyl-*t*RNA into the 30 S ribosomal subunit of the amebae (Figure 33.5). Erythromycin inhibits translocation in protein synthesis (see Figure 3.7).

To the third class, called *tissue amebicides,* belong metronidazole, emetine, dehydroemetine, and chloroquine and derivatives. Metronidazole exerts its amebicidal action at all sites where amebae are usually found. It is metabolized in the liver to 2-methyl-5-nitroimidazole-2-acetic acid, which is responsible for the amebicidal action. Emetine and dehydroemetine act against amebae mainly in the intestinal wall and the liver. Chloroquine and hydroxychloroquine act against amebae chiefly in the liver; they inhibit DNA replication, as was seen in Chapter 28 (Figure 28.5).

III. ANTITRYPANOSOMAL AGENTS

Trypanosomiasis is caused by flagellate parasites of the genus *Trypanosoma.* Those that infect man are *T. rhodesiense, T. gambiense,* and *T. cruzi.* The first two species are transmitted by the tsetse fly (*Glossina morsitans* and *G. palpalis,* respectively); they cause so-called African trypanosomiasis or sleeping sickness and are responsible for the infection, both in man and domestic animals, in about 4,5 million square miles of tropical Africa. The last species causes Chagas' disease, whose vectors are called "kissing bugs," mainly of the genuses *Triatoma* (principally *T. infestans*) and *Pastrongylus* (specially *P. megistus*). In Brazil those vectors are known by different popular names; the most common is "barbeiro," which means barber. Chagas' disease affects about

12 million people in Central and South America and Mexico.

Several other species of trypanosomes infect only domestic animals, such as cattle, sheep, swine, horses, mules, goats, cats, and dogs; among several others are the following ones: *T. brucei,* the etiological agent of nagana disease; *T. congolense,* which causes a chronic wasting disease; *T. vivax,* responsible for souma; *T. evansi,* which produces surra; *T. equinum,* the etiological agent of mal de caderas; *T. hippicum,* which causes murrina or derrengadera; *T. equiperdum,* responsible for dourine or mal de coit; *T. simiae,* which produces a chronic wasting disease.

T. gambiense and *T. rhodesiense* present two stages in their life cycle: epimastigote, formerly called crithidia stage, which occurs in the tsetse fly and in whose salivary glands it is changed into the infective tripomastigote, formerly called trypanosome stage, which is introduced into the host by the bite of the tsetse fly in search of its blood meal. Tripomastigotes eventually invade the central nervous system, causing the sleeping sickness.

T. cruzi has three stages in its life cycle: amastigote, formerly called leishmania, epimastigote, and tripomastigote. Transmission occurs during the bite of the vector: at the end of the blood meal, the insect expels with the feces some forms of tripomastigotes, and these enter the wound produced by the bite. Enlargement of several organs is one of the symptoms of the disease.

For the treatment of African trypanosomiasis there are several drugs available. All cause toxic effects. For this reason patients should be hospitalized during therapy.

For Chagas' disease only one drug, nifurtimox, is presently available, but it is effective only in some cases of the acute stage of the disease in children and adolescents.

A. Classification

Antitrypanosomal agents presently used either in human or veterinary medicine belong to one of the following classes:

1. Urea derivative: suramin
2. Arsenicals: tryparsamide, melarsoprol, melarsonyl
3. Diamidines: pentamidine, diminazene (Berenil)
4. Quinoline derivative: quinapyramine (Antrycide)
5. Phenanthridinium derivatives: dimidium (Trypadine), homidium (Ethidium), pyrithidium (Prothidium), isometamidium (Samorin), carbidium ethanesulfonate
6. Nitrofurans: furaltadone (Altafur), nitrofurazone, nifurtimox, nifurmazole
7. Antibiotics: puromycin, vermicillin, trypanomycin

Years ago, but now obsolete, were also used the following drugs: *(a)*

Table 29.3 Antitrypanosomal Drugs Most Widely Used

Official Name	Proprietary Name	Chemical Name	Structure
tryparsamide	⋯	see Table 26.1	
melarsoprol	⋯	see Table 26.1	
suramin sodium	Antrypol Bayer 205 Germanin Moranyl	hexasodium salt of 8,8′-(ureylenebis[m-phenylenecarbonylimino(4-methyl-m-phenylene)-carbonylimino])di(1,3,5-naphthalenetrisulfonic) acid	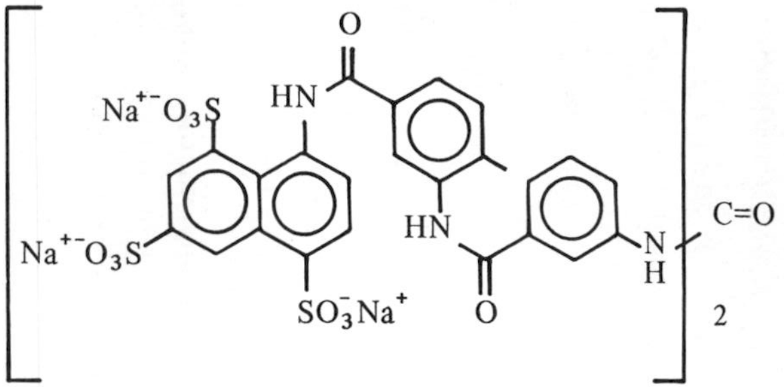
pentamidine isethionate	Lomidine	4,4′-(pentamethylene-dioxy)dibenzamidine	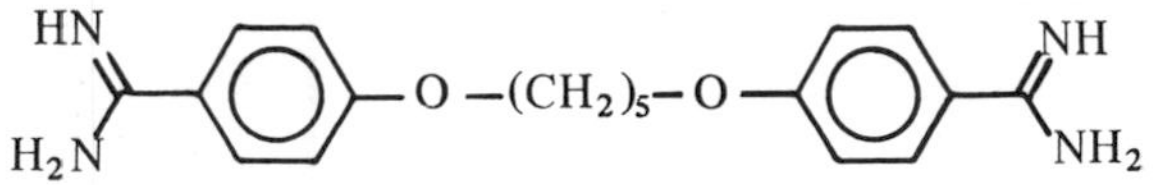
nitrofurazone[+] (nitrofural)	Furacin Yatrocin	5-nitro-2-furaldehyde semicarbazone	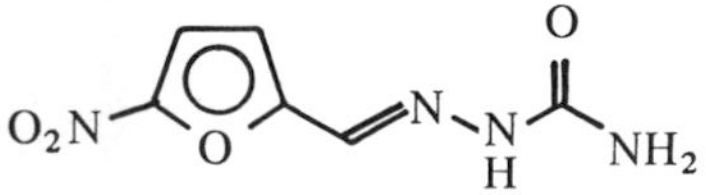
nifurtimox	Lampit	3-methyl-4-(5′-nitrofurfurylideneamino)-tetrahydro-4H-1,4-thiazine 1,1-dioxide	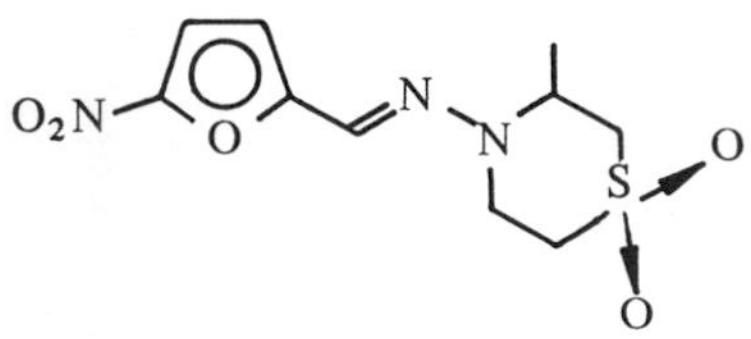

[+]In *NF XIV* (1975).

433

arsenicals and antimonials: tartar emetic, potassium antimonyl tartrate; (b) diamidines: stilbamidine, dimethylstilbamidine, propamidine, phenamidine; (c) phenanthridinium derivatives: phenidium.

The most widely used antitrypanosomal agents in human medicine are listed in Table 29.3.

Suramin Sodium

It is the drug of choice in the early stage of African trypanosomiasis. Because it does not cross the blood-brain barrier, it is not effective in the late meningoencephalitic stage of the disease. Among toxic reactions some are immediate (nausea, vomiting, shock, and syncope); others (albinuria, hematuria, and cylindruria) appear later. The structure of suramin is critical for its chemotherapeutic action. For instance, removal of the two methyl groups results practically in loss of activity. Suramin is also used as salt or complex with other trypanocides: diminazene, quinapyramine, phenanthridium types, Tozocide.

Pentamidine Isethionate

Because it does not cross the blood-brain barrier, it is effective only in the early stage of the disease. It is also used in prophylaxis of Gambian trypanosomiasis. Subcutaneous injection results in severe pain, because this drug is a potent liberator of histamine.

Nifurtimox

It is effective in the acute form of Chagas' disease in children and adolescents; however, it should not be used in children less than 1 year of age. Convulsions, nervousness, amnesia, tremor, and other untoward effects are common.

B. Mechanism of Action

The mechanism of action of most drugs of this class has already been studied in preceding chapters.

As to suramin, it acts primarily on cell division of trypanosomes. It may also sensitize the parasites to phagocytosis in the reticuloendothelial system of the host, but this hypothesis is not endorsed by some authors.

Pentamidine isethionate inhibits several enzyme systems, and it binds to DNA and nucleoproteins; but its mode of action has not been completely elucidated. Recent molecular orbital calculations performed by Korolkovas and Tamashiro on phenamidine, stilbamidine, and diminazene have shown that these drugs have LEMO value (-0.525β, -0.315β, and -0.277β, respectively) adequate to be electron acceptors in a charge transfer process involving DNA's cytosine-guanine pair, for instance, which can act as an electron donor because its HOMO value is $+0.487\beta$. This interaction with a macromolecule of trypanosomes may be operative also in pentamidine.

Phenanthridine derivatives inhibit DNA-dependent RNA polymerase by intercalation between adjacent base pairs of the double helix and complex with DNA primer of trypanosomes. Quinapyramine acts by a similar mechanism: it binds to DNA, as chloroquine does (see Fig. 28.5).

IV. ANTILEISHMANIAL DRUGS

Leishmaniasis is a widely spread disease, being endemic in several parts of Asia, Africa, and Central and South America. It is caused by three different species of *Leishmania* that are pathogenic to man: *L. donovani,* which causes the visceral leishmaniasis, or kala-azar; *L. tropica,* which produces the cutaneous leishmaniasis, also known as Oriental sore; *L. braziliensis,* the causative agent of American leishmaniasis, also called mucocutaneous leishmaniasis or uta or espundia. These diseases afflict not only man but also some other animals. The transmission is usually by the bite of sandflies, of the genus *Phlebotamus.*

The genus *Leishmania* presents only two forms in its life cycle: promastigote form, which is developed in the intestinal tube of invertebrate hosts (insects) and introduced into the mammal by the bite of the insect, where it is transformed into the amastigote form.

A. Classification

Antileishmanial drugs now used can be classified in the following groups:

1. Antimonials: sodium stibogluconate (Solustibosan), ethylstibamine (Neo-astiban, Astaril), urea stibamine (Carbostibamide), stibophen (Fuadin), meglumine antimoniate
2. Aromatic diamidines: hydroxystilbamidine (Table 30.3), stilbamidine isethionate, pentamidine, propamidine, diminazene
3. Antibiotics: amphotericin B, vermicillin
4. Miscellaneous: dehydroemetine, cycloguanil pamoate, niridazole, quinacrine (mepacrine), pyrimethamine, berberine

Infections caused by *L. donovani* are treated with antimonials, aromatic diamidines, and amphotericin B. For the treatment of infections by *L. tropica* quinacrine, berberine, antimonials, and niridazole are preferred. In the treatment of infections by *L. braziliensis* antimonials, cycloguanil pamoate, pyrimethamine, amphotericin B, and vermicillin are used. For prophylaxis of the last type of leishmaniasis pentamidine is indicated.

B. Mechanism of Action

Most drugs studied in this section were already seen in the preceding chapters, where their mechanism of action was discussed. Amphotericin B interacts with thin lipid membranes of the parasites' cells and renders them markedly more permeable to ions.

V. ANTITRICHOMONAL AGENTS

Trichomoniasis is a parasitic infection of the intestinal or genitourinary tract of man and some animals caused by flagellates of the genus *Trichomonas: T. vaginalis* occurs in humans; *T. foetus,* in cattle; and *T. gallinae,* in domestic fowl. The first species lives usually on the vaginal mucosa, but it is found also in other places of the genitourinary tract of both females (40%) and males (10%). The transmission of the infection is either direct, through sexual contact, or indirect, through toilet articles. In females it can cause vaginitis, urethritis, and cystitis. In males, it can produce nonspecific urethritis.

The drug of choice for treatment of trichomoniasis is metronidazole, which has systemic action. As adjunct therapy, some topical agents are used: furazolidone, halogenated hydroxyquinolines, povidone-iodine. Wetting agents (nonoxynol, sodium lauryl sulfate, and dioctyl sodium sulfosuccinate) are added to local trichomonacides to assist penetration into detritus, mucus and pus.

A. Classification

Antitrichomonal agents presently used or that have shown promising activity, either in human or in veterinary medicine, belong to one of the following chemical groups:

1. Arsenicals: carbarsone, glycobiarsol
2. Quinolines: diiodohydroxyquin, iodochlorhydroxyquin
3. Nitrofurans: furazolidone, nifuroxime, nifuratel
4. Nitroimidazoles: metronidazole, tinidazole (Fasigyn, Simplotan), nimorazole (Naxigin, Nulogyl), flunidazole, dimetridazole, panidazole, ipronidazole (Ipropran)
5. Antibiotics: ikarugamycin, trichomycin
6. Miscellaneous: povidone-iodine (Betadine), aminacrine, clotrimazole (Canesten), trimonil (trimagill), triclobisonium chloride (Triburon)

The most widely used drugs are the following ones: metronidazole, diiodohydroxyquin, iodochlorhydroxyquin, povidone-iodine, and some mixtures, especially Tricofuron (furazolidone + nifuroxime), besides various adjuvants.

B. Mechanism of Action

The mode of action of antitrichomonal agents was discussed in the preceding sections of this chapter.

VI. AGENTS ACTIVE IN OTHER PROTOZOAN INFECTIONS

A. Giardiasis

Giardiasis, or lambliasis, is caused by *Giardia lamblia,* formerly called *Lamblia intestinalis,* a flagellate protozoan whose infection interferes with absorption of fat and fat-soluble vitamins. Drugs used for treatment of this parasitic disease are metronidazole, nifuratel, hydroxychloroquine (Plaquenil), furazolidone, and quinacrine.

B. Chilomastix mesnili Infections

Chilomastix mesnili, a flagellate protozoan, is considered by some authors as pathogenic, whereas others classify it as a harmless flagellate commensal. Drugs that have shown activity are carbarsone, glycobiarsol, phenarsone, trichomycin, fumagillin, erythromycin, puromycin, metronidazole, chiniofon.

C. Isosporosis

Isosporosis is caused by *Isospora belli* or *Isospora hominis,* teleosporean protozoa. It is worldwide-distributed but produces a relatively mild and transient gastrointestinal disorder in man. Of far more importance are veterinary coccidiases, caused by species of the genus *Eimeria.* Usually no drug should be taken, because the disease disappears spontaneously. Drugs recommended are bismuth salicylate, and quinacrine.

D. Toxoplasmosis

Toxoplasmosis, a worldwide-distributed disease, is caused by *Toxoplasma gondii,* a toxoplasmid protozoan. It may be either congenital or acquired. For treatment a combination of pyrimethamine with sulfadiazine or sulfones is used. Several antibiotics have shown activity: carbomycin, chlortetracycline, chloramphenicol, erythromycin, fumagillin, spiramycin, trypacidin.

E. Pneumocystosis

Pneumocystosis is a cosmopolitan infection, known as intersticial plasma cell pneumonia, caused by *Pneumocystis carinii,* a toxoplasmid organism. Drugs used are pentamidine, hydroxystilbamidine, stilbamidine.

F. Balantidiasis

Balantidiasis, which occurs in several parts of the world, is an infection produced by the cilliate *Balantidium coli,* whose natural host is swine. Tetracycline, oxytetracycline, paromomycin, and metronidazole are the drugs recommended.

VII. DRUGS USED IN VETERINARY PROTOZOAN INFECTIONS

Several species of protozoa infect not only man but domestic animals or fowl as well; other species are mainly of veterinary interest. The former were studied in the preceding sections. The latter are discussed briefly in this part.

In 1970 it was estimated that the veterinary ethical market represented more than $80,000,000. The cost of introduction of a new drug in this market varies from $150,000, if the drug is already approved for human use, to $1,500,000, if the drug is for use in a food-producing animal. Almost all veterinary drugs in the U.S. now come from one of the following sources: human drugs, foreign veterinary drugs, and over-the-counter animal drugs.

A. Histomoniasis

Histomoniasis is caused by the flagellate *Histomonas meleagridis* and infects chickens and turkeys. Drugs used for treatment of this disease include acinitrazole, dimetridazole, furazolidone, ipronidazole, nithiazide, paromomycin, ronidazole.

B. Babesiasis

Babesiasis refers to a group of diseases affecting domestic animals mainly in temperate zones. It is caused by species of the genus *Babesia.* For the treatment ·of babesiasis the drug of choice is quinuronium. Other drugs used are trypan blue, acriflavine, amicarbalide, diminazene, imidocarb, pentamidine, phenamidine, propamidine, stilbamidine.

C. Coccidiosis

Coccidiosis is caused by species of the sporozoan genus *Eimeria,* infecting domestic animals and fowl. Drugs used for the treatment of this disease include a combination of a sulfonamide (sulfaquinoxaline, sulfamethiazine, sulfaguanidine) with a folic acid inhibitor (pyrimethamine, ormetoprim, cycloguanil), nitrofurazone, furazolidone, nitrophenide, amprolium, meticlor-pindol, buquinolate, methyl benzoquate, monensin, dianemycin, beclotiamine, robenidine, clopidol.

D. Anaplasmosis

Anaplasmosis is caused by *Anaplasma marginale* and it is worldwide distributed. Chlortetracycline, oxytetracycline, and tetracycline are drugs used in the treatment of this disease.

REFERENCES

General

M. Hoffer and C. W. Perry, *Annu. Rep. Med. Chem.*, 8, 141 (1973).

E. A. Steck, *The Chemotherapy of Protozoan Diseases*, Walter Reed Army Institute of Research, Washington, D.C., 1972.

P. E. Thompson, *Annu. Rev. Pharmacol.*, 7, 77 (1967).

S. H. Hutner and A. Lwoff, Eds., *Biochemistry and Physiology of Protozoa*, 3 vols., Academic, New York, 1951, 1955, 1964.

Introduction

K. G. Grell, *Protozoology*, Springer, Berlin, 1973.

N. D. Levine, *Protozoan Parasites of Domestic Animals and of Man*, 2nd ed., Burgess, Minneapolis, 1973.

M. A. Sleigh, *The Biology of Protozoa*, Elsevier, New York, 1973.

R. B. Clark and A. L. Panchen, *Synopsis of Animal Classification*, Chapman & Hall, London, 1971.

Antiamebic Agents

K. W. Jeon, Ed., *The Biology of Amoeba*, Academic, New York, 1973.

S. J. Powell, *Bull. N.Y. Acad. Med.*, 47, 469 (1971).

D. R. Seaton, *Practitioner*, 206, 16 (1971).

F. Scott and M. J. Miller, *J. Amer. Med. Assoc.*, 211, 118 (1970).

R. Elsdon-Dew, *Advan. Parasitol.*, 6, 1 (1968).

A. P. Grollman, *Proc. Nat. Acad. Sci. U.S.A.*, 56, 1867 (1966).

R. A. Neal, *Advan. Parasitol.*, 4, 1 (1966).

G. Woolfe, *Progr. Drug Res.*, 8, 11 (1965).

J. H. Burckhalter et al., *J. Med. Chem.*, 6, 89 (1963).

J. H. Burckhalter et al., *J. Org. Chem.*, 26, 4070 (1961).

E. C. Faust, *Amebiasis*, Thomas, Springfield, Ill., 1954.

Antitrypanosomal Agents

Z. Brener, *Annu. Rev. Microbiol.*, 27, 347 (1973).

E. Grunberg and E. H. Titsworth, *Annu. Rev. Microbiol.,* **27**, 317 (1973).

J. Ford, *The Role of Trypanosomiases in African Ecology,* Clarendon, London, 1971.

A. Korolkovas *et al., Rev. Farm. Bioquim. Univ. S. Paulo,* **9**, 135 (1971).

H. W. Mulligan, Ed., *The African Trypanosomiases,* Wiley-Interscience, New York, 1971.

W. H. R. Lumsden, *Advan. Parasitol.,* **8**, 227 (1970).

M. A. Miles and J. E. Rouse, *Chagas's Disease—A Bibliography,* Bureau of Hygiene and Tropical Diseases, London, 1970.

J. R. Cançado, Ed., *Doença de Chagas,* Imprensa Oficial, Belo Horizonte, 1968.

F. Köberle, *Advan. Parasitol.,* **6**, 63 (1968).

A. R. Gray, *Bull. World Health Org.,* **37**, 177 (1967).

C. A. Hoare, *Advan. Parasitol.,* **5**, 47 (1967).

J. J. Jaffe, *Trans. N. Y. Acad. Sci.,* **29**, 1057 (1967).

Antileishmanial Agents

A. Korolkovas and K. Tamashiro, *Rev. Farm. Bioquim. Univ. S. Paulo,* **12**, 109 (1974).

G. O. Doak and L. D. Freedman, *Organometallic Compounds of Arsenic, Antimony, and Bismuth,* Wiley-Interscience, New York, 1970.

S. Adler, *Advan. Parasitol.,* **2**, 35 (1964).

G. Vianna, *Arch. Brasil. Med.,* **2**, 426 (1912).

Antitrichomonal Agents

O. Jirovec and M. Petru, *Advan. Parasitol.,* **6**, 117 (1968).

R. M. Michaels, *Advan. Chemother.,* **3**, 39 (1968).

K. Wiesner and H. Fink, *Progr. Drug Res.,* **9**, 361 (1966).

R. Cavier *et al., Ann. Pharm. Franç.,* **22**, 381 (1964).

Agents Active in Other Protozoan Infections

L. Jacobs, *Advan. Parasitol.,* **5**, 1 (1968).

Drugs Used in Veterinary Protozoan Infections

"Symposium on Problems of New Animal Drug Development," *J. Amer. Vet. Med. Assoc.,* **161**, 1375-1400 (1972).

J. R. Douglas and N. F. Baker, *Annu. Rev. Pharmacol.,* **8**, 213 (1968).

E. J. L. Soulsby, *Textbook of Veterinary Clinical Parasitology,* Vol. I, Blackwell, Oxford, 1965.

CHAPTER 30

ANTISEPTIC, ANTIFUNGAL, AND ANTIBACTERIAL AGENTS

1. INTRODUCTION

In this chapter mainly topically applied chemotherapeutic agents are studied. Also, some that have localized action or very specific use are included.

II. ANTISEPTICS

Antiseptics are agents used to destroy microbes or inhibit their reproduction or metabolism; they are principally applied to cutaneous and mucous surfaces and infected wounds in order to sterilize them. However, most authorities do not recommend their application to wounds, because they retard the healing and may injure the tissues. A better procedure is to cleanse the wounds and remove pus and necrotic tissue by mechanical means. Antiseptics applied to inanimate objects or to excreta receive the name of disinfectants. Antiseptics and disinfectants are presently drugs of wide use.

Other related terms have a clear connotation in relation to antiseptics and disinfectants: biocides, sterilizers, and sanitizers. Biocides refer to preservatives that prevent bacterial and fungal attack on all sorts of organic material, such as paper, wood, and textiles. Sterilizers are substances that destroy all forms of life; one example is ethylene oxide. Sanitizers are products that reduce the number of bacteria to relatively safe levels. All these agents destroy cells by coagulation or denaturation of protoplasm protein, or cell lysis by changing structure of cell membranes and thus causing leakage of cell contents.

Several factors, such as pH, temperature, concentration, duration of contact with the microorganisms, and the presence of organic matter—blood, pus, necrotic tissue—determine the degree of effectiveness of antiseptics and disinfectants.

Antiseptics are either used as single entities or incorporated into detergents, soaps, deodorants, sprays, dusting powers, dentifrices, preservatives, urinary antiinfectives, and several other preparations. They are extensively used to

Table 30.1 Antiseptics and Disinfectants

Official Name	Proprietary Name	Chemical Name	Structure
alcohol* isopropyl alcohol[+] iodine solution* iodine tincture*		ethanol isopropanol	C_2H_5OH $(CH_3)_2CH{-}OH$ I_2
gentian violet*	Pyoktanin	N,N,N′,N′,N″,N″-hexa-methylpararosaniline chloride	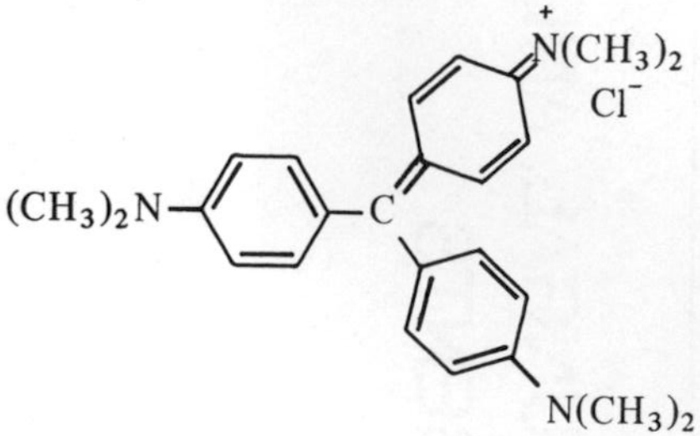
sodium hypochlorite[+]			NaOCl
halazone[+]		p-(dichlorosulfamoyl)benzoic acid	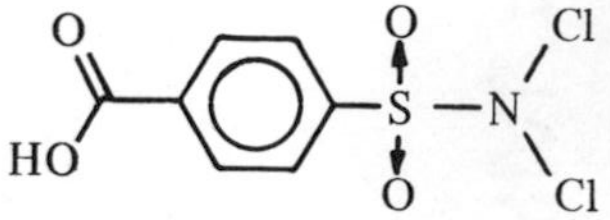
nitromersol[+]	Metaphen	5-methyl-2-nitro-7-oxa-8-mercura-bicyclo[4.2.0]octa-1,3,5-triene	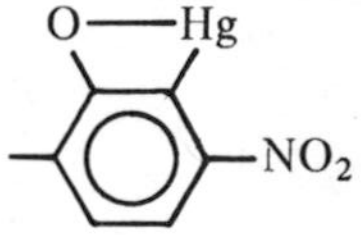
thimerosal[+]	Merthiolate	sodium [(o-carboxyphenyl)thio]ethylmercury	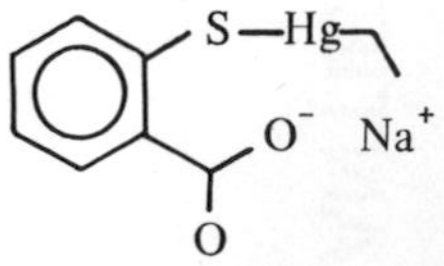

benzethonium chloride[+]	Phemerol chloride	benzyldimethyl(2-[2-(p-(1,1,3, 3-tetramethylbutyl)phenoxy)-ethoxy]ethyl)ammonium chloride 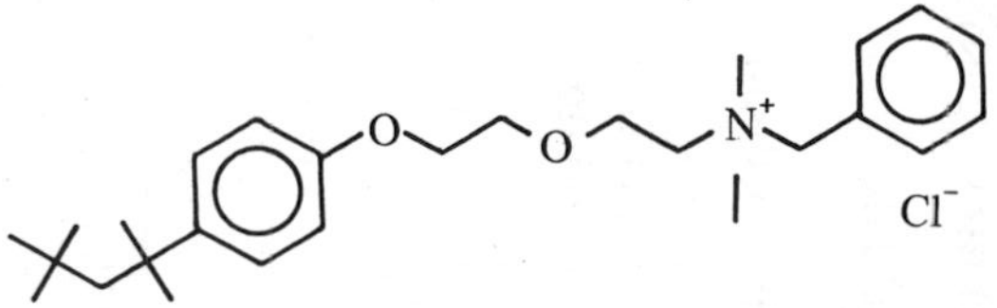
formaldehyde*	methanal	$H_2C=O$
methenamine mandelate*	hexamethylenetetramine monomandelate	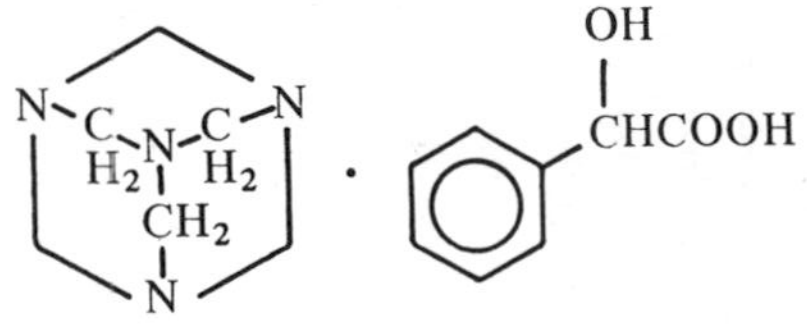
parachlorophenol*	p-chlorophenol	
hexachlorophene*	Gamophen pHisoHex	2,2'-methylenebis(3,4,6-trichloro-phenol) 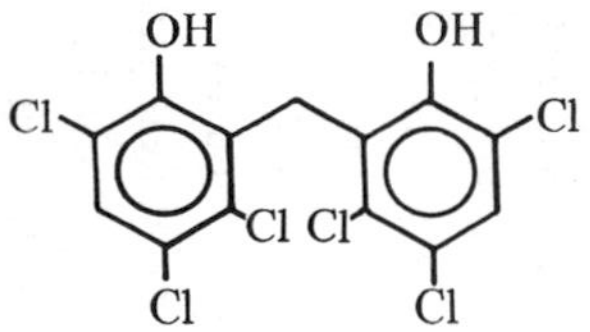

*In *USP XIX* (1975).
[+]In *NF XIV* (1975).

kill bacteria, bacterial spores, fungi, viruses, and protozoa in local infections or infestations and to prepare skin for surgical procedures.

Disinfectants are widely used for home and hospital sanitation to disinfect water, utensils, and to sterilize vaccines, blood products, and tissue grafts.

Antiseptics are not devoid of toxic effects. Topical application can irritate the skin and mucosae, causing dermatitis or allergic reactions. Absorption of these drugs produces systemic toxicity.

The first method to evaluate antiseptics was the phenol coefficient test. It was proposed by Rideal and Walker in 1903. This test was supplanted by a better one, known as use-dilution method, designed by Stuart and associates in 1963. This new method is the one officially adopted by the Association of Official Analytical Chemists.

B. Classification

Antiseptics and disinfectants most widely used are listed in Table 30.1. They belong to different chemical structures. Many of these drugs have resulted from addition, duplication, or hybridization processes of the molecular association method of drug design. Examples of addition: mercocresols, metacresyl acetate, methenamine mandelate, resorcinol monoacetate, thymol iodide, undecoylium chloride-iodine. Examples of duplication: bithionol, dichlorophene, hexachlorophene. Examples of hybridization: merbromin, thimerosal, halazone.

1. *Acids and Derivatives*

The main ones are the following: acetic acid, benzoic acid, boric acid, butyl-paraben, dehydroacetic acid, ethylparaben, mandelic acid, methylparaben, nalidixic acid, propylparaben, salicyclic acid, sodium benzoate, sodium propionate, sorbic acid.

Boric Acid

It has been used as an antiseptic, especially as an eyewash. In modern medicine it has no place, because its therapeutic value is doubtful and accidental ingestion or indiscriminate use has caused death.

Salicylic Acid

It has antiseptic, fungicidal, and keratolytic properties.

2. *Alkalies*

The most widely used are potassium hydroxide, sodium borate, and sodium hydroxide. Strong alkalies are sometimes used to disinfect excreta.

3. *Alcohols*

Examples are benzyl alcohol, chlorobutanol (Chloretone, Metaform), 2,4-

dichlorobenzyl alcohol, ethyl alcohol, ethylene glycol, isopropyl alcohol, phenethyl alcohol, phenoxyethanol (Phenoxetol), propylene glycol, trimethylene glycol.

Alcohols are also disinfectants, but their main application is as antiseptics in surgical procedures. Primary alcohols are more active than secondary ones, and these, more than tertiary ones.

Ethyl Alcohol

Used widely as a disinfectant of the skin, it has a rapid bactericidal action. Since it acts by denaturing proteins, and this process requires water, the 70% aqueous solution is more effective than undiluted ethanol. It is also used as a rubefacient and skin conditioner for bedridden patients.

Isopropyl Alcohol

It has greater bactericidal activity than ethanol, because it is more effective in reducing surface tension of bacterial cells and in denaturing proteins. It has the same uses as ethyl alcohol.

4. Phenols

Phenols are general protoplasmic poisons. Phenol itself has more historical interest than clinical use, because safer and more powerful antiseptics are available, such as halogenated phenols, which are used mainly as disinfectants, since they are too toxic and caustic for application on living tissue. The effectiveness of phenols is, therefore, enhanced by introduction of electron-withdrawing groups, such as halogens. An example is hexachlorophene, which is incorporated in soaps, creams, oils, and other vehicles for topical application. Camphorated parachlorophenol is used as a dental antiinfective.

5. Aldehydes

Those more commonly used are formaldehyde, glutaral, methenamine, methenamine hippurate, methenamine mandelate.

Methenamine

It is a polymer of formaldehyde, which is liberated by acidic sweat when this drug is used as a topical antiseptic. It can be considered, therefore, as an example of drug latentiation. It is obtained by reacting ammonium hydroxide with formaldehyde. Methenamine and its salts [hippurate, mandelate (USP)] are mainly used in urinary-tract infections, in which an acidic medium is also required. For this reason, in infections by urea-splitting microorganisms, such as *Proteus* and some *Pseudomonas* species, methenamine should be taken with acidic compounds—ammonium chloride, arginine hydrochloride, ascorbic acid, methionine, for instance—to assure suitable urinary pH (5.5 or below).

6. Halogens and Halogenophores

Halogens and halogenophores act both as antiseptics and disinfectants. Thus, elemental chlorine is used to disinfect utensils, swimming pools, drinking water, and even tissues. Chlorine-containing compounds are used mainly for disinfection of water. Iodine and iodophores are used to disinfect the skin before surgical procedures and as antiseptics on wounds. In the presence either of organic matter or of alkalies, their effectiveness is reduced.

The number of halogens and halogenophores with antiseptic activity is large: chloramine T (Chlorazene), chlorinated lime, chlorine, chloroazodin (Azochloranid), dichloramine T, halazone (Pantocid), strong iodine solution (USP), iodine tincture (USP), iodoform, oxychlorosene (Clorpactin), potassium troclosene, povidone-iodine (USP), sodium hypochlorite, symclosene, tetraiodopyrazole (Iodol), undecoylium chloride-iodine.

7. Metals and Metallic Compounds

Mercurial compounds have antiseptic and disinfectant properties. Their antibacterial activity is sharply reduced in the presence of serum and tissue proteins. Organic mercurial compounds are less toxic and caustic than inorganic salts. Poisoning with mercury compounds is treated with dimercaprol (Cf. Chapter 2, Section IV.E.4).

Silver and silver salts have antibacterial action in very low concentration (1 part in 20 million). To this action, which is found in many heavy metals, is given the name *oligodynamic action*. Silver compounds have been widely used in the past, but recently they have been replaced by more effective agents. However, silver nitrate (USP) is still used in 1% solution for instilling into the conjunctival sac of newborn infants as prophylaxis against gonorrheal ophthalmia neonatorum.

Metals and metallic compounds with antiseptic and disinfectant properties are the following: ammoniated mercury ointment, cadmium sulfide, copper, copper sulfate, merbromin (Mercurochrome), mercocresols (Mercresin), mercuric chloride, mercurobutol (Mecryl), Mercuróphen, mild silver protein, nitromersol, phenylmercuric acetate (Nylmerate), phenylmercuric nitrate (Phe-Mer-Nite), selenium sulfide, silver chloride, silver nitrate, silver picrate, sodium meralein, thimerosal, yellow mercuric oxide, zinc chloride, zinc oxide, zinc pyrithione (Zinc Omadine), zinc stearate, zinc sulfate.

8. Peroxides and Other Oxidants

Peroxides and other oxidants have some bactericidal effect as a result of the release of nascent oxygen. However, their use has been declining. Most of them are considered obsolete.

Those most widely used are hydrogen peroxide, benzoyl peroxide (USP), ozone, potassium permanganate (USP), sodium perborate, urea peroxide, zinc permanganate, zinc peroxide.

Hydrogen Peroxide (USP)

Used in 3% solution to cleanse wounds. It should never be instilled into a closed body cavity or abcess, because liberated oxygen can pass into the vascular system and cause an embolism.

9. *Quaternary Ammonium Compounds*

Quaternary ammonium compounds are active against both Gram-positive and Gram-negative bacteria, owing to their cationic surface activity. Therefore, they are inactivated by soaps, anionic substances, and inorganic matter. Structurally, they have a hydrophobic moiety (paraffinic chain, alkyl-substituted benzene, naphthalene), and a hydrophilic group (quaternary ammonium, sulfonium, phosphonium).

Some of the main ones are benzalkonium chloride (Zephiran chloride), benzethonium chloride (Phemerol chloride), cetylpyridium chloride (Ceepryn chloride).

10. *Nitrofurans*

Nitrofurans are used primarily as antibacterial agents in urinary infections. They can give rise to hypersensitivity reactions.

Examples are furazolidone (Furoxone), nifuroxime (Microfur), nitrofurantoin (Furadantin), nitrofurazone (Furacin).

11. *Dyes*

Certain dyes have bacteriostatic properties. Now their use is limited. Those still used are acriflavine (Euflavine), 9-aminoacridine, 9-amino-4-methylacridine, brilliant green, Dimazon, 4,5-dimethylproflavine, gentian violet (USP), ichthammol (Isarol), malachite green, Pyridacil.

12. *Amides and Carbanilides*

Amides and carbanilides are extensively used as bacteriostatic agents in soaps, deodorants, and lotions.

Those available are alocarban, bensalan, clofucarban, dibromsalan, fluorosalan, fursalan, metabromsalan, thiosalan, tibrofan, tribromsalan, triclocarban.

13. *Amidines and Guanidines*

Those most widely used are ambazone (Iversal), chlorhexidine, and dibrompropamidine (Brolene), as antiseptics in oropharyngeal infections.

14. *Miscellaneous Agents*

The main ones are ethylene oxide, camphor (USP), hexetidine (Sterisil), and pine oil. It should be noted that β-propiolactone (Betaprone), formerly used, is a carcinogen.

Ethylene Oxide

It is a very efficient sterilizing agent of medical and dental instruments, laboratory apparatus, pharmaceutics, food-packaging films, food, and other items. It is used in the sterilization of space ships. It is biocidal to all organisms at room temperature, being noncorrosive and readily diffusible. Since it forms explosive mixtures with air, it is usually diluted with carbon dioxide, nitrogen or a fluorohydrocarbon.

C. Mechanism of Action

Mechanisms by which antiseptics and disinfectants exert their germicidal action are varied and complex. Some of these drugs act by several mechanisms.

1. *Inactivation of Certain Enzymes*

This is the most common mechanism. It is operative in halogens and halogenophores, metals and metallic compounds, ureas, amides, carbamates, and nitrofurans.

Iodine may interact with a polypeptide chain in several ways:

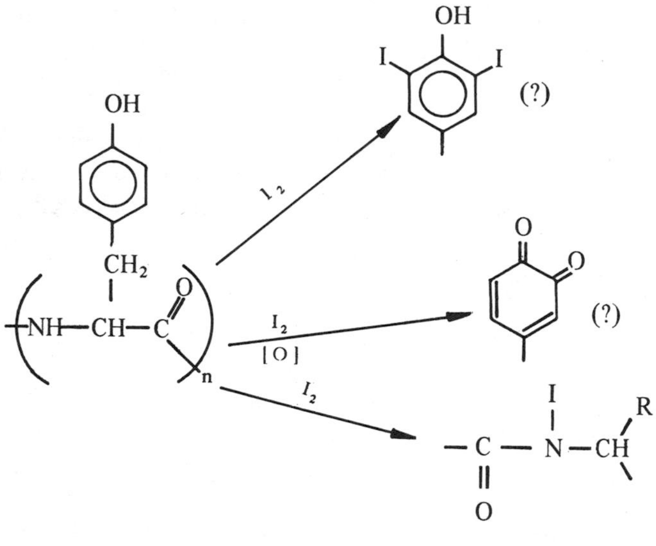

Chlorine and chlorinated compounds are first transformed to hypochlorous acid, which can (*a*) attach chlorine to a protein segment and (*b*) yield HCl and nascent oxygen, which acts as an oxidant:

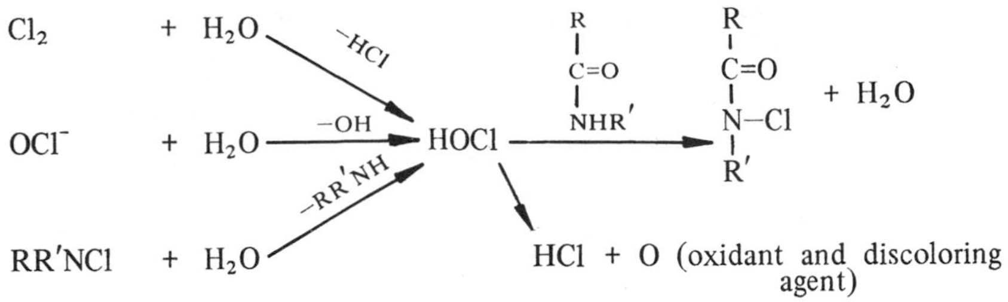

$$Cl_2 \quad + \; H_2O \xrightarrow{-HCl}$$

$$OCl^- \quad + \; H_2O \xrightarrow{-OH} HOCl$$

$$RR'NCl \quad + \; H_2O$$

$$HOCl \longrightarrow \begin{array}{c} R \\ | \\ C=O \\ | \\ NHR' \end{array} \longrightarrow \begin{array}{c} R \\ | \\ C=O \\ | \\ N-Cl \\ | \\ R' \end{array} + H_2O$$

HCl + O (oxidant and discoloring agent)

2. Denaturation of Proteins

Acids, alcohols, phenols, metals and metallic compounds, halogens and halogenophores, quaternary ammonium compounds, and some other drugs act by denaturing and coagulating proteins.

That alcohols owe their antiseptic property to protein denaturation is supported by the fact that this process requires water and that absolute alcohol, a dehydrating agent, is a less potent germicide than aqueous alcohol.

Mercury compounds first form an ion $R-Hg^+$ which then reacts covalently with cellular enzymatic thiol groups through formation of mercaptides:

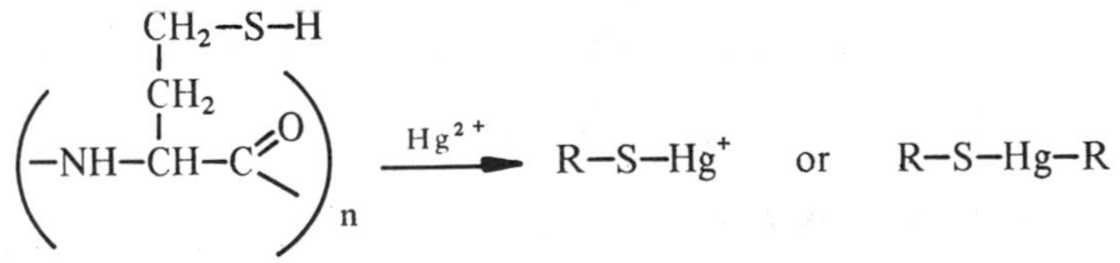

$$\left(\begin{array}{c} CH_2-S-H \\ | \\ CH_2 \\ | \\ -NH-CH-C{\overset{O}{\diagup}}_{\diagdown} \end{array} \right)_n \xrightarrow{Hg^{2+}} R-S-Hg^+ \quad \text{or} \quad R-S-Hg-R$$

Silver and silver compounds act by a similar two-step mechanism: (a) silver ions react with bacterial protein, giving an instant germicidal effect; (b) resolution and ionization of the silver proteinate which results in a milder but sustained bacteriostatic effect.

3. Alteration of Normal Permeability of Bacterial Cell Membrane

This mode of action is operative in phenols, quaternary ammonium compounds, and some other drugs. By altering membrane permeability, they allow leakage of essential bacterial cell constituents with subsequent death of the bacteria.

4. Intercalation into DNA

Some dyes may act by this mechanism. This happens, for instance, with acriflavine (Figure 3.1). In the case of acridines, their effectiveness increases with the degree of ionization. Owing to their ionization, 91 and virtually 100%, respectively, 3- and 9-aminoacridines are the most active, since they are stronger bases than other aminoacridines, as shown by resonance stabilization of the protonated forms:

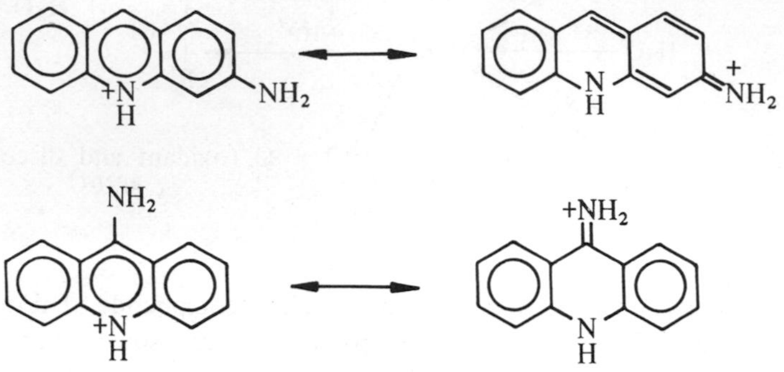

5. Chelation

Some phenols, such as oxine and related drugs, by chelation remove a metallic ion which is important for the formation of an essential metabolite.

6. Metabolic Antagonism

Hexetidine very likely acts by this mechanism, because it can be considered an antimetabolite of thiamine, replacing it in metabolic processes of the bacterial cell.

III. ANTIFUNGAL AGENTS

Antifungal agents are drugs used against infections caused by fungi. They can be either fungistatics or fungicides. They have a wide application in human and veterinary medicine. Fungitoxic agents are also used for treatment of plants, seeds, soil, and paints and as foliar protectants, industrial (pulp, leather) preservatives, wood preservatives, and some other purposes. It has been estimated that fungicides are used in the growing of one-half of the world's crops. The production of fungicides amounts to $50 million, which represents about 10% of all pesticide sales.

Human fungi infections can be divided into three groups: systemic infections, dermatophytic infections, and candidiasis (Table 30.2).

Fungitoxicity is evaluated by screening tests *in vitro* and *in vivo* similar to those used for determining antibacterial potency.

Before treatment of any infection supposedly of fungal origin, the etiological agent should be identified, because most antifungal drugs are inactive against bacteria and have narrow antifungal spectra.

Untoward effects produced by antifungal agents are rare and usually minor.

A. Classification

Hundreds of substances have antifungal activity. Those of wider use for any of the many applications in medicine, industry, and agriculture can be divided into the following chemical classes: (1) acids and derivatives; (2) alcohols; (3) phenols; (4) salicylamides; (5) sulfur compounds; (6) metallic compounds; (7) diamidines; (8) dithiocarbamates and related compounds; (9) quaternary ammonium compounds; (10) nitro compounds; (11) quinones; (12) halogenophores; (13) heterocyclic compounds; (14) antibiotics; (15) miscellaneous compounds, such as acrisorcin, carbol-fuchsin, dodine, gentian violet, glutural, hydrogen peroxide, potassium permanganate.

Some mixtures of antifungal agents are also marketed. A combination of chlordantoin and benzalkonium chloride is known as Sporostacin and is indicated for treatment of candidal infections of the vulvovaginal area.

Several new antibiotics have shown antifungal activity *in vitro* or in experimental animals: aabomycin, brassicicolin A, chlorflavonin, eulicin, gonidodomin, gougeroxymycin, hondamycin, oryzachlorin, subsporins A, B, and C.

Antifungal agents most widely used in human medicine are listed in Table 30.3.

Amphotericin B

It is the drug of choice in most fungal diseases in man. Its action is fungistatic rather than fungicidal. In systemic infections it is applied intravenously and intrathecally; in candidiasis, topically. Adverse reactions are rare but do occur.

Griseofulvin

This antibiotic is the only drug that is effective by oral route in dermatophytic infections, especially in tinea infections of the scalp and glabrous skin. It is contraindicated in cases of acute intermittent porphyria. Untoward effects are minor.

Nystatin

This fungistatic and fungicidal antibiotic is active against all species of *Candida* that infect man. Since it is too toxic for parenteral use and not absorbed from the gastrointestinal tract, its usefulness is restricted to topical infections.

Candicidin

It has both fungistatic and fungicidal action, being effective only for treatment of mycotic vaginitis, in which case the sexual partner should be treated simultaneously. It is applied topically.

Table 30.2 Fungal Infections and Their Treatment

Type of Infections	Pathogen	Drug of choice	Alternate drugs
Systemic infections			
Histoplasmosis	*Histoplasma capsulatum*	amphotericin B	
Coccidioidomycosis	*Coccidioides immitis*		
North American blastomycosis	*Blastomyces dermatitidis*		hydroxystilbamidine
Paracoccidioidomycosis (South American blastomycosis)	*Blastomyces brasiliensis*		
Sporotrichosis	*Sporotrichum schenckii*		iodides hydroxystilbamidine
Cryptococcosis	*Cryptococcus neoformans*		
Dermatophytic infections			
Athlete's foot Tinea	*Epidermophyton* species *Trichophyton* species	griseofulvin tolnaftate	haloprogin iodochlorhydroxyquin
Ringworm of skin, hair, nails	*Microsporum* species		
Tinea versicolor (pityriasis versicolor)	*Malassezia furfur*	acrisorcin	selenium sulfide tolnaftate haloprogin
Candidiasis (Moniliasis)	*Candida albicans* *Candida krusis* *Candida tropicalis* *Candida pseudotropicalis*	amphotericin B	nystatin flucytosine

Mycotic vaginitis	amphotericin B nystatin candicidin	sporostacin
Other infections Actinomycosis	*Actinomyces israeli*[a] penicillin	lincomycin tetracyclines erythromycin
Nocardiosis	*Nocardia* species[a] sulfadiazine	tetracycline streptomycin

[a]Now considered to be higher forms of bacteria.

Table 30.3 Antifungal Agents

Official Name	Proprietary Name	Chemical Name	Structure
potassium iodide*			KI
selenium sulfide*	Selsun		SeS
benzoic acid*			C_6H_5COOH
benzoic and salicylic acid ointment*			
undecylenic acid*	Desenex Enzactin Fungacetin	10-undecenoic acid	$CH_2=CH(CH_2)_8COOH$
triacetin		glyceryl triacetate	CH_2OOCCH_3 $-CHOOCCH_3$ $-CH_2OOCCH_3$
salicylanilide	Salinidol	N-phenyl salicylamide	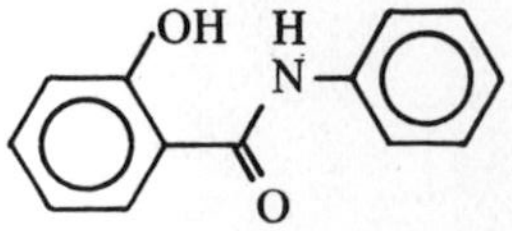
tolnaftate*	Tinactin	O-2-naphthyl m,N-dimethyl-thiocarbanilate	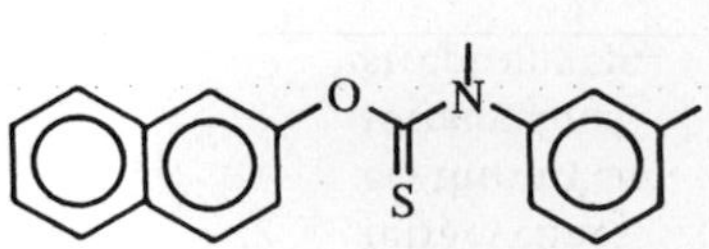
flucytosine*	Ancobon	5-fluorocytosine	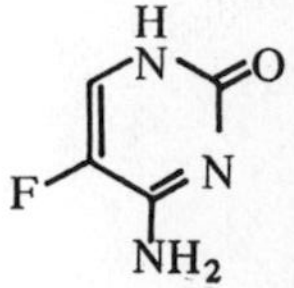

acrisorcin[+]

Akrinol

9-aminoacridinium 4-hexylresor-
cinolate

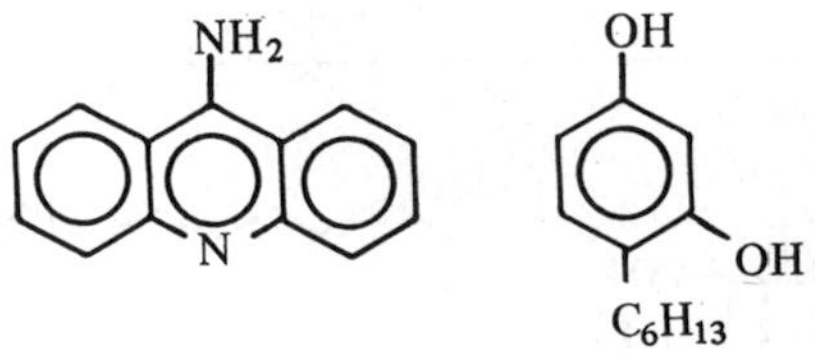

hydroxystilbamidine
isethionate*

2-hydroxy-4,4'-stilbenedicarboxamidine
bis(2-hydroxyethanesulfonate)

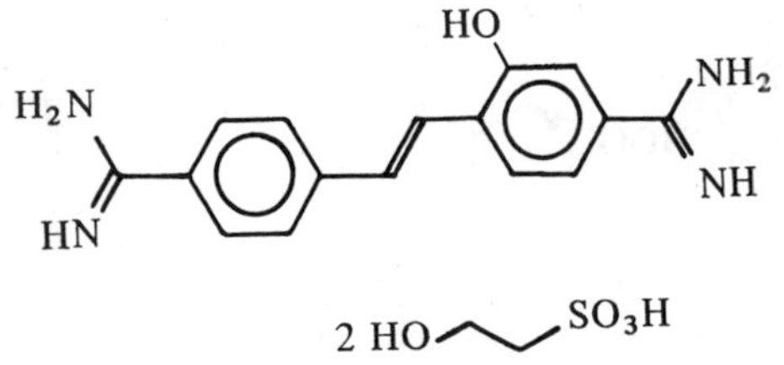

iodochlorhydroxyquin[+]   see Table 29.2

griseofulvin*
Fulvicin U/F
Grifulvin V
Grisactin

7-chloro-2',4,6-trimethoxy-
6'β-methylspiro[benzo-
furan-2(3H),1'-[2]cyclo-
hexene]-3,4'-dione

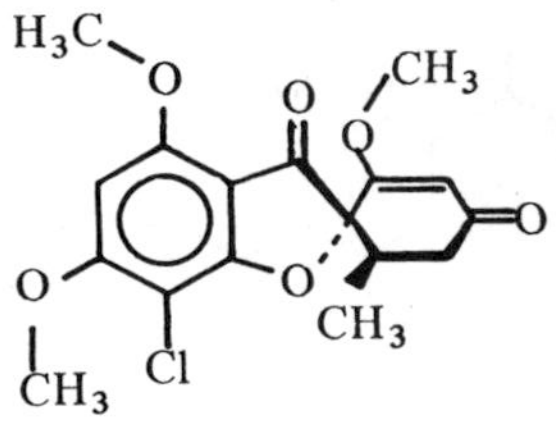

Table 30.3 Antifungal Agents (continued)

Official Name	Proprietary Name	Chemical Name	Structure
nystatin*	Mycostatin Nilstat		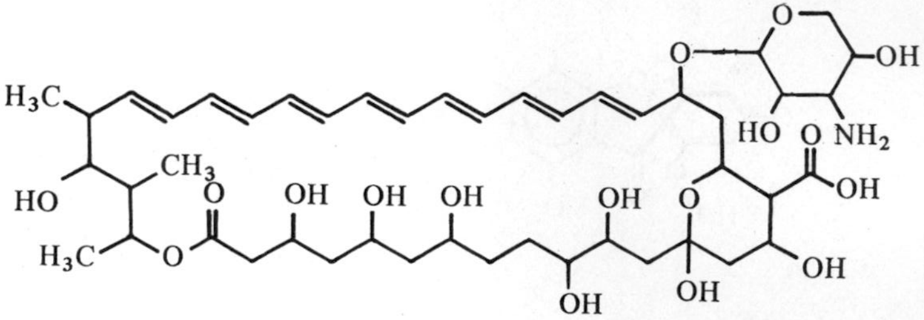
candicidin[+]	Candeptin Vanobid	heptaene macrolide of unknown structure closely related to amphotericin B	
amphotericin B*	Fungizone		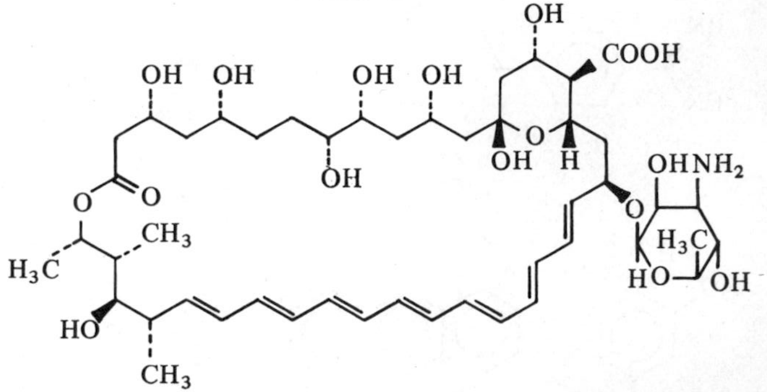

*In *USP XIX* (1975).
[+]In *NF XIV* (1975).

B. Mechanism of Action

Based on the mode of action, there are two groups of fungicides. The first group, which comprises the majority of antifungal agents, includes those drugs that react indiscriminately with functional groups of the fungal cell. The second group, which is the smaller, includes those agents that enter into a few reactions with cellular components, such as cell wall, cell membrane, mitochondria, ribosomes, or nucleus to destroy existing structures or prevent their synthesis. In other words, fungicides may either react with functional groups of the fungal cell or interfere with the synthesis or activity of enzymes by one or more mechanisms.

Some fungicides, such as fatty acids, are nonspecific. Most of them, however, are structurally specific. Several groups of fungicides owe their activity to reaction with functional groups of fungal cells, such as thiol (found in glutathione, cysteine, lipoic acid, coenzyme A, thiamine), amino (present in asparagine, glutamine), carboxyl, and hydroxyl. With these groups some can form a covalent bond and thus exert their action for a prolonged period. This happens with quinones, organometallic compounds, dithiocarbamates, captan, and analogues.

Quinones, α,β-unsaturated ketones, difolatan, haloprogin, and several other fungicides react with thiol groups either by oxidation, conjugate addition, or elimination of chlorine, e.g.,

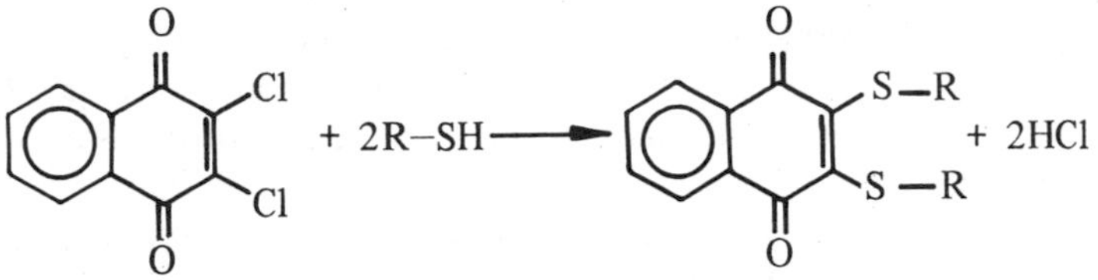

Mercury compounds and organometallic fungicides apparently owe their fungitoxicity primarily to reaction with essential thiol groups of the cell wall and cell membrane:

$$R-Hg^+ + R'-SH \rightarrow R-Hg-S-R'$$

Dithiocarbamates inhibit a number of enzymes, some of which contain metals, suggesting that their sites of action are the metal atoms in the cell membrane; they might form cationic complexes, such as

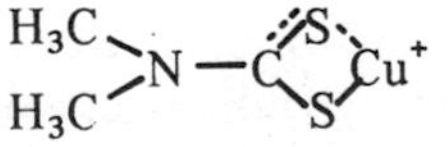

1:1 complex   or   1:2 complex

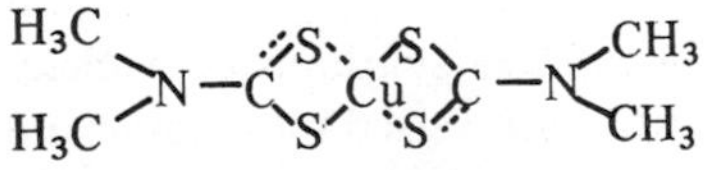

However, it seems that these fungicides interact with sulfhydryl-containing enzymes or coenzymes and in this way inhibit metabolism of fungi at many sites and in many processes. The mechanism involves two steps:

$$(a) \quad Me_2N-\overset{\overset{\displaystyle S}{\|}}{C}-S-S-\overset{\overset{\displaystyle S}{\|}}{C}-NMe_2 \underset{}{\overset{R-S^-}{\rightleftharpoons}} Me_2N-\overset{\overset{\displaystyle S}{\|}}{C}-S-S-R + Me_2N-\overset{\overset{\displaystyle S}{\|}}{C}-S^-$$

$$(b) \quad Me_2N-\overset{\overset{\displaystyle S}{\|}}{C}-S-S-R \underset{}{\overset{R-S^-}{\rightleftharpoons}} R-S-S-R + Me_2N-\overset{\overset{\displaystyle S}{\|}}{C}-S^-$$

It is likely that tolnaftate, which is a thionocarbamate, may act by a similar mechanism.

Captan and other compounds of the general formula $R-SCCl_3$, where $-SCCl_3$ is the toxophoric group, react with thiol groups also by a two-step mechanism:

$$(a) \quad R-SCCl_3 + 2R'-SH \rightarrow Cl-\overset{\overset{\displaystyle S}{\|}}{C}-Cl + R-H + R'-S-S-R' + HCl$$

$$(b) \quad Cl-\overset{\overset{\displaystyle S}{\|}}{C}-Cl + 2R'-SH \rightarrow R'-S-\overset{\overset{\displaystyle S}{\|}}{C}-S-R' + 2HCl$$

In the first step thiophosgene is formed; in the second step thiophosgene reacts with thiol groups. The fungitoxicity results from inhibition of synthesis of certain amino compounds and inactivation of CoA and other thiol-containing organic substances.

Dyrene is a powerful reactive agent. It acts by forming a covalent bond with amino compounds of the fungal cell membrane:

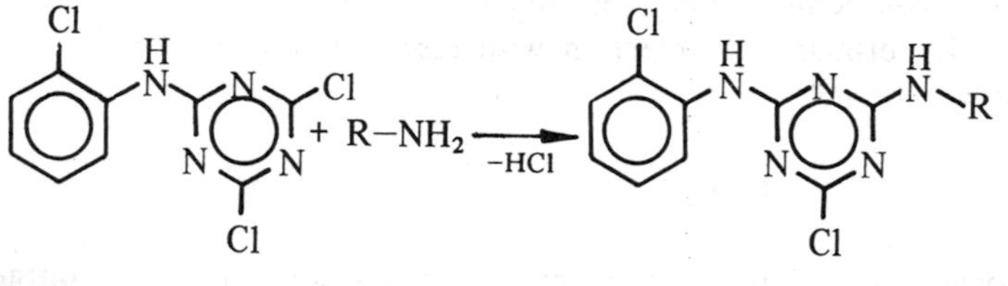

The fungitoxic effect of sulfur results from excessive removal of protons, very likely of dehydrogenase systems of respiration, from fungal protoplast. The consequence of this oxidation-reduction reaction is loss of energy by the fungal cell.

Glyodin inhibits synthesis of nucleic acids and proteins. Since this inhibition is competitively reversed by guanine and also xanthine, glyodin is an antimetabolite, because it is a bioisostere of cytosine.

Some fungicides owe their action to a chelation mechanism. This happens with oxine and iodochlorhydroxyquin, for instance. It is likely that they compete with coenzymes for metal-binding sites on enzymes or by inserting copper into macromolecules where other metals normally function.

Acrisorcin may act by intercalation between the base pairs of DNA, the same way as other acridines (see Chapter 28).

Phenols act by interference with specific metabolic pathways, probably by denaturation of fungal proteins through reaction of the acidic OH group with basic centers in the protein molecule located on cell wall.

Salicylamides cause impairment of cellular respiration of fungi. They may react through their hydrogen atom of the anilide with electron donors of polypeptides, especially thiol groups of the enzymes, and thus inactivate them.

The primary site of action of thiabendazole is the terminal electron-transfer system of the mitochondria.

Antibiotics act by several mechanisms. Thus, polyene antibiotics, which are characterized by a macrolid with a β-hydroxylated part and a conjugated double-bond system in the lactone ring, owe their action to combination with sterols in the cell membrane. This results in reorientation of the sterol molecules, which causes alteration in permeability and leakage of essential fungal cell constituents: K^+, inorganic phosphates, carboxylic acids, amino acids, phosphate esters. Such a mechanism is operative in both heptaenes—amphotericin B, nystatin, candidin, candicidin, fungimycin—and tetraenes—pimaricin, filipin, etruscomycin, rimocidin, and fungichromin. Tetraenes are stronger lysing agents, but heptaenes are better growth inhibitors. Polyenes have no action on bacteria, because membranes of these microorganisms lack sterols. Fungi, on the other hand, whose membranes contain ergosterol, are susceptible to these antibiotics.

Griseofulvin owes its fungicidal action to inhibition of synthesis of cell wall material as well as to inhibition of synthesis of nucleic acids and proteins, since it was seen that it binds to RNA.

Cycloheximide acts by inhibiting synthesis of DNA and protein, but it is not clear if it prevents the binding of aminoacyl-tRNA to mRNA or the peptide formation.

The primary site of action of pyrrolnitrin is the blockade of the electron transfer between the flavoprotein of NADH dehydrogenase and the cytochrome B of the respiratory chain.

IV. ANTIBACTERIAL AGENTS

Antibacterial agents are drugs used in the treatment of infections caused by bacteria. According to the effect produced, antibacterial agents can be bacteriostatic or bactericidal. Bacterial infections can be systemic or topical. The most widely used drugs for systemic bacterial infections are antibiotics and sulfonamides; these agents are studied in the next three chapters. In this chapter we study only other antibacterial agents that are applied topically or have localized action in various infections: skin, eyes, ears, genitourinary tract. Actually, most of them are used in infections of the genitourinary tract.

Most drugs used in these infections have been already studied in the preceding chapters or preceding sections of this chapter, especially as antiseptics.

Antibiotics, which are widely used also as systemic antibacterial agents, are seen in Chapter 33. Here mention should be made that several antibiotics are used in topical or localized bacterial infections. Among other ones, the following:

1. In topical bacterial infections: bacitracin, gentamicin, kanamycin, neomycin, novobiocin, spectinomycin, streptomycin, troleandomycin, tyrothricin, vancomycin

2. In ophthalmology: ampicillin, bacitracin, carbenicillin, cephalexin, cephaloridine, cephalothin, chloramphenicol, colistimethate, erythromycin, gentamicin, kanamycin, lincomycin, methicillin, neomycin, penicillin, polymixin, streptomycin, tetracycline, vancomycin

3. In otic preparations: colistin, neomycin, polymixin

A. Classification

Antibacterial agents with localized action most widely used can be divided into the following classes:

1. Antibiotics: those already mentioned; they are discussed in Chapter 33

2. Sulfonamides: they are studied in Chapter 31

3. Nitrofurans: furazolidone, nifuratrone, nifurpipone, nifurprazine (Strensol), nitrofurantoin, nitrofurazone

4. Cyclic amines: methenamine (Uritone), methenamine hippurate (Hiprex), methenamine mandelate (Mandelamine)

5. Naphthyridine derivatives: amfonelic acid, nalidixic acid, oxolynic acid, pyromidic acid

6. Xylenol derivatives: xibornol (Nanbacine)

The most useful antibacterial agents are listed in Table 30.4.

Furazolidone

It has wide antibacterial and antiprotozoal activity, being used for the treatment of bacterial enteritis and dysentery and for giardiasis. In some patients it can cause a disulfiram type of reaction.

Table 30.4 Antibacterial Agents with Topical or Localized Action[a]

Official Name	Proprietary Name	Chemical Name	Structure
furazolidone	Furoxone	3-[(5-nitrofurfurylidene)amino]-2-oxazol-idinone	
nitrofurantoin*	Cyantin Furadantin Macrodantin	1-[(5-nitrofurfurylidene)amino]hydantoin	
nitrofurazone[+] methenamine[+]	Uritone Urotropin	see Table 29.3 hexamethylenetetramine	
nalidixic acid[+]	NegGram	1-ethyl-1,4-dihydro-7-methyl-4-oxo-1,8-naphthyridine-3-carboxylic acid	

*In *USP XIX* (1975).
[+]In *NF XIV* (1975).
[a]Sulfonamides and antibiotics are not included because they are discussed in the next three chapters.

Nitrofurazone

It acts against many Gram-positive and Gram-negative microorganisms present in surface infections. It is used in otitis externa, impetigo, infected lacerations, and other infections of the skin and mucous membranes.

Nalidixic Acid

It is effective only against many Gram-negative bacteria, such as *Proteus* and *E. coli.* It is used by oral route in the treatment of urinary-tract infections. Resistance to it appears rapidly.

B. Mechanism of Action

Methenamine and its salts act by releasing formaldehyde in acidic pH. Formaldehyde is very reactive toward several functional groups in bacterial cells.

Nalidixic acid inhibits selectively and reversibly DNA synthesis of bacteria, without effect on DNA synthesis in mammalian cells. This inhibition may be due to its structural similarity to purine nucleosides, such as guanosine.

Nitrofurans inhibit some enzymes, especially those involved in the formation of acetylcoenzyme A from pyruvic acid, thus interfering in the production of energy (Figure 30.1). According to molecular-orbital calculations, this activity depends greatly on the nitro group that, *in vivo*, is reduced to $-NHOH$ or $-NH_2$.

Figure 30.1 Site of action of nitrofurans in citric acid cycle.

The mechanism of action of some other antibacterial agents has already been discussed in preceding sections. The mode of action of sulfonamides and antibiotics is studied in the next chapters.

REFERENCES

Antiseptics

C. A. Lawrence and S. S. Block, Eds., *Disinfection, Sterilization, and Preservation,* Lea & Febiger, Philadelphia, 1968.

H. S. Bean, *J. Appl. Bacteriol.*, **30**, 6 (1967).

A. Albert, *Acridines*, 2nd ed., Arnold, London, 1966.

H. L. Davis, Ed., "Mechanism and Evaluation of Antiseptics," *Ann. N.Y. Acad. Sci.*, **53**, 1-219 (1950).

Antifungal Agents

S. Shadomy and G.E. Wagner, *Annu. Rep. Med. Chem.*, **10**, 120 (1975).

A. Korolkovas, *Rev. Bras. Farm.*, **55**, 147 (1974).

S. Shadomy, *Annu. Rep. Med. Chem.*, **9**, 107 (1974).

J. A. Waitz and C. G. Drube, *Annu. Rep. Med. Chem.*, **8**, 116 (1973).

C. Hansch and E. J. Lien, *J. Med. Chem.*, **14**, 653 (1971).

R. J. Lukens, *Chemistry of Fungicidal Action*, Chapman & Hall, London, 1971.

N. N. Melnikov, *Chemistry of Pesticides*, Springer, New York, 1971.

W. B. Turner, *Fungal Metabolites*, Academic, London, 1971.

S. Kinsky, *Annu. Rev. Pharmacol.*, **10**, 119 (1970).

C. da S. Lacaz *et al.*, Eds., *O Grande Mundo dos Fungos*, Poligono e Universidade de São Paulo, São Paulo, 1970.

D. C. Torgeson, Eds., *Fungicides*, 2 vols., Academic, New York, 1967, 1969.

A. Kreutzberger, *Progr. Drug Res.*, **9**, 356 (1968).

Antibacterial Agents

E. Grunberg and E. H. Titsworth, *Annu. Rev. Microbiol.*, **27**, 317 (1973).

A. Bauerfeind, *Antibiot. Chemother. (Basel)*, **17**, 122 (1971).

F. M. Harold, *Advan. Microbiol. Physiol.*, **4**, 46 (1970).

R. Cavier *et al.*, *Chim. Ther.*, **5**, 270 (1970).

E. B. Winshell and H. S. Rosenkranz, *J. Bacteriol.*, **104**, 1168 (1970).

CHAPTER 31

SULFONAMIDES

I. INTRODUCTION

Sulfonamides are drugs extensively used for treatment of bacterial infections. They are effective in diseases caused by the following microorganisms: *Escherichia coli, Klebsiella pneumoniae, Pasteurela pestis, Diplococcus pneumoniae, Salmonella typhi, S. enteritidis, S. typhimurium, S. schottmulleri, Staphylococcus pyogenes aureus, Streptococcus pyogenes, Shigella sonnei, S. flexneri, Listeria monocytogenes, Pseudomonas aeruginosa, Brucella melitensis, B. abortus, B. suis, Proteus vulgaris, Vibrio cholerae, Bacillus anthracis, B. subtilis, Corynebacterium diphtheriae, Hemophilus influenzae, H. ducreyi, Francisella tularensis.* They are also effective against most strains of *Neisseria gonorrheae* and *N. meningitidis.* They are drugs of choice in the treatment of nocardiosis and chancroid. Long-acting sulfonamides have shown activity in leprosy. Certain species of protozoa are also susceptible. Thus, some sulfonamides, in combination with other types of agents, are active against chloroquine-resistant strains of plasmodia and infections caused by other microorganisms. The treatment of choice against *Toxoplasma gondii* is a combination of sulfadiazine or trisulfapyrimidines with pyrimethamine.

In usual doses (2 to 4 g daily for short-acting and 500 mg daily for long-acting agents), sulfonamides exert only bacteriostatic effect. However, when high concentrations are reached in the body, as is the case in the treatment of urinary infections, they exert bactericidal action. Since sulfonamides are bacteriostatic and not bactericidal agents in customary sulfonamido therapy, it is important to maintain bacteriostatic concentrations during sufficient time for the mammal organism to produce antibodies. In treatment with penicillins, which are bactericidal agents, it is important to keep high concentration of the antibiotics.

Sulfonamides are administered almost always by oral route. Topical application is usually ineffective, because pus and cellular debris inhibit their action. However, topical sulfonamides should be applied in the conjunctival sac, otic canal, and vagina.

Urinary-tract infections are treated with sulfamethizole, sulfisoxazole, or sulfachlorpyridazine, because they are excreted in high antibacterial concentration—largely in the active form, that is, without undergoing metabolism—have adequate solubility at acidic pH values, and maintain good antibacterial levels in the blood and tissues during urinary excretion.

As a class, sulfonamides are rapidly absorbed from the gastrointestinal tract, with the exception of intestinal sulfonamides. Effective concentrations in the blood vary between 6 and 15 mg/100 ml.

Sulfonamides undergo inactivation primarily in the liver either by acetylation or by conjugation of the free aromatic amino group with glucuronic acid. They are excreted mainly in the urine. Owing to their low solubility in water, they can accumulate in the kidneys, producing crystalluria. The incidence of this hazard may be reduced by using trisulfapyrimidines instead of a single sulfonamide and also by administering large amounts of fluids to the patients. Concomitant administration of methenamine, which releases formaldehyde *in vivo*, with sulfamethizole and sulfathiazole should be avoided, because these sulfonamides form insoluble derivatives with formaldehyde in the urine.

Other untoward effects are nausea, dizziness, hypersensitivity reactions—such as urticaria, purpura, photosensitization, and erythema nodosum—blood dyscrasias—for example, leukopenia, granulocytopenia, agranulocytosis—hepatitis, and several other adverse reactions. A serious one is hemolytic anemia, especially in those patients with congenital erythrocytic glucose-6-phosphate dehydrogenase (G6PD) deficiency, which is most common in Negroes and people of Mediterranean ethnic groups. Sulfonamides may cause Stevens-Johnson syndrome, especially in children, which is fatal in approximately 25% of susceptible patients. These drugs are contraindicated for pregnant women and nursing mothers and for infants less than 2 months of age.

Owing to the widespread and indiscriminate use of sulfonamides, some strains of bacteria resistant to these drugs and cross-resistance among different types of sulfonamides have been reported. *Pseudomonas* is commonly resistant. Some strains of *Klebsiella, Enterobacter,* and *Proteus* are totally resistant. Resistance is found also among the following susceptible microorganisms: *H. influenzae, Diplococcus pneumoniae, S. faecium,* staphylococci, anaerobic streptococci, clostridiae, spirochetes.

II. HISTORY

The first sulfonamide to be synthesized was sulfanilamide. As early as 1908 Gelmo prepared it as part of a program in search of new azo dyes. For many years it was used only as an intermediate product in the dye industry. Its antibacterial activity was first discovered a quarter of a century later, after a long run of interesting circumstances.

In 1932 Domagk reported that prontosil rubrum, synthesized by Mietzsch and Klarer, was active in experimental infections by β-hemolytic streptococci. For this discovery Domagk received the Nobel prize in 1938.

In 1935, a group of investigators—Tréfouël, Tréfouël, Nitti, and Bovet—working under Fourneau at the Pasteur Institute in Paris reported that *in vivo* the azo linkage of prontosil rubrum is reduced by azo reductase, as it is now known, yielding sulfanilamide, and this is the active moiety:

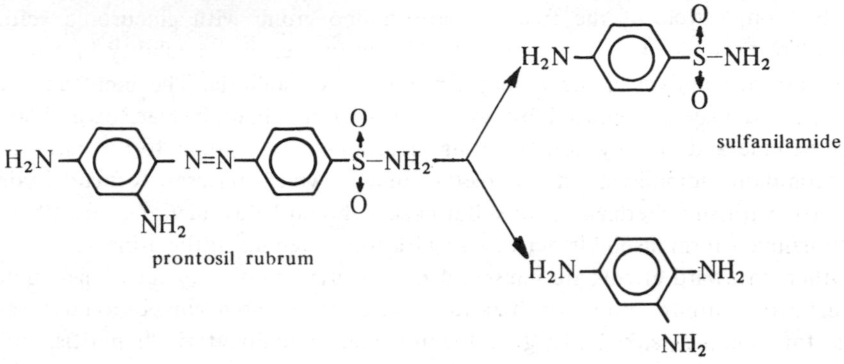

sulfanilamide

prontosil rubrum

A retrospective look at sulfonamides leaves no doubt that, besides providing the first efficient treatment of bacterial infections, they unleashed a revolution in chemotherapy by introducing and substantiating the concept of metabolic antagonism, which has been very useful in medicinal chemistry to explain the mechanism of action of many drugs and to design rationally new therapeutic agents.

With the purpose of obtaining new and hopefully better sulfonamides, until this date about 15,000 sulfanilamide derivatives, analogues, and related compounds, especially those related to *p*-aminobenzoic acid, were synthesized in the most extensive and intensive program of molecular modification. It included position isomers, analogues and isosteres of the benzene ring, additional substituents on the ring, substitution on the two functional groups, and modification of the functional groups with others. This program resulted in the discovery that some of these compounds had other pharmacological properties besides, or instead of, antibacterial activity. Further modification of the lead compound with the newly discovered pharmacological activity led to the introduction of many new drugs: antibacterial agents (sulfanilamides), leprostatic agents (sulfones), diuretics (heterocyclic sulfonamides, benzenedisulfonamides, thiazides), hypoglycemic agents (sulfonylureas), antimalarial agents (chlorguanide, cycloguanil), antithyroid drugs (propylthiouracil, methimazole), and one useful drug for the treatment of gout (probenecid) (Figure 2.7).

III. CLASSIFICATION

A. General Structure

Sulfonamides, also called sulfanilamides or sulfa drugs, have the following general structure:

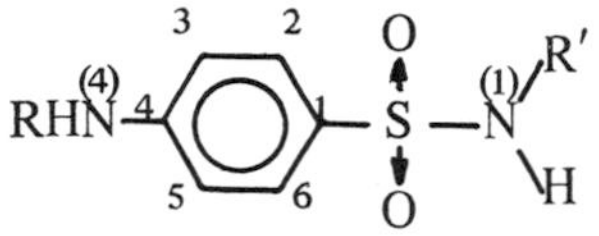

They are derivatives of sulfanilic acid, which exists both in molecular form and in ionized form:

Sulfonamido nitrogen and anilino nitrogen are designated N^1 and N^4, respectively.

The group $p-NH_2-C_6H_4SO_2-$ is called sulfanilyl; the group $p-NH_2-C_6H_4-SO_2NH-$, sulfanilamido; and the group $p-NH_2-SO_2-$, sulfamyl.

As a general principle for maintenance of antibacterial action, the fundamental structure of sulfonamides cannot be modified. This is explained by the fact that sulfonamides are competitive antagonists of p-aminobenzoic acid, and, therefore, their structure should be closely similar to that of this essential metabolite; that is, it should include the following features: benzene ring with only two substituents oriented para to each other, 4-amino group (or a group such as azo, nitro, or substituted amino that *in vivo* yields 4-amino group), and the singly substituted 1-sulfonamido group (except for the sulfones and the ring N-methyl derivatives). Drugs related to sulfonamides, although not having its fundamental structure, may exert antibacterial action by a different mechanism.

B. Chemical Properties

Sulfonamides are white crystalline powders, usually sparingly soluble in water. Being weak acids, they form salts with bases:

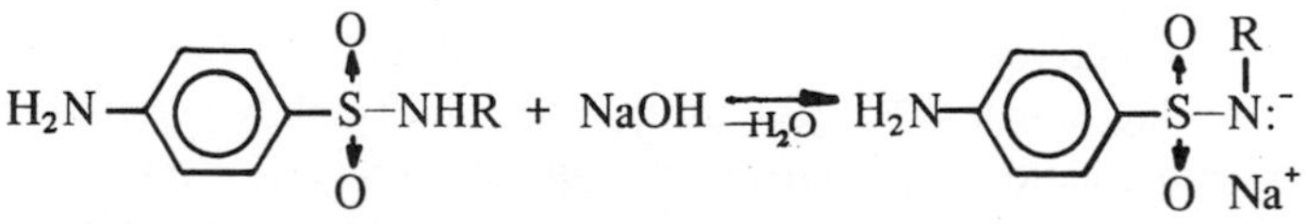

These salts are more soluble in water than the free sulfonamides. Since sodium salts of sulfonamides, with the exception of sodium sulfacetamide, have high pH, when injected they precipitate in the blood as the free acids, causing damage to the tissues.

C. Structure-Activity Relationships

Various physical and chemical parameters have been correlated with chemotherapeutic activity of sulfonamides: pK_a, protein binding, and electronic charge distribution, among others. Unfortunately, no single parameter can explain the action of sulfonamides.

1. pK_a

Bell and Roblin postulated that the antibacterial activity of sulfonamides is related to pK_a. Maximal activity would be found in those having pK_a lying between 6.0 and 7.5.

The pK_a is calculated from the following equation:

$$pK_a = pH + \log \frac{[HA]}{[A^-]}$$

When the sulfonamide is 50% dissociated, $[HA]/[A^-] = 1$. Since $\log 1 = 0$, in this case $pK_a = pH$. According to some authors, this is the pK_a for maximal activity, because sulfonamides penetrate cell wall as undissociated molecules but act in ionized forms.

However, the hypothesis that acidity or ionization alone explains the degree of activity of sulfonamides does not apply to all series of these drugs.

According to Cammarata and Allen, there is little evidence to support the assumption that, in order to exert maximal activity, a sulfonamide must have a definite pK_a. Rather, they point out that many sulfonamides lying outside of the values of pK_a (6.0 to 7.5), postulated by Bell and Roblin as being connected with maximal activity, are potentially capable of high activities, especially those having electron withdrawing substituents on an N^1-heterocyclic ring.

2. *Protein Binding*

Sulfonamides bind differently to plasmatic proteins. The binding sites in the protein might be the basic centers of arginine, histidine, and lysine. In sulfonamides the groups involved in binding are alkyl, alkoxy, alkylthio, halo, aryl, or heteroaryl at N^1, as well as substances on an N^1-phenyl or heterocyclic ring, and an N^4-acetyl group. The forces mainly involved in protein-sulfonamide interactions are hydrophobic binding and ionic attraction.

The percentage of binding varies greatly. Thus, sulfadimethoxine is 99.5 to 99.8% bound, whereas sulfaguanidine is only 5% bound. For activity, toxicity, metabolism, and glomerular filtration, however, it is the unbound part that is significant. There is no correlation between the degree of protein binding and the potency of chemotherapeutic activity.

3. Electron-Charge Distribution

Several electronic parameters show significant relationships to antibacterial action of sulfonamides. For instance, Foernzler and Martin observed a qualitative relationship between activity and the formal π charge on N^1: the greater this charge, the greater the biological activity. A similar relationship was obtained by recent molecular-orbital calculations by Korolkovas and Tamashiro.

D. Classes of Sulfonamides

Various criteria can be used to classify sulfonamides: for instance, spectrum of activity, duration of action, therapeutic uses, chemical structure. For brevity's sake, we shall classify them according to therapeutic application, although some have more than one therapeutic use. Thus, there are systemic sulfonamides, intestinal sulfonamides, urinary sulfonamides, ophthalmic sulfonamides, and sulfonamides for special uses.

1. Systemic Sulfonamides

They are used in systemic infections. According to duration of action there are two main groups of these sulfonamides: short-acting and long-acting.

a. Short-Acting Sulfonamides

They are rapidly absorbed and rapidly excreted. Their half-lives vary from 5 to 15 hours. They are administered every 4 to 6 hours. These sulfonamides are preferred for systemic infections, because application can be halted if a serious untoward effect arises. The most important are sulfachlorpyridazine, sulfadiazine, sulfadimidine (sulfamethazine), sulfaethidole, sulfafurazole (sulfisoxazole), sulfamerazine, sulfamethizole, sulfamethoxazole, trisulfapyrimidines.

b. Long-Acting Sulfonamides

They are rapidly absorbed and slowly excreted. Their half-lives are 35 to 40 hours; that of sulfadoxine, however, is 179 hours. They can be given once or twice a day. Slow excretion is related to an optimal lipophilicity. Groups that impart this characteristic are lower alkyl (methyl, but especially ethyl), lower alkoxy (methoxy and ethoxy), halogen, and occasionally aromatic ring. Slow elimination probably results from a high rate of tubular reabsorption.

Long-acting sulfonamides should be used only under extraordinary circumstances, for the following reasons: (a) they have no clinical advantages over the short-acting sulfonamides; (b) they may not cross the brain-blood barrier as well as the short-acting sulfonamides; (c) since they are slowly eliminated, they can reach a dangerous concentration in the blood, especially in patients with impaired renal function.

The most widely used are sulfadimethoxine, sulfadoxine, sulfalene (Kelfizina), sulfameter, sulfamethoxydiazine (Bayrena, Durenat), sulfamethoxypyridazine, sulfamethoxypyridazine acetyl, sulfametomidine (Deposulf), sulfamoxole (Tardamide), sulfaperin (Pallidin), sulfaphenazole (Sulfabid, Orisul), sulfasomizole (Amidozol), sulfasymazine.

2. Intestinal Sulfonamides

These drugs are used in intestinal infections. They were designed by drug latentiation: hydrophylic moieties were attached to the free amino group with the purpose of obtaining highly water-soluble latent forms of sulfonamides. Owing to the presence of these moieties (succinyl, phthalyl) that are strongly hydrophylic, they are poorly absorbed from the gastrointestinal tract and thus reach a high concentration in the colon lumen, where bacterial hydrolysis releases the sulfanilamido moiety.

The most commonly used are phthalylsulfacetamide, phthalylsulfathiazole, succinylsulfathiazole, sulfaguanol (Enterecura), sulfaloxic acid (Enteromide).

3. Urinary Sulfonamides

These sulfonamides are used in urinary infections, because they are rapidly absorbed but slowly excreted by the kidneys and thus reach a high concentration there.

The main ones are sulfacetamide, sulfachlorpyridazine, sulfacytine, sulfadimethoxine, sulfaethidol, sulfafurazole, sulfameter, sulfamethizole, sulfamethoxazole, sulfamethoxypyridazine, sulfaphenazole, sulfaurea (Urosulfan), sulfisomidine.

4. Ophthalmic Sulfonamides

They are topically applied. The main ones are sulfacetamide-sodium, sulfisoxazole diolamine.

5. Sulfonamides for Special Uses

Among others there are (a) sulfapyridine (Dagenan), in dermatitis herpetiformis; (b) salicylazosulfapyridine (salazosulfapyridine) and nitrosulfathiazole (Nisulfazole), in ulcerative colitis; (c) mafenide acetate (Sulfamylon acetate) and sulfathiourea (Fontamide), topically for burn and wound antisepsis; (d) sulfamonomethoxine, effective not only in chloroquine sensitive malaria but also in chloroquine-resistant malaria; (c) silver sulfadiazine (Flamazine), used for prevention and treatment of burn infections.

E. New Sulfonamides

Some new sulfonamides have been introduced recently. Among several others, the following deserve mention: sulfabenzamide, sulfacecol, sulfaclorazole, sulfaclozine, sulfanitran, sulfapyrazole, sulfatroxazole, sulfatrozole, sulfazamet.

F. Obsolete Sulfonamides

Because of their low efficacy and high proportion of side effects many sulfonamides once used are now obsolete; among them the following ones: sulfaguanidine (once used in intestinal infections), sulfanilamide, sulfathiazole.

G. Sulfonamides Clinically Used

Sulfonamides now most widely used are listed in Table 31.1.

Sulfacetamide Sodium

It is used in ophthalmology for the treatment of acute catarrhal conjunctivitis by local instillation into the conjunctival sac.

Sulfadiazine

Used in free form and also as sodium salt. It has a wide application as systemic antibacterial agent. In ophthalmology it is used in the treatment of trachoma and ocular toxoplasmosis.

Sulfisoxazole

It is marketed both as free form and as acetyl and diolamine. It is less prone to cause crystalluria, because its solubility is relatively high. It is better than tetracycline for the treatment of chancroid. The acetyl derivative is located on the amide of the sulfonamide rather than on the free amino group. The diolamine salt is used topically in the eye and as a vaginal cream. In cases of recrudescence of resistant malaria one of the recommended regimes is to give orally sulfisoxazole in combination with pyrimethamine and concomitant with continuous administration of intravenous infusion of quinine dihydrochloride.

Sulfadimethoxine

This long-acting sulfonamide is used in the treatment of several bacterial infections, especially of the urinary and respiratory tracts. The dose is 500 mg once daily.

Sulfadoxine

Formerly called sulformethoxine, it has a broad spectrum of action, being effective against bacteria, fungi, and plasmodia. Since it is a very long-acting

Table 31.1 Sulfonamides

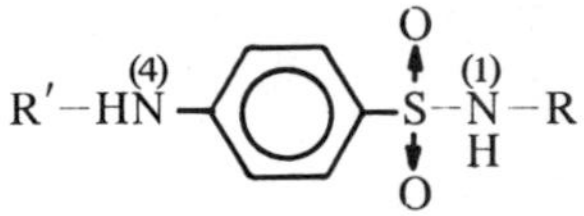

$$R'-HN-\overset{(4)}{\underset{}{\bigcirc}}-\overset{O}{\underset{O}{\overset{\uparrow}{\underset{\downarrow}{S}}}}-\overset{(1)}{\underset{H}{N}}-R$$

Official Name	Proprietary Name	Chemical Name	R	R'
sulfacetamide*	Cetamide Sulamyd	N-sulfanilylacetamide	—COCH₃	H
sulfapyridine*	Dagenan	N¹-2-pyridylsulfanilamide	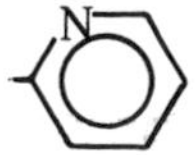 (2-pyridyl)	H
sulfadiazine*	Coco-Diazine Microsulfon Pyrimal	N¹-2-pyrimidinylsulfanila- mide	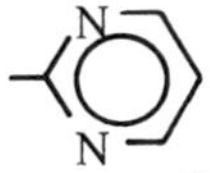 (2-pyrimidinyl)	H
sulfamerazine*+		N¹-(4-methyl-2-pyrimidin- yl) sulfanilamide	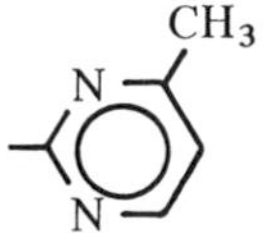 (4-methyl-2-pyrimidinyl)	H
sulfamethazine* (sulfadimidine)		N¹-(4,6-dimethyl-2-pyri- midinyl) sulfanilamide	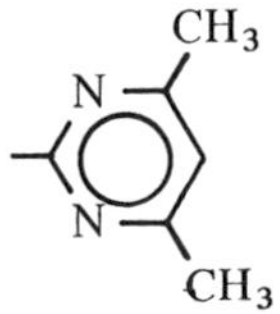 (4,6-dimethyl-2-pyrimidinyl)	H
sulfachlorpyridazine	Sonilyn	N¹-(6-chloro-3-pyridazinyl)- sulfanilamide	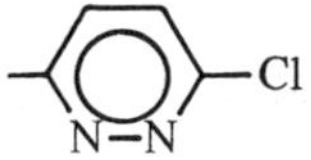 (6-chloro-3-pyridazinyl)	H

sulfamethizole[+]	Thiosulfil Ultrasul	N^1-(5-methyl-1,3,4-thiadiazol-2-yl) sulfanilamide	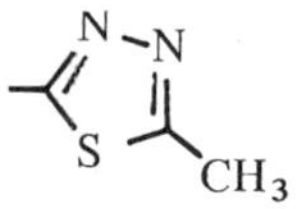	H
sulfamethoxazole[+]	Gantanol	N^1-(5-methyl-3-isoxazolyl)-sulfanilamide	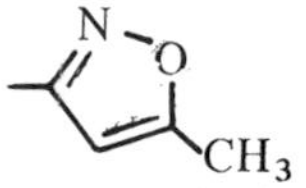	H
sulfaethidole[+]	Sul-Spansion Sul-Spantab	N^1-(5-ethyl-1,3,4-thiadiazol-2-yl) sulfanilamide	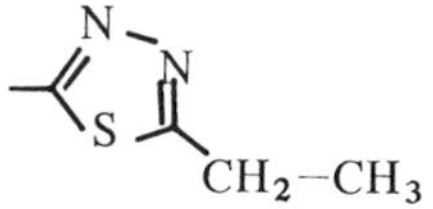	H
sulfisoxazole diolamine[+]	Gantrisin SK-Soxazole Sodizole Sosol Soxomide Sulfisocon	N^1-(3,4-dimethyl-5-isoxazolyl)-sulfanilamide with 2,2'-imino-bis(ethanol)	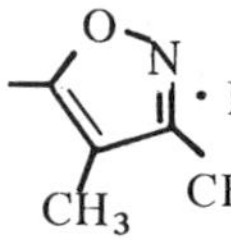 · $HN(CH_2CH_2OH)_2$	H
sulfamethoxypyridazine	Kynex Midicel	N^1-(6-methoxy-3-pyridazinyl)-sulfanilamide	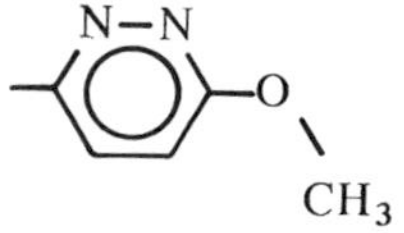	H
sulfalene	Kelfizina	N^1-(3-methoxy-2-pyrazinyl)-sulfanilamide	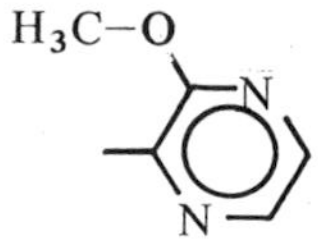	H

Table 31.1 Sulfonamides (continued)

Official Name	Proprietary Name	Chemical Name	R	R'
sulfameter	Sulla	N^1-(5-methoxy-2-pyrimidin-yl)sulfanilamide	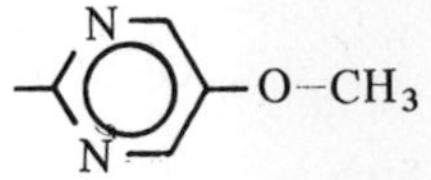 (5-methoxy-2-pyrimidinyl)	H
sulfadoxine	Fanasil Fanasulf	N^1-(5,6-dimethoxy-4-pyrimidinyl) sulfanilamide	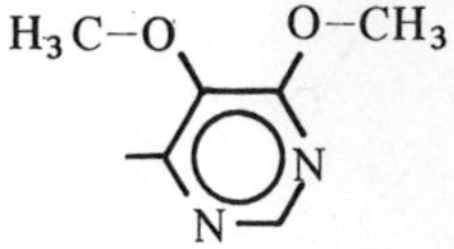 (5,6-dimethoxy-4-pyrimidinyl)	H
sulfadimethoxine[+]	Madribon	N^1-(2,6-dimethoxy-4-pyrimidinyl) sulfanilamide	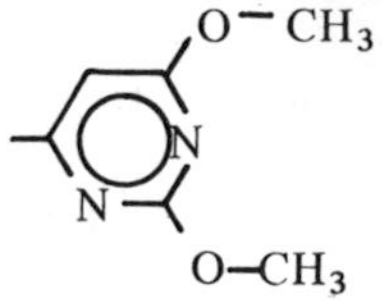 (2,6-dimethoxy-4-pyrimidinyl)	H
sulfamonomethoxine		N^1-(6-methoxy-4-pyrimidinyl) sulfanilamide monohydrate	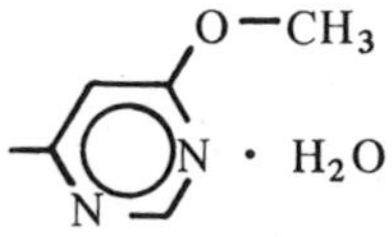 (6-methoxy-4-pyrimidinyl) · H_2O	H
sulfaclomide		N^1-(5-chloro-2,6-dimethyl-4-pyrimidinyl)sulfanilamide	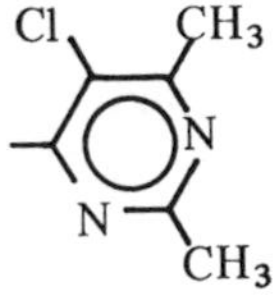 (5-chloro-2,6-dimethyl-4-pyrimidinyl)	H

sulfasymazine		N^1-(4,6-diethyl-s-triazin-2-yl)-sulfanilamide	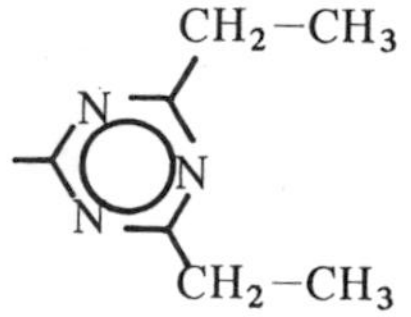 [s-triazine ring with two CH_2-CH_3 groups]	H
sulfisoxazole acetyl*	Gantrisin Acetyl	N-(3,4-dimethyl-5-isoxazolyl)-N-sulfanilyl-acetamide	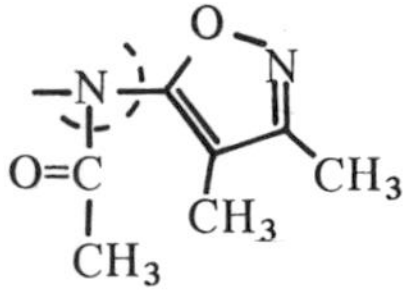 [isoxazole ring with CH_3, CH_3 and $O=C-CH_3$]	H
sulfasalazine[+] (salazosulfapyridine)	Azulfidine	5-[p-(2-pyridylsulfamoyl)-phenylazo] salicylic acid	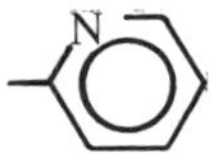 [pyridine ring]	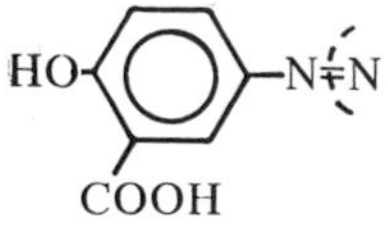 [HO— and COOH-substituted benzene with $N=N$ azo linkage]
phthalylsulfacetamide		N^1-acetyl-N^4-phthaloylsulfanil-amide	—$COCH_3$	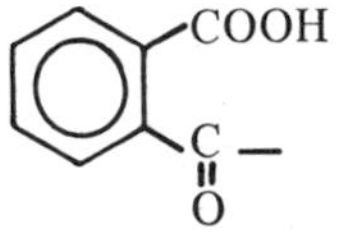 [phthaloyl group, benzene ring with two C=O]
phthalylsulfathiazole[+]		4'-(2-thiazolylsulfamoyl)phthal-anilic acid	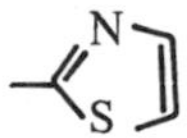 [thiazole ring]	same as above
succinylsulfathiazole	Sulfasuxidine	4'-(2-thiazolylsulfamoyl)-succinanilic acid	same as above	$HOOCCH_2CH_2CO-$
mafenide*	Sulfamylon	α-amino-p-toluenesulfonamide	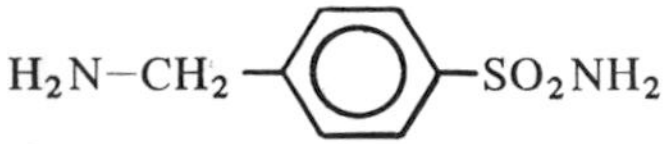 H_2N-CH_2-	[benzene ring with SO_2NH_2]

*In *USP XIX* (1975).
[+]In *NF XIV* (1975).

sulfonamide—its half-life averages 179 hours—weekly doses are sufficient to maintain effective therapeutic levels. In combination with pyrimethamine or quinine, it is used in the treatment of infections caused by resistant strains of malaria.

Acetyl Sulfamethoxypyridazine

It is tasteless and, for this very reason, is used in liquid pediatric preparations.

Sulfalene

In combination with trimethoprim, it is used in cases of recrudescence of resistant malaria.

H. Mixtures

Several combinations of sulfonamides are marketed with the purpose of reducing the incidence of crystalluria. The most common is trisulfapyrimidines, a combination of equal parts of sulfadiazine, sulfamerazine, and sulfamethazine. The basis of use is that each compound, administered in one-third the total dosage, behaves independently from the standpoint of solubility.

Another mixture is the combination of sulfadiazine with sulfamerazine, also in equal parts. Another one, marketed as Sultrin, contains sulfathiazole, sulfacetamide, and nitrobenzoylsulfanilamide, for the prophylaxis or treatment of cervical and vaginal infections.

Some pharmaceutical preparations contain a sulfa drug combined with an inhibitor of dihydrofolate reductase, such as pyrimethamine or trimethoprim. This combination potentiates the bacteriostatic effect of the sulfonamide, because it blocks the metabolic pathways of the microorganism at two different sites. A good synergistic combination is sulfamethoxazole with trimethoprim, marketed under the name of Bactrim, now considered the drug of first choice in the treatment of several urogenital, renal, and respiratory infections.

IV. MECHANISM OF ACTION

Sulfonamides owe their chemotherapeutic activity to competition with p-aminobenzoic acid (PABA) in the synthesis of folic acid, as first suggested by Woods and Fildes in 1940. Mammals also require folic acid as an essential growth factor, but they get it from food intake. At physiological pH, folic acid exists as a dianion, which cannot cross the bacterial cell wall by passive diffusion, because its negative charges are either attracted or repelled by opposite or equal charges of the bacterial wall. Therefore, an energy-requiring active transport mechanism should be involved. However, with the exception of *Streptococcus faecium* (formerly called *S. faecalis*), bacteria lack this mechanism.

For this reason these bacteria have to synthesize folic acid *de novo* from *p*-aminobenzoic acid, and, consequently, this process is inhibited by sulfonamides. The antagonism between PABA and sulfa drugs is a fine example of exploitation of biochemical differences between parasite and host.

Sulfonamides are active only at the multiplication phase of bacteria, after a certain period of lag. This period depends on the amount of accumulated PABA: in certain species it lasts six generations.

Two bacterial enzyme systems capable of performing the synthesis of folic acid have been described by Weisman and Brown: (*a*) the first one, called dihydropteroate synthase, catalyzes the synthesis of a precursor of folic acid, namely, dihydropteroic acid, through condensation of *p*-aminobenzoic acid with a pteridine derivative; with the subsequent addition of dihydropteroic acid to glutamic acid the biosynthesis of folic acid is completed; (*b*) the second one catalyzes the synthesis of folic acid itself by direct coupling of *p*-aminobenzoyl glutamate with a pteridine derivative. Both routes of folic acid biosynthesis can be antagonized by sulfonamides, because these chemotherapeutic agents are structurally similar to both *p*-aminobenzoic acid (Figure 31.1) and *p*-aminobenzoic acid-glutamate.

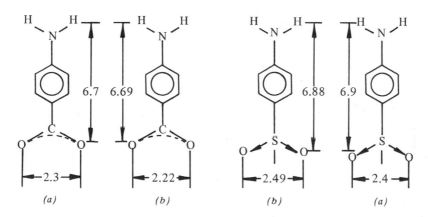

Figure 31.1 Interatomic distances, in angströms, in *p*-aminobenzoic acid and in sulfanilamide: (*a*) after Bell and Roblin; (*b*) after Korolkovas. (See References, Mechanism of Action.)

The site of action of sulfonamides, according to the Woods-Fildes original proposition, is shown in Figure 28.4.

Shefter and coworkers proposed recently that N^1 substituents in sulfonamides may play the role of competing for a site on the enzyme surface reserved for the glutamate residue in *p*-aminobenzoic acid-glutamate through one of the two following ways: (*a*) either by direct competition in the linking of *p*-aminobenzoic acid-glutamate with the pteridine derivative; (*b*) or by indirect interference with the coupling of glutamate to dihydropteroic acid.

Another theory on the mode of action of sulfonamides is that of Seydel, who modified slightly the hypothesis originally proposed by Tschesche in 1947. It assumes that sulfonamides and PABA do not compete in the same reaction and that the competitive antagonism between them is not a direct one. Rather, the bacteriostatic effect of sulfonamides results from inhibition of biosynthesis of coenzyme F by forming Schiff bases with formyltetrahydropteridine:

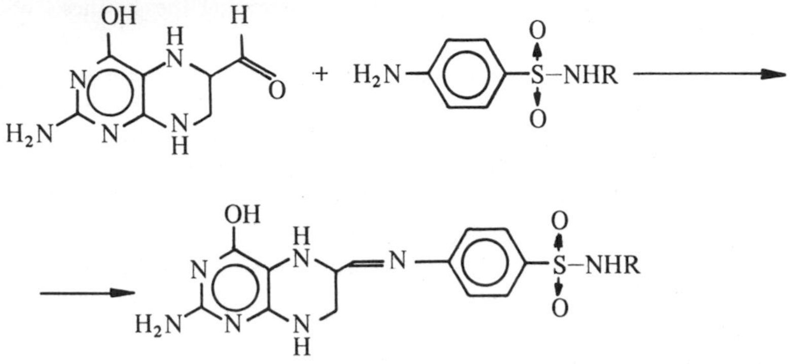

Schiff base

Inhibition by sulfonamides of the enzyme systems involved in folic acid biosynthesis can be reversed by addition of a small amount of p-aminobenzoic acid: 10^{-4} M of PABA to 1 M of a sulfonamide. Since PABA competes with sulfa drugs, local anesthetics with a p-aminobenzoic moiety—procaine, for instance—should not be applied during treatment with sulfonamides, because esterase splits these local anesthetics yielding PABA.

In the case of the sulfamethoxazole-trimethoprim mixture (Bactrim), it produces a double block in folic acid biosynthesis: sulfamethoxazole, owing to its structural similarity to PABA, inhibits the synthesis of dihydropteroic acid; trimethoprim, owing to its structural analogy with the dihydrofolic acid molecule, blocks the conversion of dihydro- to tetrahydrofolic acid (Figure 28.4).

REFERENCES

General

M. Finland, *Clin. Pharmacol. Ther.*, **13**, 469 (1972).

J. K. Seydel, *J. Pharm. Sci.*, **57**, 1455 (1968).

T. Struller, *Progr. Drug Res.*, **12**, 389 (1968).

E. H. Northey, *The Sulfonamides and Allied Compounds*, Reinhold, New York, 1948.

Introduction

J. O. Cohen, Ed., *The Staphylococci,* Wiley-Interscience, New York, 1972.

A. K. Miller, *Advan. Appl. Microbiol.,* **14**, 151 (1971).

E. Reimerdes and J. H. Thumin, *Arzneim.-Forsch.,* **20**, 1171 (1970).

History

American Chemical Society, *Advan. Chem. Ser.,* **45** (1964).

G. Domagk, *Deut. Med. Wochenschr.,* **61**, 250 (1935).

J. Tréfouël *et al., Compt. Rend. Soc. Biol.,* **120**, 756 (1935).

Classification

A. Korolkovas and K. Tamashiro, *Rev. Farm. Bioquim. Univ. S. Paulo,* **11**, 15 (1973).

S. S. Yang and J. K. Guillory, *J. Pharm. Sci.,* **61**, 26 (1972).

J. K. Seydel, "Physicochemical Approaches to the Rational Development of New Drugs," in E. J. Ariëns, Ed., *Drug Design,* Vol. I, Academic, New York, 1971, pp. 343-379.

M. Yamazaki *et al., Chem. Pharm. Bull.,* **18**, 702 (1970).

G. Domagk, *Antibiot. Chemother. (Basel),* **4**, 1 (1957).

Mechanism of Action

J. K. Seydel, *Top. Infec. Dis.,* **1**, 25 (1975).

G. L. Biagi *et al., J. Med. Chem.,* **17**, 28 (1974).

L. Bock *et al., J. Med. Chem.,* **17**, 23 (1974).

A. Korolkovas and K. Tamashiro, *Rev. Farm. Bioquim. Univ. S. Paulo,* **12**, 37 (1974).

A. Korolkovas, *Rev. Paul. Med.,* **82**, 185 (1973).

G. H. Miller *et al., J. Med. Chem.,* **15**, 700 (1972).

E. Shefter *et al., J. Pharm. Sci.,* **61**, 872 (1972).

P. H. Bell and R. O. Roblin, Jr., *J. Amer. Chem. Soc.,* **64**, 2905 (1942).

P. Fildes, *Lancet,* **I**, 955 (1940).

D. D. Woods, *Brit. J. Exp. Pathol.,* **21**, 74 (1940).

CHAPTER 32

ANTITUBERCULOUS AND ANTILEPRAL AGENTS

I. INTRODUCTION

Antituberculous and antilepral agents are drugs used for the treatment of parasitic diseases caused by mycobacteria.

Mycobacteria are Gram-positive acid-fast bacilli. There are several species of mycobacteria, but the most important clinically are those responsible for tuberculosis and leprosy. Chemotherapy does not cure these diseases in the strict sense of the word. Mycobacterial bacilli are very difficult to kill and are eliminated from the body extremely slowly after chemotherapeutic treatment of the infections. The success of chemotherapy depends on the duration of treatment: the more prolonged, the more effective. Since it is so long (6 months to 4 years), 15 to 20% of patients abandon it. Furthermore, although several useful agents are presently available, the chemotherapy of mycobacterial infections is still inadequate.

Tuberculosis (TB) is caused by several species of mycobacteria. The main etiological agent is *Mycobacterium tuberculosis,* first isolated by Koch in 1882; it takes 18 to 24 hours to reproduce. To a small extent, tuberculosis is caused also by *M. bovis (M. tuberculosis* var. *bovis).* Atypical pathogenic species of mycobacteria, now divided into five groups, that are responsible for animal diseases similar to tuberculosis are *M. kansasii, M. marinum (balnei), M. scrofulaceum, M. phlei, M. fortuitum, M. smegmatis, M. avium, M. ulcerans, M. intracellulare, M. microti.* Another species, but not belonging to any atypical group, is *M. paratuberculosis,* which infects cattle and sheep.

In developed countries there has been a steady decrease in phthisis. In the United States the death rate, which was 71.0 in 1930, declined to 22.0 in 1950 and 4.1 in 1965. However, according to a recent report of the WHO, in the world there are still 10 to 20 million contagious cases, to which 2 to 3 million new cases are added annually, and the mortality index is approximately 5 to 10%. As a public-health problem, tuberculosis continues to plague all countries, especially those underdeveloped and those in the process of development.

Leprosy in man, also known as Hansen's disease, is caused by *Mycobacterium leprae,* first described by Hansen in 1874. Another species, *M. lepraemurium,* causes leprosy in rats. The leprosy bacillus takes about a month to reproduce, and the incubation time for human leprosy is 3 to 5 years.

In spite of the availability of antilepral agents, the number of infected people has not decreased since 1965; actually, there are still 11 million leprotics in the world, and the incidence of leprosy is about 250,000 new cases per year.

Often leprosy does not require any medical care, because, as it happens also with tuberculosis, it may be self limiting and self healing, being only a temporary localized infection. Usually, however, leprosy manifests itself in two different types: lepromatous and tuberculoid. The lepromatous type, also called cutaneous or nodular leprosy, is the most serious and most infectious. It gives a negative lepromine reaction. The first symptoms are thickened, glossy, and corrugated skin, owing to diffuse infiltration of *M. leprae,* followed by atrophy of skin and muscle and absorption of small bones in consequence of the invasion of the bacilli into the larger nerve trunks. This form of leprosy should be treated during 4 years or more.

The tuberculoid type, also called neural or macroanesthetic leprosy, is a milder form. It is characterized by macules, which are discolored patches on the skin, and anesthesia consequent to early nerve invasion. It gives a positive lepromine reaction. This form of leprosy should be treated for at least 2 years.

II. CLASSIFICATION

Although tuberculosis and leprosy have several characteristics in common, it is more convenient to study antituberculous and antilepral agents separately, because their spectrum of activity is different: the first are usually inactive in leprosy, and the second ones do not exert antituberculous activity.

A. Antituberculous Agents

A serious problem of antituberculous chemotherapy is the rapid emergence of drug-resistant tubercle bacilli. Even with the very potent isoniazid, monotherapy results in 11% of resistant bacilli after 1 month, 52% after 2 months, and 71% after 3 months. On the other hand, combination therapy (isoniazid + streptomycin + aminosalicylic acid) does not permit the emergence of more than 1% of resistant bacilli after 3 months.

In combination therapy, preferentially mycobactericidal drugs having equal activity and acting by different mechanisms should be used at the proper dosage. Furthermore, these drugs must not have cross-resistance. Mutual or cross-

Table 32.1 Antituberculous Agents

Official Name	Proprietary Name	Chemical Name	Structure
isoniazid*	Hyzyd Niconyl Nydrazid	isonicotinic acid hydrazide	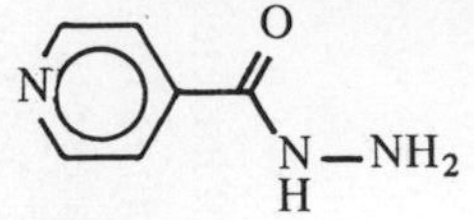
pyrazinamide*		pyrazinecarboxamide	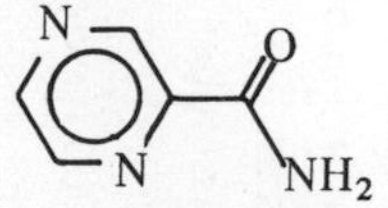
ethionamide*	Trecator-SC	2-ethylthioisonicotinamide	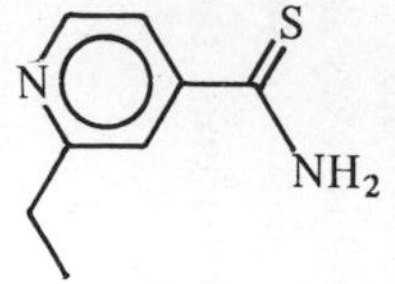
aminosalicylic acid$^+$ (PAS)	Pamisyl Parasal	4-aminosalicylic acid	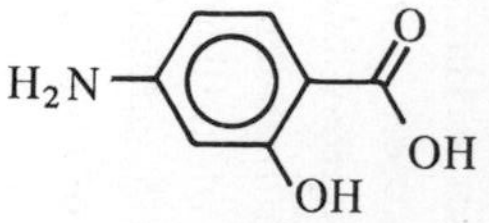
benzoylpas calcium$^+$ (calcium benzamido-salicylate)	Benzapas	calcium 4-benzamidosalicylate	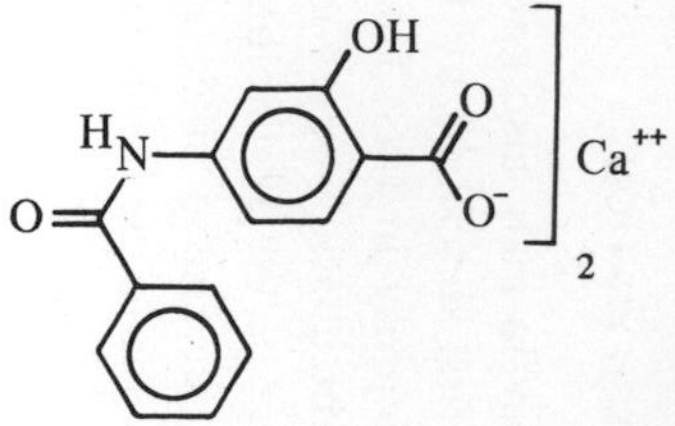

482

ethambutol* Myambutol (+)-2,2'-(ethylenediimino)-di-1-butanol

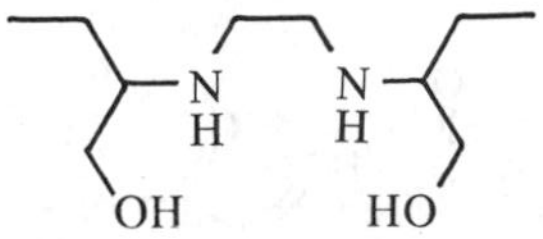

streptomycin
sulfate*

2,4-diguanidino-3,5,6-trihydroxycyclohexyl-
5-deoxy-2-O-(2-deoxy-2-methylamino-α-L-
glucopyranosyl)-3-formyl-β-L-*lyxo*pentano-
furanoside sulfate

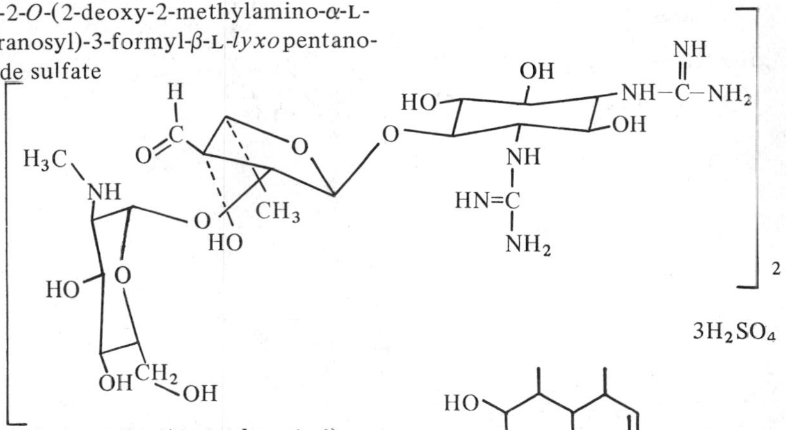

$3H_2SO_4$

rifampin*
(rifampicin) Rifadin
Rimactane 3(((4-methyl-1-piperazinyl)imino]methyl)-
rifamycin

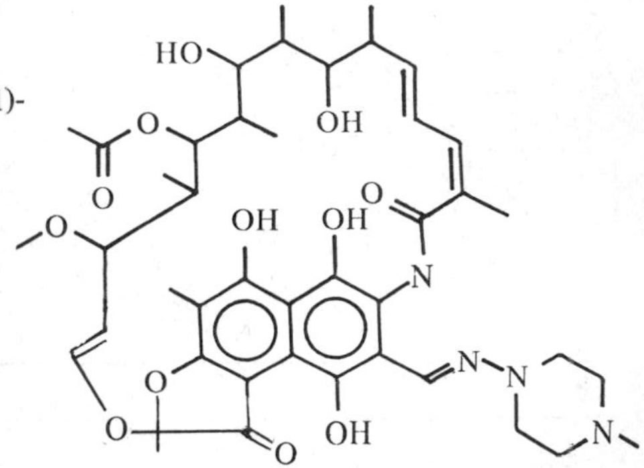

Table 32.1 Antituberculous Agents (continued)

Official Name	Proprietary Name	Chemical Name	Structure
cycloserine*	Seromycin	R(+)-4-amino-3-isoxazolidinone	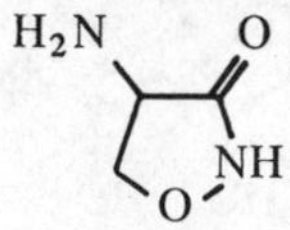
kanamycin sulfate*	Kantrex	4,6-diamino-2-hydroxy-1,3-cyclohexene 3,6′-diamino-3,6′-dideoxydi-α-D-glucoside sulfate	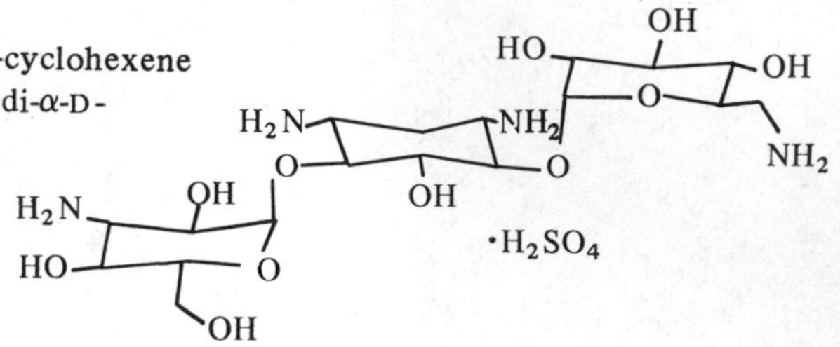
viomycin sulfate*	Viocin sulfate		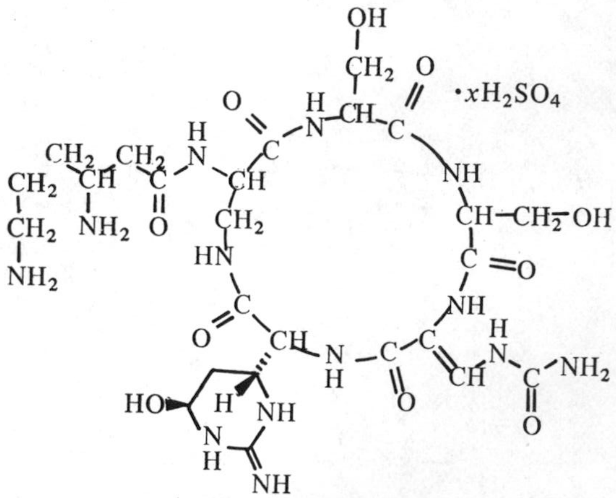

$\cdot 2H_2SO_4$

capreomycin
sulfate*

Capastat
sulfate

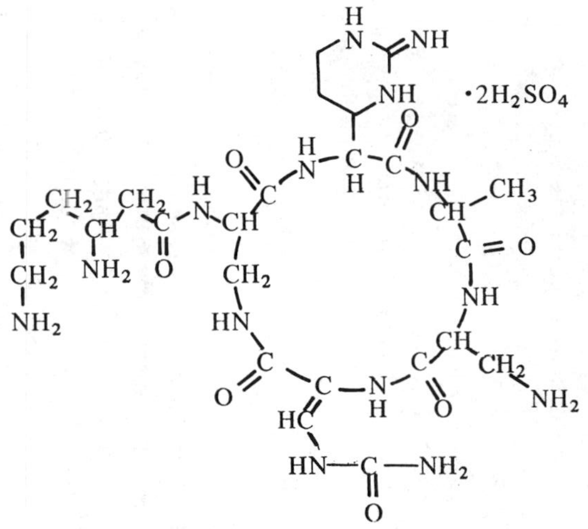

*In *USP XIX* (1975).
†In *NF XIV* (1975).

resistance occurs with PAS and ethionamide, isoniazid and cycloserine, aminoglycosidic antibiotics (kanamycins, neomycins, paromomycins) and peptidic antibiotics (viomycin, capreomycin).

Most of the antituberculous agents are bactericidal agents; they are effective only against actively growing tubercle bacilli. This explains why dormant bacilli become *persistors,* that is, still sensitive to the antituberculotic present. Since the peak concentration of the drug determines the therapeutic response, antituberculotics are more effective in a single daily dose than in divided doses.

Until recently the *primary* antituberculous drugs were streptomycin, isoniazid, and aminosalicylic acid. They were used as drugs of choice in combinations of two or three. All other antituberculous drugs were regarded as *secondary* agents, being reserved for cases of retreatment when tubercle bacilli were already resistant to one or more of the primary drugs. From 1970 on, however, the chemotherapy of tuberculosis has been undergoing a thorough reevaluation. According to the American Medical Association (AMA), now the primary agents are isoniazid, streptomycin, ethambutol, and aminosalicylic acid. The secondary drugs, because they are considered less effective and more toxic, include capreomycin, cycloserine, ethionamide, kanamycin, pyrazinamide, and viomycin. When retreatment is necessary, as the regimen of choice AMA recommends rifampin plus one or two other agents (isoniazid, ethambutol, streptomycin). The reason for not recommending rifampin as a primary drug is, according to the AMA, its high cost. Some authors, however, include this antibiotic among the primary drugs. Freerksen and associates consider rifampin and ethambutol as *basic* or *fundamental* drugs; other agents are regarded as *reserve* or *supplementary* drugs. At present the combination of isoniazid and ethambutol is considered the most suitable one. However, the combination of isoniazid, rifampin, and ethambutol is expected to prove more effective than all previous forms of treatment for the initial stages of tuberculosis, as well as for relapses and resistant cases.

Antituberculous drugs, especially aminosalicylates, cause hypersensitivity reactions.

The most widely used antituberculotics are listed in Table 32.1. However, many other drugs are active.

Antituberculous agents can be divided into the following classes:

1. Aminosalicylates and derivatives: aminosalicylic acid, aminosalicylate calcium, aminosalicylate potassium, aminosalicylate sodium, PAS resin (Rezipas), fenamisal (Tebamin), calcium benzamidosalicylate (Benzapas)

2. Hydrazides and derivatives: isoniazid, ftivazide, salinazid (Acozide), sulfoniazid (Isorilone), verazide (Isoverazide), streptoniazid (Streptoidrazide, Strazide), Glucazid, Menazone, Pasdrazide

3. Thiosemicarbazones: thioacetazone (amithiozone, Conteben, Tibione)

4. Thioamides: ethionamide, protionamide (Trevintix)

5. Thioureas: thiocarbanidin (Thioban), thiocarlide (Isoxyl)
6. Nicotinamides: pyrazinamide, morfazinamide
7. Aliphatic diamines: ethambutol
8. Thiols: ditophal (Etisul)
9. Surfactants: macrocyclon
10. Antibiotics: streptomycin, kanamycin, viomycin, capreomycin, cycloserine, rifampin, terizidone

Isoniazid

It is an odorless, white crystalline powder which is very soluble in water. Owing to its low toxicity, high effectiveness, ease of administration, low dosage (5 mg/kg of body weight daily), good tolerance, and low cost, isoniazid is nearly the ideal drug in the treatment of active tuberculosis. However, in order to prevent the emergence of resistant strains, it is advisable to use isoniazid in combination with other drugs. For prophylaxis it is used alone, but it should be administered for at least 1 year. Usually there is no cross-resistance between isoniazid and PAS or streptomycin, but it does occur occasionally. Among adverse reactions caused by isoniazid, the most common are those affecting the nervous system, some of them resulting because isoniazid acts as an antimetabolite of pyridoxine. To prevent these effects, pyridoxine should be administered concomitantly.

Ethambutol

It is a white, water-soluble, odorless, crystalline powder. Three isomers of this drug are possible. Ethambutol is the dextro isomer, which is 12 times more active than the meso isomer and 200 to 500 times more active than the levo isomer. However, the toxicities of the three isomers are identical. Considered for some years a secondary drug, now ethambutol, in combination with other drugs (isoniazid, or PAS, or streptomycin sulfate), is used as a primary agent in the treatment of pulmonary tuberculosis. Adverse reactions are dose related and include optic neuropathy, peripheral neuritis, anaphylactic shock, dermatitis, fever, nausea, dizziness, and other side effects. The usual dose is 15 mg/kg of body weight daily. No cross-resistance has been observed between ethambutol and some other antituberculous drugs.

Aminosalicylic Acid

For being less effective than isoniazid, streptomycin, or ethambutol in pulmonary tuberculosis, PAS or its salts should always be used in combination with one or two of those drugs and never alone. This combination has synergistic effect and prevents the rapid emergence of resistant strains. Large doses (8 to 15 g daily) are required for maximal benefit. When fever, rash, malaise, headache, and other hypersensitivity reactions occur, the drug should be discontinued immediately.

Rifampin

It is a semisynthetic antibiotic. The original antibiotic, of the family of rifamycins, is produced by *Streptomyces mediterranei*. Rifampin exerts bacteriostatic and bactericidal action. It is now considered a primary drug in the treatment of tuberculosis. However, its high cost prevents its wider administration. Rifampin is also the drug of choice in cases of retreatment but should be used with another antituberculous agent. No cross-resistance was found with any other antituberculous drug. Adverse reactions are rare, but it may have lethal effect on patients with preexisting liver disease. It causes teratogenic effects in experimental animals. Rifampin imparts orange-red color to urine, feces, sputum, saliva, tears, and sweat.

Streptomycin

It is an aminoglycosidic antibiotic produced by *Streptomyces griseus* and composed of streptidine, α-streptose, and *N*-methyl-L-glucosamine. It is used as the sulfate, which is soluble in water, and occurs as odorless white granules or powder, having a bitter taste. Among several other adverse reactions, streptomycin causes damage to the vestibular nerve, which may result, although rarely, in deafness. For this reason streptomycin is being replaced by less toxic drugs.

B. Antilepral Agents

Antilepral agents can be divided into the following classes:

1. Sulfones: dapsone, acedapsone, amidapsone, solasulfone (solapsone, Sulphetrone), sodium sulfadiasulfone (Promacetin)
2. Thioureas: thiambutosine, thiocarbanidin, 4,4'-diethyoxydiphenylthiourea
3. Phenazines: clofazimine
4. Hydrazine derivatives: isoniazid, Vadrine
5. Thiols: ditophal (Etisul)
6. Thiosemicarbazones: thioacetazone
7. Amides: thalidomide
8. Antibiotics: dihydrostreptomycin, cycloserine, oxytetracycline, chlorotetracycline, capreomycin, rifampin
9. Sulfonamides: sulfamethoxypyridazine, sulfamethoxypyrimidine, sulfadimethoxine, sulfamonomethoxine, sulfadoxine, sulfalene
10. Surfactants: macrocyclon

Antilepral agents most widely used are listed in Table 32.2.

Dapsone

It occurs as an odorless, white crystalline substance, which is almost insoluble in water. It should be protected from light, because in the presence of water or other impurities it undergoes photodecomposition. Dapsone is characterized by a conjugated system of single and double bonds. *In vivo* it undergoes metabolism to a great extent, yielding substances more soluble in water but still active, provided one free amino group is retained. Dapsone exerts bacteriostatic effect. It is the drug of choice in the treatment of all forms of leprosy. Treatment should continue for 5 years or more and even for life. Adverse reactions are not frequent: transient cyanosis, discoloration of the skin, dizziness, nausea, headache, vomiting. In a few cases, hemolysis and hemolytic anemia have been observed, especially in patients with glucose-6-phosphate dehydrogenase (G6PD) deficiency. High doses may cause psychoses.

Acedapsone

Known also as DADDS, this repository or latent form of dapsone is less toxic than the parent compound. Administered by intramuscular injection in the form of a suspension every 77 days for 48 weeks in the dose of 225 mg in 40% benzyl benzoate and 60% castor oil, acedapsone releases slowly either dapsone or its monoacetylated derivative, which are the active substances.

III. MECHANISM OF ACTION

The mechanism of action of certain antimycobacterial agents is already fairly well understood but not unquestionably established. Of other agents little is known. Many antituberculous agents—isoniazid, ethambutol, and thioacetazone—have chelating properties, but chelation alone cannot always explain the mechanism of action of these drugs.

For the mechanism of action of isoniazid the following theories have been advanced: (*a*) chelation of heavy metal ions, especially copper, which are essential for *M. tuberculosis;* (*b*) disturbance in the biosynthesis of catalase and peroxidase of the mycobacterium; (*c*) biochemical malfunction as a result of hydrazone formation with pyridoxal, pyruvic acid, or other physiological carbonylic compounds which are important in the metabolism of *M. tuber-culosis;* (*d*) injury to mycobacterium by the oxidation of isoniazid by hydrogen peroxide in the presence of manganese ions; (*e*) formation of a defective enzyme by replacement of nicotinamide by isoniazid in the biosynthesis of NAD^+; (*f*) selective binding of isoniazid to tyrosine residues of nucleic acids; (*g*) transformation of isoniazid to isonicotinic acid, which is incorporated into the diphosphopyridine nucleotide molecule, now called nicotinamide adenine dinucleotide, thereby originating a fraudulent coenzyme I in mycobacterium; the hydrazino group of isoniazid would act, therefore, only as a transport moiety; isoniazid would be, consequently, a latent form of isonicotinic

Table 32.2 Antilepral Agents

Official Name	Proprietary Name	Chemical Name	Structure
dapsone*	Avlosulfon	4,4'-diaminodiphenylsulfone	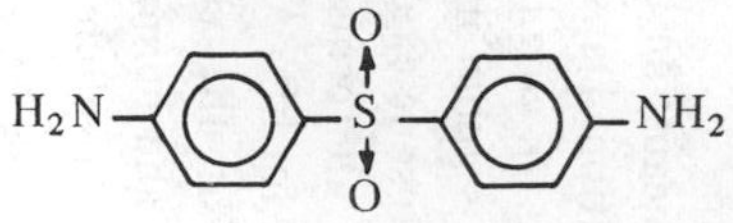 $H_2N-C_6H_4-SO_2-C_6H_4-NH_2$
acedapsone	Hansolar	N,N'-diacetyl-4,4'-diamino-diphenylsulfone	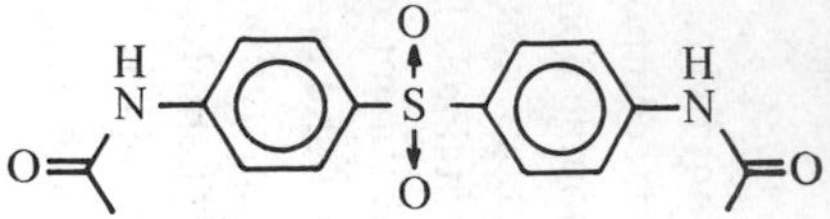 $CH_3CONH-C_6H_4-SO_2-C_6H_4-NHCOCH_3$
acetosulfone sodium	Promacetin	N-(6-sulfanilyl metanilyl)acet-amide sodium	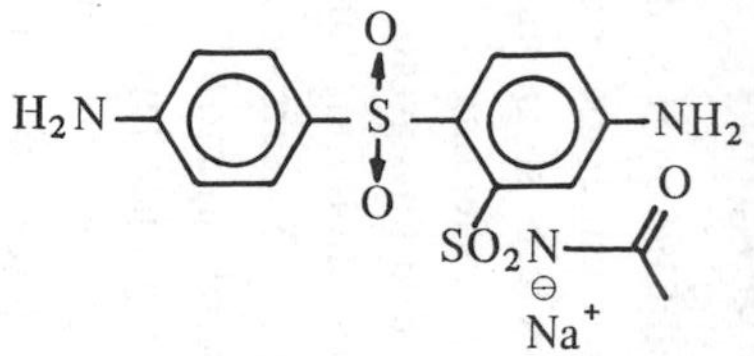 $H_2N-C_6H_4-SO_2-C_6H_3(SO_2N^{\ominus}Na^+)(NHCOCH_3)$
sulfoxone sodium⁺	Diasone sodium	disodium [sulfonylbis(p-phenyl-eneimino)] dimethanesulfinate	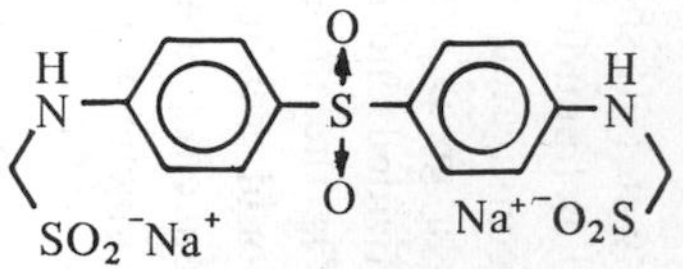 $Na^+{}^-O_2S-CH_2-NH-C_6H_4-SO_2-C_6H_4-NH-CH_2-SO_2{}^-Na^+$

| glucosulfone sodium | Promin | disodium 4,4'-diaminodiphenyl-sulfone-*N,N'*-di(dextrosesulfonate) 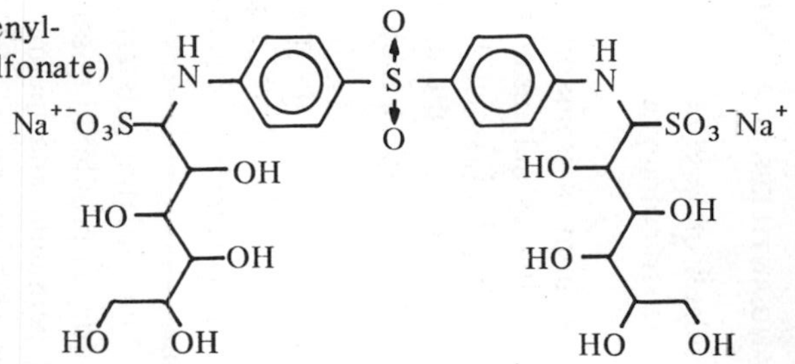 |
| clofazimine | Lamprene | 3-(*p*-chloroanilino)-10(*p*-chlorophenyl)-2,10-dihydro-2-(isopropylimino)phen-azine 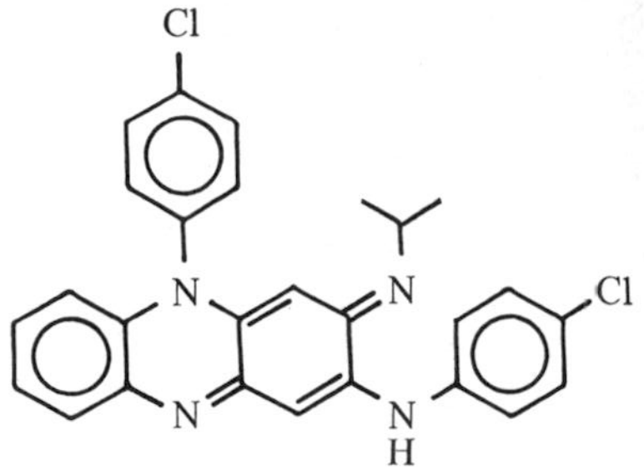 |

acid; (*h*) inhibition of the synthesis of cell wall mycolates of *M. tuberculosis;* the speculation is that mycolate-depleted cells might be structurally weak and therefore susceptible to rupture.

There is evidence that ethionamide acts by the mechanism of isoniazid explained in *g*. In unaltered form it penetrates the cell membranes and, inside the cell, it is enzymatically split or simply hydrolyzed, yielding H_2S, which has a slight inhibitory effect on the mycobacteria, and isonicotinic acid, the active moiety. In other words, the thioamide group serves as a transport moiety, because ethionamide would be also a latent form of isonicotinic acid.

Pyrazinamide presumably interferes in nicotinamide metabolism. Its mode of action may be similar to that of isoniazid. However, since there is no cross-resistance between pyrazinamide and isoniazid, the mechanism of action of both compounds must differ in some essential point.

Ethambutol is structurally similar to polyamines, and it has chelating properties. These characteristics explain its antimycobacterial action on proliferating cells. It interferes with a function of cellular polyamines (spermidine and spermine) and divalent cations involved in the synthesis or stabilization of RNA, possibly forming a chelate such as that of Figure 32.1.

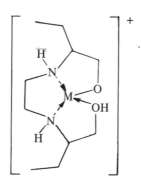

Figure 32.1 Chelate formed by ethambutol.

The mechanism of action of thioacetazone is not completely clarified, but among several other hypotheses one of the most accepted is that it owes its antimycobacterial action to chelating properties, which is common to thiosemicarbazones, including thiambutosine, which is a thiourea.

Rifampin acts by inhibiting DNA-dependent RNA polymerase of *M. tuberculosis,* forming with this enzyme a rather stable complex, but without involvement of covalent linkages. Formation of the complex prevents initiation of RNA synthesis of mycobacterium.

Streptomycin and other aminoglycosidic antibiotics—kanamycins, neomycins, paromomycin and gentamycin, for instance—interact with mRNA, enhancing ambiguity or misreading of the genetic message (Figure 3.7). It was once thought that this miscoding was responsible for the antituberculous action of streptomycin and kanamycin. However, recent studies suggest that, in addition to this action, streptomycin inhibits all aspects of protein synthesis—initiation, elongation and termination. The activity of streptomycin against *M. tuberculosis* results, therefore, from inhibition of the translation process.

Cycloserine acts by inhibiting competitively two of the three enzymes involved in the incorporation of D-alanine into the cell wall precursor UDP-MurNAc-pentapeptide, that is, alanine racemase and D-alanine:D-alanine synthetase (Figure 32.2).

Most authors believe that aminosalicylic acid and its derivatives, as well as dapsone and its derivatives, act by a competitive antagonism with *p*-aminobenzoic acid in a way similar to sulfonamides. However, Seydel advanced the hypothesis that aminosalicylic acid and its derivatives act by being incorporated in a precursor of coenzyme F by forming Schiff bases (see Chapter 31, Section IV).

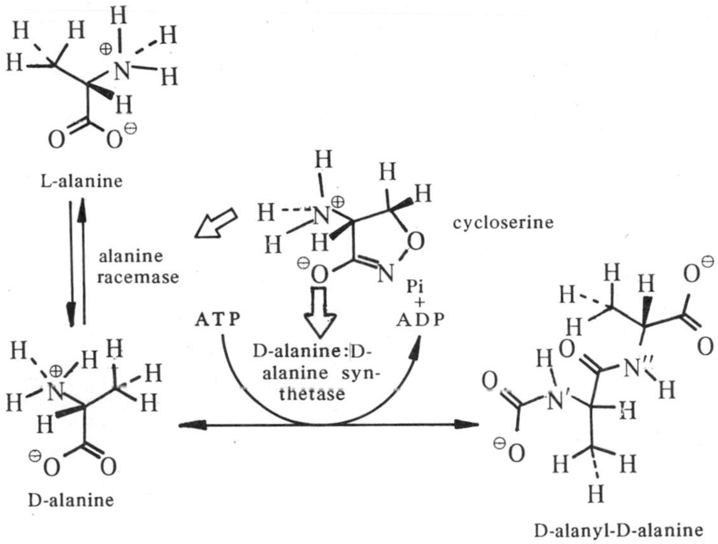

Figure 32.2   Mechanism of action of cycloserine.

REFERENCES

General

R. F. Johnston, *Top. Med. Chem.*, 3, 203 (1970).

W. Loeffler, *Schweiz. Med. Wochenschr.*, 100, 1790 (1970).

S. A. Waksman, *The Conquest of Tuberculosis,* University of California Press, Berkeley, 1964.

Classification

R. Benveniste and J. Davies, *Annu. Rev. Biochem.*, **42**, 471 (1973).

W. Lester, *Annu. Rev. Microbiol.*, **26**, 85 (1972).

G. Binda *et al.*, *Arzneim.-Forsch.*, **21**, 1907 (1971).

R. Protivinsky, *Antibiot. Chemother. (Basel)*, **17**, 87 (1971).

E. Freerksen *et al.*, Eds., "Experimental and Clinical Evaluation of the Tuberculostatics," *Antibiot. Chemother. (Basel)*, **16**, 1-536 (1970).

Mechanism of Action

A. Korolkovas, *Rev. Bras. Clin. Ter.*, **4**, 197 (1975).

K. Takayama, *Ann. N. Y. Acad. Sci.*, **235**, 426 (1974).

P. Sensi, *Pure Appl. Chem.*, **35**, 383 (1973).

S. Riva and L. G. Silvestri, *Annu. Rev. Microbiol.*, **26**, 199 (1972).

S. Pestka, *Annu. Rev. Microbiol.*, **25**, 487 (1971).

R. Protivinsky, *Antibiot. Chemother. (Basel)*, **17**, 101 (1971).

CHAPTER 33

ANTIBIOTICS

I. INTRODUCTION

Antibiotics are specific chemical substances derived from or produced by living organisms that even in small concentrations are capable of inhibiting the life processes of other organisms.

The word *antibiotic,* which comes from two Greek words ($\overset{,}{\alpha}\nu\tau\iota$ + $\overset{,}{\beta}\iota o \varsigma$) and literally means against life, was coined by Paul Vuillemin in 1889. In 1942 Waksman introduced this word in the presently accepted sense.

The first antibiotics were isolated from microorganisms, but some antibiotics are now obtained even from higher plants and animals. Presently, the sources of antibiotics, with approximate percentages, are *Pseudomonales,* 1.2: *Eubacteriales,* especially *Bacilli,* 7.7; *Actinomycetales,* 58.2; fungi, 18.1; algae and lichens, 0.9; higher plants, 12.1; animals, 1.8.

To date, in the literature more than 3100 antibiotics, about 2400 originating from microorganisms, have been described. Only a few dozen, however, are used in medicine, because, besides other reasons, most antibiotics are devoid of selective toxicity, being almost equally toxic to both parasite and host. Most antibiotics are produced by microorganisms, but some have resulted from chemical modification of known antibiotics or microbial metabolites; for instance, the semisynthetic penicillins and cephalosporins, the modified tetracyclines and rifamycins, dihydrostreptomycin, clindamycin, and troleandomycin, to name just a few. Furthermore, some antibiotics are now completely prepared by synthesis, for example, chloramphenicol.

Antibiotics are the most widely prescribed class of drugs. More than 12% of all recent medical prescriptions in the United States contain an antibiotic. They are used for several purposes: (*a*) treatment of systemic, circulatory, respiratory, genitourinary, gastrointestinal, ophthalmic, soft-tissue, skeletal, and topical infections; (*b*) prophylaxis of infections, either in healthy or in ill persons; (*c*) diagnosis of malignancies—tetracycline has been used for this.

Some clinically used antibiotics are potentially nephrotoxic or hepatotoxic. For instance, amphotericin B, bacitracin, cephaloridine, colistin, gentamicin, kanamycin, neomycin, paromomycin, polymyxins, streptomycin, tetracyclines, vancomycin, and viomycin are nephrotoxic. On the other hand, benzylpenicillin, chloramphenicol, erythromycin esters, novobiocin, oleandomycin esters, rifampin, and tetracyclines are hepatotoxic.

Owing to the widespread and indiscriminate use of antibiotics, resistant strains of pathogenic microorganisms have emerged. Broad-spectrum antibiotics can cause the phenomenon of superinfection, discussed in Chapter 25, Section IV.E.

II. HISTORY

The empirical application of molds, now known to contain antibiotics, is very old. Centuries before our era, folk medicine used them rather frequently. But the phenomenon of antibiosis was studied scientifically only after Pasteur laid the foundation of bacteriology. As early as 1877, Pasteur and Joubert reported that a liquid culture of aerobic bacteria inhibited the growth of *B. anthracis.*

Fleming, in 1929, observed by chance and then reported the inhibitory action of the *Penicillium notatum* on the growth of staphylococci. His discovery did not have immediate practical application, because he could neither isolate penicillin, the substance responsible for this action, nor produce it in great yields. This task was accomplished 10 years later by Florey and Chain. In 1944 another landmark was set in the history of natural penicillins: at Northern Regional Research Laboratories, Moyer and Coghill discovered that deep culture in corn steep liquor increased yields of penicillins by more than 1000%. In 1959, Beecham Laboratories obtained 6-aminopenicillanic acid (6-APA) from *P. chrysogenum* molds. The availability of 6-APA made possible the introduction of semisynthetic penicillins, with new and better properties than the natural ones.

In 1939 Dubos. at the Rockefeller Institute, isolated from a strain of *Bacillus brevis* an antimicrobial substance which he named tyrothricin. Actually, this substance contains two antibiotics: gramicidin and tyrocidine. Waksman, who was Dubos's professor at Rutgers University, in his systematic search for antimicrobial substances from actinomycetes, was successful in isolating several new antibiotics: actinomycin (1940), streptothricin (1942), streptomycin (1943), neomycin (1949).

The development of basic knowledge and experimental techniques, related to the newly introduced field of antibiotics coupled with the important chemotherapeutic uses of penicillins and streptomycin, prompted several investigators in many parts of the world to search for new antibiotics. This effort resulted in the introduction of the following ones in a rapid succession: bacitracin in 1943 by Johnson and Meleney; chloramphenicol in 1947 by Burkholder; chlortetracycline in 1947 by Duggar, then 73 years old; oxytetracycline in 1950 by Finlay; erythromycin in 1952 by McGuire; oleandomycin in 1954 by Sobin; kanamycin in 1957 by Umezawa; rifamycins in 1959 by Sensi. For this reason the period 1940-1959 has been justly called the golden era of antibiotic discovery.

The next decade (1960-1970) saw a decline in the isolation of therapeutically useful antibiotics: fusidic acid in 1961 by Godtfredsen and associates; lincomycin in 1962 by Mason and coworkers; gentamicin and capreomycin in 1963. However, during this period known antibiotics were chemically or biochemically modified. Some molecular modifications were very successful, resulting in new therapeutic agents, such as semisynthetic penicillins, cephalosporins, rifamycins, tetracyclines, and lincomycins, for instance.

In this decade efforts are being made to discover new antibiotics by the following approaches: (a) molecular modification of useful antibiotics; (b) isolation of new antibiotics; (c) structural variation of toxic or poorly efficacious antibiotics. These approaches have been and are being fruitful, as will be seen in the next sections of this chapter.

III. CLASSIFICATION

Among the several criteria used to classify antibiotics, the main ones are (a) biosynthesis; (b) spectrum of activity; (c) chemical structure.

A. Biosynthesis

According to biosynthesis, antibiotics can be divided into the following classes:

1. Antibiotics derived from amino acids: cycloserine, chloramphenicol, penicillins, cephalosporins, gramicidin, tyrocidin, bacltracin, polymyxin, colistine, cholistimethate, viomycin, capreomycin

2. Antibiotics derived from carbohydrates: streptomycin, kanamycin, neomycin, paromomycin, gentamicin, lincomycin, spectinomycin

3. Antibiotics derived essentially from acetate and propionate: fusidic acid, griseofulvin, macrolide antibiotics, polyene antibiotics, tetracyclines

4. Miscellaneous antibiotics: novobiocin, puromycin, rifamycins, ristocetin, streptovaricins, vancomycin

B. Spectrum of Activity

1. Broad-spectrum antibiotics: some penicillins (e.g., ampicillin, carbenicillin), some cephalosporins (cephaloridine), chloramphenicol, macrolides, rifamycins, tetracyclines, edeine, cuprimyxin, fosfomycin, nucleocidin, puromycin, sparsomycin, some aminoglycosides (neomycin, kanamycin, kanendomycin, gentamicin, paromomycin), negamycin, pikromycin, butirosins, narbomycin

2. Antibiotics active predominantly against Gram-positive bacteria: bacitracin, most penicillins (e.g., benzylpenicillin, cloxacillin, dicloxacillin, phenbenicillin, pheneticillin, oxacillin), fusidic acid, gramicidin, erythromycin, lincomycin, clindamycin, novobiocin, ristocetin, spiramycin, tyrothricin, vancomycin, some cephalosporins (cephalotin), neopluramycin, maridomycins, rosamıcin, ribostamycin, oxazinomycin, rabelomycin

3. Antibiotics active predominantly against Gram-negative bacteria: colistin, polymyxin B, sulfomyxin, bicyclomicin

4. Antibiotics active predominantly against mycobacteria: cycloserine, capreomycin, decoynine, lividomycin A, rifampin, streptomycin, viomycin

5. Antibiotics active against fungi: polyene antibiotics (e.g., amphotericin B, candicidin, fumagillin, hamycin, nystatin, trichomycin), glutarimides for fungal diseases of plants (cycloheximide, streptovitacins, streptimidones), actidin, griseofulvin, azalomycin, pyrrolnitrin, variotin, toyocamycin

6. Antiamebic antibiotics: fumagillin, paromomycin, puromycin

7. Antineoplastic antibiotics: adriamycin, azaserine, bleomycin, dactinomycin, daunorubicin, mithramycin, mitomycin C, porfiromycins, pactamycin, carzinophilin, chromomycin A, sarkomycin, psicofuranine, mycophenolic acid, phleomycin, streptonigrin, actinomycins, olivomycin, anthracyclines

C. Chemical Structure

Taking into account the chemical structure, many authors have presented chemical classifications of antibiotics. None, however, is satisfactory. For didactic reasons, we shall divide antibiotics of clinical interest into the following classes: penicillins, cephalosporins, chloramphenicol and derivatives, tetracyclines, polypeptide antibiotics, polyene antibiotics, macrolide antibiotics, aminoglycoside antibiotics, ansamycins, anthracyclines, lincomycin group, nucleoside antibiotics, glutarimide antibiotics, miscellaneous antibiotics.

D. Penicillins

1. *Structure*

Penicillins, as well as cephalosporins, which are globally called β-lactam antibiotics, are characterized by three fundamental structural requirements: (a) the fused β-lactam structure, (b) a free carboxyl group, and (c) one or more properly substituted amino side-chain groups.

Hou and Poole pointed out that all β-lactam antibiotics with "high antibacterial activity possess a continuous −C−CO−NH−C−CO−N−C−COO− linkage, beginning with the side chain (−C−CO−NH−) and continuing along the β-lactam and the thiazolidine or dihydrothiazine nucleus."

All penicillins have the same β-lactam-thiazolidine general structure that contains three asymmetric centers. Therefore, theoretically this structure could present eight optically active forms. However, the natural isomer, presumably the only one with biological activity, has the following stereochemistry:

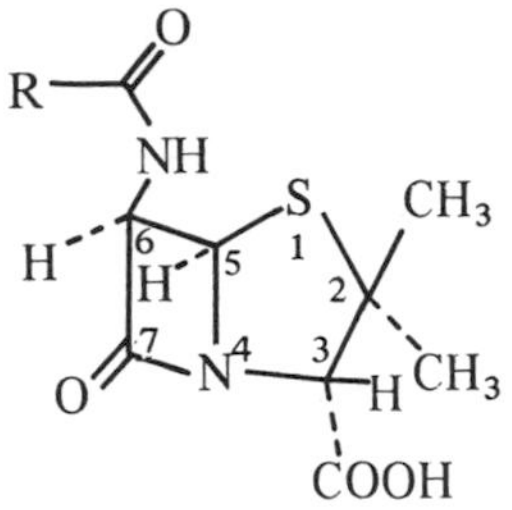

The fused rings are not coplanar but folded along the C–5,N–4 axis. This nonplanarity suppresses normal amide resonance. The amide-carrying carbon atom (C–6) has the L configuration, whereas the carbon atom bearing the carboxyl group (C–3) has the D configuration. The natural penicillins are derived from a cysteinylvaline precursor.

The intensity of antibacterial activity depends also on the stereochemistry of the side chain. For instance, of the three diastereoisomers of phenethicillin, the L epimer is the most active and the D epimer is the least active, whereas the DL form has an intermediate activity against Gram-positive bacteria.

2. Nomenclature

Penicillins are named in the following different ways:

1. 4-Thia-1-azabicyclo[3.2.0]heptanes—adopted in *Chemical Abstracts*. According to this nomenclature, ampicillin is (-)-6-(2-amino-2-phenylacetamido)-3,3-dimethyl-7-oxo-4-oxo-thia-1-azabicyclo[3.2.0]heptane-2-carboxylic acid.

2. Penam derivatives—penam is the name given to unsubstituted bicyclic lactam. Thus, penicillin G is 6-phenylacetamido-2,2-dimethylpenam-3-carboxylic acid

R = R′ = H   penam

R = COOH
R′ = CH$_3$ }   penicillanic acid

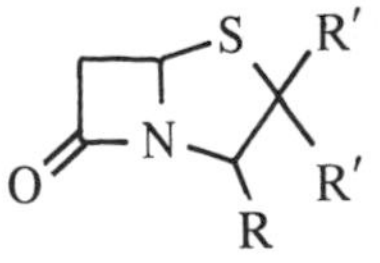

3. Penicillanic acid derivatives—penicillanic acid is the duly substituted penam. Thus, methicillin is 2,6-dimethoxybenzamidopenicillanic acid.

4. Penicillin derivatives—it is the most commonly used nomenclature; penicillin is the name given to the general structure without R. Accordingly, suncillin is [D-α-(sulfoamino)benzyl]penicillin. Usually the side-chain R is an aralkyl, aryl or heterocyclic grouping (Table 33.1).

Table 33.1 Penicillins

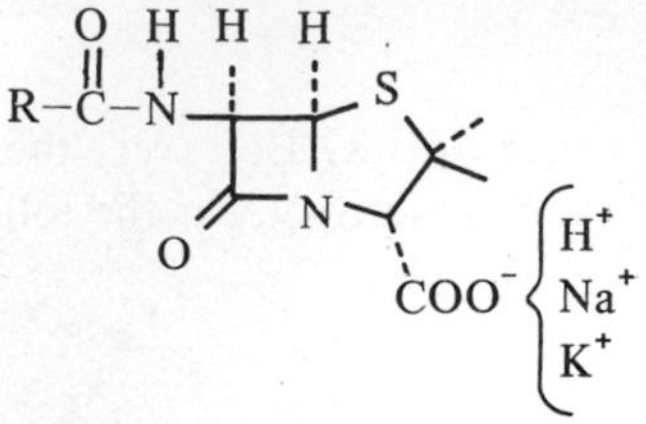

Official Name	Proprietary Name	R
penicillin G*+ (benzylpenicillin)	...	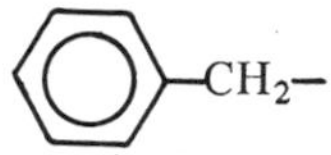
ampicillin*	Omnipen Penbritin	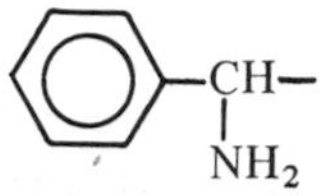
carbenicillin*	Geopen Pyopen	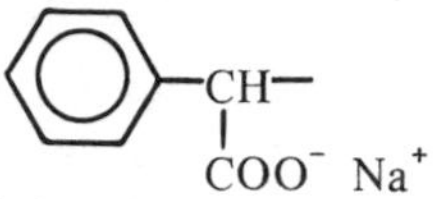
cloxacillin*	Tegopen	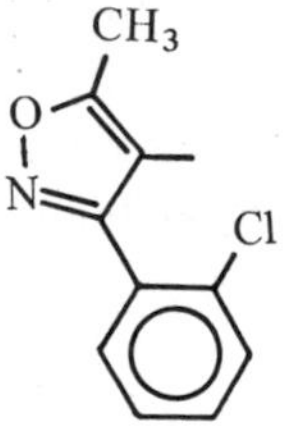
dicloxacillin*	Dynapen Pathocil Veracillin	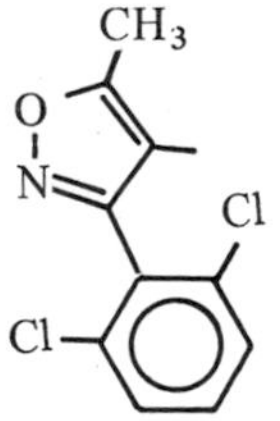
hetacillin	Penplenum Versapen	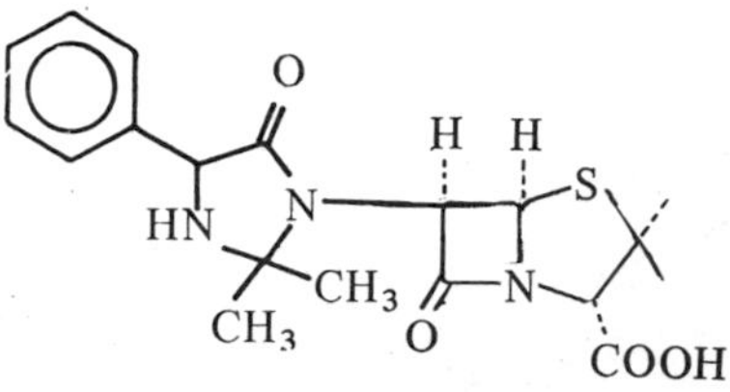
methicillin*	Staphcillin	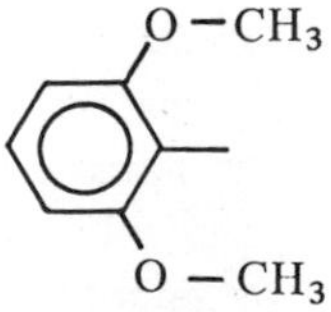

Table 33.1 Penicillins (continued)

Official Name	Proprietary Name	R
nafcillin*	Unipen	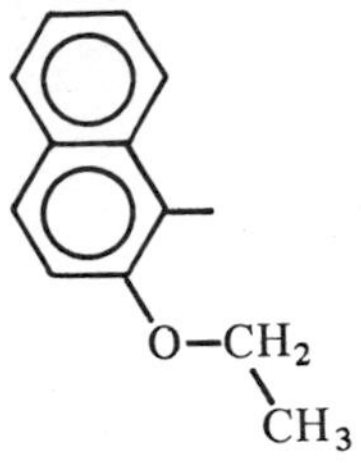
oxacillin*	Bactocill Prostaphlin	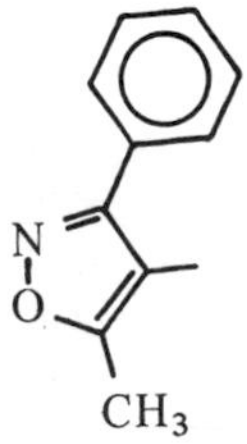
phenethicillin[+]	Darcil Maxipen Syncillin	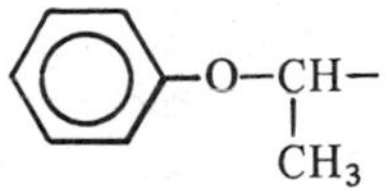
penicillin V*[+]	Pen-Vee V-Cillin	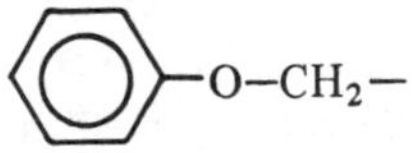
carindacillin	Geocillin	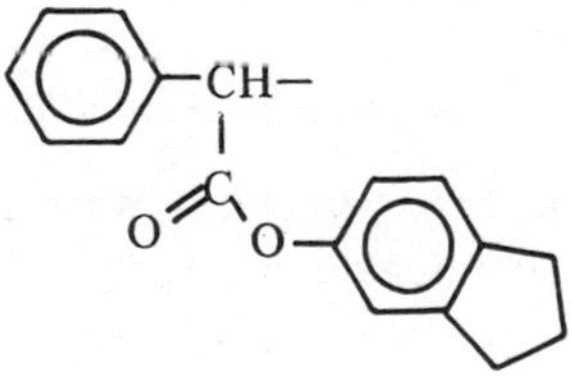
metampicillin	Suvipen	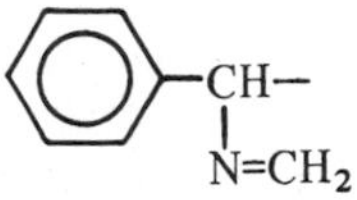
sulbenicillin	Lilacillin	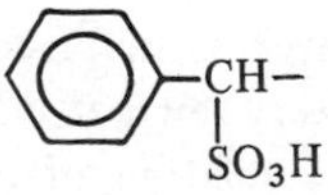

Table 33.1 Penicillins (continued)

Official Name	Proprietary Name	R
amoxycillin	Almoxil	HO—⟨O⟩—CH— | NH$_2$

*In *USP XIX* (1975).
$^+$In *NF XIV* (1975).

3. Physicochemical Properties

Penicillins are white or slightly yellowish-white crystalline powders, strongly dextrorotatory. Owing to the carboxyl group attached to the fused ring, all penicillins are relatively strong acids with pK_a values around 2.65. However, those that contain a basic group in the side chain exist as zwitterions; such is the case of ampicillin whose pK_a is 7.4. Therefore, penicillins can be classified as monobasic and zwitterionic. Most penicillins are used as sodium, potassium, or other salts, which are water soluble, whereas free penicillins are sparingly soluble in water. Penicillins and their salts have a strong tendency to form crystalline hydrates.

Owing to the strained amide bond in the fused β-lactam of the nucleus, penicillins are very reactive. They are extremely susceptible to nucleophilic attack, as well as to electrophilic attack. They are inactivated by hydrolysis, particularly in the presence of heavy metal salts, acids, and especially bases and also by the catalytic action of enzymes: acylases (amidases) and β-lactamases.

4. Semisynthetic Penicillins

The first penicillins were isolated from cultures of *Penicillium notatum* and *P. chrysogenum* species of fungi. Later on new penicillins were prepared by adding precursors, such as carboxylic acids or related compounds, to the fermentation mixture. Under certain conditions those precursors were incorporated into the side chain of the new penicillins. Unfortunately, this method had limited success.

A third approach, which also proved to yield scarce results, was to modify the available penicillins by chemical synthesis. An apparently bright perspective was opened by Sheehan and Henery-Logan in 1957, when the first total chemical synthesis of phenoxymethyl penicillin in significant yield was reported. However, it involved several steps, and the yields were low; for these reasons, only very few penicillins have been prepared by total synthesis.

The isolation of 6-aminopenicillanic acid (6–APA) in the Beecham Research

Laboratories in 1957 permitted facile preparation of a great number of the so-called semisynthetic penicillins. These penicillins are synthesized by making 6—APA react with one of the following groups of compounds: (*a*) carboxylic acids using *N,N'*-dicyclohexylcarbodiimide as condensing agent, at room temperature; (*b*) acyl chlorides, dissolved in inert organic solvent, in the presence of a proton-acceptor substance, such as triethylamine; (*c*) acid anhydrides, particularly cyclic ones.

By one or more of these or other synthetic ways of molecular modification, new penicillins having better properties were prepared:

1. Acid-stable penicillins, which have a powerful electron-attracting group attached to the amino side chain, such as phenethicillin, propicillin, phenbenicillin, and epicillin, in which acidic rearrangement is prevented. This rearrangement proceeds in the following way:

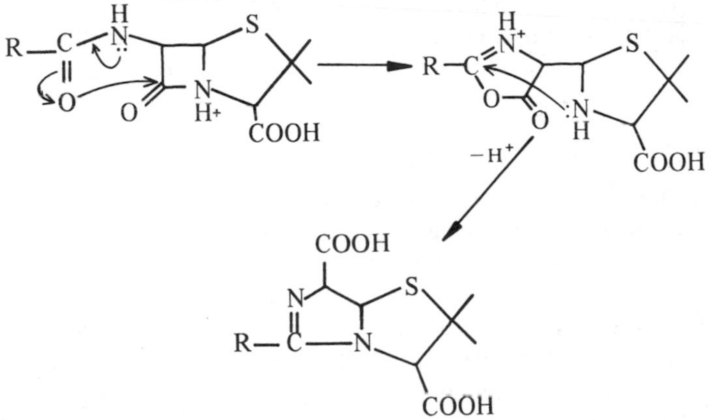

2. Resistant-to-β-lactamase penicillins, which have a bulky group attached to the amino side chain. By steric hindrance this group interferes with the enzyme attachment to the penicillins and causes a conformational change in the enzyme and loss of activity. Examples are azidocillin, flucloxacillin, methicillin, and pirazocillin.

3. Acid-stable and resistant-β-lactamase penicillins, in which the group attached to the amino side chain has electron-attracting properties and is bulky. Examples are oxacillin, cloxacillin, dicloxacillin, nafcillin, diphenicillin, quinacillin, and a new penicillin, cyclacillin.

4. Broad-spectrum penicillins, in which the hydrophilic character of the amino side chain imparts greater activity against Gram-negative bacteria. Examples are ampicillin, carbenicillin, carfecillin, carindacillin, epicillin, sulbenicillin, Suncillin, ticarcillin, and BL-P1654.

5. Latent Penicillins

Intramuscularly injected penicillins are rapidly eliminated from the body. For this reason, when penicillin was scarce and very expensive, it was usual to collect urine and recover penicillin. Later on, in order to increase the persistence of penicillin in the body, the Romansky formula and buffered tablets of Al^{3+}-penicillin, and, more recently, probenecid, a drug that blocks renal tubular excretion of penicillin, have been used. Now, semisynthetic penicillins, some of which stay longer in the body, and repository preparations, which are latent forms of penicillins, are preferred. Some of the latent forms are the following ones: (a) salts which are sparingly soluble in water: procaine benzylpenicillin, benzathine benzylpenicillin, pheniracillin (Zinacillin = 2,5-diphenylpiperazine dibenzylpenicillin), clemizole penicillin (Anergopen = clemizole benzylpenicillin, which has antiallergic action); (b) esters that hydrolyze in vivo, yielding penicillin: penamecillin (acetoxymethyl ester of benzylpenicillin), pivampicin (pivaloyloxymethyl ester of benzylpenicillin), pivampicillin (pivaloyloxymethyl ester of ampicillin); (c) hybridization of a penicillin with other antibiotics: penimocycline (Intraxium, a hybrid of metampicillin with tetracycline).

6. Classes of Penicillins

1. Penicillins with a broad spectrum of activity but not resistant to β-lactamase: ampicillin, ampicillin guanylureido derivative, carbenicillin, carindacillin, hetacillin, cyclacillin, metampicillin, sulbenicillin, amoxycillin, pivampicillin, BL-P 1597

2. Penicillins with a narrow spectrum of activity but resistant to β-lactamase: cloxacillin, oxacillin, dicloxacillin, nafcillin, diphenicillin, methicillin, pirazocillin, flucloxacillin, azidocillin, quinacillin, DA 2047

3. Penicillins with an intermediate spectrum of activity but highly resistant to acidic hydrolysis: phenoxymethylpenicillin, pheneticillin, propicillin, phenbenicillin, clometocillin, flucloxacillin

7. Therapeutic Uses

Penicillins are bactericidal agents, being more effective against Gram-positive bacteria, because their cell wall structure is more susceptible to inhibitory action than Gram-negative bacteria. They kill dividing rather than resting microorganisms. Their binding to serum depends on the hydrophobic character. Penicillins are excreted by both the liver and the kidneys, either unaltered or metabolized.

Most penicillins are active only against Gram-positive microorganisms. They are indicated for treatment and prophylaxis of infections caused by cocci (gonococci, β-hemolytic streptococci, and group A β-hemolytic streptococci), Treponema pallidum, Bacillus anthracis, Clostridium, Corynebacterium diphther-

iae, and several species of *Actinomyces.* Other penicillins are effective against Gram-negative bacteria: *Hemophilus influenzae,* most strains of *Escherichia coli,* indole-negative *Proteus mirabilis,* and some species of *Salmonella, Shigella,* and *Pseudomonas.*

Penicillins should not be used topically, because by this route they are inactive and are often prone to cause hypersensitization.

8. Adverse Reactions

Penicillins have a tendency to induce allergic reactions of two types: (*a*) an immediate, profound, and often lethal reaction called anaphylactic shock; (*b*) a delayed serum sickness type of response. It is believed that hypersensitivity responses are not caused by unchanged penicillins but by their degradation products that form antigenic complexes by reacting with proteins or their components.

Penicillinase (Neutropen) was once used in the treatment of penicillin-induced hypersensitivity reactions, but it is of little value and, paradoxically, it itself can cause allergic responses. Since the incidence of allergic reactions is relatively high (1 to 5%), extreme caution should be taken with patients with a history of hypersensitivity. The usual test for hypersensitivity is the intracutaneous injection of a small amount of benzylpenicillin. According to the AMA, "this procedure is dangerous and is contraindicated since even the small amount used can cause serious reactions or death in susceptible patients." For this test much safer preparations, injected intradermally, are now available: benzylpenicilloyl-polylysine conjugate (BPL) and a minor determinant mixture (MDM).

9. Most Widely Used Penicillins

The most widely used penicillins are listed in Table 33.1. Besides these, several others are commercially available: azidocillin (Globacillin), carfecillin, clometocillin (Rixapen), cyclacillin (Ultracillin), diphenicillin, epicillin (Omnisan), fibracillin, flucloxacillin (Floxapen), phenbenicillin, pirazocillin, pivampicillin (Pondocillin), propicillin, quinacillin, suncillin.

Penicillin G

Also called benzylpenicillin, it is the drug of choice and the most widely used penicillin for treating infections caused by β-hemolytic streptococci, gonococci, meningococci, pneumococci, *Clostridium, Treponema pallidum, Corynebacterium diphtheriae, Bacillus anthracis,* and some species of *Actinomyces.* It is administered usually by parenteral route, because gastric juice inactivates it. The intravenous route is used when massive doses are indicated. Penicillin G is also marketed in the form of salts of sodium, potassium, procaine, benzathine, and aluminium monostearate, besides combinations of two or more of these salts.

Phenoxymethylpenicillin

It is resistant to hydrolysis by gastric juice and, therefore, is used by oral route. Its spectrum of activity is the same as that of penicillin G. It is used in the free form and also as benzathine, hydrobamine, and potassium salts. The usual dosage is 125 to 500 mg four to six times daily.

Methicillin Sodium

It is resistant to β-lactamase. It is used, by parenteral route, in the treatment of infections caused by staphylococci, pneumococci, and β-hemolytic streptococci, as the drug of second choice (the first choice is penicillin G). Its most frequent use is in penicillin G-resistant staphylococci infections.

Nafcillin Sodium

It is resistant both to acids and β-lactamase. It is effective against pneumococci, β-hemolytic streptococci, and staphylococci, including penicillin G-resistant strains. It is used both orally and parenterally. The usual dosage is 250 mg to 1 g every 4 to 6 hours, preferably 2 hours before meals.

Ampicillin

It is a broad-spectrum penicillin, because it acts also against Gram-negative bacteria. This broader activity is ascribed to the amino group which presumably confers cell wall penetration power to this penicillin. It is often preferable to tetracyclines or chloramphenicol when a broad-spectrum antibacterial agent is needed. Ampicillin is available also in the forms of trihydrate and sodium salt. The usual dosage is 250 to 500 mg four times daily.

Carindacillin

It is the indanyl derivative of carbenicillin with the advantage of being active by oral route, because it is well absorbed from the gastrointestinal tract. It has a broader spectrum than any other commercially available penicillin. Its main indication is the treatment of urinary infections caused by Gram-negative bacteria, especially *Pseudomonas aeruginosa.*

E. Cephalosporins

1. *Structure and Properties*

Cephalosporins are β-lactam antibiotics, with the same fundamental structural requirements as penicillins. They have the same β-lactam-dihydrothiazine structure that contains two asymmetrical centers. Thus four optically active forms are possible. The natural isomer has the following stereochemistry:

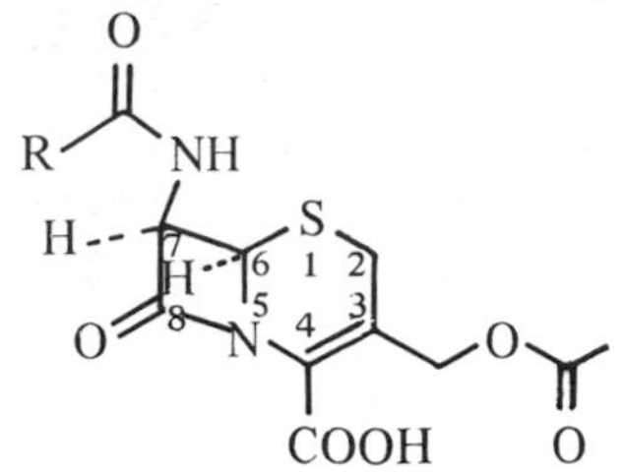

The fused rings are not coplanar but folded along the C−6,N−5 bond but less markedly than in penicillins. The amide-bearing carbon atom (C−7) has the L configuration.

The physicochemical properties of cephalosporins are very similar to those of penicillins. For instance, most cephalosporins are available as soluble salts or as zwitterions. Those with an asymmetric carbon atom are stereospecific.

2. Nomenclature

Cephalosporins are named in the following ways:

1. 5-Thia-1-azabicyclo[4.2.0]octanes−adopted in *Chemical Abstracts*. Accordingly, cephalothin is 3-(acetoxymethyl)-8-oxo-7-[2-(2-thienyl)acetamido]-5-thia-1-azabicyclo[4.2.0]oct-2-ene-2-carboxylic acid.

2. Cepham derivatives−cepham is the name given to the unsubstituted bicyclic lactam:

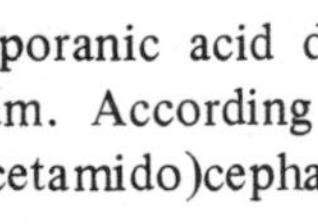

cepham 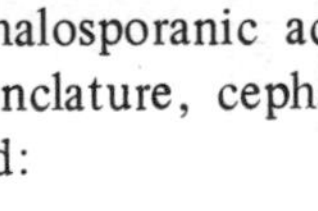   cephem

Thus, cephaloridine is 3-pyridinomethyl-7-(2-thiophene-2-acetamido)-3-cephem-4-carboxylate.

3. Cephalosporanic acid derivatives−cephalosporanic acid is the duly substituted cepham. According to this nomenclature, cephaloglycin is 7-(D-α-aminophenylacetamido)cephalosporanic acid:

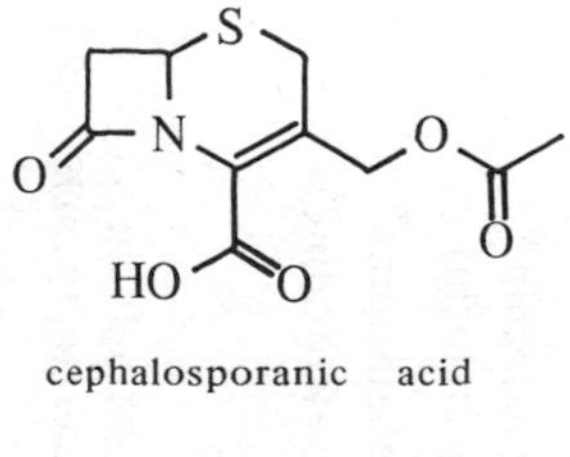

cephalosporanic   acid

Table 33.2 Cephalosporins

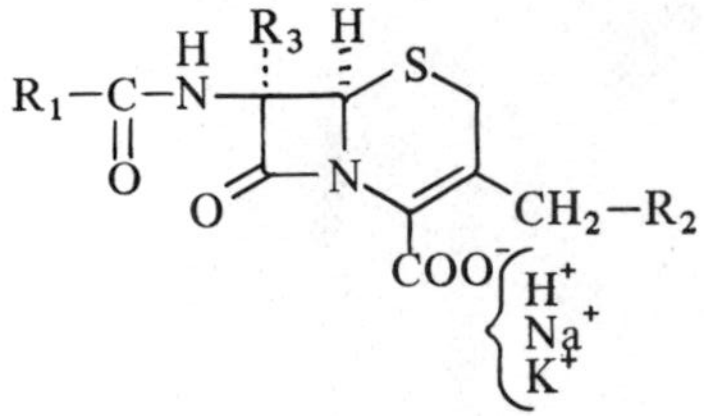

Official Name	Proprietary Name	R_1	R_2	R_3
cephalothin* (cefalotin)	Keflin	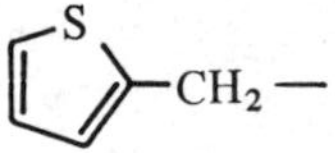 (2-thienyl)$-CH_2-$	$-O-CO-CH_3$	H
cephaloridine+ (cefaloridine)	Ceporan Ceporin Keflordin Loridine	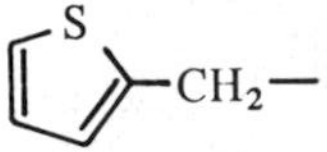 (2-thienyl)$-CH_2-$	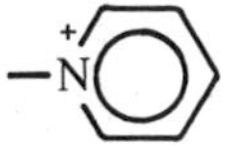 pyridinium	H
cephaloglycin+ (cefaloglycin)	Kafocin	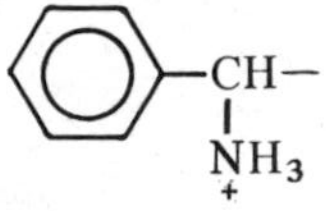 phenyl$-\overset{\underset{NH_3^+}{\shortmid}}{CH}-$	$-O-CO-CH_3$	H
cephalexin* (cefalexin)	Ceporex Keflex	 phenyl$-\overset{\underset{NH_2}{\shortmid}}{CH}-$	$-H$	H
cephacetrile (cefacetril)	Celospor	$N\equiv C-CH_2-$	$-O-CO-CH_3$	H

cephanone (cefanone)	(1-oxidopyridin-1-ium-2-yl)–CH$_2$–	$-S-\!\!\begin{smallmatrix}N-N\\ \|\ \|\end{smallmatrix}\!\!S{-}C{-}CH_3$	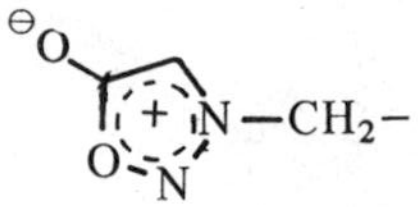 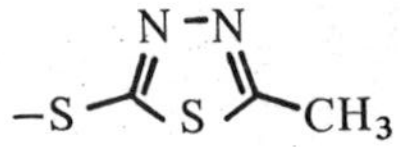 H
cephapirin (cefapirin) Brisfirina Cefatrex Cefatrexil	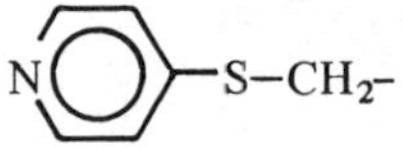 (pyridin-4-yl)S–CH$_2$–	–O–CO–CH$_3$	H
cephradine (cefradine) Eskacef Velosef	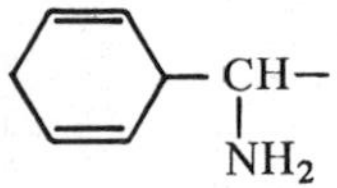 (cyclohexa-1,4-dien-1-yl)CH(NH$_2$)–	H	H
cefazolin Cefamezin Kefazol	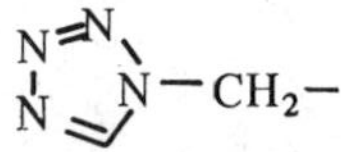 (tetrazol-1-yl)–CH$_2$–	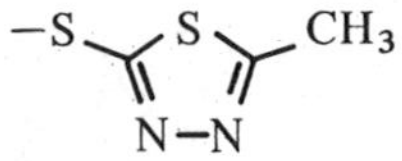 $-S-\!\!\begin{smallmatrix}S\\ \|\end{smallmatrix}\!\!\begin{smallmatrix}CH_3\\ \|\\ N-N\end{smallmatrix}$	H
cefoxitin	(thiophen-2-yl)–CH$_2$–	$-O-C(=O)-NH_2$	O–CH$_3$

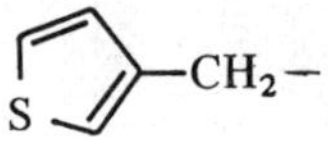

*In *USP XIX* (1975).
†In *NF XIV* (1975).

3. Semisynthetic Cephalosporins

Only cephalosporin C is found in nature, isolated from cultures of fungi, especially *Cephalosporium* species. In 1966 Woodward and associates reported the first total chemical synthesis of cephalosporin C, but their process is not industrially feasible. The discovery of 7-aminocephalosporanic acid (7-ACA) as a product of hydrolysis of cephalosporin C resulted in the introduction of several better cephalosporins, which were obtained by molecular modification of the two side chains: 7-acylamide and 3-acetoxymethyl. Variation of the first side chain is performed by the same processes used for the preparation of semisynthetic penicillins, but the method of choice is acetylation by acyl chlorides.

4. Therapeutic Uses and Adverse Reactions

Cephalosporins are used for treatment of infections caused by most Gram-positive cocci and many Gram-negative bacteria, especially *Escherichia coli, Proteus mirabilis,* and *Klebsiella.* They are not effective in infections caused by *Pseudomonas,* most species of indole-positive *Proteus* and *Enterobacter.* Most cephalosporins are resistant to cephalosporinase, an enzyme that inactivates them by cleaving the β-lactam ring. This resistance is ascribed to the side chain at C-3.

   Adverse reactions include tubular necrosis, diarrhea, thrombophlebitis, and rarely anaphylaxis.

5. Classes of Cephalosporins

Therapeutically used cephalosporins are mainly those listed in Table 33.2. Besides these, the following ones are also available: cefaloram, cefamandole, BLS 151, BLS 217, BLS 339, CMT. In the United States only cephalothin, cephaloridine, cephaloglycin, and cephalexin are currently marketed. These four cephalosporins have a broad spectrum of activity, being effective against both Gram-positive and Gram-negative bacteria. The first two are poorly absorbed, whereas the remaining two are well absorbed.

   Cephalosporins are usually grouped into two classes:

   1. Injectable cephalosporins: cephalothin, cephaloridine, cephacetrile, cefazolin, cephanone, cephapirin, BLS 217, CMT

   2. Oral cephalosporins: cephaloglycin, cephalexin, cephradine, cefoxitin, cephalexin

Cephalothin Sodium

It should be reserved only for serious infections. It is the drug of choice in staphylococcal infections when the patient is hypersensitive to penicillin and when the parenteral route is indicated. The usual dosage by intramuscular injection is 0.5 to 1 g four to six times daily; by intravenous route, 4 to 6 g daily in divided doses.

Cephalexin

Used as the monohydrate, primarily for the treatment of respiratory, urinary, skin, and soft-tissue infections. The usual dosage is 250 mg every 6 hours.

Cephanone Sodium

It has a broader spectrum of activity than most cephalosporins, especially against Gram-negative microorganisms.

Cephapirin Sodium

It is better tolerated and less painful on injection than cephalothin.

Cefoxitin Sodium

It has a broader spectrum of activity against Gram-negative microorganisms compared with older cephalosporins. It is effective also against indole-positive *Proteus* and *Serratia,* being highly resistant to β-lactamase hydrolysis.

Cephradine

It is active against most strains of β-lactamase producing staphylococci.

F. Chloramphenicol and Derivatives

Chloramphenicol (USP) is a broad-spectrum antibiotic. Structurally, it is D(−)-threo(−)-2,2-dichloro-N-[(β-hydroxy-α-(hydroxymethyl)-p-nitrophenethyl] acetamide:

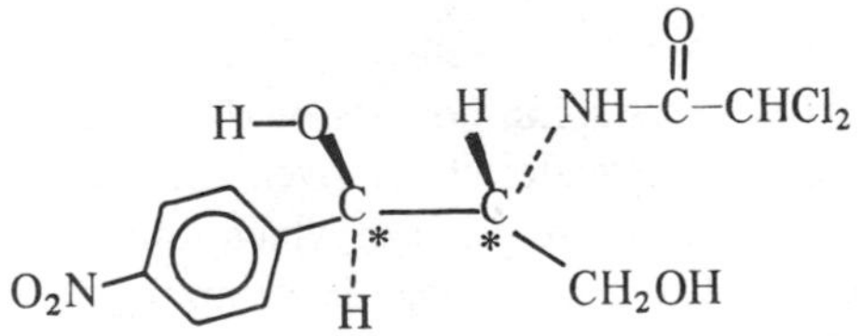

$$O_2N-C_6H_4-\underset{H}{\overset{H-O}{C^*}}-\underset{CH_2OH}{\overset{H\ NH-C(=O)-CHCl_2}{C^*}}$$

It was first isolated from *Streptomyces venezuelae,* a microorganism that was found by Paul Burkholder in a sample of soil collected close to Caracas, Venezuela, and brought to Parke, Davis Laboratories of the United States through Oliver Kamm. Since 1949 chloramphenicol has been obtained by total synthesis, which is relatively simple and less expensive than extraction from fermentation molds.

Stereochemistry plays an important role in its action. Since it has two asymmetric centers, four isomers are possible: (−)-threo, (+)-threo, (+)-erythro, and (-)-erythro. However, only the natural isomer has marked antibacterial activity. Its fundamental structure is essential for activity. Extensive molecular modification has not resulted in better products. The nitro group can be replaced without significant loss of activity by similar very strong electron-attracting moieties, such as acetyl (CH_3CO, as in cetophenicol) and methylsulfonyl (CH_3SO_2, as in thiamphenicol, the dextro isomer, and racefenicol, the racemic mixture). These drugs, however, are not better than the parent compound, but in certain cases they can be used instead of chloramphenicol. The dichloroacetyl group has been replaced by azidoacetyl (CH_2N_3), but the resulting product, called azidamphenicol, is used mainly in ophthalmology in Europe.

In order to mask its bitter taste and improve its physicochemical properties, several esters of chloramphenicol have been prepared and marketed. These esters are inactive per se, but in vivo they release the parent compound. Therefore, they are latent forms of chloramphenicol. Some esters are insoluble in water but are devoid of the bitter taste of the parent compound, which indicates them for pediatric use: palmitate (USP), stearate, cinnamate. Other esters are water soluble, suitable for parenteral administration: glycinate hydrochloride, sodium succinate (USP), arginine succinate.

Chloramphenicol is the most effective drug for treating acute typhoid fever and several other serious infections produced by various *Salmonellae*. It is also effective against many strains of Gram-positive and Gram-negative bacteria, *Rickettsia* and the psittacosis-lymphogranuloma group. However, owing to its serious adverse reactions—blood dyscrasias, including aplastic anemia with pancytopenia—chloramphenicol should be reserved for those infections not susceptible to less dangerous drugs. It is not recommended for both premature and full-term infants during the first 2 weeks of life, because it can cause gray-baby syndrome, which may be fatal.

Chloramphenicol

It is a white, very stable, crystalline powder, with no odor but very bitter taste. It is only slightly soluble in water. It is given preferably by oral route. Parenteral route is reserved for those cases in which oral administration is contraindicated or impractical. The usual dosage is 50 mg/kg of body weight daily in divided doses every 6 to 8 hours.

G. Tetracyclines

1. Structure and Nomenclature

Tetracyclines (from the Greek τετρα = four and κύκλος = circle) are characterized by a common octahydronaphthacene skeleton, a system formed by four fused rings, and their broad spectrum of action. The first two members were isolated from *Streptomyces aureofaciens* (chlortetracycline) and *S. rimosus* (oxytetracycline). Tetracycline, the prototype of this class of antibiotics, was first obtained by hydrogenolysis of 7-chlorotetracycline. In 1957 the 6-demethyltetracycline family was described. More recently, new tetracyclines, with improved properties, have been prepared by semisynthetic procedures. Some tetracyclines have been totally synthesized. In 1970 quelocardine, a new antibiotic of the tetracycline family, was described.

The general structure of tetracyclines and the absolute configuration are as represented in Figure 33.1. They have many asymmetric centers: tetracycline

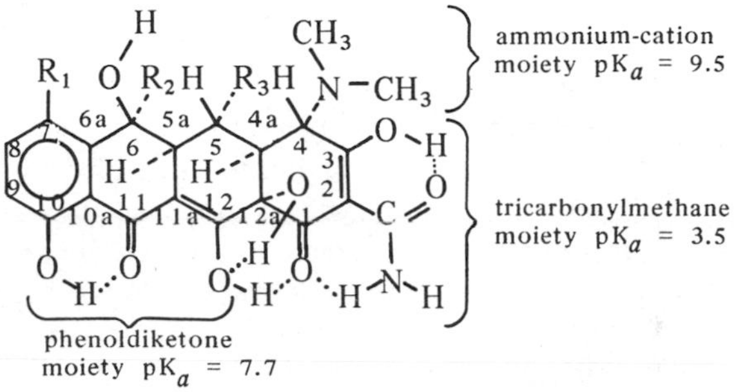

Figure 33.1 Absolute configuration of tetracyclines, showing the sections responsible for the different pK_a's and the groups involved in hydrogen bonding.

itself has 5 of these centers. The important features for chemotherapeutic activity are:

1. The 2-amide group—one of the hydrogen atoms can be substituted without loss of activity.

2. The 4-methylamino moiety—removal of this moiety results in a substantial loss of activity.

3. The proper stereochemistry of the aforementioned moiety—the 4-epitetracyclines are less active than the natural tetracyclines.

4. The proper stereochemistry of substituents at carbon 5a—epimerization or dehydrogenation causes a marked decrease in activity.

5. The conjugated system from carbons 10 to 12, in which the oxygen functions at positions 10, 11, and 12 appear to be essential for biological activity—alteration of this chromophoric moiety results in nearly or totally inactive compounds.

Tetracycline is chemically named 4-(dimethylamino)-1,4,4a,5,5a,6,11,12a-octahydro-3,6,10,12,12a-pentahydroxy-6-methyl-1,11-dioxo-2-naphthacenecarboxamide.

2. Physicochemical Properties

Tetracyclines are amphoteric compounds, and many form water-soluble salts with strong acids and strong bases. Acid salts are formed through protonation of the dimethylamino group on carbon atom 4; they are stable. Basic salts are formed with sodium or potassium hydroxides; they are unstable. The common chromophoric system imparts to them a yellow color. In each tetracycline three ionizable groups can be distinguished, with the approximate following pK_a's: 3.5, 7.7, and 9.5.

Owing to the presence of groups that can form several intramolecular hydrogen bonds, tetracyclines have chelating properties, forming insoluble complexes with iron, calcium, magnesium, and aluminum salts. Therefore, for better absorption they should not be administered with milk or milk products or antacids or other agents containing these salts.

In solutions of intermediate pH (2 to 6) tetracyclines undergo epimerization at carbon atom 4 and reach an equilibrium when approximately equal amounts of isomers are formed. Epitetracyclines are much less active than the natural isomers. This fact explains the decrease of potency of aged solutions.

Strong acids and strong bases inactivate tetracyclines having a 6-hydroxyl group by forming anhydrotetracyclines and isotetracyclines, respectively. Efforts to overcome this inactivation resulted in the development of 6-deoxytetracyclines, which are more stable and have more prolonged action. Examples are methacycline, doxycycline, sancycline.

3. Classes of Tetracyclines

Tetracyclines can be grouped into three classes:

1. Natural tetracyclines: chlortetracycline, oxytetracycline, tetracycline, demeclocycline.

2. Semisynthetic tetracyclines with unchanged carboxamide group: doxycycline, methacycline, minocycline, sancycline.

3. Latent forms of tetracycline. This latentiation was accomplished in several ways: (a) by a Mannich reaction at the amide function: apicycline (Traserit), clomocycline (Megaclor), guamecycline, lymecycline, meglucycline, mepycycline (Boniciclina), penimepicycline, pipacycline (Meplion), rolitetracycline; these tetracyclines are highly water soluble and suitable for both parenteral and oral administration; (b) by formation of salts: tetracycline lactate, tetracycline lauryl sulfate (Teloril)—two latent forms; tetracycline phosphate complex (Tetrex), which is insoluble in water; and tetracycline cyclohexyl sulfamate (Sifacycline), which is a new pleasant-tasting salt of tetracycline; (c) by molecular association: etamocycline (addition of two molecules of tetracyclines through an ethylenediamine bridge) and penimocycline (hybridization of tetracycline with metampicillin).

4. Therapeutic Uses

Tetracyclines are broad-spectrum bacteriostatic agents. They are effective in treatment of infections caused by many species of Gram-positive and Gram-negative bacteria, spirochets, *Rickettsiae,* and some of the larger viruses.

Tetracyclines are also used for nontherapeutic purposes: (*a*) promotion of growth rate of farm livestock; (*b*) conservation of food; (*c*) microbiological control of fermentation.

5. Adverse Reactions

At usual dosages tetracyclines are relatively safe drugs. Nausea, vomiting, anorexia, flatulence, pyrosis, and diarrhea are the most common side effects. Other minor reactions (glossitis, enterocolitis, stomatitis) result from suppression of normal enteric flora with overgrowth of other microorganisms. Hypersensitivity and photosensitivity reactions are common. Superinfection may occur occasionally, especially moniliasis, caused by *Candida albicans;* this complication is inhibited by simultaneous administration of antifungal agents, such as nystatin or amphotericin B. A serious effect is the permanent staining of teeth of children during calcification caused by deposition of the tetracyclines taken by young children or pregnant mothers, since these drugs readily cross the placenta and localize in calcium-containing structures by forming a chelate with calcium phosphate.

Prolonged treatment with tetracyclines can give rise to resistant strains of microorganisms. Cross-resistance among these antibiotics usually is complete.

6. Clinically Used Tetracyclines

The most widely used tetracyclines are listed in Table 33.3. Others, however, are available: etamocycline, penimocycline, Sifacycline, guamecycline, meglucycline, mepycycline, apicycline, clomocycline.

Tetracycline

It is the most popular member of the tetracyclines. It is used both in the free form and as either hydrochloride or phosphate salt. Dosages vary according to the route used: oral (all forms), 1 g daily in four divided doses; intramuscular (as hydrochloride or phosphate salt), 200 mg to 1 g daily in two or three divided doses; intravenous (hydrochloride), 500 mg twice daily.

Doxycycline

It is used both as monohydrate and as hyclate. The absence of the 6-hydroxyl group, which was replaced by the methyl group, imparts greater stability and longer action. In fact, it is more completely absorbed and more slowly excreted than most other tetracyclines. These properties allow smaller and less frequent doses and recommend its use in patients with impaired renal function. The usual dosage is 200 mg on the first day, followed by 100 mg daily.

Table 33.3 Tetracyclines

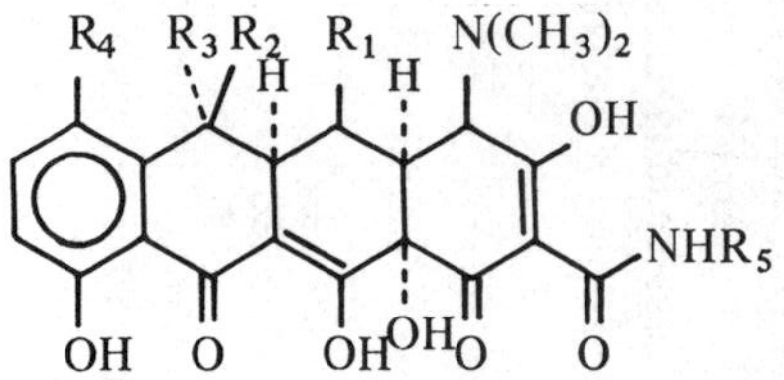

Official Name	Proprietary Name	R_1	R_2	R_3	R_4	R_5
tetracycline*+	Achromycin Panmycin Tetracyn Tetrex-S	H	OH	CH_3	H	H
chlortetracycline+	Aureomycin	H	OH	CH_3	Cl	H
oxytetracycline*+	Terramycin	OH	OH	CH_3	H	H
demeclocycline+	Declomycin	H	OH	H	Cl	H
rolitetracycline+	Syntetrin	H	OH	CH_3	H	CH_2-N 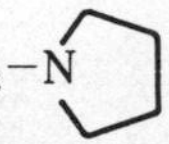 (pyrrolidine)
lymecycline	Tetralysal	H	OH	CH_3	H	$CH_2NHCH(CH_2)_4NH_2$ $\mid$ COOH
doxycycline*	Vibramycin	OH	H	CH_3	H	H
sancycline	Bonomycin	H	H	H	H	H
minocycline*+	Minocin	H	H	H	$N(CH_3)_2$	H
methacycline+	Rondomycin	OH	$=CH_2$		H	H

*In *USP XIX* (1975).
+In *NF XIV* (1975).

Rolitetracycline

Its great water solubility, conferred by the substituent at the 2-amide function, facilitates its parenteral administration. The usual dosages vary: intramuscular, 150 mg every 8 to 12 hours; intravenous, 350 to 700 mg every 12 hours.

H. Polypeptide Antibiotics

1. *Structure*

Polypeptide antibiotics have a polypeptide structure, which is usually very complicated, being predominantly cyclic. Although some hundreds of polypeptide antibiotics have been described, only a few have clinical use, because most of them are extremely toxic, especially to the kidneys. Their main sources are the genera *Bacillus* and *Streptomyces*.

These antibiotics can be divided according to several criteria. According to their components they may be either homeomeric (if comprised only of amino acids) or heteromeric peptides (if constituted of amino acids plus other moieties). With consideration of ring components, they may be divided into homodetic (amino acids in peptide linkages in the ring) and heterodetic peptides (ring containing other linkages). Thus, bacitracin, gramicidin S, and tyrocidines are homeomeric homodetic peptides; colistins and polymyxins are heteromeric homodetic peptides; actinomycins and mikamycins are heteromeric heterodetic peptides (Figure 33.2).

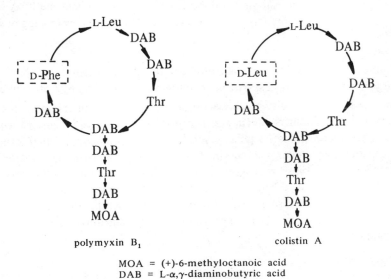

polymyxin B₁   colistin A

MOA = (+)-6-methyloctanoic acid
DAB = L-α,γ-diaminobutyric acid

Figure 33.2 Structures of two polypeptide antibiotics.

Depsipeptide is the name given to peptide-related antibiotics that contain hydroxy and amino acid residues linked by amide and ester bonds. Examples are valinomycin and enniatin B.

2. *Physicochemical Properties*

Polypeptide antibiotics can be of acidic, basic, or neutral types. The acidic antibiotics have free carboxyl groups, indicating a partially noncyclic structure. The basic antibiotics have free amino groups and, therefore, a partially noncyclic structure. The neutral antibiotics have no free carboxyl or amino group, because they have a cyclic structure or because their reactive groups are neutralized by formylation.

3. *Therapeutic Uses*

Polypeptide antibiotics of medical interest have narrow action. For instance, gramicidins are active only against Gram-positive bacteria. Polymyxins are effective against Gram-negative microorganisms.

These antibiotics have other applications. Thus, thiostrepton is used in veterinary medicine; nisins are used in food preservation, and bacitracin has been used in animal supplements.

4. *Clinically Used Polypeptide Antibiotics*

Clinically used polypeptide antibiotics are mainly the following six: bacitracin (Baciguent), polymyxin B sulfate (Aerosporin), sodium colistimethate (USP) (Coly-Mycin M), colistin sulfate (Coly-Mycin S), tyrothricin, and methoxydine. Besides these, three—capreomycin, cycloserine, and viomycin—were already seen in Chapter 32 (Table 32.1), and bleomycin and dactinomycin are studied in the next chapter.

Several others, however, have useful chemotherapeutic action: actinomycins, albomycin, amphomycin, aspartocin, bacillomycin, bleomycins, bottromycins, cactinomycin, circulins, distamycins, echinomycin, edeine, enduracidins, enniatins, enramycin, epidermidin, etamycin, ferrimycins, fungistatin, fusafungine, gatavalin, glumamycins, gramicidins (NF), ilamycins, janiemycin, jolipeptin, licheniformins, lupincolistin, malformin, micrococcin P, mikamycins, mycobacillin, negamycin, neocarzinostatin, nisins, ostreogrycins, peptidolipine NA, polypeptin, pristina-mycins, quinomycin, saramycetin, siomycin, stendomycins, streptogramins, subtilins, succinimycin, suzukacillin, telomycin, thiopeptin, thiostrepton, trio-stin, tuberactinomycins, tyrocidines, valinomycin, vancomycin, vernamycins, virginiamycin.

Bacitracin (USP)

It is a mixture at least of 10 different polypeptides. The official product contains mainly bacitracin A and small amounts of B, D, E, and F bacitracins.

It is a white to pale-buff, stable, almost odorless powder. Bacitracin and its zinc salt are used topically against Gram-positive microorganisms and *Neisseria*.

Polymyxin B Sulfate (USP)

Polymyxin, first isolated in 1947 almost simultaneously in three laboratories, is actually a mixture of at least 12 different cyclic basic polypeptides. Polymyxin B is a mixture of B_1 and B_2 polymyxins. It is active against several Gram-negative bacteria. Its main use is for treatment of infections caused by *Pseudomonas* species, especially meningitis. It can be administered by all routes. However, it is nephrotoxic.

Colistin Sulfate (USP)

Colistine is a mixture of A, B, and C colistins, which belong to the polymyxin group (polymyxin E). Colistin sulfate is used orally to treat bacterial enteritis in infants and children. The usual dosage is 3 to 5 mg/kg of body weight daily in three divided doses.

I. Polyene Antibiotics

Polyene antibiotics are produced by various strains of *Streptomyces*. They are characterized by a large ring containing a lactone function and a sequence of conjugated double bonds, which is their chromophoric part. Owing to their macrolidic structure, the polyenes could be studied as a subclass of macrolide antibiotics and not as a separate class. All are poorly water soluble, and in the presence of light many undergo autooxidation. They occur as pale-yellow, crystalline compounds or powders. As to chemical reactivity, they can be acidic, basic, amphoteric and nonionic.

According to the number of double bonds present in the conjugated system, polyenes are classified as tri-, tetra-, penta-, hexa-, and heptaenes.

Polyene antibiotics have no antibacterial or antirickettsial activity but they are active against molds and yeasts.

The most widely used polyene antibiotics were studied in Chapter 30, as antifungal agents: amphotericin B, candicidin, nystatin. A few others have been approved for clinical use: natamycin (= pimaricin), hachimycin (= trichomycin). Several others have been extensively studied and found active: aliomycin, ascosin, cabacidin, candidin, capacidin, chromin, endomycins, eurocidin, etruscomycin, filipin, fumagillin, fungichromin, gerobrucin, hamycin, lagosin, α-lipomycin, β-lipomycin, manumycin, lucimycin, mycoticin, oleficin, perimycin, protocidin, rimocidin, tetrin.

J. Macrolide Antibiotics

Macrolide antibiotics are produced by *Streptomyces* species. They are character-

Table 33.4 Macrolide Antibiotics

Official Name	Proprietary Name	Structure
erythromycin*+	E-Mycin Erythrocin Ilocytin	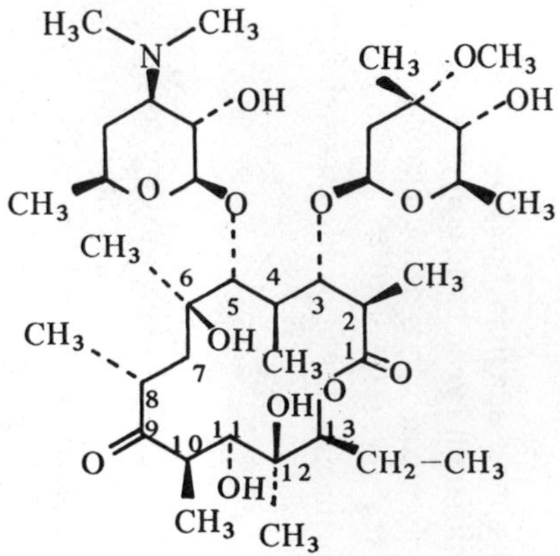
carbomycin	Magnamycin	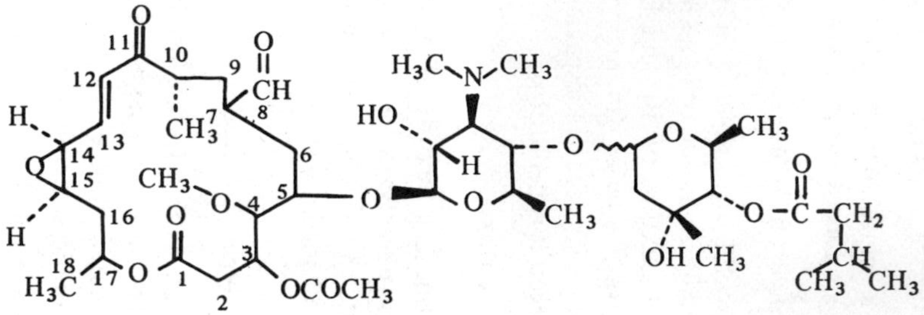

oleandomycin — Matromycin

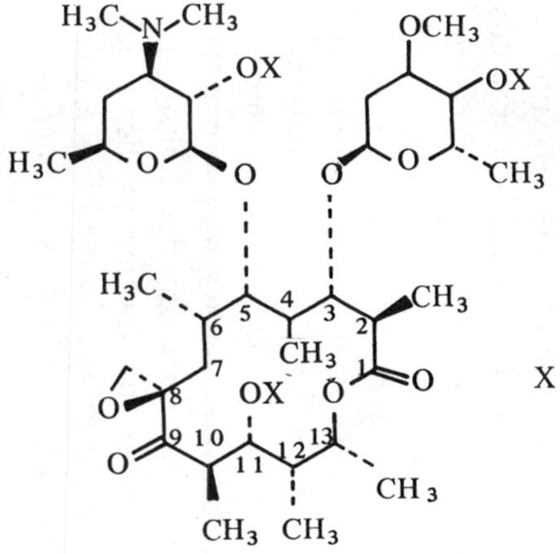

X = H

X = COCH₃ → $X = COCH_3$

troleandomycin — Cyclamycin
TAO

spiramycin — Rovamycin

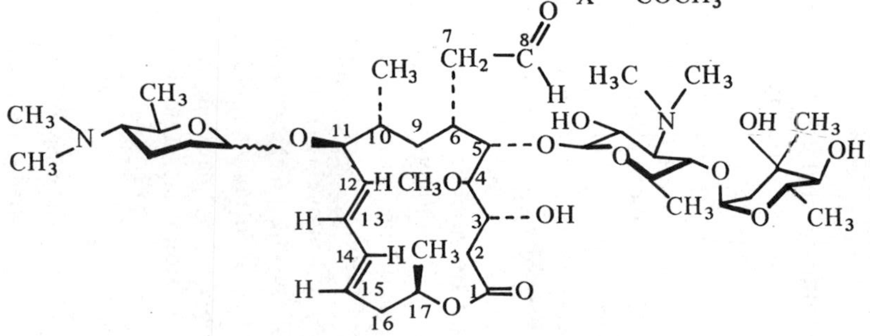

kitasamycin — Syneptine — a mixture of at least six fractions

*In USP XIX (1975).
†In NF XIV (1975).

ized by five common chemical features: (*a*) a macrocyclic lactone, usually having 12 to 17 atoms—hence the name *macrolide* (from the Greek μακρος, large); (*b*) a ketone group; (*c*) one or two amino sugars glycosidically linked to the nucleus; (*d*) a neutral sugar linked either to amino sugar or to nucleus; (*e*) the presence of the dimethylamino moiety on the sugar residue, which explains the basicity of these antibiotics and affords the possibility of preparing their salts.

The pure macrolide antibiotics are colorless and usually crystalline. In neutral solutions they are stable, but in acidic media they are hydrolyzed at the glycosidic bonds, and in basic media the lactone bonds are saponified.

The first member of this class to be identified was pikromycin in 1950. Now, more than 60 are described. Those marketed are listed in Table 33.4, but only erythromycin, oleandomycin, and troleandomycin are available in the United States. Many others, however, have shown promising activity: acumycin, aldgamycin, amaromycin, angolamycin, bandamycins, carbomycin, chalcomycin, cirramycins, erythromycins (B and C), espinomycins, kitasamycins, lankamycin group, maridomycins, megalomycin, methymycin, midecamycin, narbomycin, neomethymycin, neutramycin, niddamycin, pikromycin, platenomycins, proactinomycin, 9-propionylmaridomycin, relomycin, rosamycin, shincomycin A, spiramycin, tertiomycins, tylosin.

Erythromycin

It is a very bitter, white or yellowish-white, sparingly water-soluble crystalline powder. It is used both as free base and in the form of salts (gluceptate, lactobionate, and stearate) or esters (estolate, ethylcarbonate, and ethylsuccinate). Gluceptate and lactobionate are water soluble and are administered parenterally. The stearate salt is insoluble in water and tasteless; it is used in tablets and suspensions. The esters are tasteless and are used in oral suspensions; they cannot be considered latent forms of erythromycin, because they are biologically active themselves; that is, their activity does not depend on their hydrolysis. Erythromycin is active against most Gram-positive bacteria and some Gram-negative organisms. It is an alternate drug to penicillin when high blood levels are required. Its main use is in the treatment of group A β-hemolytic streptococcal, staphylococcal, and pneumococcal infections. Adverse serious reactions occur rarely, but the estolate can produce jaundice. The usual dosage is calculated on the basis of the free erythromycin: 250 mg four times daily.

Oleandomycin

It is used as the phosphate, a white, water-soluble, crystalline powder. Routes of administration are intravenous, intramuscular, and oral. Its main use is in treatment of infections refractory to other antibiotics.

K. Aminoglycoside Antibiotics

Aminoglycoside antibiotics are a group of basic carbohydrate substances that are water-soluble and form crystalline salts. They are used as salts, especially the sulfate.

These antibiotics are active against *Escherichia coli* and most species of *Enterobacter, Klebsiella, Salmonella, Shigella,* and *Proteus.* They have no effect on fungi or viruses. Cross-resistance is developed among all members of this class of antibiotics. A serious adverse reaction is permanent damage to cochlear and vestibular parts of the eighth cranial nerve. Most of these antibiotics are also used as local or topical antibacterial agents.

The most widely used antibiotics of this class are listed in Table 33.5. Several others are either marketed or under study: amicetin, amikacin, bluensomycin, butirosin A, chromomycins, dideoxykanamycin B, dideoxyneamine, hygromycin S, kanendomycin, kasugamycin, lividomycins, nebramycin, olivomycins, oxamicetin, ribostamycin, ristocetin, sisomycin, spectinomycin, streptonicozid, tobramycin. Some of these resulted from molecular modification of natural antibiotics and have better properties than the parent compounds.

Kanamycin

It is basic but used as sulfate, which is very water soluble and stable to heat and chemicals. It is used as an antituberculous, antibacterial, and antiinfective agent.

Gentamicin Sulfate

It is a mixture of three different gentamicins, which occurs as a water-soluble, white to buff-colored substance. Its main systemic use is for the treatment of serious infections by Gram-negative organisms. By injection, the usual dosage is 5 mg/kg of body weight daily. Topical use is not advisable.

L. Ansamycins

Ansamycins are macrolide antibiotics containing an aliphatic ansa bridge, a bridge connecting two nonadjacent positions of an aromatic nucleus. This class consists of geldanamycins, rifamycins, streptovaricins, tolypomycins, and miathansine. These antibiotics, which are notably active against Gram-positive bacteria, are produced by different strains of *Streptomyces* species.

Rifamycins have chemotherapeutic activity. Thus rifamycin (Rifocin) is a broad-spectrum antibiotic. Rifampin, which is a rifamycin, is already marketed for the treatment of tuberculosis (Table 32.1). Streptovaricins also have antituberculous activity, but they are too toxic. Furthermore, certain rifamycin derivatives have broad-spectrum activity, and streptovaricins are active against

Table 33.5 Aminoglycoside Antibiotics

Official Name	Proprietary Name	Structure
streptomycin sulfate*		see Table 32.1
kanamycin sulfate*		see Table 32.1
gentamicin sulfate*	Garamycin	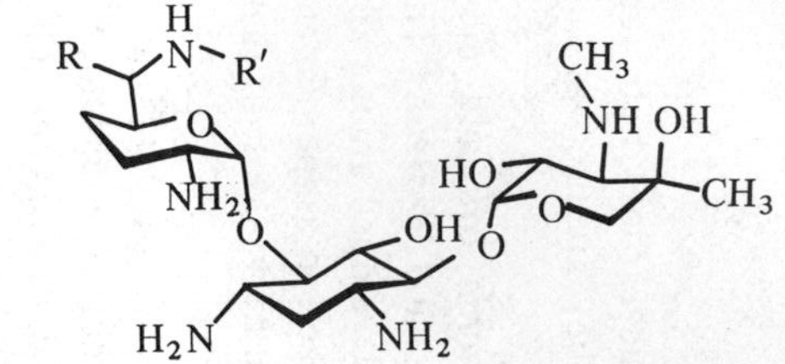
neomycin sulfate*	Mycifradin sulfate Myciguent Neobiotic	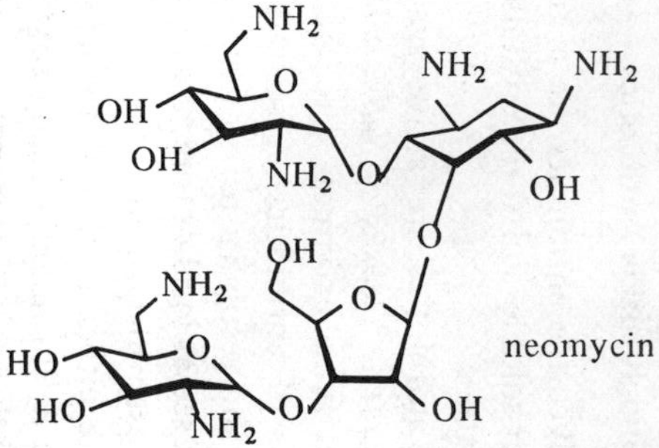

gentamicin C₁: R, R′ = CH₃
gentamicin C₂: R = CH₃, R′ = H
gentamicin C₁A: R, R′ = H

paromomycin sulfate[+]   Humatin

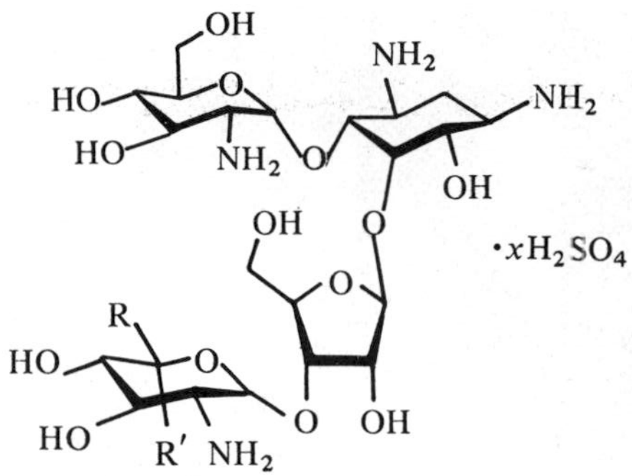

·xH_2SO_4

paromomycin I: R=H, R'=CH_2NH_2
paromomycin II: R=CH_2NH_2, R'=H

spectinomycin*   Trobicin

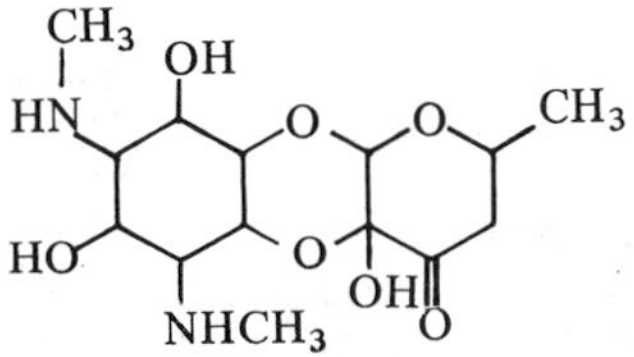

vancomycin
hydrochloride*[+]   Vancocin
Hydrochloride

glycopeptide of unknown structure

*In *USP XIX* (1975).
[+]In *NF XIV* (1975).

525

Gram-negative bacteria. Geldanamycins are active against protozoa. Miathansine has shown antineoplastic activity.

M. Anthracyclines

Anthracyclines are antibiotics characterized by a tetrahydrotetracenquinone chromophore consisting of three flat, coplanar six-membered rings:

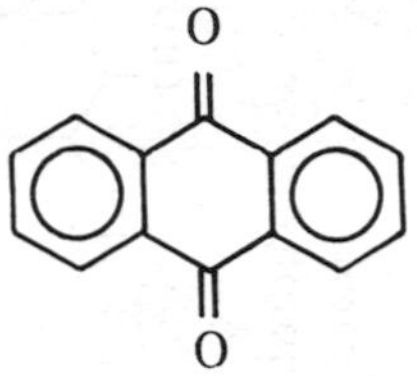

Of special interest to medicinal chemistry are the following anthracyclines: acetyladriamycin, N-acetyldaunomycin, adriamycin, cinerubin, daunorubicin, doxorubicin, nogalamycin, piperazinodaunomycin, rabelomycin, rhodomycins, rubomycins. Most of these antibiotics have antineoplastic activity (Chapter 34).

N. Lincomycin Group

The lincomycin group of antibiotics includes cilesticetin, clindamycin, and lincomycin. The last two are clinically used (Table 33.6). Lincomycin is produced

Table 33.6 Lincomycin Group

Official Name	Proprietary Name	Structure
lincomycin*	Lincocin	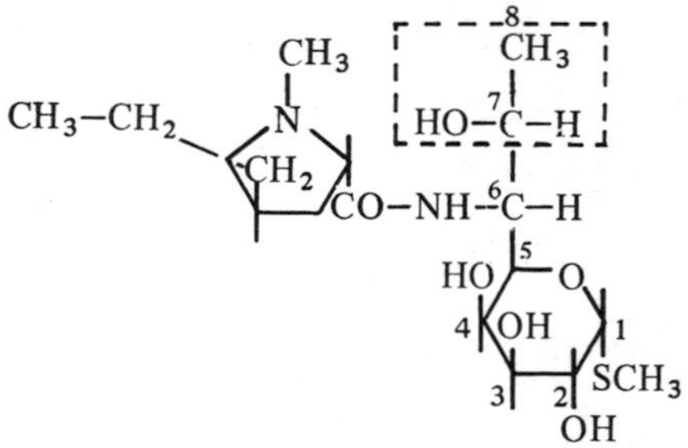
clindamycin*+	Cleocin	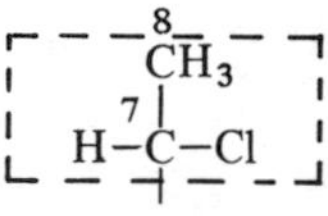

*In *USP XIX* (1975).
+In *NF XIV* (1975).

by *Streptomyces lincolnensis* var. *lincolnensis*. They are active against the common Gram-positive pathogens. Some halogenated lincomycins have shown activity in experimental malaria.

This class of antibiotics is characterized by a 4-alkylsubstituted hygric acid bound to an alkyl 6-amino-α-thiooctapyranoside by an amide linkage.

Lincomycin

Used as hydrochloride monohydrate, a white, water-soluble, crystalline and stable solid in the dry state. It is primarily used against microorganisms resistant to the penicillins and erythromycin but susceptible to this antibiotic or in patients who do not tolerate other antibacterial agents.

Clindamycin

It is available as phosphate, hydrochloride hydrate, and palmitate. Clindamycin is a derivative of lincomycin, being obtained by substituting the 7-hydroxyl group by chlorine. It has better intestinal absorbability, greater potency, and lower adverse effects than the parent antibiotic but the same indications.

O. Nucleoside Antibiotics

In this class are included cordycepin, decoynine, nebularine, nucleocidin, psicofuranine, puromycin, sangivamycin, toyocamycin, tubercidin. Most are produced by different strains of *Streptomyces* species. Their chemotherapeutic activities vary. Some are broad-spectrum antibiotics, whereas others are active against bacteria, and still others have shown activity in tumor cells. However, none of this class of antibiotics is clinically used.

P. Glutarimide Antibiotics

Glutarimide antibiotics are characterized by a common β-(2-hydroxyethyl)-glutarimide moiety attached to a cyclic or acyclic ketone. They are produced by *Streptomyces* species. All have antifungal activity, and most show antineoplastic action. Some are being used as antifungal agents. The prototype of this class is cycloheximide (Actidione), used as an antifungal agent. Other important members of this class include acetoxycycloheximide, actiphenol, fermicidin, inactone, mycophenolic acid, neocycloheximide, niromycin, protomycin, streptimidones, streptovitacins.

Q. Miscellaneous Antibiotics

In this class we include all other antibiotics that have useful action. Some are already being used as chemotherapeutic agents, especially against fungi and tumors. Novobiocin, as calcium or sodium salt, is used as antibacterial agent, but the high incidence of adverse reactions indicate that its use should be abandoned. Others are used chiefly for agricultural purposes. Most are being investigated.

Among several other antibiotics that are placed in this class, the following ones deserve mention: alazopeptin, albofungin, antimycin A, azalomycins, azaserine, bicyclomycin, blasticidins, carzinophilin, cellocidin, cuprimyxin, dianemycin, diazomycins, DON, fosfomycin, fusidic acid, griseofulvin, ikaruga-mycin, mitosanes, monensin, nigericin, novobiocin, oxazinomycin, pecilocin, polyoxins, pyrrolnitrin, sarkomycin, thermorubin A, variotin, xantocillin.

IV. MECHANISM OF ACTION

According to their mechanism of action, antibiotics can be divided into the following classes (Gale *et al.*, Vazquez):

1. Antibiotics affecting bacterial cell wall synthesis
2. Antibiotics affecting the function of the cytoplasmic membrane
3. Antibiotics affecting nucleic acid synthesis
4. Antibiotics inhibiting protein synthesis

The mechanism of action of some antibiotics has already been studied in Chapters 30 and 32, and that of antineoplastic antibiotics will be seen in the next chapter. In this section the mode of action of the most widely used antibiotics not studied thus far in this book is discussed.

A. Inhibitors of Bacterial Cell Wall Synthesis

Antibiotics acting by this mechanism can be further divided into the following groups:

1. Antibiotics which inhibit biosynthetic enzymes: fosfomycin, cycloserine, penicillins, cephalosporins
2. Antibiotics which combine with carrier molecules: bacitracin
3. Antibiotics which combine with substrates: vancomycin, ristocetin, ristomycin
4. Antibiotics whose site of action on cell wall synthesis has not been localized: novobiocin, prasinomycin, enduracidin, moenomycin

1. β-Lactam Antibiotics

Resemblances between a segment of penicillins and cephalosporins structures and certain segments of *N*-acetylmuramic acid, D-alanyl-D-alanine, and L-alanyl-D-glutamic acid have been pointed out by different authors. These structural analogies were invoked to explain the mechanism of action of β-lactam antibiotics.

The last step in the bacterial cell wall synthesis is a cross-linking reaction between two nascent peptidoglycan units catalyzed by transpeptidase. Owing to their structural resemblance to the D-alanyl-D-alanine terminal group of the pentapeptide part of the nascent peptidoglycan units, both penicillins and cephalosporins, according to the hypothesis advanced by Tipper and Strominger in 1965, inhibit this enzyme by reacting covalently with it, thus preventing the formation of the bacterial cell wall (Figure 33.3). As a result, the high internal

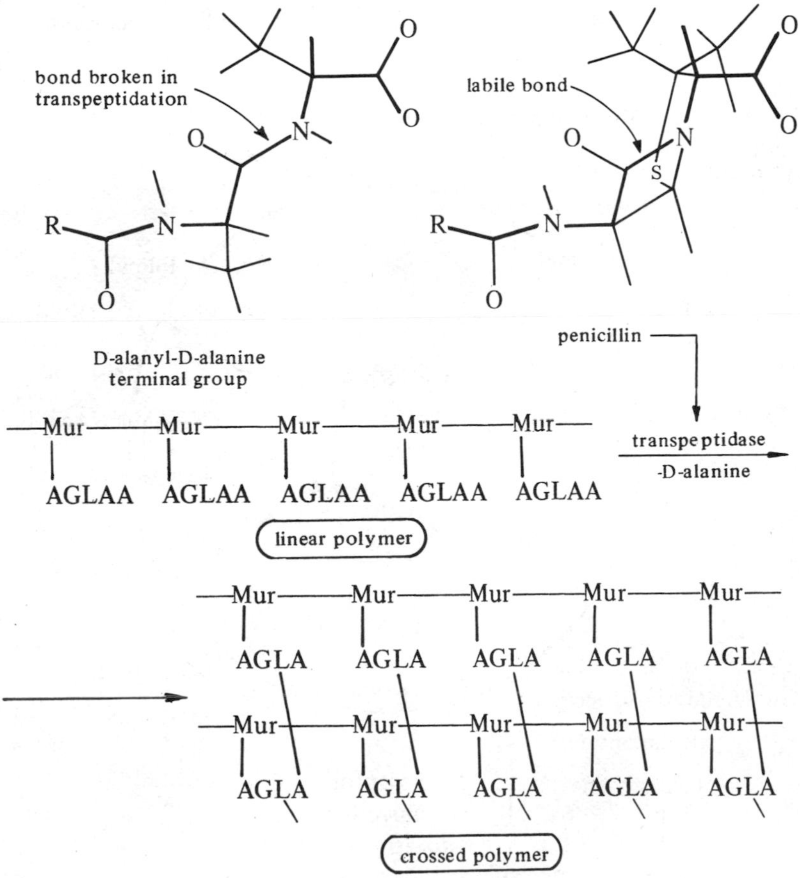

Mur = N-acetylglycosyl-N-acetylmuramic acid

AGLAA = L-alanyl-D-glutamyl-L-lysyl-D-alanyl-D-alanine

Figure 33.3 Mechanism of action of β-lactam antibiotics. Owing to their structural resemblance to the D-alanyl-D-alanine terminal group of the bacterial cell wall nascent peptidoglycan units, they inhibit transpeptidase by attaching to this enzyme through a covalent bond. As a result, the linear polymer is not transformed into the crossed polymer, and the bacterial cell wall fails to form. [After Strominger et al., Top. Pharm. Sci., 1, 53 (1968).]

turgor pressure of bacteria (about 20 atm for Gram-positive and 5 for Gram-negative types) causes a rupture of the bacterial cell wall and a bursting of the bacterial cytoplasm, with subsequent death of the microorganism. Therefore, these antibiotics kill only growing bacteria; they do not affect dormant or persisting forms. Since mammal cells have no walls, β-lactam antibiotics, as well as other antibiotics that inhibit bacterial cell wall synthesis (Figure 33.4), are highly specific.

It has been found that three enzymes are sensitive to penicillins: peptidoglycan transpeptidase, D-alanine carboxypeptidase, and endopeptidase.

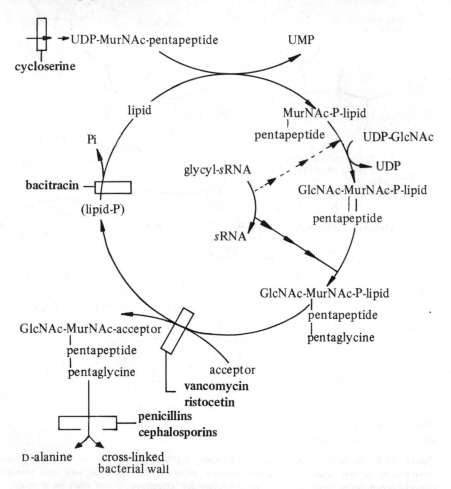

Figure 33.4 Site of action of some antibiotics that inhibit bacterial cell wall synthesis. [After M. Matsuhashi, C. P. Dietrich, and J. L. Strominger, Proc. Nat. Acad. Sci., *U.S.A.*, 54, 587 (1965).]

2. Vancomycin and Ristocetin

Vancomycin and ristocetin inhibit the incorporation of amino acids into peptidoglycan, which is a component part of the cell wall of Gram-positive bacteria. They accomplish this by binding to the D-alanyl-D-alanine termini of nascent peptidoglycan and thus inhibit glycopeptide synthetase.

B. Inhibitors of the Function of the Cytoplasmic Membrane

Antibiotics acting by this mechanism can be further divided into the following groups:

1. Antibiotics causing disorganization of the cytoplasmic membrane: tyrocidins, polymyxins, polyenes

2. Antibiotics producing specific changes in cation permeability: gramicidins, valinomycin, nonactin, enniatins, actins, nigericin, monensin, alamethicin

3. Antibiotics which inhibit membrane-bound enzymes involved in energy transfer: oligomycin

Polypeptide (tyrocidins, colistin, polymyxins), as well as polyene (amphothericin B, nystatin, candidin), antibiotics owe their antibacterial action to disorganization they cause to the structure of membranes, with the result that they lose their properties as permeability barriers, permitting passage into the cell of ions that are normally excluded. In polypeptide antibiotics the cyclic structure and the basic groups are apparently related to antibacterial activity. Polyene antibiotics bind to membranes containing sterols, as in fungi (see Chapter 30). As to gramicidins, they promote rapid loss of K^+ from the cell by exchange for other ions: Na^+, Li^+ or NH_4^+.

Depsipeptide antibiotics (valinomycin, enniatins) have no medicinal use. However, their mechanism of action deserves mention. They form well-defined complexes with potassium in which the potassium atom occupies the central part, being enclosed by a cage, and either transport this ion across cell membrane, which is usually not permeable to this ion, or become incorporated into the membrane to form pores specifically permeable to potassium.

Macrotetrolides (nonactin) have a similar mode of action.

C. Inhibitors of Nucleic Acid Synthesis

Antibiotics acting by this mechanism can be further divided into the following groups:

1. Antibiotics which interfere with nucleotide metabolism: (a) inhibitors of nucleotide synthesis: azaserine, DON; (b) inhibitors of nucleotide interconversion: hadacidin, psicofuranine, decoyinine, showdomycin, mycophenolic acid; (c) antibiotics which become incorporated into polynucleotides: tubercidin, for-

mycin, cordycepin

2. Antibiotics which impair the template formation of DNA: (*a*) by intercalation: anthracyclines (nogalamycin, daunorubicin, cinerubin, doxorubicin), dactinomycin; (*b*) by noncovalent interaction with DNA: chromomycin, mithramycin, olivomycin, netropsin, distamycin, phleomycin; (*c*) by covalent interaction with DNA: mitomycin C, carzinophilin, anthramycin; (*d*) by strand-breaking DNA: bleomycin, streptonigrin

3. Antibiotics which inhibit RNA polymerase in nucleic acid synthesis: rifamycins, streptovaricins, streptolydigin, cycloheximide

Most antibiotics acting as inhibitors of nucleic acid synthesis have antineoplastic activity and, therefore, are studied in the next chapter. Such is the case, for instance, with anthracyclines. They owe their antineoplastic action to binding to DNA by intercalation through the chromophore, possibly by a charge-transfer mechanism. The sugar residues on the antibiotic chromophores assist in this binding by stabilizing the complexes by hydrophobic or electrostatic and hydrogen-bonding interactions.

Other antibiotics acting by this mechanism were seen in the preceding chapters, and their mechanism was explained there. It is the case of rifamycins and streptovaricins, which are antituberculous agents which act by inhibiting DNA-dependent RNA polymerase more strongly than any other known inhibitor.

D. Inhibitors of Protein Synthesis

Antibiotics acting by this mechanism can be further divided into the following groups:

1. Inhibitors of the initiation phase: edeine, streptomycin and related aminoglycoside antibiotics (dihydrostreptomycin, gentamicin, kanamycin, paromomycin), spectinomycin, kasugamycin, pactamycin. Other possible inhibitors of this phase are anisomycin, clindamycin, lincomycin, carbomycin, spiramycin, streptogramin A group, and chloramphenicol.

2. Inhibitors of the elongation cycle: (*a*) inhibitors of aminoacyl-*t*RNA binding: edeine, tetracyclines, siomycin and related antibiotics (sporangiomycin, thiopeptin, thiostrepton), fusidic acid. Other possible inhibitors of aminoacyl-*t*RNA binding to the larger ribosome subunit are anisomycin, clindamycin, lincomycin, carbomycin, spiramycin, streptogramin A group and chloramphenicol; (*b*) inhibitors of peptide bond formation: puromycin, chloramphenicol, lincomycin group (celesticetin, clindamycin, lincomycin), macrolides (angolamycin, carbomycin, spiramycin), streptogramin A group, actinobolin, amicetin, blasticidin S, gougerotin, sparsomycin, anisomycin, trichodermin, tenuazonic acid; (*c*) inhibitors of translocation: siomycin and related antibiotics (sporangiomycin, thiopeptin, thiostrepton), fusidic acid, erythromycin, cycloheximide;

(*d*) inhibitors causing breakdown of bacterial polysomes: streptomycin group (dihydrostreptomycin, gentamicin, kanamycin, paromomycin, streptomycin); (*e*) other inhibitors of the elongation cycle: macrolide antibiotics (chalcomycin, erythromycin, lankamycin, methymycin, oleandomycin), streptogramin B group (staphylomycin S, streptogramin, viridogrisein).

3. Inhibitors of the termination phase: (*a*) inhibitors of interaction of the nonsense-terminating codon: streptomycins, tetracyclines; (*b*) inhibitors of the release reaction: puromycin, chloramphenicol, lincomycin group (celesticetin, clindamycin, lincomycin), macrolides (angolamycin, carbomycin, spiramycin), streptogramin A group, actinobolin, amicetin, blasticidin S, gougerotin, sparsomycin, anisomycin, trichodermin, tenuazonic acid.

1. Tetracyclines

Owing to the chelating properties of tetracyclines, it had been suggested that their antibacterial activities could be ascribed to their ability to remove essential metallic ions, such as Mg^{2+}. Now it is believed that chelation may facilitate the transport of tetracyclines to the site of action, but this phenomenon does not play a fundamental role in the mechanism of action of these antibiotics.

Tetracyclines inhibit bacterial protein synthesis by blocking the attachment of aminoacyl-*t*RNA to the smaller subunit of the ribosomal A (acceptor) site, although they may also bind elsewhere on the ribosome. More precisely, they are inhibitors of codon-anticodon interaction at the A site of the smaller ribosomal subunit, either 30 S or 40 S, which explains their lack of selectivity and adverse reactions.

The inhibitory properties of tetracyclines are apparently related to the electronic structure, according to recent results of quantum-chemical studies performed by Peradejordi and coworkers. They found that the action of tetra-

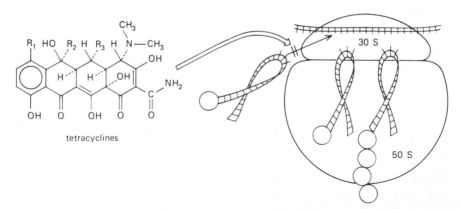

Figure 33.5 Mechanism of antibacterial action of tetracyclines. These antibiotics prevent the attachment of aminoacyl-*t*RNA to the 30 S ribosomal subunit.

cyclines seems to involve a direct interaction of carbon atom 6 and the phenoldiketone region with the proposed ribosome receptor site (Figure 33.5).

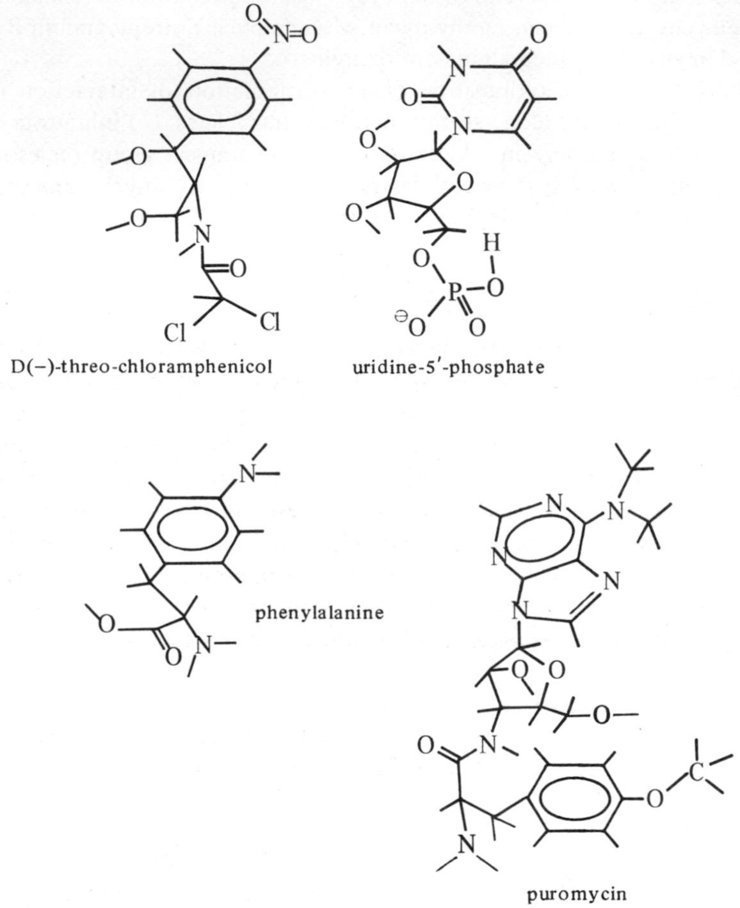

D(−)-threo-chloramphenicol   uridine-5′-phosphate

phenylalanine

puromycin

Figure 33.6 Structural similarities among chloramphenicol and some metabolites and puromycin. As an analogue of uridine-5′-phosphate, it could inhibit the binding of template RNA by competing specifically with the uridylic residues for the attachment to ribosomes. Phenylalanine has the property of decreasing the toxicity of this antibiotic. Puromycin is perhaps a competitive antagonist of chloramphenicol.

2. Chloramphenicol

Chloramphenicol, in one way, presents structural similarity to several metabolites, such as tryptophan, phenylalanine, and uridine-5′-phosphate (Figure 33.6). On the basis of this analogy it was postulated that it might act by metabolic antagonism. In another way, it is also a structural analogue of puromycin. This antibiotic has a resemblance to $3'$-aminosyladenosine moiety of aminoacyl-tRNA. Therefore, by acting on the A site of the peptidyl transferase center of ribosomes, puromycin forms a peptide bond with the initiator amino acid. In this way, it blocks the correct peptide-bond formation. However, chloramphenicol is a total or partial competitive inhibitor of puromycin reaction in various systems. Consequently, it does not act like puromycin.

Although the precise mechanism of its action at the molecular level is not well understood, it is accepted that chloramphenicol inhibits peptide bond formation possibly by blocking the peptidyl transferase center, thus preventing the interaction between this enzyme and the substrate in the 50 S subunit of the ribosomal A site. In other words, it inhibits peptide chain elongation and the movement of ribosomes along mRNA. This inhibition is stereospecific, not being observed with the three stereoisomers of chloramphenicol.

3. Macrolide Antibiotics

Erythromycin, oleandomycin, troleandomycin, and some other macrolide antibiotics owe their antibacterial action to inhibition of the elongation cycle; that is, they inhibit translocations and consequently protein synthesis.

4. Lincomycin Group

The lincomycin group of antibiotics inhibit bacterial protein synthesis, especially in Gram-positive bacteria, without affecting the synthesis of RNA or DNA. They act by blocking the peptidyl transferase center of the 50 S subunit. Therefore, they are not only inhibitors of peptide-bond formation but also inhibitors of aminoacyl-tRNA binding at the level of the larger ribosomal subunits.

REFERENCES

General

C. da S. Lacaz, Ed., *Antibióticos,* 3rd ed., Edgard Blücher and Universidade de São Paulo, São Paulo, 1975.

R. Gallien, *Antibiot. Chemother. (Basel),* **17**, 137 (1971).

D. Gottlieb and P. D. Shaw, Eds., *Antibiotics,* 2 vols., Springer, New York, 1967.

T. Korzybski *et al., Antibiotics: Origin, Nature and Properties,* 2 vols., Pergamon, Oxford, 1967.

Introduction

G. H. Wagman and M. J. Weinstein, *Annu. Rep. Med. Chem.,* **10**, 109 (1975).

F. Leitner and C. A. Claridge, *Annu. Rep. Med. Chem.,* **9**, 95 (1974).

K. E. Price and F. Leitner, *Annu. Rep. Med. Chem.,* **8**, 104 (1973).

F. C. Sciavolino, *Annu. Rep. Med. Chem.*, 7, 99 (1972).

O. Gsell, *Antibiot. Chemother. (Basel)*, 14, 1 (1968).

H. Knothe, *Antibiotic. Chemother. (Basel)*, 14, 217 (1968).

H. Umezawa, *Index of Antibiotics from Actinomycetes*, University of Tokyo Press, Tokyo, 1967.

S. A. Waksmann, *Advan. Appl. Microbiol.*, 5, 235 (1963).

History

L. H. Conover, *Advan. Chem. Ser.*, 108, 33 (1971).

O. K. Sebek and D. Perlman, *Advan. Appl. Microbiol.*, 14, 123 (1971).

S. A. Waksmann, *Advan. Appl. Microbiol.*, 11, 1 (1969).

S. J. Childress, *Top. Med. Chem.*, 1, 109 (1967).

E. B. Chain, *Proc. Roy. Soc. Med.*, 58, 85 (1965).

Classification

Biosynthesis

Z. Vaněk et al., *Pure Appl. Chem.*, 34, 463 (1973).

H. Zähner and W. K. Maas, *Biology of Antibiotics*, Springer, New York, 1972.

D. Perlman and M. Bodansky, *Annu. Rev. Biochem.*, 40, 449 (1971).

G. Sermonti, *Genetics of Antibiotic–Producing Microorganisms*, Wiley-Interscience, New York, 1969.

L. A. Mitscher, *J. Pharm. Sci.*, 57, 1633 (1968).

J. F. Snell, Ed., *Biosynthesis of Antibiotics*, Vol. I, Academic, New York, 1967.

Spectrum of Activity

J. Klastersky, Ed., *Clinical Use of Combinations of Antibiotics*, Wiley Biomedical, New York, 1975.

AMA Department of Drugs, *AMA Drug Evaluations*, 2nd ed., Publishing Sciences Group, Acton, Mass., 1973

A. Kucers, *The Use of Antibiotics*, Lippincott, Philadelphia, 1972.

L. D. Sabath et al., *Clin. Pharmacol. Ther.*, 11, 161 (1970).

J. Crofton, *Brit. Med. J.*, 2, 137, 209 (1969).

Chemical Structure

R. M. Evans, *The Chemistry of Antibiotics Used in Medicine*, Pergamon, Oxford, 1965.

M. W. Miller, *The Pfizer Handbook of Microbial Metabolites*, McGraw-Hill, New York, 1961.

H. S. Goldberg et al., *Antibiotics: Their Chemistry and Non-Medical Uses*, Van Nostrand, Princeton, 1959.

Penicillins

D. B. Boyd, *J. Med. Chem.*, 16, 1195 (1973).

G. N. Rolinson and R. Sutherland, *Advan. Pharmacol. Chemother.*, 11, 151 (1973).

D. B. Boyd, *J. Amer. Chem. Soc.*, **94**, 6513 (1972).

E. Brandl, *Sci. Pharm.*, **40**, 1 (1972).

A. E. Bird and J. H. C. Nayler, "Design of Penicillins," in E. J. Ariëns, Ed., *Drug Design*, Vol. II, Academic, New York, 1971, pp. 277-318.

E. Brandl, *Sci. Pharm.*, **39**, 267 (1971).

J. P. Hou and J. W. Poole, *J. Pharm. Sci.*, **60**, 503 (1971).

M. S. Manhas and A. K. Bose, *Beta-Lactams: Natural and Synthetic*, Part I, Wiley, New York, 1971.

M. S. Manhas and A. K. Bose, *Synthesis of Penicillin, Cephalosporin C and Analogs*, Dekker, New York, 1969.

K. E. Price, *Advan. Appl. Microbiol.*, **11**, 17 (1969).

M. A. Schwartz, *J. Pharm. Sci.*, **58**, 643 (1969).

E. P. Abraham, *Top. Pharm. Sci.*, **1**, 1 (1968).

K. Heusler, *Top. Pharm. Sci.*, **1**, 33 (1968).

J. Ploquin *et al.*, *Prod. Probl. Pharm.*, **23**, 258 (1968).

G. I. Stewart, *The Penicillin Group of Drugs*, Elsevier, Amsterdam, 1965.

F. P. Doyle and J. H. C. Nayler, *Advan. Drug Res.*, **1**, 1 (1964).

J. C. Sheehan, *Advan. Chem. Ser.*, **45**, 15 (1964).

Cephalosporins

E. H. Flynn, Ed., *Cephalosporins and Penicillins: Chemistry and Biology*, Academic, New York, 1972.

G. Nomine, *Chim. Ther.*, **6**, 53 (1971).

R. B. Morin and B. G. Jackson, *Progr. Chem. Nat. Prod.*, **28**, 343 (1970).

M. L. Sassiver and A. Lewis, *Advan. Appl. Microbiol,*, **13**, 163 (1970).

R. M. Sweet and L. P. Dahl, *J. Amer. Chem. Soc.*, **92**, 5489 (1970).

Chloramphenicol and Derivatives

V. S. Malik, *Advan. Appl. Microbiol.*, **15**, 297 (1972).

T. D. Brock, *Bacteriol. Rev.*, **25**, 32 (1961).

Tetracyclines

R. E. Hughes *et al.*, *J. Amer. Chem. Soc.*, **93**, 1037 (1971).

A. I. Laskin, *Antibiot. Chemother. (Basel)*, **17**, 1 (1971).

R. K. Blackwood and A. R. English, *Advan. Appl. Microbiol.*, **13**, 237 (1970).

D. L. J. Clive, *Quart. Rev. (London)*, **22**, 435 (1968).

Polypeptide Antibiotics

P. De Santis *et al.*, *Nature New Biol.*, **237**, 94 (1972).

D. Perlman and M. Bodanszky, *Annu. Rev. Biochem.*, **40**, 449 (1971).

R. Protivinsky, *Antibiot. Chemother. (Basel)*, **17**, 87 (1971).

C. Toniolo, *Farmaco, Ed. Sci.*, **26**, 741 (1971).

A. Taylor, *Advan. Appl. Microbiol.*, **12**, 189 (1970).

M. Bodanszky and D. Perlman, *Science*, **163**, 352 (1969).

M. M. Shemyakin *et al.*, *Angew. Chem. Int. Ed. Engl.*, 8, 492 (1969).

S. A. Waksman, Ed., *Actinomycins: Nature, Formation and Activities*, Wiley, New York, 1968.

Polyene Antibiotics

W. Oroshnik and A. D. Mebane, *Progr. Chem. Nat. Prod.*, 21, 18 (1963).

Macrolide Antibiotics

W. D. Celmer, *Pure Appl. Chem.*, 28, 413 (1971).

E. E. Schmid, *Antibiot. Chemother. (Basel)*, 17, 52 (1971).

R. S. Griffith and H. R. Black, *Med. Clin. North Amer.*, 54, 1199 (1970).

M. Berry, *Quart. Rev. (London)*, 17, 343 (1963).

Aminoglycoside Antibiotics

F. E. Hahn, *Antibiot. Chemother. (Basel)*, 17, 29 (1971).

K. L. Rinehart, Jr., *The Neomycins and Related Antibiotics*, Wiley, New York, 1964.

Ansamycins

V. von Prelog and W. Oppolzer, *Helv. Chim. Acta*, 56, 2279 (1973).

W. Lester, *Annu. Rev. Microbiol.*, 26, 85 (1972).

K. L. Rinehart, Jr., *Accounts Chem. Res.*, 5, 57 (1972).

S. Riva and L. G. Silvestri, *Annu. Rev. Microbiol.*, 26, 199 (1972).

G. Binda *et al.*, *Arzneim.-Forsch.*, 21, 1907 (1971).

K. L. Rinehart, Jr. *et al.*, *J. Amer. Chem. Soc.*, 93, 6273, 6275 (1971).

N. Bergamini *et al.*, *Arzneim.-Forsch.*, 20, 1546 (1970).

Anthracyclines

K. Yamamoto *et al.*, *J. Med. Chem.*, 15, 872 (1972).

I. H. Goldberg and P. A. Friedman, *Annu. Rev. Biochem.*, 40, 775 (1971).

Lincomycin Group

B. J. Magerlein, *Advan. Appl. Microbiol.*, 14, 185 (1971).

Nucleoside Antibiotics

A. Bloch, Ed., "Chemistry, Biology, and Clinical Uses of Nucleoside Analogs," *Ann. N. Y. Acad. Sci.*, 255, 1-610 (1975).

R. J. Suhadolnik, *Nucleoside Antibiotics*, Wiley-Interscience, New York, 1970

Glutarimide Antibiotics

F. Johnson *et al.*, *J. Amer. Chem. Soc.*, 86, 118 (1964).

Miscellaneous Antibiotics

A. Morris and A. D. Russell, *Progr. Med. Chem.*, 8, 39 (1971).

E. Freerksen et al., Antibiot. Chemother. (Basel), 6, 303 (1959).

Mechanism of Action
General

J. W. Corcoran and F. E. Hahn, Eds., "Mechanism of Action of Antimicrobial and Antitumor Agents," Antibiotics, Vol. III, Springer, Berlin, 1975.

J. Drews and F. E. Hahn, Eds., "Drug Receptor Interactions in Antimicrobial Chemotherapy," Top. Infect. Dis., 1 (1975).

A. Korolkovas, Rev. Bras Farm., 55, 57 (1974).

E. F. Gale et al., The Molecular Basis of Antibiotic Action, Wiley-Interscience, London, 1972.

J. H. Hash, Annu. Rev. Pharmacol., 12, 35 (1972).

E. Muñoz et al., Eds., Molecular Mechanisms of Antibiotic Action on Protein Biosynthesis and Membranes, Elsevier, Amsterdam, 1972.

V. Lorian, Arch. Intern. Med., 128, 623 (1971).

H. Schönfua, Ed., Antibiotics and Chemotherapy: Mode of Action, Karger, Basel, 1971.

R. M. Swenson and J. P. Sanford, Advan. Intern. Med., 16, 373 (1970).

D. Vazquez, Annu. Rep. Med. Chem., 1969, 156 (1970).

N. S. Beard, Jr. et al., Pharmacol. Rev., 21, 213 (1969).

T. Bücher and H. Sies, Eds., Inhibitors: Tools in Cell Research, Springer, New York, 1969.

A. D. Russell, Progr. Med. Chem., 6, 135 (1969).

R. A. Cox, Quart. Rev. (London), 22, 499 (1969).

H. J. Rogers, "The Mode of Action of Antibiotics," in E. E. Bittar, Ed., The Biological Basis of Medicine, Vol. II, Academic, London, 1968, pp. 421-448.

Inhibitors of Bacterial Cell Wall Synthesis

D. B. Boyd et al., J. Med. Chem., 18, 408 (1975).

J. M. Indelicato et al., J. Med. Chem., 17, 523 (1974).

M. R. J. Salton and A. Tomasz, Eds., "Mode of Action of Antibiotics on Microbial Walls Membranes," Ann. N.Y. Acad. Sci., 235, 1-620 (1974).

W. C. Topp and B. G. Christensen, J. Med. Chem., 17, 342 (1974).

H. R. Perkins and M. Nieto, Pure Appl. Chem., 35, 371 (1973).

R. Hartmann et al., Nature (London), 235, 426 (1972).

B. Lee, J. Mol. Biol., 61, 463 (1971).

A. Morris and A. D. Russell, Progr. Med. Chem., 8, 39 (1971).

Inhibitors of the Function of the Cytoplasmic Membrane

D. S. Feingold et al., Ann. N.Y. Acad. Sci., 235, 480 (1974).

H. W. Huang, J. Theor. Biol., 32, 351, 363 (1971).

S. C. Kinsky, Annu. Rev. Pharmacol., 10, 119 (1970).

Inhibitors of Nucleic Acid Synthesis

H. Kersten and W. Kersten, Inhibitors of Nucleic Acid Synthesis, Springer, Berlin, 1974.

M. A. Apple, *Annu. Rep. Med. Chem.,* 8, 251 (1973).

Y. Miura, "Drugs Affecting Nucleic Acid and Protein Synthesis," in S. Dikstein, Ed., *Fundamentals of Cell Pharmacology,* Thomas, Springfield, Ill., 1973, pp. 43-66.

P. Sensi, *Pure Appl. Chem.,* 34, 383 (1973).

W. J. Pigram *et al., Nature New Biol.,* 235, 17 (1972).

I. G. Goldberg and P. A. Friedman, *Annu. Rev. Biochem.,* 40, 775 (1971).

F. E. Hahn, Ed., "Complexes of Biologically Active Substances with Nucleic Acids and Their Modes of Action," *Progr. Mol. Subcell. Biol.,* 2, 1-400 (1971).

W. Wehrli and M. Staehelin, *Bacteriol. Rev.,* 35, 290 (1971).

B. A. Newton, *Advan. Pharmacol. Chemother.,* 8, 247 (1970).

M. Waring, *J. Mol. Biol.,* 54, 247 (1970).

K. G. Lark, *Annu. Rev. Biochem.,* 38, 569 (1969).

G. Hartmann *et al., Angew. Chem., Int. Ed. Engl.,* 7, 693 (1968).

Inhibitors of Protein Synthesis

H.-D. Höltje and L. B. Kier, *J. Med. Chem.,* 17, 814 (1974).

D. Vazquez, *Pure Appl. Chem.,* 35, 355 (1973).

H. B. Bosmann and R. A. Winston, *Chem.-Biol. Interactions,* 4, 113, 129 (1972).

A. Korolkovas, *Rev. Bras. Clin. Ter.,* 1, 729 (1972).

J. R. Smythies *et al., Experientia,* 28, 1253 (1972).

F. Peradejordi *et al., J. Pharm. Sci.,* 60, 576 (1971).

S. Pestka, *Annu. Rev. Microbiol.,* 25, 487 (1971).

Z. Vogel *et al., J. Mol. Biol.,* 60, 339 (1971).

CHAPTER 34

ANTINEOPLASTIC AGENTS

I. INTRODUCTION

Antineoplastic agents are drugs used for the treatment of cancer. Increasingly, modern treatment involves combinations of drugs.

Cancer or *neoplasm* (from the Greek, νεος, new and πλασμα, formation) refers not to a single disease but to a group of diseases probably caused by several agents, such as certain chemical compounds, radiant energy, certain viruses, and cellular mutation of unknown origin. Cancer is characterized by an abnormal and uncontrolled division of cells exhibiting varying degrees of malignancy which produce tumors and invade adjacent normal tissue, affecting one or more different organs and systems of the body.

Often cancer cells separate themselves from the primary tumor and, carried by the lymphatic system or blood stream, reach distant sites of the organism where they divide and form secondary tumors. This phenomenon is called *metastasis* (from the Greek, μετα, after, and στασίς, standing).

The main characteristics of malignant tumors are, therefore: (*a*) *autonomous growth,* insensitive to the normal control mechanisms that limit cell growth and division in differentiated tissues, and (*b*) *invasiveness,* of adjacent capillaries and lymph channels and, metastatically, of remote parts of the organism; this invasion of normal tissues by growing in between normal cells is what makes cancer lethal.

Neoplasms occupy second place as the *causa mortis,* next to cardiovascular disorders. In 1970 the death rate per 100,000 population was 360.3 for diseases of the heart and 162.0 for cancer and other malignant neoplasms. In 1972 the latter figure rose to 166.8. Now, in the United States, cancer causes over 250,000 fatalities every year. Cancer authorities claim that 1.3 million Americans may be found to have cancer for the first time in 1974.

According to their localization and shape, tumors receive different names: carcinoma (glandular tissue), sarcoma (connective tissue), lymphoma (lymphatic ganglia), leukemia (blood cells; it can be myeloid leukemia and lymphoid leukemia). Tumors can be classified also according to their anatomic location.

Leukemias receive also various names, depending on the blood-cell picture and symptoms presented.

Neoplastic diseases affect both children and adults. In children the most common of these diseases are acute leukemias, usually of the lymphoblastic or stem-cell types, and four solid tumors: Wilm's tumor, neuroblastoma, retinoblastoma, and embryonal rhabdomyosarcoma; in adults the following ones: acute leukemia (mainly of the myeloblastic granulocytic type), chronic lymphocytic leukemia, chronic myelocytic leukemia, Hodgkin's disease, multiple myeloma, polycythemia vera, cell sarcoma, lymphosarcoma, choriocarcinoma, carcinomas (of the prostate, testis, breast, ovary, endometrium, cervix, lung, colon, stomach, liver, pancreas), melanoma.

Antineoplastic agents available today are in general only palliative, especially in the case of leukemia. Usually they do not cure cancer but only effect its temporary remission. However, 10 disseminated human neoplasms are highly responsive to chemotherapy: acute lymphocytic leukemia, Burkitt's lymphoma, choriocarcinoma, embryonal testicular cancer, Ewing's sarcoma, Hodgkin's disease, lymphosarcoma, reticulum cell sarcoma, retinoblastoma, Wilm's tumor. About 50% of patients with these diseases who are treated with modern chemotherapy achieve normal life expectancy.

Most antineoplastic agents are essentially antigrowth drugs, designed on the rationale that cancer cells multiply at a rate greater than all normal cells, and, therefore, anticancer agents must somehow interfere with cellular mitosis. Yet tumor cells do not undergo mitosis more rapidly than all normal cells. For instance, cells of the hemopoietic system, internal mucosa, oral mucosa, hair follicles, and skin divide faster than cancer cells. For this reason, drugs which act by destroying rapidly dividing cells mainly attack normal tissues and, hence, can be very toxic or even lethal to the patient. Nevertheless, they are resorted to as adjuvants to surgery and radiotherapy, which still remain the primary approaches to the cancer problem.

In order to obtain therapeutic synergism, combinations of different antineoplastic agents are now used, usually offering better results than in single-drug therapy, with reduction of toxic effects, enhancement of antitumor action, and delay of onset of drug resistance. One of these regimes (vincristine, mercaptopurine, methotrexate and prednisone) is used for treatment of acute leukemia in children. Other examples of combination therapy are prednisone and vincristine, methotrexate, mercaptopurine, and cyclophosphamide, successively, and prednisolone, vincristine, methotrexate, and mercaptopurine.

Since most antineoplastic drugs interfere with cell division, they are highly toxic, particularly to organs and systems characterized by rapid cell proliferation, such as bone marrow and the gastrointestinal tract. Common symptoms of intoxication with these drugs are leukopenia, thrombocytopenia, anorexia, nausea, vomiting, alopecia, thrombophlebitis, and cystitis.

II. HISTORY

Until recently the only approaches to treatment of cancer were surgery, for localized and accessible tumors, and radiotherapy. Immunotherapy has been tried but with limited success. Thirty years ago a new weapon was added, chemotherapy, with the introduction of antineoplastic agents, such as hormones, nitrogen mustards, and folic acid antagonists. Before then, however, potassium arsenite, known as Fowler's solution, had been used for the treatment of leukemia. Now, there are several classes of compounds with antitumor activity. Unfortunately, notwithstanding the fact that about a quarter million chemical compounds have been tested as potential antineoplastic agents and cures have been achieved in certain forms of cancer, chemotherapy has not solved the total problem of cancer. There are several reasons for this failure, the first two being the principal ones:

1. Biochemical and morphological differences between normal and neoplastic cells are slight. Therefore, antineoplastic agents are devoid of selective toxicity to tumor cells. One of the few exceptions known today is that certain leukemic cells lack the enzyme L-asparagine synthetase and, therefore, require L-asparagine from nutrients. Furthermore, since tumors differ one from another and each one may have its own distinct metabolism, it is unlikely that a drug active against all types of neoplasms can ever by developed.

2. Most neoplastic cells are not truly foreign to the host. Hence, they do not arouse an immunological response, in contrast to what happens in microbial infections, in which immunological defenses play a very important role in assisting the chemotherapeutic agent. Lately, however, foreign antigens were detected in several human neoplasms—carcinoma of the colon, neuroblastoma, and Burkitt's lymphoma, for instance—which suggests that host defense mechanisms may be involved in long-term remissions.

3. Cancer cells very rapidly develop resistance to antineoplastic drugs. One attempt to overcome this is combination therapy. But even this extreme therapeutic resource does not solve the problem; in fact, it can worsen the disease by allowing the most malignant types of cancer cells to emerge from the treated tumor.

4. Some tumors are poorly irrigated by blood, and this hampers the easy access of drugs to the cancer cells. With the aim of increasing the concentration of drugs in the sites desired, infusion and regional perfusion techniques are being used.

5. No ideal way of assessing the therapeutic usefulness of a potential antineoplastic agent is available. The findings with experimental transplanted animal tumors, the most important type of screening test used nowadays, are not necessarily extrapolated to human tumors.

6. Most antineoplastic agents produce very toxic untoward effects, such as the suppression of the immune response. Incidentally, study of the phenomenon of immunosuppression caused by anticancer drugs has resulted in the introduction of drugs which have permitted organ transplantations. Among other antineoplastic drugs, the following ones are being used also as immunosuppressive agents: methotrexate, cyclophosphamide, mercaptopurine, azathioprine, fluorouracil.

7. Under certain conditions most, if not all, known antineoplastic agents are also carcinogenic.

The first antineoplastic drugs introduced in chemotherapy were the alkylating agents. They are analogues of so-called mustard gas, also known as sulfur mustard (bis-2-chloroethyl sulfide), a compound known since the middle of the last century. In 1919 Krumbhaar and Krumbhaar observed that it causes leukopenia, a drastic reduction of the white cells. Similar results were observed in British soldiers during World War I and American troops in World War II who were exposed accidentally to this substance.

Considering that in the treatment of certain leukemias a decrease of leucocytes is taken as improvement, thousands of structural analogues of mustard gas have since then been synthesized—for example, nitrogen mustards—and tested as potential antileukemic agents, but only a few of them have been introduced into clinical use.

Another rational approach led to introduction of cyclophosphamide and related compounds. These drugs were designed and synthesized as potential substrates for the phosphoramidases, enzymes in which cancer cells are reportedly richer than normal cells. In the specific case of cyclophosphamide, the inactive transport form of nor-mechloretamine, it was hoped that those enzymes would cleave that molecule at the site of its action, releasing thus the active moiety at the desired spot. The expected activation actually occurs, but neither in the desired way nor in the tumor. It does occur in the liver by an oxidase system of its microsomes.

Purine antagonists were developed by Hitchings, Elion and coworkers, starting in 1942, in a rational attempt to discover selective inhibitors of nucleic acid biosynthesis. Those researchers were stimulated and guided by the then recently ennunciated antimetabolite theory. Their efforts were directed to the synthesis of potential antimetabolites of the purine and pyrimidine bases of the nucleic acids. Preparation of isosteric analogues of hypoxanthine resulted in the introduction of mercaptopurine in 1952.

Synthesis and screening of numerous analogues of natural purine bases, nucleosides, and nucleotides supplied other antineoplastic agents, such as 6-thioguanine and azathioprine, besides drugs having other activities: allopurinol, used mainly in the treatment of gout and hyperuricemia and as coadjuvant in anticancer therapy, and trimethoprim and pyrimethamine, which are antimalarial agents.

In 1948 Farber observed that certain analogues of folic acid caused temporary remissions in acute leukemia in children. This observation led to introduction of folic acid antagonists as antineoplastic agents: aminopterin in 1948 and methotrexate in 1949.

The pyrimidine antagonists resulted from a rational design of drugs by Heidelberger and colleagues, starting in 1957, after the observation made by Rutman and coworkers in 1954 that a rat hepatoma incorporated uracil in nucleic acid synthesis to a much greater extent than normal liver cells. Knowing that by catalytic action of thymidylate synthetase uracil of deoxyuridylic acid is converted to thymine of thymidylic acid by 1-carbon transfer and that fluoroacetate is a powerful metabolic poison, they reasoned that replacement of hydrogen atom in the 5 position of uracil or deoxyuridine by the stable fluorine atom might alter drastically DNA biosynthesis. Their hopes came true. Isosteric replacement worked. 5-Fluorouracil and 5-fluorodeoxyuridine, which were synthesized, proved to be pyrimidine antagonists.

The antibacterial and antiprotozoal activities of antibiotics prompted the screening of these agents in cancer chemotherapy. Several antibiotics have shown antitumor activity.

Hormones were introduced as antineoplastic agents on the assumption that neoplasms arising in organs or tissues normally susceptible to hormone control would very likely respond to hormone treatment, mainly during the earlier stages of malignancy. This concept proved its validity about 40 years ago, when hormonal agents were used successfully for the first time in the treatment of carcinomas of the brest and prostate.

III. CLASSIFICATION

Antineoplastic agents can be divided into the following classes: alkylating agents, antimetabolites, antibiotics, plant products, hormones, miscellaneous agents, and radioactive isotopes. Owing to their peculiar action, some of them are also called cytotoxics, cytostatics, antiblastics, oncolytics, and antimitotics.

A. Alkylating Agents

Alkylating agents comprise a heterogeneous group of chemical compounds:

1. Nitrogen mustards: carmustine, chlorambucil, chlornaphazine (Aleukon), chloroethylaminouracil (Dopan), cyclophosphamide, Cytostasan, defosfamide (Mitarson), extramustine (Estracyt), iphosphamide, lomustine (CCNU), mannomustine (Degranol), mechlorethamine (= chlormethine), melphalan, merofan, methyl-CCNU, mitoclomine, mitotenamine, Nitromin, phenesterin, phosphoramide mustard, trichlormethine (= trimustin, Leukamin), trophosphamide, uramustine

2. Aziridines: azatepa, benzodepa (Dualar), inproquone, meturedepa (Turloc), thiotepa, triaziquone (Trenimon), triethylenemelamine (= tetramine), triethylenephosphoramide (TEPA), uredepa (Avinar)

3. Methanesulfonate esters: bromacrylide, busulfan, mannitol busulfan, mannosulfan, piposulfan (Ancyte)

4. Epoxides: epipropidine (Eponate), epoxypiperazine, etoglucid (Epodyl)

5. Miscellaneous agents: dibromodulcitol, 3,3-dimethyl-1-phenyltriazene, 5-(3,3-dimethyl-1-triazeno)-imidazole-4-carboxamide (dacarbazine, DTIC), hexamethylmelamine, mitobronitol, pipobroman, trimethylolmelamine

Some of the most active alkylating agents resulted from the application, intentional or empirical, of drug latentiation. With the purpose of assisting in the active transport of a pharmacophoric moiety, several naturally occurring substances—for example, amino acids, carbohydrates, peptides, proteins, steroids, and numerous heterocyclic compounds—were attached to it as carriers. Such is the case with the following drugs: melphalan (the carrier is L-phenylalanine), uramustine (the carrier is uracil, which not only assists in the transport of the drug to the site of action but is also incorporated into the RNA nucleus; for this reason, uramustine has a dual mode of action: as alkylating agent and as antimetabolite), mitobronitol (the carrier is mannitol), dibromodulcitol (the carrier is dulcitol), chloroquine and quinacrine nitrogen mustards (the carriers are chloroquine and quinacrine, respectively, and were used as such because they localize in cell nuclei), steroid esters of p-[N,N-bis(2-chloroethyl)amino]-phenylacetic acid (the carriers are cholesterol, testosterone, and deoxycorticosterone, which may impart to the drugs a dual mode of action, that is, as alkylating agents and as hormones, active per se).

The most widely used alkylating agents are listed in Table 34.1

Cyclophosphamide

It is used as monohydrate, which is quite stable. It is one of the most versatile and dependable alkylating agents, being effective in lymphosarcomas, reticulum cell sarcomas, disseminated Hodgkin's disease. It is one of the primary drugs used for childhood neuroblastoma. Cyclophosphamide is a pro-drug; one (or more) of its products of metabolism is active.

Busulfan

It occurs as practically water-insoluble crystals. It is the drug of choice in chronic myelocytic leukemia. The usual dosage is 4 to 8 mg daily.

B. Antimetabolites

Since cancer is characterized by an abnormal cellular metabolism and mitosis, and mitosis is controlled by the nucleic acids, it is easily understandable that most of the antimetabolites used as antineoplastic agents are structural analogues of the metabolites involved in the biosynthesis of nucleic acids and the purine- and pyrimidine-containing cofactors.

Antimetabolites used in cancer chemotherapy can be divided into the following groups:

1. Amino acids antagonists: azaserine, azotomycin, DON, fenclonine
2. Folic acid antagonists: aminopterin, methasquin, methotrexate, pteropterin (Teropterin), citrovorum factor
3. Pyrimidine antagonists: 5-azacytidine, azaribine, azauracil, azauridine, bromouracil, cyclocytidine, cytarabine, 5-diazauracil, fluorouracil, fluxoridine, ftorafur, idoxuridine, iodouracil
4. Purine antagonists: azaguanine, azathioprine (Imuran), broxuridine, butiopurine (Cytogran), mercaptopurine, psicofuranine, thiamiprine (Guaneran), thioguanine, tioinosine

Antimetabolites most widely used are listed in Table 34.2.

Fluorouracil

It is a sparingly water-soluble, white crystalline powder, which is stable to heat. However, its solution is light-sensitive. Since it is highly reactive drug, inhalation of the powder and exposure to the skin should be avoided. Fluorouracil and floxuridine are the only effective drugs in the treatment of gastrointestinal cancer. Leukopenia is the primary side effect, but is is mild. The usual dosage, by intravenous route, is 12 mg/kg of body weight daily.

Mercaptopurine

It is a water-insoluble, odorless crystalline powder. It is effective in acute lymphoblastic and stem-cell leukemia of children. The usual dosage is approximately 2.5 mg/kg of body weight daily.

Methotrexate

It is indicated for the treatment of trophoblastic tumors, such as choriocarcinoma, chorioadenoma destruens, and hydatidiform mole, carcinoma of the testes, leukemic meningitis, mycosis fungoides, and several other malignant diseases. It is contraindicated in the presence of existing liver, bone-marrow, or renal damage, especially the last, since it is excreted by the kidney. Owing to its abortive effect, it should not be administered during the first trimester of pregnancy.

C. Antibiotics

Many antibiotics have antineoplastic activity. Those mostly used are listed

Table 34.1 Alkylating Agents

Official Name	Proprietary Name	Chemical Name	Structure
mechloretamine* (chlormetin)	Mustargen	2,2'-dichloro-N-methyldiethyl-amine	$CH_3-N(CH_2CH_2Cl)_2$
chlorambucil*	Leukeran	4-(p-[bis(2-chloroethyl)amino]-phenyl)butyric acid	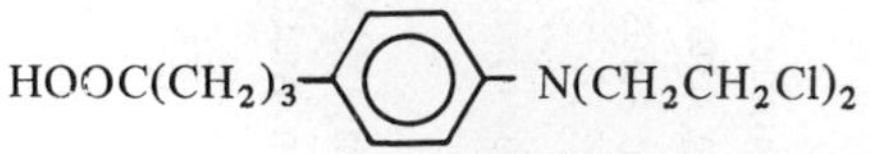 $HOOC(CH_2)_3-$⟨benzene ring⟩$-N(CH_2CH_2Cl)_2$
melphalan*	Alkeran	L-3-(p-[bis(2-chloroethyl)-amino]phenyl)alanine	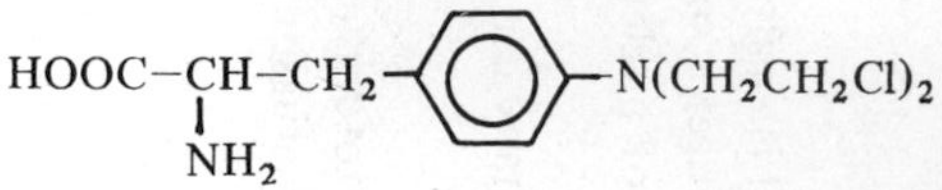 $HOOC-\underset{\underset{NH_2}{\vert}}{CH}-CH_2-$⟨benzene ring⟩$-N(CH_2CH_2Cl)_2$
cyclophosphamide*	Cytoxan Endoxan	2-[bis(2-chloroethyl)amino] tetrahydro-2H-1,3,2-oxazaphos-phorine 2-oxide	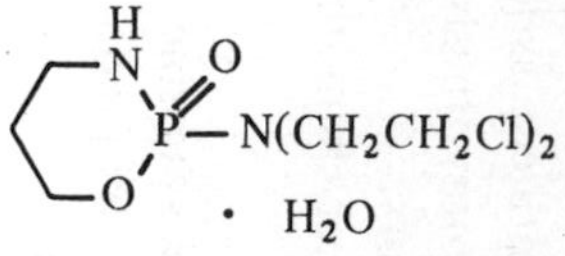 ⟨oxazaphosphorine ring, $O=P-N(CH_2CH_2Cl)_2$⟩ $\cdot\ H_2O$
uracil mustard+ (uramustine)		5-[bis(2-chloroethyl)amino]-uracil	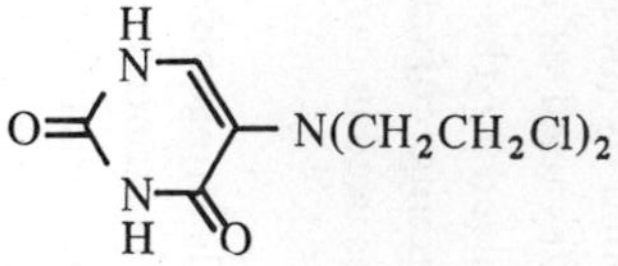 ⟨uracil ring with $N(CH_2CH_2Cl)_2$⟩
thiotepa+	Tifosyl	tris(1-aziridinyl)phosphine sulfide	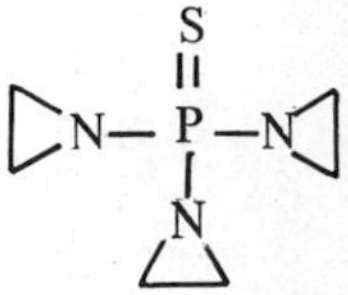 ⟨$S=P$ with three aziridinyl (N) rings⟩

triethylenemelamine[+]	TEM Triameline	2,4,6-tris(1-aziridinyl)-s-triazine	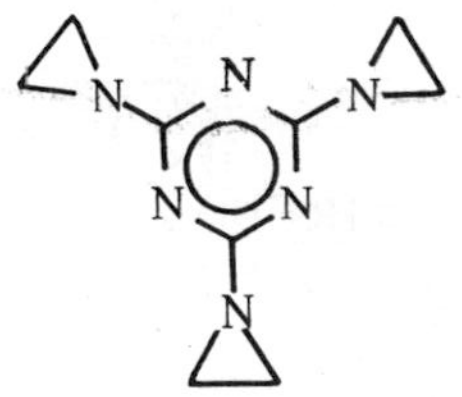
busulfan*	Myleran	1,4-butanediol dimethanesulfonate	$CH_3SO_2-O-CH_2CH_2CH_2CH_2-O-SO_2CH_3$
mitobronitol	Myelobromol	1,6-dibromo-1,6-dideoxy-D-mannitol	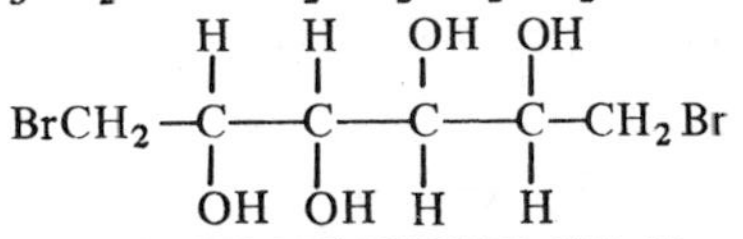 $BrCH_2-C-C-C-C-CH_2Br$ (with OH, H, OH, OH, H, OH, H, OH substituents)
carmustine	BCNU	1,3-bis(2-chloroethyl)-1-nitro-sourea	$ClCH_2CH_2NCONHCH_2CH_2Cl$ NO
pipobroman[+]	Vercyte	1,4-bis(3-bromopropionyl)-piperazine	 $BrCH_2CH_2CO-N \diagup \diagdown N-COCH_2CH_2Br$

*In USP XIX (1975).
[+]In NF XIV (1975).

Table 34.2 Antimetabolites

Official Name	Proprietary Name	Chemical Name	Structure
methotrexate*	Amethopterin	L-(+)-N-[p-[[(2,4-diamino-6-pteridinyl)-methyl]methylamino]benzoyl]-glutamic acid	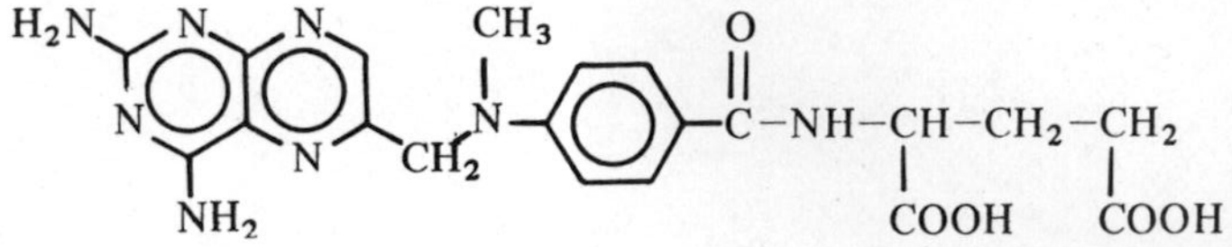
fluorouracil*	Fluracil Fluril	5-fluorouracil	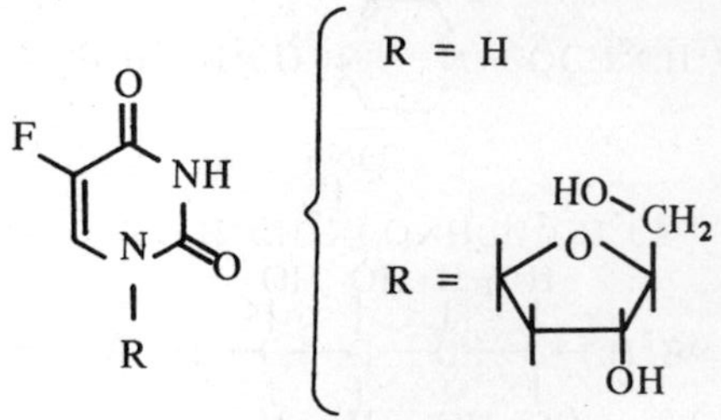
floxuridine+	FUDR	2'-deoxy-5-fluorouridine	R = H R =
cytarabine* (ara-C)	Cytosar	1-β-D-arabinofuranosylcy-tosine	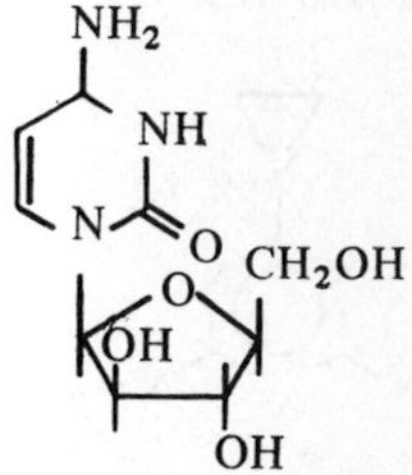

mercaptopurine*   Leukerin   6-*H*-purine-6-thione monohydrate
                   Purinethol

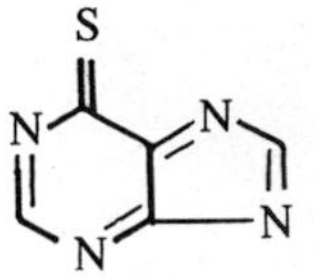

S ... $\cdot$ H$_2$O

thioguanine*   Lamis   2-aminopurine-6(1*H*)-thione
                         hemihydrate

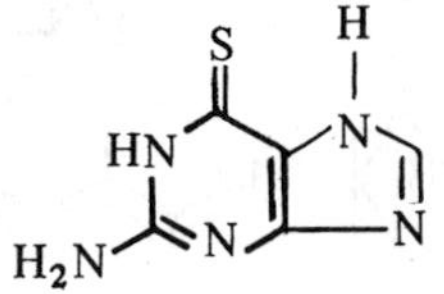

S ... $\cdot$ ½H$_2$O

*In *USP XIX* (1975).
⁺In *NF XIV* (1975).

Table 34.3 Antibiotics

Official Name	Proprietary Name	Structure
dactinomycin*	Cosmegen	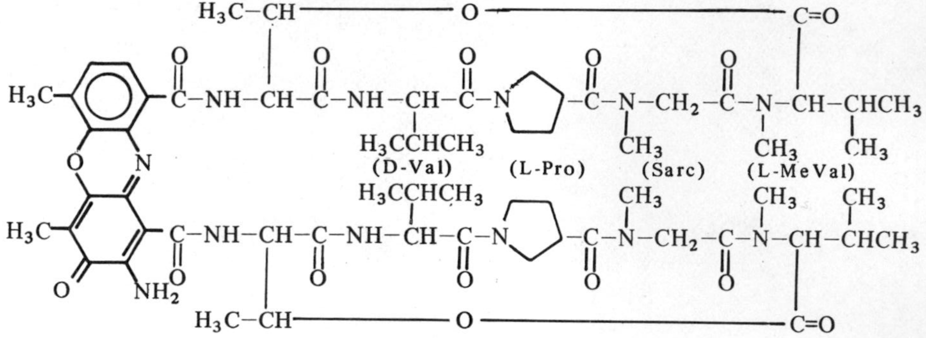
mithramycin*	Mithracin	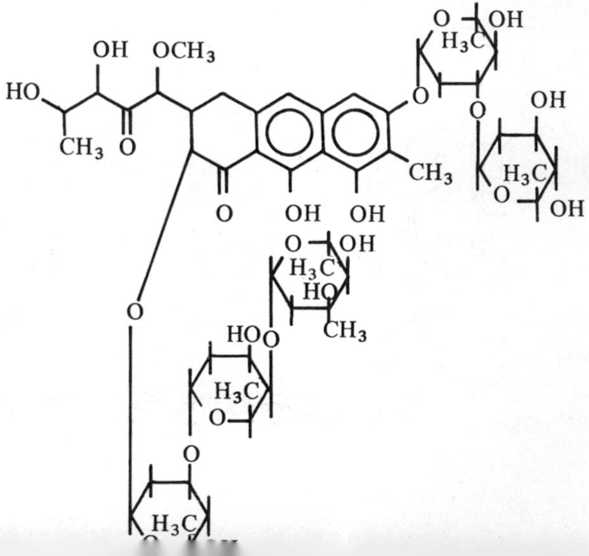

Structure labels for dactinomycin (Cosmegen):

H_3C-CH — O — $C=O$

C—N—CH$_2$—C—N—CH—CH—CHCH$_3$

CH$_3$ (Sarc), CH$_3$ CH$_3$ (L-MeVal)

C—N—CH—C (L-Pro)

H_3CCHCH$_3$ (D-Val)

C—NH—CH—CH—C

C—NH—CH—C

N

H_3C

O

H_3C

O

NH$_2$

H_3C-CH — O — $C=O$

Structure labels for mithramycin (Mithracin):

CH$_3$, OH, OCH$_3$, HO, CH$_3$, O, H_3C, HO, HOO, OH, HO, CH$_3$

bleomycin sulfate   Blenoxane

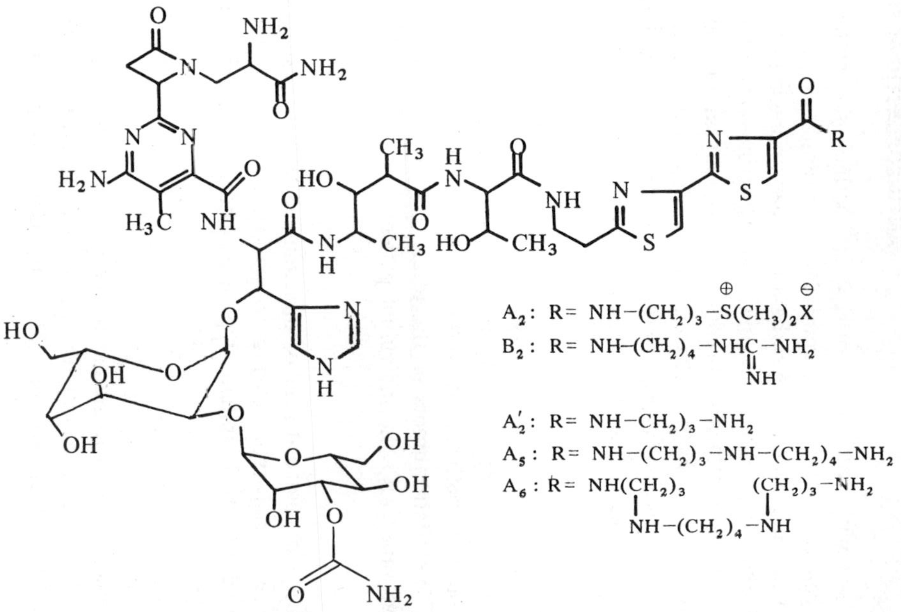

A_2: R= $NH-(CH_2)_3-\overset{\oplus}{S}(CH_3)_2 \; X^{\ominus}$

B_2: R= $NH-(CH_2)_4-NH\overset{\|NH}{C}-NH_2$

A_2': R= $NH-CH_2)_3-NH_2$

A_5: R= $NH-(CH_2)_3-NH-(CH_2)_4-NH_2$

A_6: R= $NH(CH_2)_3-NH-(CH_2)_3-NH_2$
$\qquad\qquad\quad$ $NH-(CH_2)_4-NH$ $\;\;(CH_2)_3-NH_2$

*In *USP XIX* (1975).

553

in Table 34.3. Others with anticancer activity are actinobolin, actinogen, adriamycin, anthracyclines, anthramycin, aranoflavin, azaserine, carcinostatin, carzinophilin, chaetocin, chromomycin, cycloheximide, daunorubicin (Toyomycin), largomycin, miathansine, mitocromin, mitomalcin, mitomycins, mitosper, mycophenolic acid, neocarcinostatin, neopluramycin, oxazinomycin, olivomycins, pactamycin, peptinogan, phleomycin, pillaromycin, porfiromycins, psicofuranine, puromycin, rufochromomycin, sarkomycin, sparsomycin, streptonigrin (Nigrin), streptozotocin, tenuazonic acid, tubercidin, verticillin A.

Dactinomycin

It occurs as a water-soluble, bright-red, crystalline powder that is slightly hygroscopic, being sensitive to heat and light. It is highly effective in Wilm's tumor. In combination with methotrexate and chlorambucil, it is of major value in the treatment of metastatic tumors of the testes. It causes severe adverse reactions. The dosage varies with the individual patient.

Mithramycin

It is indicated primarily for the treatment of disseminated testicular carcinoma, especially the embryonal cell type. Owing to its severe toxic reactions, it should be administered only to hospitalized patients under the supervision of experienced physicians.

Bleomycin

It is used in some squamous cell carcinomas in the head and neck region. It is also effective in lymphomas and in testicular carcinomas. It has the advantage of being only slightly myelosuppressive, in contrast to many anticancer agents.

D. Plant Products

Dozens of plant products have shown antineoplastic activity. They can be divided into the following groups:

1. Vinca alkaloids: vinblastine, vincristine
2. Podophylotoxin and analogues: Burseran, epipodophylotoxin, mitopozide (Proresipar), Proresidor
3. Colchicine, analogues and derivatives: Citostal, demecolcine (Colcemid), Thiocolciran
4. Phenanthrene derivatives: aristolochic acid, criptobleurine, tylocrebrine, tylophorine
5. Ellipticine and related compounds: demethylellipticine, ellipticine, methoxyellipticine, olivacine, oxaellipticine, thiaellipticine, thiaovalicine
6. Benzylisoquinoline alkaloids: berberine, coralyne, fagaronine, nitidine, pseudolycorine, tetrandine, thalidasine

7. Aporphine alkaloids: hernandaline, thalicarpine

8. Camptothecin and other quinoline alkaloids: acronin, camptothecin, hydroxycamptothecin, 9-methoxycamptothecin

9. Pyrrolizidine antibiotics: eliotrine, fulvine, lasiocarpine, monoerotaline, senecionine, spectabiline

10. Naphthoquinones: arnebin, lapachol and derivatives, shikonin

11. Proteins: ricin

12. Miscellaneous alkaloids: acronycin, cephalotaxin, coptisine, deoxyharringtonine, emetine, harringtonine, homoharringtonine, isoharringtonine, liriodenine, narciclasine, oxopurpureine, β-solamarine, solapalmitenin

13. Other plant products: costunolide, crotepoxide, cucurbitacins, damsine, datiscoside, elatericin, elebrigenin, elephantin, elephantopin, epoxidone, eupachlorin, euparotin, eupatine, eupatoretin, jatrophone, matrine, maytanisine, oximatrine, primulagenin A, sofogiaponicin, taxodione, taxodone, taxol, tripdiolide, triptolide, tulipinolide, vernodalin, vernolepin, withacnistin, withaferin

Plant alkaloids most used are listed in Table 34.4.   •

Vincristine Sulfate

It is effective primarily in acute leukemias in children. The adverse reactions produced by this drug are reversible.

E. Hormones

Some neoplasms are susceptible to hormone control. Unfortunately, not all are. Furthermore, most of the hormone-responsive neoplasms eventually are reactivated, becoming refractory to further treatment. However, some cancer cells do show some sensitivity to hormones, as evidenced by their remission when treated with hormonal agents. Usually they constitute only an adjuvant therapy to surgery.

Their main clinical indications are in the management of patients with disseminated neoplasias: androgens, for breast carcinoma; estrogens, for prostatic and mammary carcinoma; progestins, for carcinoma of uterus; adrenocorticoids, for acute leukemia, particularly in children, chronic lymphocytic leukemia, multiple myeloma, and lymphomas—they are usually used in combination or in sequence with other classes of anticancer drugs.

As to untoward reactions, androgens produce masculinizing effects; estrogens, feminilizing effects. Adrenocorticoids can cause disturbance of electrolyte balance.

Hormones can be divided into the following groups:

1. Androgens: calusterone (Methosarb), dromostanolone propionate, fluoxymesterone, testolactone (Teslac), testosterone, testosterone propionate

Table 34.4 Plant Products

Official Name	Proprietary Name	Structure
vinblastine sulfate*	Velban	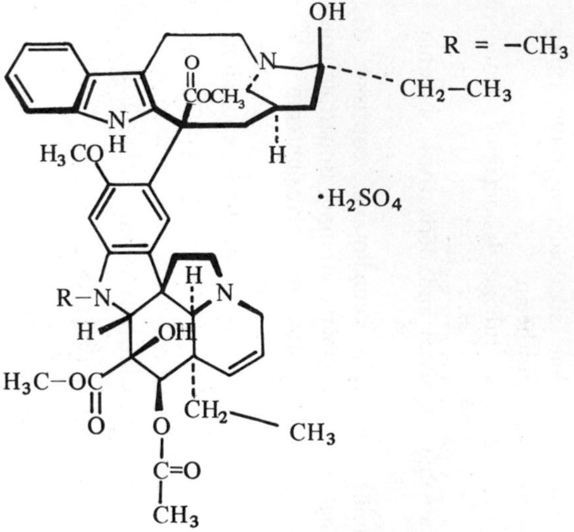 R = —CH₃
		(structure diagram)
vincristine sulfate*	Oncovin	R = —C(=O)—H

*In *USP XIX* (1975).

Table 34.5 Antineoplastic Hormones

Official Name	Proprietary Name	Chemical Name	Structure
Androgens			
testosterone*+	Andronate Maserate Neo-Hombreol Oreton Propionate Perandren	17β-hydroxyandrost-4-en-3-one	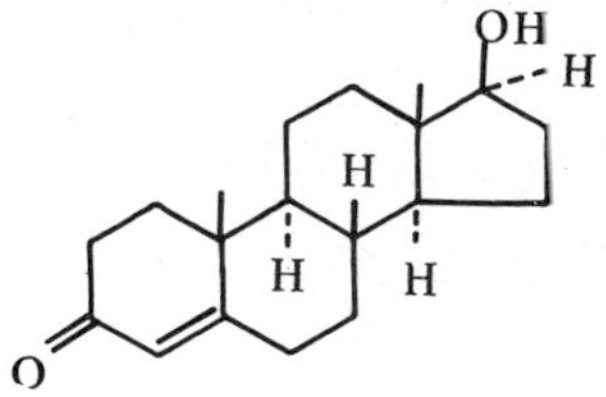
fluoxymesterone*	Halotestin Ultandren	9-fluoro-11β,17β-di-hydroxy-17-methylan-drost-4-en-3-one	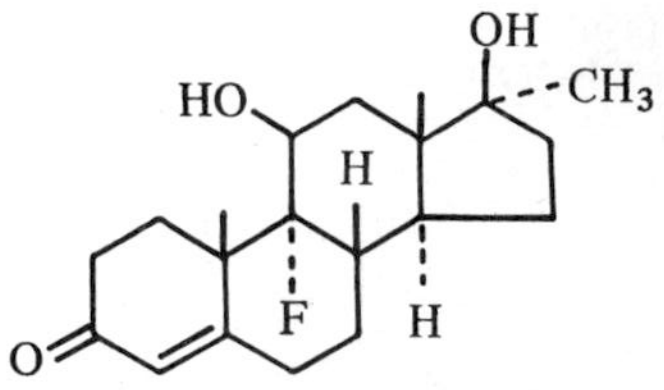
dromostanolone propionate+	Drolban Masterone	2α-methyl-17β-hyd-roxy-5α-androstan-3-one propionate	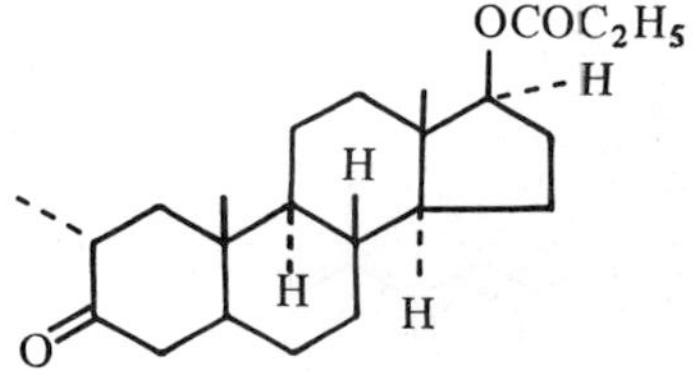
testolactone+	Teslac	D-homo-17α-oxaan-drosta-1,4-diene-3,17-dione	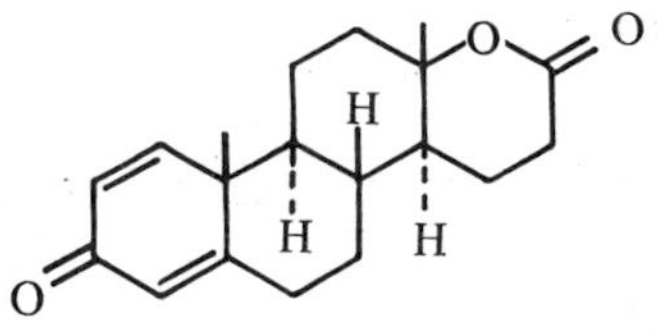

Table 34.5 Antineoplastic Hormones (continued)

Official Name	Proprietary Name	Chemical Name	Structure
Estrogens			
diethylstilbestrol*[+] diethylstilbestrol dipropionate[+]		α,α'-diethyl-(*E*)-4,4'-stilbenediol	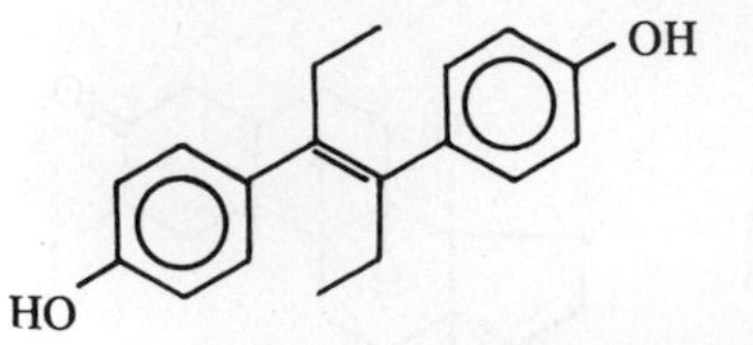
ethinyl estradiol*	Esteed Estinyl Eticylol Lynoral	19-nor-17α-pregna-1,3,5-(10)-trien-20-yne-3,17-diol	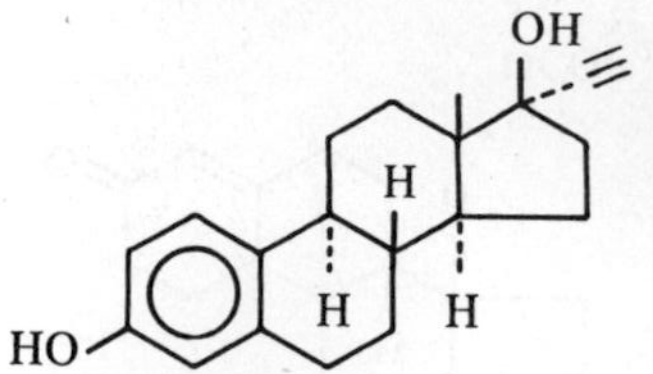
Progestagens			
hydroxyprogesterone caproate*	Delalutin	17-hydroxypregn-4-ene-3,20-dione hexanoate	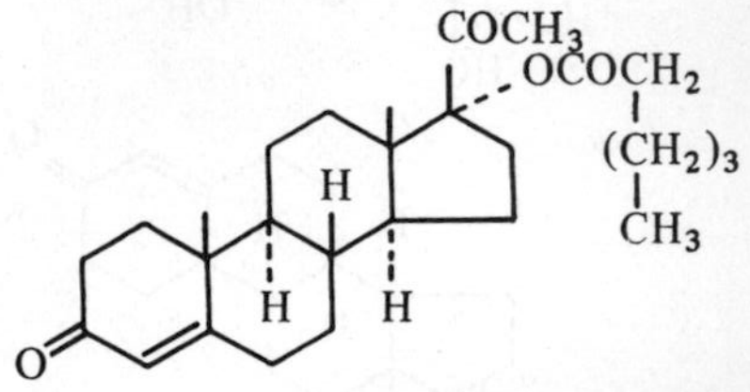

medroxyprogesterone acetate*	COCH₃ / OCOCH₃	17-hydroxy-6α-methyl-pregn-4-ene-3,20-dione acetate	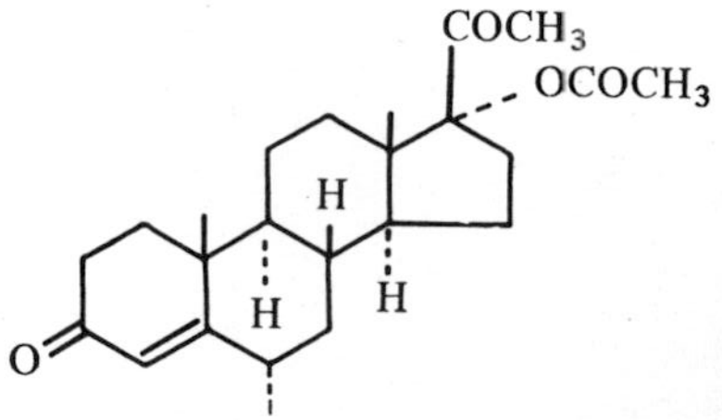 Lutoral Oragest Provera

medroxyprogesterone acetate*

$COCH_3$ / $OCOCH_3$

17-hydroxy-6α-methyl-pregn-4-ene-3,20-dione acetate

Lutoral
Oragest
Provera

Adrenocorticoids
prednisone*

$COCH_2OH$ / OH

17,21-dihydroxypregna-1,4-diene-3,11,20-trione

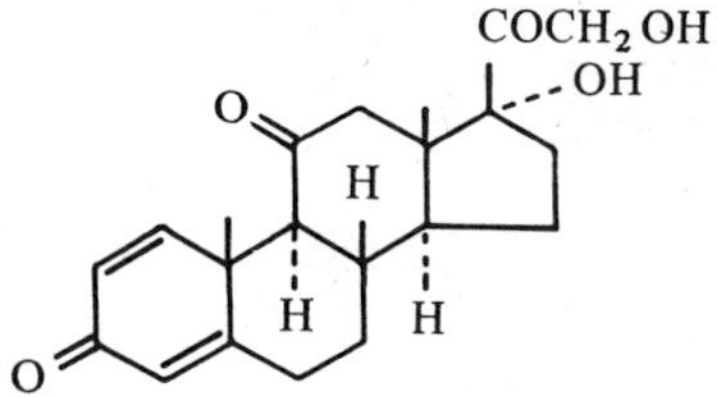

Deltasone
Deltra
Meticorten
Paracort

prednisolone*

$COCH_2OH$ / OH

11β,17,21-trihydroxy-pregna-1,4-diene-3, 20-dione

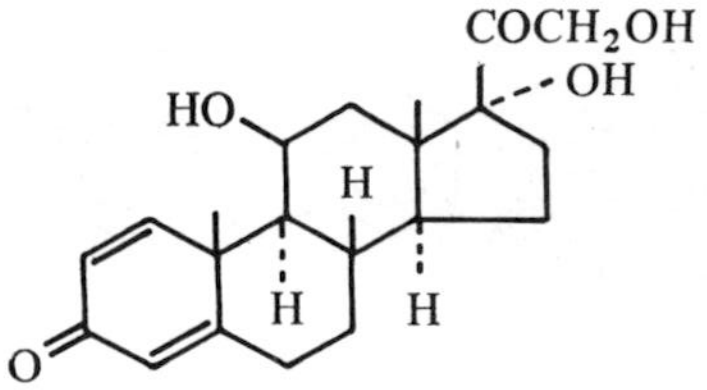

Delta Cortef
Hydeltra
Meticortelone
Metiderm
Paracortol
Sterane
Sterolone

*In *USP XIX* (1975).
†In *NF XIV* (1975).

Table 34.6 Miscellaneous Antineoplastic Agents

Official Name	Proprietary Name	Chemical Name	Structure
hydroxyurea*	Hydrea	hydroxyurea	$H_2N-CO-NH-OH$
mitotane*	Lysodren	1,1-dichloro-2-(o-chlorophenyl)-2-(p-chlorophenyl)ethane	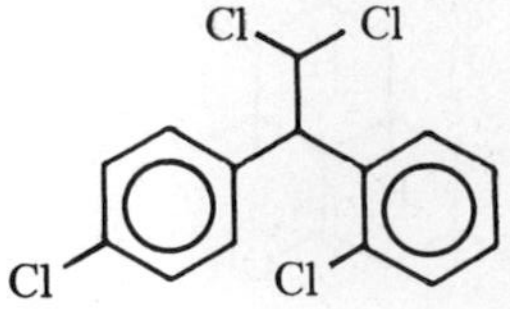
procarbazine*	Matulane Natulan	N-isopropyl-α-(2-methylhydrazino)-p-toluamide	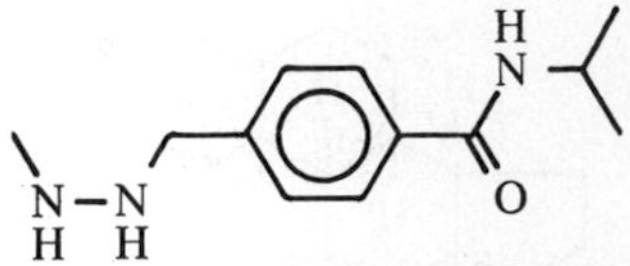
urethan		ethyl carbamate	$H_2N-CO-O-C_2H_5$
quinacrine*	Atabrine	see Table 28.2	see Figure 28.2

*In *USP XIX* (1975).

2. Estrogens: conjugated estrogens (Premarin), diethylstilbestrol, esterified estrogens (Amnestrogen, Menest), estramustine (Estracyl), ethinyl estradiol

3. Adrenocorticoids: prednisolone, prednisone

4. Progestagens: hydroxyprogesterone caproate, medroxyprogesterone acetate, megestrol acetate

Hormones most widely used as anticancer agents are listed in Table 34.5.

F. Miscellaneous Agents

Several other agents, besides those just discussed, have antineoplastic activity. Furthermore, new ones are constantly being sought, either from natural sources or by synthetic means. Those most widely used in medical practice are listed in Table 34.6. They belong to different groups of chemical structures.

G. Radioactive Isotopes

Owing to the ionizing radiations which have destructive action on living cells, the radioactive isotopes listed in Table 34.7 arc used for treatment of neoplasms.

Table 34.7   Antineoplastic Radioisotopes

Radioactive Isotopes	Half-Life (Days)	Usual Dose (mCi)
sodium iodide I 131 solution* capsules*	8.08	1-200 oral or intravenous
sodium phosphate P 32 solution*	14.3	1-12 oral or intravenous
gold Au 198 injection[+]	2.70	35-150 intracavitory

*In *USP XIX* (1975).
[+]In *NF XIV* (1975).

For being selectively incorporated into thyroid tissue, sodium iodide I 131 is used both as diagnostic and antineoplastic agent in the management of metastatic carcinoma of thyroid gland. Sodium phosphate P 32 tends to concentrate selectively in cells that undergo rapid division; it has shown good activity in polycythemia rubra vera. Gold Au 198 injection is administered into the serous cavity in the treatment of serous effusions developed in cancer.

IV. MECHANISM OF ACTION

Most antineoplastic agents act by interfering at some stage of nucleic acid (particularly DNA) or protein biosynthesis, as shown in Figure 34.1. Furthermore,

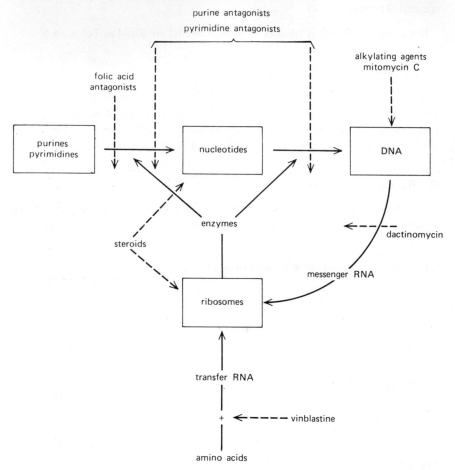

Figure 34.1 Sites of action of some antineoplastic agents. (Adapted from T. A. Connors, in G. Mathé, Ed., *Scientific Basis of Cancer Chemotherapy*, Springer, New York, 1969, pp. 1-17.)

this action depends on the cell cycle. Some drugs are active only in the stationary phase; others act in one or more phases of the logarithmic growth: mitotic phase, postmitotic phase, synthetic phase, or premitotic phase. In general, drugs that form covalent bonds with their receptors and thus inhibit irreversibly are lethal at all phases; such is the case of alkylating agents, which are used against solid tumors whose cells are largely in stationary phase. On the other hand, drugs that complex with their receptors through weaker interactions, such as ionic and hydrogen bonds, charge transfer, and hydrophobic and van der Waals interactions, allowing just reversible inhibition, are active only to rapidly metabolizing cells, for instance, the antimetabolites, drugs of choice in leukemias whose cells are totally or almost totally in the logarithmic phase. The action of antimetabolites is more intense when the cells are in the synthetic phase.

A. Alkylating Agents

All alkylating agents probably act in the same way. Initially they form highly reactive intermediates (Figure 34.2). These intermedicates then react with sulfur,

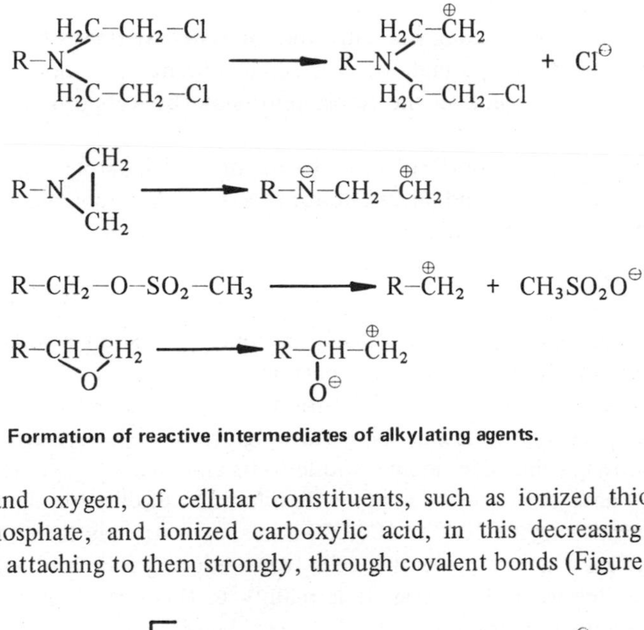

Figure 34.2 Formation of reactive intermediates of alkylating agents.

nitrogen, and oxygen, of cellular constituents, such as ionized thiol, amine, ionized phosphate, and ionized carboxylic acid, in this decreasing order of preference, attaching to them strongly, through covalent bonds (Figure 34.3).

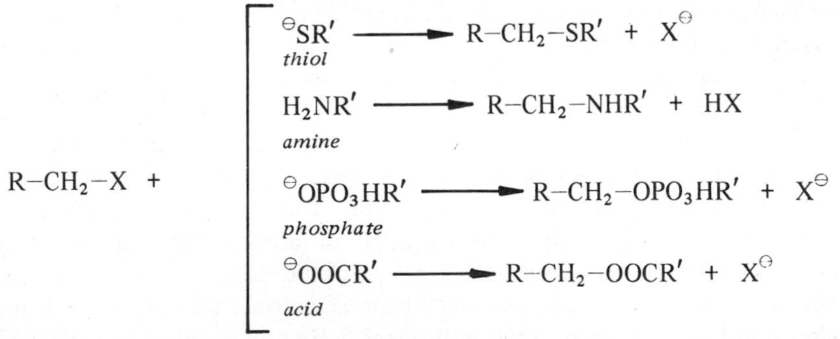

Figure 34.3 Reaction of alkylating agents with major nucleophiles.

As highly chemically reactive species, alkylating agents are nonselective. They react irreversibly with various nucleophiles and not necessarily with those which are contained in tumor cells. Besides their therapeutic activity, certain bifunctional nitrogen and sulfur mustards mimic the effects produced by X-rays, such as mutation, chromosomal aberrations, degeneration of the bone marrow, sterility, decrease or suppression of the immune response. For this reason, they are often called *radiomimetics.* It is realized that similarities between both effects are superficial, and, therefore, it may be misleading to call alkylating agents by the name radiomimetics.

The cell component most sensitive to the action of alkylating agents is DNA. The primary site of attack is atom 7 of guanine; then, in decreasing order of sensitivity, atom 3 of adenine and atoms 1 of adenine and cytosine. Alkylation with subsequent cross-linking not only inhibits DNA synthesis but also causes deletion of guanine (depurination) with concomitant formation of an easily hydrolyzable apurinic link in the ribose-phosphate backbone of DNA (Figure 3.6).

Although there are mono-, bi-, and trifunctional alkylating agents, greatest anticancer activity is found in bifunctional agents: they contain two centers separated by an optimal distance.

B. Antimetabolites

Glutamine antagonists (azaserine, DON, azotomycin) inhibit several metabolic processes in which glutamine takes part as a cofactor. Their antineoplastic activity is ascribed, however, to inhibition of the most sensitive enzyme to their action, namely, phosphoribosylformylglycinamidine synthetase, which converts formylglycinamide ribonucleotide to its amidine.

Folic acid antagonists act by nonspecific inhibition of dihydrofolate reductase and bind to this enzyme 3000 to 100,000 times more strongly than its substrate. Hence, antifolics act by preventing cells from reducing folic acid and dihydrofolic acid to tetrahydrofolic acid. It is mainly to this action, and not to their interference with 1-carbon transfers, that is ascribed their biological activity.

Pyrimidine antagonists owe their action to products of anabolism which interfere in the early phases of DNA synthesis by depletion of thymidylic acid, which leads to "thymineless death." For instance, both fluorouracil and floxuridine become active after being anabolyzed to 5-fluoro-2'-deoxyuridylic acid: this product, as a competitive inhibitor of thymidylate synthetase, is responsible for their antineoplastic activity. It is interesting that catabolic degradation of fluorouracil takes place only in normal cells but not in cancer cells; this phenomenon may explain the antineoplastic action of this drug.

Similarly, bromouracil and iodouracil become active after their anabolism to 2'-deoxyuridine derivatives. Their anticancer activity is ascribed to their incorporation into DNA in place of thymine or to their suppression of normal DNA biosynthesis.

Azauridine owes its action to its product of anabolism, 6-azauridylic acid, which is a competitive inhibitor of orotidylate decarboxylase, the enzyme that catalyzes the conversion of orotidylic acid to uridylic acid.

Cytarabine is a nucleoside obtained synthetically and distinguished by having D-arabinose as the pentose moiety. D-Arabinose presents an important feature that may play an essential role in the mechanism of action of cytarabine: by steric hindrance it prevents the free rotation of the cytosine moiety around the nucleosidic bond, because the spatial orientation of the hydroxyl group on the 3' position is opposite to that found in deoxynucleotides. Cytarabine appears to act either by inhibiting competitively the conversion of uridine to 2'-deoxycytidylic acid, but not to cytidylic acid, or by being incorporated into nucleic acids. Recently it was observed that this drug and its 5-fluoroderivative kill only cells in the synthetic phase of the logarithmic growth.

Purine antagonists become active after their anabolization to nucleotides and sometimes thence to di- and triphosphate derivatives.

Mercaptopurine acquires activity after its conversion to 6-thioinosinic acid. In this form it interferes with many metabolic processes vital to the growth and mitosis of cells by inhibiting the biosynthesis of purine nucleotides.

The mechanism of thioguanine is similar. It is a potent inhibitor of *de novo* purine nucleotide synthesis. But it acts also by incorporation into DNA, after being anabolyzed to its nucleotide, phosphorylated, and reduced to a derivative.

As to azathioprine and thiamiprine, designed to act as latent forms of mercaptopurine and thioguanine, respectively, they are cleaved by sulphydryl groups, as in glutathione, regenerating slowly the parent compounds, which are responsible for their action, after undergoing anabolism to their respective ribonucleotides.

An interesting feature of mercaptopurine and thioguanine is that both these compounds are catabolyzed by xanthine oxidase to thiouric acid, a noncarcinostatic metabolite. This process can be important to their antineoplastic activity, because many cancer cells are poorer in xanthine oxidase than certain normal cells.

C. Antibiotics

Mitomycin C and porfiromycin owe their anticancer action to the cross-linking of complementary DNA strands. This is made possible by a prior enzymatic activation *in vivo*, in which the quinone ring is reduced, and by losing the tertiary methoxy group the antibiotics acquire the protonated active form capable of acting as bifunctional alkylating agents having a 4-carbon-atom chain separating the reactive centers. Recent evidence, however, indicates that most mitomycins molecules attach to a single base of nucleic acid by monofunctional alkylation.

Dactinomycin and other actinomycins intercalate their chromophore between base pairs (very likely guanine-cytosine) of DNA, and the peptide lactone rings interact with each strand of the DNA double helix, but the exact mode of this complexation is not definitely determined. Dactinomycin is a selective inhibitor of DNA-dependent RNA synthesis.

Daunorubicin blocks synthesis of RNA and, therefore, of proteins by inhibiting both DNA and RNA polymerase, through intercalation between the base pairs of both DNA and RNA. Apparently this interference is during the premitotic phase of the logarithmic growth and it causes a delay in the onset of mitosis in cells that have already biosynthesized DNA.

Mithramycin, chromomycin, and olivomycin inhibit DNA and RNA synthesis by binding specifically to guanine residues in helical DNA.

Bleomycin inhibits DNA synthesis, causing the formation of single-stranded scissions in the sugar-template backbone of DNA. Streptonigrin acts in the same way.

D. Plant Products

In spite of very similar structures, vinca alkaloids exhibit impressive differences in biological activity. This may be ascribed to their greater or lesser ability to penetrate different types of cells. The most striking feature, however, is the lack of cross-resistance among them; this may depend on cell altered permeability.

Three main different effects, atributed to a common metabolite, may be responsible for their antineoplastic activity: metaphase arrest; inhibition of synthesis of soluble or transfer RNA by an acylating mechanism involving $-COOCH_3$ at positions 3 and 18'; inhibition of incorporation of acetate into phospholipids.

E. Hormones

Since steroids take part in or mediate protein and nucleic acid metabolism, their mechanism of antineoplastic action may be ascribed to interference with nucleic acid metabolism.

F. Miscellaneous Agents

The antineoplastic action of hydroxyurea may derive either from disturbance in the structure of DNA or from inhibition of DNA synthesis as a result of interference with ribonucleotide reductase, the enzyme that converts ribonucleotides to deoxyribonucleotides. Probably hydroxyurea acts as a chelator of ferrous ions, perhaps after metabolization.

The action of procarbazine may derive from inhibition of DNA synthesis, presumably by splitting DNA molecule into small fragments caused by hydrogen proxide formed in the process of oxidation. On the other hand, it has been suggested that the active metabolite is the azo derivative.

G Radioactive Isotopes

Radioactive isotopes destroy neoplastic cells by ionizing radiation.

REFERENCES

Introduction

M. J. Lacher, Ed., *Hodgkin's Disease,* Wiley Biomedical, New York, 1976.

C. Pochedly and D. Miller, Eds., *Wilms' Tumor,* Wiley Biomedical, New York, 1976.

F. Rutledge, R. C. Boronow and J. T. Wharton, *Gynecologic Oncology,* Wiley Biomedical, New York, 1976.

A. Bloch, *Annu. Rep. Med. Chem.,* **10,** 131 (1975); 9, 139 (1974).

H. G. Seydel, Ed., *Tumors of the Nervous System,* Wiley Biomedical, New York, 1975.

H. G. Seydel, A. Chait and J. T. Gmelich, *Cancer of the Lung,* Wiley Biomedical, New York, 1975.

J. C. Arcos and M. E. Argus, *Chemical Induction of Cancer,* Vols. IIA and IIB, Academic, New York, 1974.

H. Busch, Ed., *The Molecular Biology of Cancer,* Academic, New York, 1973.

C. C. Cheng and K. Y. Zee-Cheng, *Annu. Rep. Med. Chem.,* 8, 128 (1973).

Committee on Professional Education of UICC, Eds., *Clinical Oncology,* Springer, Berlin, 1973.

C. W. Parker, *Pharmacol. Rev.,* **25,** 325 (1973).

H. S. Rosenkranz, *Annu. Rev. Microbiol.,* **27,** 383 (1973).

R. Süss *et al., Cancer: Experiments and Concepts,* Springer, Berlin, 1973.

J. Tooze, Ed., *The Molecular Biology of Tumor Viruses,* Cold Spring Harbor, N. Y., 1973.

R. L. Nelson and H. S. Mason, *J. Theor. Biol.,* **37,** 197 (1972).

V. S. Shapot, *Advan. Cancer Res.,* **15,** 253 (1972).

J. M. Vaeth, Ed., *Frontiers of Radiation Therapy and Oncology,* 6 vols., Karger, Basel, 1968-1972.

G. Mathé, *Advan. Cancer Res.,* **14,** 1 (1971).

K. Nishioka, *Advan. Cancer Res.,* **14,** 231 (1971).

E. D. Bergmann and B. Pullman, Eds., *Physico-Chemical Mechanisms of Carcinogenesis,* Academic, New York, 1970.

M. Green, *Annu. Rev. Biochem.,* **39,** 701 (1970).

C. Heildelberger, *Cancer Res.,* **30,** 1549 (1970).

O. Jarret, *Advan. Cancer Res.,* **13,** 39 (1970).

J. A. Miller, *Cancer Res.,* **30,** 559 (1970).

P. Vigier, *Progr. Med. Virol.,* **12,** 240 (1970).

H. Hamperl and L. V. Ackerman, *Illustrated Tumor Nomenclature,* 2nd ed., Springer, Berlin, 1969.

W. J. Harrington, *Advan. Int. Med.,* **15,** 317 (1969).

History

G. W. Camiener and W. J. Wechter, *Progr. Drug Res.*, **16**, 67 (1972).

T. Makinodan *et al.*, *Pharmacol. Rev.*, **22**, 189 (1970).

K. Habel, *Advan. Immunol.*, **10**, 229 (1969).

"Symposium: A Critical Evaluation of Cancer Chemotherapy," *Cancer Res.*, **29**, 2255-2485 (1969).

G. H. Hitchings and G. B. Elion, *Pharmacol. Rev.*, **15**, 365 (1963).

Classification

T. H. Maugh II, *Science*, **184**, 970 (1974).

A. C. Sartorelli and D. G. Johns, Eds., *Antineoplastic and Immunosuppressive Agents*, Springer, Berlin, 1974.

G. Brulé *et al.*, *Drug Therapy of Cancer*, World Health Organization, Geneva, 1973.

A. R. Jones, *Drug Metabolism Rev.*, **2**, 71 (1973).

M. Artico, *Farmaco, Ed. Sci.*, **27**, 683 (1972).

I. Brodsky and S. B. Kahn, Eds., *Cancer Chemotherapy II*, Grune and Stratton, New York, 1972.

S. K. Carter *et al.*, *Advan. Cancer Res.*, **16**, 273 (1972).

I. Gresser, *Advan. Cancer Res.*, **16**, 97 (1972).

K. Jewers *et al.*, *Progr. Med. Chem.*, **9**, 1 (1972).

M. Artico, *Farmaco, Ed. Sci.*, **26**, 156 (1971).

J. R. Bertino, Ed., "Folate Antagonists as Chemotherapeutic Agents," *Ann. N.Y. Acad. Sci.*, **186**, 1-519 (1971).

F. Elkerbout *et al.*, Eds., *Cancer Chemotherapy*, Leiden University Press, Leiden, 1971.

B. Goodell *et al.*, *Clin. Pharmacol. Ther.*, **12**, 599 (1971).

J. L. Hartwell, *Lloydia*, **34**, 310 (1971).

J. A. Stock, "The Design of Tumor-Inhibitory Alkylating Agents," in E. J. Ariëns, Ed., *Drug Design*, Vol. II, Academic, New York, 1971, pp. 531-571.

F. Bergel, *Ergebn. Physiol.*, **62**, 91 (1970).

R. L. Capizzi *et al.*, *Annu. Rev. Med.*, **21**, 433 (1970).

W. H. Cole, Ed., *Chemotherapy of Cancer*, Lea & Febiger, Philadelphia, 1970.

S. M. Kupchan, *Trans. N.Y. Acad. Sci.*, **32**, 85 (1970).

R. B. Livingston *et al.*, *Advan. Pharmacol. Chemother.*, **8**, 57 (1970).

E. Boessen and W. Davis, *Cytotoxic Drugs in the Treatment of Cancer*, Arnold, London, 1969.

J. L. Hartwell and B. J. Abbott, *Advan. Pharmacol. Chemother.*, **7**, 117 (1969).

Symposium on Vincristine, *Cancer Chemother. Rep.*, **52**, 453-535 (1968).

I. Brodsky and S. B. Kahn, Eds., *Cancer Chemotherapy*, Grune and Straton, New York, 1967.

C. Heidelberger, *Annu. Rev. Pharmacol.*, **7**, 101 (1967).

M. B. Shimkin, *Top. Med. Chem.*, **1**, 79 (1967).

G. M. Timmis and D. C. Williams, *Chemotherapy of Cancer: The Antimetabolite Approach*, Butterworth, London, 1967.

L. F. Larionov, *Cancer Chemotherapy*, Pergamon, Oxford, 1965.

J. A. Montgomery, *Progr. Drug. Res.*, 8, 431 (1965).

A. Goldin, *et al.*, Eds., *Advances in Chemotherapy*, Vols. 1 and 2, Academic, New York, 1964, 1965.

P. Emmelot, "The Molecular Basis of Cancer Chemotherapy," in E. J. Ariëns, Ed., *Molecular Pharmacology*, Vol. II, Academic, New York, 1964, pp. 53-198.

Mechanism of Action

A. Bloch, Ed., "Chemistry, Biology, and Clinical Uses of Nucleoside Analogs," *Ann. N. Y. Acad. Sci.*, 255, 1-610 (1975).

J. W. Corcoran and F. E. Hahn, Eds., "Mechanism of Action of Antimicrobial and Antitumor Agents," *Antibiotics*, Vol. III, Springer, Berlin, 1975.

G. B. Grindey *et al.*, "Approaches to the Rational Combination of Antimetabolites for Cancer Chemotherapy," in E. J. Ariëns, Ed., *Drug Design*, Vol. V, Academic, New York, 1975, pp. 169-249.

T. J. Bardos, *Top. Curr. Chem.*, 52, 63 (1974).

T. A. Connors, *Top. Curr. Chem.*, 52, 141 (1974).

H. Kersten and W. Kersten, *Inhibitors of Nucleic Acid Synthesis*, Springer, Berlin, 1974.

W. A. Remers and C. S. Schepman, *J. Med. Chem.*, 17, 729 (1974).

A. Bloch, "The Design of Biologically Active Nucleosides," in E. J. Ariëns, Ed., *Drug Design*, Vol. IV, Academic, New York, 1973, pp. 285-378.

E. F. Gale *et al.*, *The Molecular Basis of Antibiotic Action*, Wiley-Interscience, London, 1972.

B. Goodell *et al.*, *Clin. Pharmacol. Ther.*, 12, 599 (1971).

A. Korolkovas, *Rev. Farm. Bioquim. Univ. S. Paulo*, 9, 5 (1971).

G. V. Gursky, *Mol. Biol.*, 3, 592 (1970).

P. Roy-Burman, *Analogues of Nucleic Acid Components*, Springer, Berlin, 1970.

R. D. Wells and J. E. Larson, *J. Mol. Biol.*, 49, 319 (1970).

K. Y. Zee-Cheng and C. C. Cheng, *J. Pharm. Sci.*, 59, 1630 (1970).

T. A. Connors, "Anti-Cancer Agents: Their Detection by Screening Tests and Their Mechanism of Action," in G. Mathé, Ed., *Scientific Basis of Cancer Chemotherapy*, Springer, New York, 1969, pp. 1-17.

CHAPTER 35

ANTIVIRAL AGENTS

I. INTRODUCTION

Antiviral agents are substances used in the treatment and prophylaxis of diseases caused by viruses.

Until recently viruses were considered to be infective agents that passed through bacterial filters and were unable to grow in the usual culture media devoid of cells. This concept is no longer accepted. Now viruses are defined as those infective agents which during their evolutive cycle have only one of the nucleic acids, either DNA or RNA, but not both and which multiply from nucleic acids but do not have ATP-generating enzymes. According to this modern concept, rickettsiae and the psittacosis-lymphogranuloma-trachoma group are no longer regarded as viruses because they have both DNA and RNA.

Viruses belong, therefore, to two great classes: DNA viruses and RNA viruses. The main groups of DNA viruses are poxviruses (vaccinia, smallpox, cowpox), adenoviruses (adenovirus 5), herpesviruses (Herpes simplex, Herpes zoster, varicella), papovaviruses (papieloma, polyoma). The principal groups of RNA viruses are picornaviruses (poliomyelitis, echovirus, rhinovirus), myxoviruses (influenza A, B, and C, mumps), arboviruses (yellow fever, equine encephalomyelitis), leukoviruses (Rous sarcoma), and arenoviruses (lymphocytic choriomeningitis).

Both DNA and RNA viruses lack a cell wall and related enzyme systems, such as those involved in internal metabolism. For this reason they are not susceptible to antibiotics, whose action results from the inhibition of certain phases of cellular metabolism, as was seen in the preceding chapter. For their existence and growth, viruses need a host cell on whose enzymatic activity they are dependent. They reproduce not by fission, as bacteria, but by bursting the host cell open.

Antiviral chemotherapy is confronted with two major obstacles which explain its slow progress and scanty achievements: (*a*) the intimate relationship that exists between multiplying viruses and mammal cells; thus, most antiviral agents lack selectivity, being equally toxic to both virus and host; (*b*) many viral diseases are diagnosed too late for effective treatment; usually, the first symptoms of viral diseases appear at the final stage of viral multiplication.

Despite the great effort toward discovering useful antiviral agents, only a few are presently available. None has broad-spectrum antiviral activity, although drugs of this sort are badly needed. Therefore, antiviral chemotherapy is at the same stage as antibacterial chemotherapy before the introduction of sulfa drugs.

For prophylaxis of most serious viral diseases, the only available and dependable resources are vaccines. For treatment of others, sera are the recommended preparations.

The rational design of antiviral agents which is now being sought involves selective inhibitors of one or more unique steps of the replicative cycle of viruses. With some small variations this process generally includes adsorption, penetration and uncoating, *de novo* synthesis of messenger RNA (by DNA viruses only), binding of viral *m*RNA to ribosomes, synthesis of viral enzymes (including the viral RNA polymerase), synthesis of viral nucleic acids and structural proteins, assembly of subunits, and release of viral particles. Antiviral agents act by inhibiting one or more of these steps, as will be seen in Section III.

II. CLASSIFICATION

Many substances have shown antiviral activity either *in vitro* or *in vivo* or both. However, most of them cannot be and are not used clinically because of their toxic effects. These substances belong to the following classes:

1. Adamantanes: amantadine, rimantadine, tromantadine
2. Nucleosides: azaribine, azathymidine, azauridine, cytarabine (ara-C), floxuridine, formycin, idoxuridine, pyrazomycin, showdomycin, trifluorothymidine, vidarabine (ara-A), virazole
3. Thiosemicarbazones: methisazone
4. Inosine derivatives: isoprinosine (Viruxan)
5. Amidines and guanidines: canavanine, guanidine, moroxydine (Virustat), D-penicillamine
6. Isoquinolines: famotine, memotine
7. Benzimidazoles: (*S,S*)-1,2bis(5-methoxy-2-benzimidazolyl)-1,2-ethanediol, 5,6-dichloro-1-β-D-ribofuranosylbenzimidazole (DRB), 5-fluoro-2-(α-hydroxybenzyl)-1-propylbenzimidazole, hydroxybenzylbenzimidazole (HBB)
8. Epidithiadiketopiperazine derivatives: acetylaranotin, aranotin, chetomin, gliotoxin, melanacidins, oryzachlorin, sporidesmins
9. Amino acid analogues: fluorophenylalanine, selenocystine
10. Enzymes: L-asparaginase
11. Antibiotics: alanosine, α-amanitin, anisomycin, cactinomycins, cephalomycin, congocidin (= netropsin), cycloheximide, cylindrochlorin, daunorubicin, distamycins, ehrlichin, filipin, fusidin, helenine, heliomycin, mycophenolic acid, myxoviromycin, novobiocin, phleomycin, propionin, puromycin, quinomycin, rifamycins, streptothricin, streptovaricins, tenuazonic acid, Thiolutin, topolymycin, tunicamycin, virocidin, viscosin

Table 35.1 Antiviral Agents

Official Name	Proprietary Name	Chemical Name	Structure
amantadine[+]	Symmetrel	1-adamantanamine	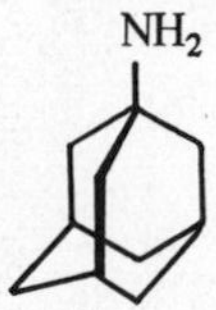
methisazone	Marboran	N-methylisatin β-thiosemicarbazone	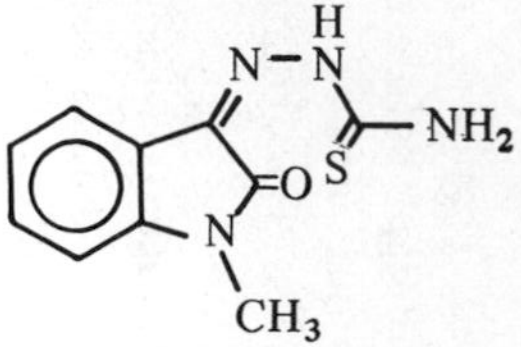
idoxuridine*	Dendrid Herplex Stoxil	2'-deoxy-5-iodouridine	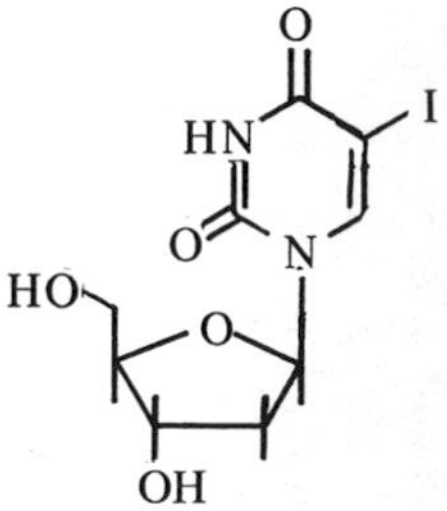
cytarabine* (ara-C)		see Table 34.2	

vidarabine (ara-A)	9-β-D-arabinofuranosyladenine 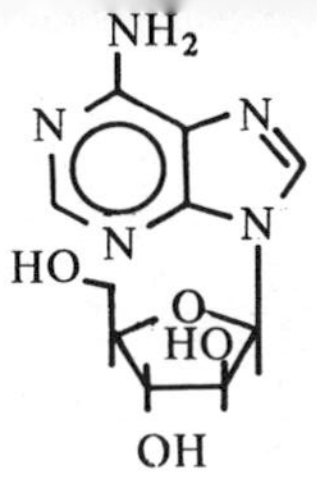
cyclooctylamine	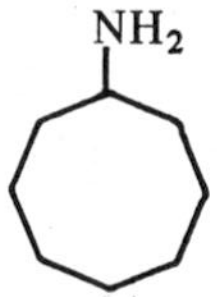
Viruxan	
isoprinosine	acetamidobenzoate salt of inosine-dimethylaminoisopropanol complex (1:3) 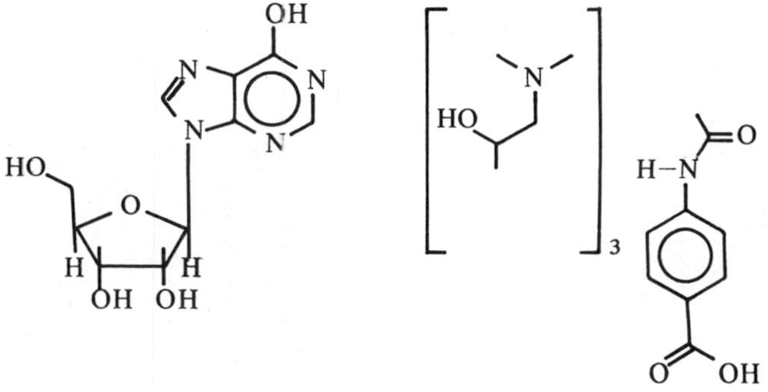

12. Plant products: colchicine, demecolcine, elenolate, emetine, gymnenic acid, vinblastine, vincristine

13. Miscellaneous agents: aurintricarboxylic acid, caprochlorone, cyclo-octylamine, ethidium bromide, kethoxal, oxoline, quinacrine, RSV (a mixture of diphenyl glyoxal and diphenyl glyoxal superoxide), tebrofen, xenalamine, xenaldial

14. Interferon and interferon inducers: antivirin, cycloheximide, helenine, interferon, kanamycin, phage double-stranded RNA, phagicin, phosphomannans, phosphorylated polysaccharides, polyacetal carboxylic acid, polyacrylic acid, polymethacrylic acid, polyriboinosinic-polyribocytidylic acid (poly rI-rC), pyran-copolymer, statolon, tilorone, vaccines, BL-20803

Antiviral agents most widely used in medicine are listed in Table 35.1.

Amantadine Hydrochloride

It is a freely water-soluble white crystalline powder, with a bitter taste. It is used both for prophylaxis and treatment of infections caused by A_2 strain of influenza virus, being effective in most cases. It is used also as antiparkinsonism agent. Adverse reactions include irritability, ataxia, tremor, slurred speech, insomnia, lethargy, dizziness, and several other minor discomforts.

Idoxuridine

It occurs as a white crystalline powder which is slightly soluble in water. It is active against some DNA and RNA viruses, but it is used topically only for the treatment of herpes simplex infections of the lids, cornea, and conjunctiva. This drug may cause slight local irritation, photophobia, and other minor adverse reactions.

Cytarabine

It has the same spectrum of activity as idoxuridine.

Vidarabine

Its spectrum of activity is the same as idoxuridine and cytarabine, but it may have certain advantages over them.

Methisazone

This drug occurs as a deep-orange-yellow chemical. It has very wide antiviral spectrum. However, its use is restricted to the DNA viruses, especially those responsible for smallpox, alastrim, adenoviruses, and varicella.

III. MECHANISM OF ACTION

Antiviral agents act at different sites and processes of the viral replicative cycle.

Amantadine and derivatives block the penetration of certain strains of RNA viruses into mammal cells and inhibit the uncoating of these viruses in the host cells. Their antiviral activity results primarily from this prevention of uncoating.

Interferon acts probably by inhibiting the attachment of viral mRNA to ribosomes. This inhibition is selective, because the initiation of viral protein synthesis is prevented without interference with the translation of mRNA by the host cell. Aurintricarboxylic acid (ATA) acts by a similar mechanism.

Compounds having the epidithiapiperazinedione moiety—acetylaranotin, for instance—inhibit viral RNA polymerase, and consequently multiplication of RNA viruses is prevented.

Idoxuridine prevents viral replication as a result of its incorporation into viral DNA. The same mechanism is responsible for the action of other antiviral nucleosides, such as cytarabine and vidarabine, although they are not incorporated as extensively as idoxuridine.

Rifamycins, streptovaricins, cactinomycins, distamycins, and thiosemicarbazones—for example, methisazone—owe their antiviral action primarily to blockade of normal cellular DNA → RNA transcription. In the case of thiosemicarbazones, their ability for selective chelation plays an important role in the mechanism of action.

REFERENCES

General

J. B. G. Kwapinski, Ed., *Molecular Microbiology,* Wiley, New York, 1974.

R. A. Bucknall, *Advan. Pharmacol. Chemother.,* **11**, 295 (1973).

W. A. Carter, *Selective Inhibitors of Viral Function,* Chemical Rubber Co. Press, Cleveland, 1973.

G. Werner, *Rev. Port. Farm.,* **23**, 143 (1973).

D. J. Bauer, Ed., *Chemotherapy of Virus Diseases,* Vol. I, Pergamon, Oxford, 1972.

E. C. Herrmann and W. R. Stinebring, Eds., "Antiviral Substances," *Ann. N.Y. Acad. Sci.,* **173**, 1-844 (1970).

Introduction

M. R. Hilleman, *Science,* **164**, 506 (1969).

H. B. Levy, Ed., *The Biochemistry of Viruses,* Dekker, New York, 1969.

S. E. Luria and J. Darnell, Jr., *General Virology,* 2nd ed., Wiley, New York, 1967.

Classification

S. Baron and G. Galasso, *Annu. Rep. Med. Chem.,* **10**, 161 (1975).

E. De Clercq, *Top. Curr. Chem.,* **52**, 173 (1974).

A. R Schwartz, *Annu. Rep. Med. Chem.,* **9**, 128 (1974).

D. J. Bauer, *Brit. Med. J.,* 3, 275 (1973).

H. B. Levy, *Annu. Rep. Med. Chem.,* 8, 150 (1973).

R. W. Sidwell *et al., Science,* 177, 705 (1972).

E. De Clercq and T. C. Merigan, *Annu. Rev. Med.,* 21, 17 (1971).

M. R. Hilleman and A. A. Tytell, *Sci. Am.,* 255(1), 26 (1971).

D. L. Swallow, *Progr. Med. Chem.,* 8, 119 (1971).

T. C. Merigan, Ed., "Symposium on Interferon and Host Response to Viral Infection," *Arch. Intern. Med.,* 126, 49-158 (1970).

J. Vilcěk, *Interferon,* Springer, Vienna, 1969.

T. S. Osdene, *Top. Med. Chem.,* 1, 137 (1967).

Mechanism of Action

P. Chandra, *Top. Curr. Chem.,* 52, 99 (1974).

M. A. Apple, *Annu. Rep. Med. Chem.,* 8, 251 (1973).

E. A. C. Follett and T. H. Pennington, *Advan. Virus Res.,* 18, 105 (1973).

A. P. Grollman and S. B. Horwitz, "The Rational Design of Antiviral Agents," in E. J. Ariëns, Ed., *Drug Design,* Vol. II, Academic, New York, 1971, pp. 261-276.

B. Goz and W. H. Prusoff, *Annu. Rev. Pharmacol.,* 10, 143 1970.

PART 6

VITAMINS

Vitamins are substances that in trace quantities are essential for normal metabolism of living beings. Many vitamins are integral parts of coenzymes; this explains their essential role in living processes. Owing to differences in metabolism, a substance may be a vitamin for man without being a vitamin for bacteria or protozoans, and vice versa. Except for vitamin D, which may be synthesized in the skin, and nicotinamide, which is a metabolic product of tryptophan, all vitamins needed by humans must be supplied to the body from exogenous sources, mainly vegetables. All vitamins, with the exception of A and D, are synthesized by plants. Nowadays, however, most are obtained by industrial synthetic processes.

The name *vitamine* was coined by Funk in 1912. It derives from the Latin *vita,* which means life, and the word *amine,* because he believed that all such substances were amines. Soon new vitamins were described that contain no nitrogen. For this reason, the name vitamine was changed to *vitamin.*

Usually vitamins are supplied by a well-balanced diet, and healthy individuals have no need to ingest extra amounts of them as medicines. In certain conditions, however, extra amounts are needed in order to cure or to prevent specific deficiency syndromes, such as beriberi, scurvy, pellagra, rickets, and night blindness.

Avitaminosis may result from (*a*) primary deficiency—inadequate diet, due to poverty, ignorance, food fads, traumatic stresses, poor dentition, and other causes; (*b*) secondary conditioned deficiency—malabsorption (intestinal abnormality or chronic diarrhea), increased demand (during periods of pregnancy, lactation, growth, and certain diseases), or reduced storage facilities (protein binding and transport to the site of action).

Substances are known that are antagonists of vitamins. Thus, pyrithiamine (3.V.C.3) and oxythiamine antagonize thiamine; coumarins and indandiones are antagonists of vitamin K; deoxypyridoxine antagonizes pyridoxine; glucoascorbic acid is an antagonist of vitamin C.

According to their solubility in fats or in water, vitamins are classifed as fat soluble or water soluble.

Carefully conducted studies by the Food and Nutrition Board of the National Research Council have established the recommended dietary allowances for individual vitamins. These allowances are usually greater than the minimal daily requirements. Unfortunately, excessive intake of vitamins is a common practice in some countries, especially in the United States. Preparations containing a single vitamin or a group of vitamins, sometimes with mineral salts included, are consumed in large quantities, in most cases without medical prescription. This practice is not only wasteful but dangerous, because excessive ingestion of certain vitamins, especially the fat-soluble vitamins, may produce adverse effects.

We exclude from this discussion dietary factors, such as inositol, choline, and essential fatty acids, as well as substances that are vitamins to other living beings (bacteria, protozoans) but not to mammals: p-aminobenzoic acid and lipoic acid (thioctic acid).

REFERENCES

M. M. Hashmi, *Assay of Vitamins in Pharmaceutical Preparations,* Wiley, New York, 1973.

R. J. Kutsky, *Handbook of Vitamins and Hormones,* Van Nostrand Reinhold, New York, 1973.

A. E. Axelrod, *Amer. J. Clin. Nutr.,* **24**, 265 (1971).

H. Baker and O. Frank, *Clinical Vitaminology: Methods and Interpretation,* Wiley-Interscience, New York, 1969.

J. Marks, *The Vitamins in Health and Diseases,* Churchill, London, 1968.

N. S. Scrimshaw, *Vitamins Hormones,* **26**, 705 (1968).

M. Freed, Ed., *Methods of Vitamin Assay,* 3rd ed., Wiley, New York, 1966.

S. F. Dyke, *The Chemistry of Vitamins,* Wiley-Interscience, New York, 1965.

A. F. Wagner and K. Folkers, *Vitamins and Coenzymes,* Wiley-Interscience, New York, 1964.

CHAPTER 36

FAT-SOLUBLE VITAMINS

I. INTRODUCTION

Fat-soluble vitamins are those vitamins that are soluble in lipids but not in water: A, D, E, and K. They are usually stored in the liver. Excessive intake of these vitamins may result in toxic manifestations. Deficiency produces several diseases. Thus, deficiency of vitamin A causes hyperkeratosis, xerophthalmia, keratomalacia, and night blindness; of vitamin D, rickets in growing children, infantile tetany, and osteomalacia; of vitamin E, kwashiorkor, macrocytic, and hemolytic anemia in infants; of vitamin K, hypoprothrombinemia.

II. CLASSIFICATION

Only four fat-soluble vitamins are known: A, D, E, and K (Table 36.1). Vitamin K and analogues were studied in Chapter 23.

A. Vitamin A

Preformed vitamin A occurs in animal fats, fish (cod, tuna, shark, halibut, turbot, percomorph) oils, liver, milk, cheese, butter, eggs, and other diet sources. Several yellow and green vegetables and carrots contain carotenoids, some of which have provitamin A activity; that is, they are transformed into vitamin A, a reaction that takes place in the small intestine and liver of mammals. Carotenoids that have vitamin A activity are α-carotene, β-carotene, γ-carotene, α-carotene epoxide, aphanin, citroxanthin, cryptoxanthin, echinenone, myxoxanthin, and torularhodin. They are structurally characterized by the presence of a chromophore, which occupies the central part of the molecule, linked to two terminal groups, usually α- or β-ionone. This structure allows the existence of several stereoisomers. Thus, of one of the two vitamins A, namely, retinol or vitamin A_1, 16 isomers are possible, but only 4 are strain-free and have been found in nature: the one with greatest vitaminic activity is all-trans. Vitamin A_2 is 3,4-dehydroretinol; its stereochemistry is also all trans, and its vitaminic activity is about 40% of that of vitamin A. A new derivative of vitamin

Table 36.1 Fat-Soluble Vitamins

Official Name	Proprietary Name	Chemical Name	Structure
vitamin A*	Alphalin Anatola	all-trans-3,7-dimethyl-9-(2,6,6-trimethyl-1-cyclohexen-1-yl)-2,4,6,8-nonatetraen-1-ol	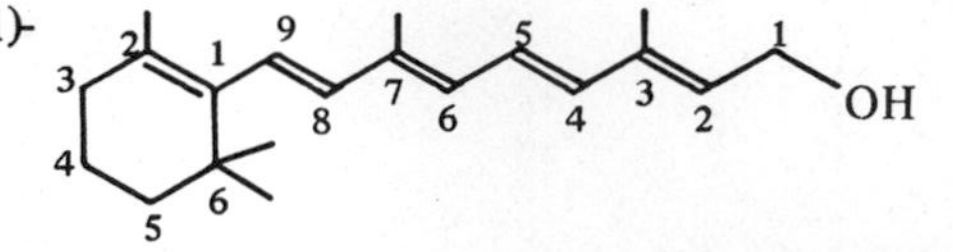
cholecalciferol*	Calciferol	(3β)-9,10-secocholesta-5,7,10(19)-trien-3-ol	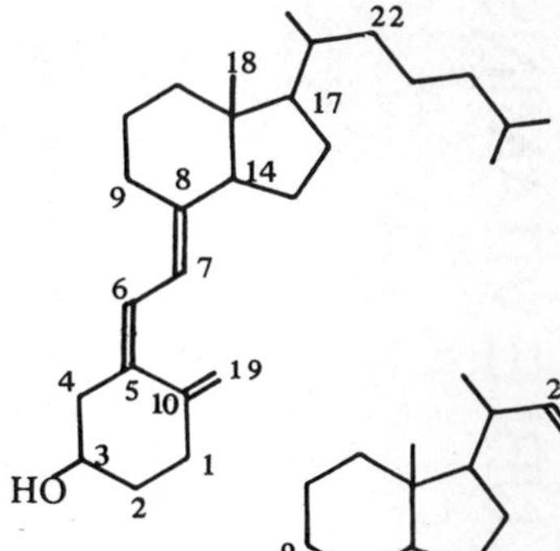
ergocalciferol*	Deltalin Drisdol	(3β)-9,10-secoergosta-5,7,10(19),22-tetraen-3-ol	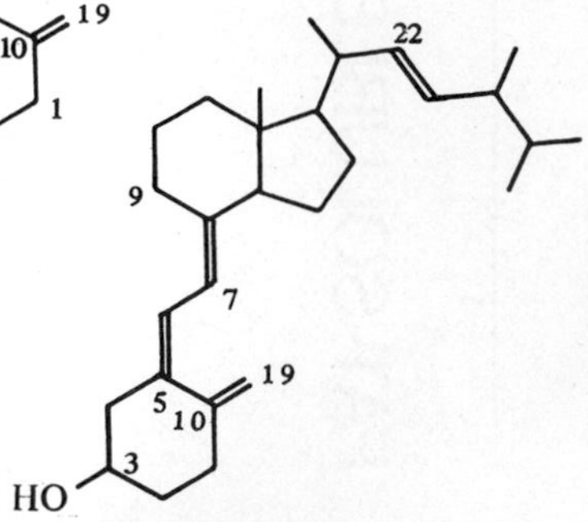

dihydrotachysterol*

A.T. 10
Hytakerol

(3β)-9,10-secoergosta-5,7,22-trien-3-ol

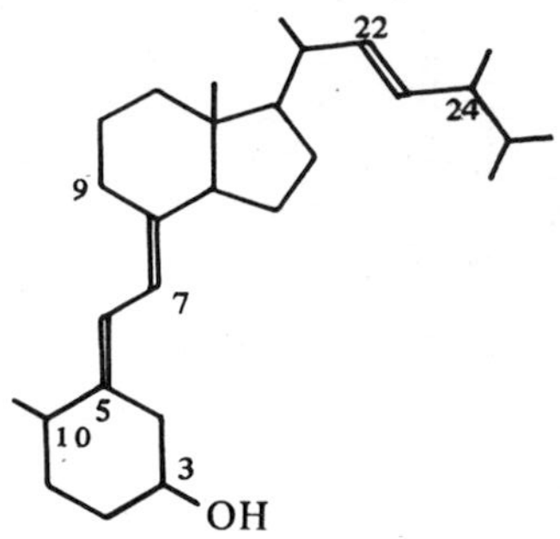

vitamin E+

Ecofrol
Tocopherex

(+)-2,5,7,8-tetramethyl-2-(4,8,12-trimethyltridecyl)-6-chromanol

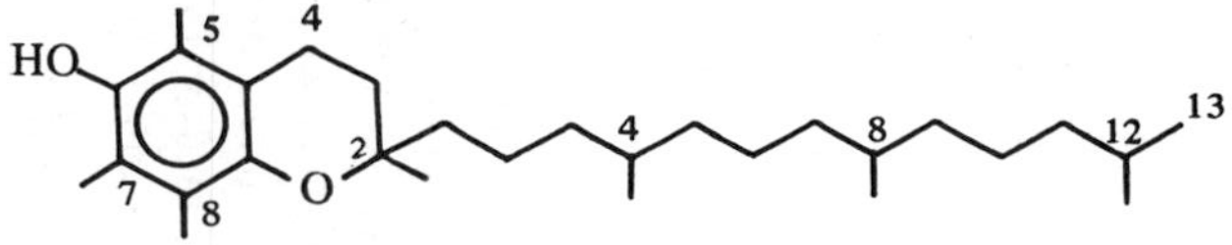

vitamin K and analogues

see Table 25.3

*In *USP XIX* (1975).
+In *NF XIV* (1975).

A is tretinoine, or all-trans retinoic acid.

Pure crystalline vitamin A occurs as pale-yellow plates or crystals, which are insoluble in water but soluble in ethanol, other organic solvents, and fixed oils. It is unstable in the presence of air or light, as well as in oxidized or readily oxidized fats and oils. The main alteration results from oxidation, which is prevented by an association of ($\pm$)-α-tocopherol, vitamin C, and nordihydroguaiaretic acid.

Owing to its polyene structure, vitamin A gives colored reactions with many reagents. Thus, it imparts an intense blue color to dry chloroform to which a chloroform solution of antimony chloride is added (Carr-Price reaction).

The recommended daily dietary allowance for adults is 5000 universal units. Excessive intake of vitamin A (but not of carotene) produces a toxic syndrome known as hypervitaminosis A, which is characterized by irritability, anorexia, weigth loss, itching, fatigue, alopecia, gingivitis, abdominal discomfort, insomnia, menstrual irregularities, hyperostoses, premature closure of the epiphyses, and other adverse reactions. Withdrawal of the vitamin causes regression of most of the symptoms within a week, but the hyperostoses remain for months.

Vitamin A is marketed both in the free form and in the form of esters, especially acetate, palmitate, and propionate.

Vitamin A

It occurs as a yellow to red oily liquid, almost odorless or with a fishy odor. It is insoluble in water but soluble in absolute ethanol, vegetable oils, ether, and chloroform. It is unstable to air and light. The usual dosage in cases of mild to moderate deficiency is 25,000 to 50,000 IU daily; in severe deficiency, 50,000 to 100,000 IU daily.

B. Vitamin D

Several vitamins D are found in nature, but only vitamin D_2 (cholecalciferol) and vitamin D_3 (ergocalciferol) have equal antirachitic activity in man. Both are steroid derivatives and are obtained by ultraviolet irradiation: the first, of ergosterol; the second, of 7-dehydrocholesterol. A third type of vitamin D is dihydrotachysterol, but its antirachitic potency is only 0.25% of that of calciferol; for this reason it is not used in rickets; it is, however, used for the treatment of all forms of parathyroid tetany. One of the biologically active metabolites of vitamin D is 25-hydroxycholecalciferol (25-HCC); it is called supervitamin D, and it may eventually be used for the treatment of avitaminosis D.

The recommended daily dietary allowance is 400 IU.

Although the unsaturated conjugated part of the molecule is of primary importance, activity depends also on the presence and the proper stereochemistry of C_3—OH. Epimerization of this group or its conversion to a ketone group greatly decreases, although not completely, activity of vitamins D_2 and D_3.

Hypervitaminosis D is characterized by hypercalcemia, ectopic calcification in soft tissues, nausea and vomiting, headache, and other minor adverse reactions. Other symptoms are osteoporosis, hypertension, and decreased renal function.

Vitamins D used clinically are white, odorless, crystalline powders that are insoluble in water but soluble in fatty oils and many organic solvents. The usual dosage in infants and children for treatment of rickets is 1000 to 4000 IU daily (1 mg = 4000 IU); for familial hypophosphatemia, 50,000 to 200,000 IU daily. In adults, the usual dosage for osteomalacia, 10,000 to 50,000 IU; for renal osteodystrophy, 50,000 to 500,000 IU daily.

C. Vitamin E

Vitamin E is comprised of a group of naturally occurring α-, β-, γ-, and δ-tocopherols, which are distributed widely in nature. The best dietary sources are legumes (soybean), cereal products (rice, corn), vegetable oils, liver, eggs, butter. Clinically used vitamin E is predominantly α-tocopherol, especially the (+) isomer, and the racemic mixture. It is obtained either by extraction from vegetable oils or by chemical synthesis from trimethylhydroquinone or 2,5-dimethyl-4-nitrophenol as starting material.

The recommended daily dietary allowance is 20 to 30 IU (IU = 1 mg).

Based on studies in animals, hypovitaminosis E occurs rarely, and only vitamin E is effective for the treatment or prevention of this condition. The usual dosage is four to five times the recommended dietary allowance.

Hypervitaminosis E, which is observed only with large doses in animals, causes certain reversible symptoms: skeletal-muscle weakness, gastrointestinal disorder, and disturbances of reproductive functions.

Vitamin E

It occurs as light-yellow, viscous, odorless oils that are insoluble in water but soluble in organic solvents and fixed oils. These oils are a mixture of α-tocopherol (mainly (+) isomer) and their acetates or their succinates. Tocopherols are readily oxidized by air, but their acetates and benzoates, which are equally active, are more resistant to oxidation.

III. MECHANISM OF ACTION

Vitamin A plays an essential role in general metabolism. It is involved in control regulation of the transport of metabolites across cell membranes and in the maintenance of the integrity of epithelial tissue by preventing metaplasia of the stratified squamous type. It participates in the visual process by promoting resynthesis of rhodopsin, which is the light-sensitive pigment present in the retina. Details on this mechanism can be found in textbooks on biochemistry.

Vitamin D promotes the absorption of calcium from the intestine, increases renal tubular reabsorption of phosphate, and is involved in the mobilization of bone calcium and maintenance of calcium levels in the serum. These activities are closely related to that of the parathyroid hormone and of calcitonin and probably result from the mediation of metabolic products of vitamin D, especially 25-HCC and 1,25-DHCC, because vitamin D itself is inert in the synthesis of proper enzymes. There is evidence that vitamin D is involved in DNA-mediated RNA synthesis and the subsequent translation into proteins.

Vitamin E is an antioxidant, and its activity is apparently related to heme synthesis. Its antioxidant properties may be involved in the stabilization of vitamin A and unsaturated fatty acids, preventing random free-radical reactions in the body. Its primary mechanism, however, is not known.

REFERENCES

General

H. F. DeLuca and J. W. Suttie, *The Fat-Soluble Vitamins,* University of Wisconsin Press, Madison, 1970.

R. A. Morton, Ed., *Fat-Soluble Vitamins,* Pergamon, Oxford, 1970.

Nutrition Society Symposium, "Recent Developments in the Fat-Soluble Vitamins," *Federation Proc.,* **28**, 1670-1701 (1969).

Classification

O. Isler, Ed., *Carotenoids,* Wiley, New York, 1971.

D. R. Threlfall, *Vitamins Hormones,* **29**, 153 (1971).

H. F. DeLuca, *New Engl. J. Med.,* **281**, 1103 (1969).

G. M. Sanders *et al., Progr. Chem. Nat. Prod.,* **27**, 131 (1969).

M. S. Seelig, *Ann. N.Y. Acad. Sci.,* **147**, 539 (1969).

W. J. Darby, *Vitamins Hormones,* **26**, 685 (1968).

C. D. Fitch, *Vitamins Hormones,* **26**, 501 (1968).

J. A. Olson, *Vitamins Hormones,* **26**, 1 (1968).

J. A. Olson, *Pharmacol. Rev.,* **19**, 559 (1967).

R. S. Harris *et al.,* "International Symposium on Recent Advances in Research on Vitamins K and Related Quinones," *Vitamins Hormones,* **24**, 293-689 (1966).

Mechanism of Action

E. Kodicek, *Acta Vitaminol. Enzymol.,* **25**, 153 (1971).

W. Boguth, *Vitamins Hormones,* **27**, 1 (1969).

H. F. DeLuca, *Vitamins Hormones,* **25**, 315 (1967).

C. Martins, *Vitamins Hormones,* **24**, 441 (1966).

CHAPTER 37

WATER-SOLUBLE VITAMINS

I. INTRODUCTION

Water-soluble vitamins are those vitamins that in general are soluble in water but not in lipids; however, some of them are slightly soluble in certain organic solvents. These vitamins include ascorbic acid, niacin, riboflavin, thiamine, pyridoxine, pantothenic acid, folic acid, and vitamin B_{12}. The last two were studied in Chapter 23. Excessive intake of water-soluble vitamins, although economically wasteful, is not harmful to the body, because their toxicity is low, owing probably to rapid excretion of the excess amounts.

Deficiency of water-soluble vitamins produces several diseases. Thus, deficiency of ascorbic acid causes scurvy, which consists in degeneration of collagen and intercellular ground substance, resulting in disturbances of bone growth, hemorrhages of the gums and other parts of the body, loosening of the teeth, capillary fragility with consequent cutaneous hemorrhages, and other abnormalities. Deficiency of thiamine produces beriberi, which manifests itself in two main forms: (*a*) dry beriberi, whose principal symptom is polyneuropathy; (*b*) acute wet beriberi, whose predominant symptoms are edema and serous effusions. Deficiency of riboflavin causes loss of hair (alopecia), lesions of the skin, eyes, lips, mouth, and genitalia. Deficiency of pyridoxine causes seborrheic and desquamative dermatitis of the eyes and mouth, glossitis and stomatitis, intertrigo of breasts and inguinal region of women, and many other clinical changes. Deficiency of niacin produces pellagra, whose manifestations are symmetrically distributed erythematous lesions on exposed surfaces of the body, red swelling of the tongue and oral mucous membranes, and central-nervous-system and gastrointestinal disturbances. Deficiency of pantothenic acid is responsible for several discomforts, such as malaise, fatigue, headache, nausea, sleep disturbances, and abdominal cramps.

II. CLASSIFICATION

Water-soluble vitamins known thus far are those listed in Table 37.1, besides folic acid and vitamin B_{12}, which were already studied in Chapter 23. Niacin,

586

pyridoxine, riboflavin, pantothenic acid, and thiamine belong to the B complex group.

A. Ascorbic Acid

Ascorbic acid, also called vitamin C, is distributed very widely in higher plants, especially citrus fruits, tomatoes, peppers, paprikas, blackberries, and other fruits. Only primates and the guinea pig require it as a dietary factor.

The recommended daily dietary allowance for adults is 40 to 60 mg.

Since ascorbic acid is practically nontoxic, excessive intake does not cause adverse reactions. After saturation of the tissues, excess is rapidly excreted in the urine. However, administration of large doses concomitantly with sulfa drugs and PAS may cause crystalluria. Claims that large doses prevent or abort the common cold are apparently not supported by well-controlled clinical trials.

Four official forms of vitamin C are available: ascorbic acid, sodium ascorbate, ascorbic acid injection, and ascorbyl palmitate.

Ascorbic Acid

It occurs as an odorless, white or slightly yellow powder or crystals. At room temperature, dry crystals are stable on exposure to air, but moisture makes them darken gradually. The usual dosage for treatment of scurvy is 300 mg daily for at least 2 weeks.

B. Thiamine

Thiamine, also called vitamin B_1 and, formerly, aneurine, occurs in moderate quantities in egg yolks, peas, brans, rice, beans, nuts, yeast extracts, and some other vegetables. Since the isolation of thiamine in the crystalline form from these sources is too expensive, only concentrates from these sources are marketed. Pure thiamine for use in pharmaceutical preparations is obtained by synthesis. Thiamine is marketed as the hydrochloride, mononitrate (which is not hygroscopic and more stable than hydrochloride), bromide hydrobromide, and napadisilate.

The recommended daily dietary allowance for adults is 1.0 to 1.5 mg.

Some derivatives of vitamin B_1 are also marketed: (a) those having both heterocycles of thiamine: cocarboxylase (which is thiamine pyrophosphate chloride), thiamine monophosphoric ester chloride, thiamine triphosphoric ester; (b) those in which the thiadiazolium ring is opened: acetiamine (Thianeuron), bentiamine (Névriton), dicetiamine (Dicetamin), benfotiamine (Vitanévril), cyclocarbotiamine (Cometamine), prosultiamine (Proneurin).

Thiamine Hydrochloride

It occurs either as white crystals or crystalline powder, with a slight yeastlike

Table 37.1 Water-Soluble Vitamins

Official Name	Proprietary Name	Chemical Name	Structure
ascorbic acid*	Cebione Cecon Cevalin	3-oxo-L-gulofuranolactone (enolic form)	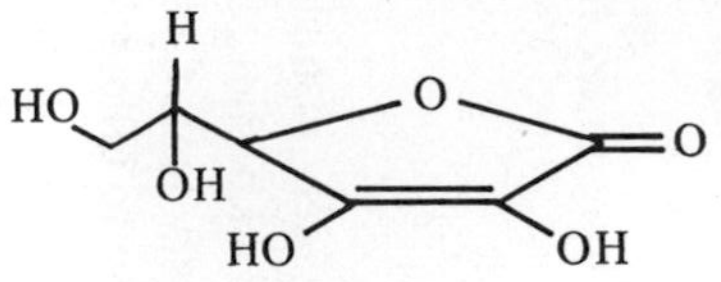
thiamine*	Betalin S Bewon Thiabev	3-[(4-amino-2-methyl-5-pyrimidinyl)-methyl]-5-(β-hydroxyethyl)-4-methylthiazolium chloride	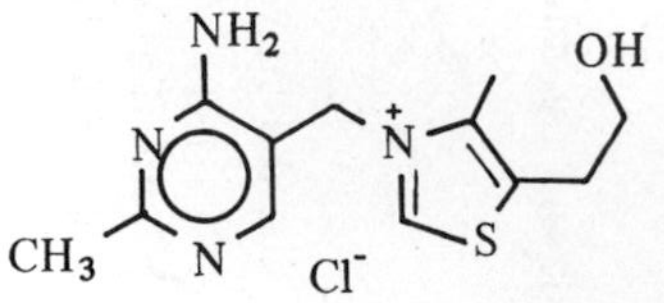
riboflavin*	Beflavin Beflavit Flavaxin	7,8-dimethyl-10-(D-ribit-1-yl) isoalloxazine	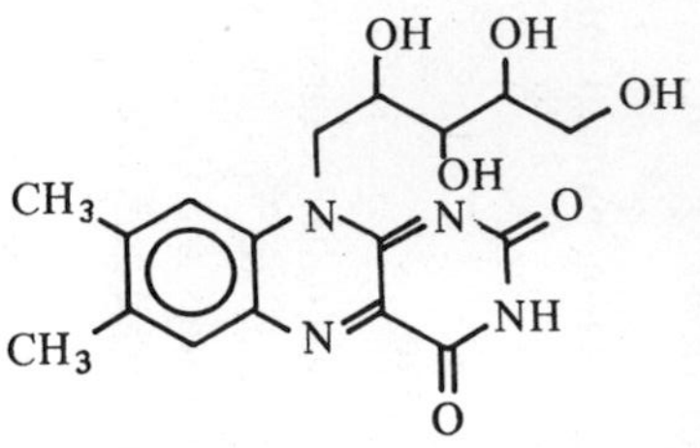
pyridoxine*	Beesix Hexa-betalin Hydoxin	3-hydroxy-4,5-di(hydroxymethyl)-2-methylpyridine	See text

niacin[+]	Nicobid Nicotene	nicotinic acid	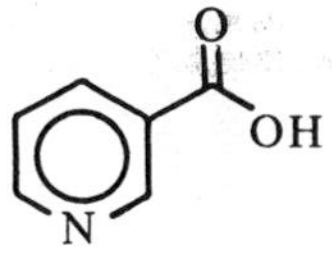
niacinamide[*]	Niacimid Nicotamide	nicotinamide	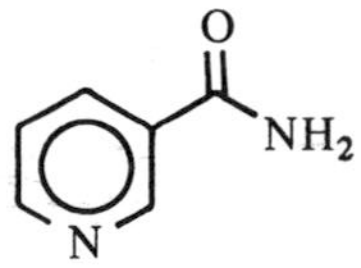
calcium pantothenate[*]	Pantholin	calcium salt of N-(2,4-dihydroxy-3,3-dimethylbutyryl)-β-alanine	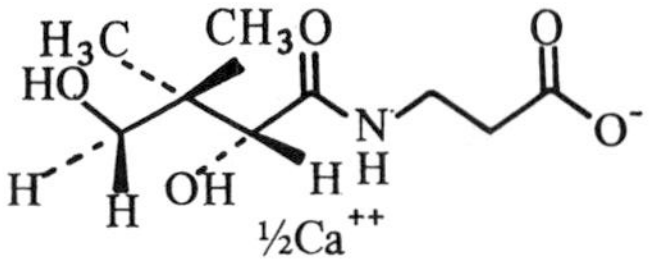

*In USP XIX (1975).
[+]In NF XIV (1975).

odor. In the dry state it is stable to heat, but when exposed to air it absorbs 4% of water and it is oxidized. For this reason it should be kept in well-closed containers and away from light. Its aqueous solutions are stable in acidic media, but at pH 5 or 6 thiamine is inactivated by chemical cleavage. With permanganate or alkaline potassium ferricyanide it yields thiochrome, a vivid blue flourescent compound: this reaction is used for colorimetric assay of thiamine. The usual dosage, by oral, intramuscular, or intravenous routes, is 5 to 10 mg three times daily.

C. Riboflavin

Riboflavin occurs widely in animal and plant foodstuffs, such as liver, milk, kidneys, beef, eggs, oysters, wheat germ, turnip greens, beet greens, rice polishings. In the intestinal mucosa riboflavin is transformed into riboflavin mononucleotide (FNM), which in the liver is converted into flavin adenine dinucleotide (FAD).

The recommended daily dietary allowance for adults is 1.3 to 1.7 mg.

In the presence of oxygen, riboflavin is irreversibly transformed by light to lumiflavin, lumichrome, and minor compounds.

Three different crystalline forms of riboflavin are known, each one having a different solubility in water, varying from 1 part in 3000 to 1 part in 15,000. This low solubility in water made it necessary to search for solubilizing agents; one of the most widely used is sodium 3-hydroxynaphthoate.

Several derivatives of riboflavin have been prepared, but most of them are not as active as the natural vitamin B_2. However, those with equal activity can be divided into two classes: (a) water-soluble derivatives: sodium riboflavin phosphate (Hyryl) and methylolriboflavin (Hyflavin); (b) water-insoluble derivatives, designed to be long-acting: riboflavin O-tetrabutyrate (Hibon), marketed in Japan. In the United States, vitamin B_2 is marketed as riboflavin and methylol riboflavin (Hyflavin), which is a mixture of methylol derivatives of riboflavin in weakly alkaline solution.

Riboflavin

It occurs as a yellow to orange-yellow crystalline powder, with a bitter taste and a slight odor. In the dry state it is stable to diffused light; however, in solution, especially alkaline, it undergoes decomposition. The usual dosage is 5 to 10 mg daily.

D. Pyridoxine

Pyridoxine, also called vitamin B_6, occurs widely in nature. The dietary sources are liver, cereal brans, yeast, crude cane molasses, wheat germ. It is a mixture of pyridoxine, pyridoxal, and pyridoxamine, which are interconverted in the body:

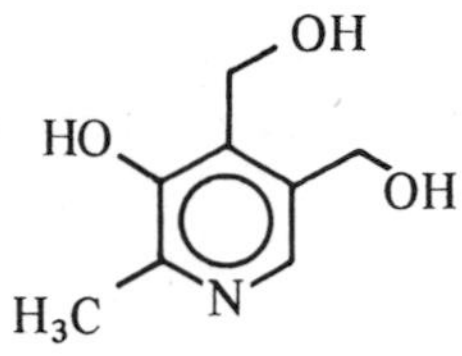

pyridoxine 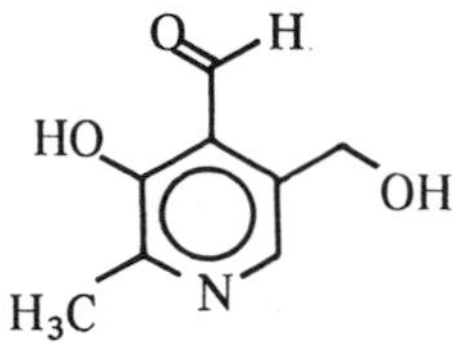   pyridoxal 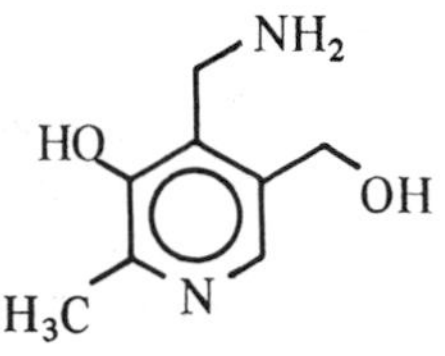   pyridoxamine

The most stable and the most common analogue used in pharmaceutical preparations is pyridoxine, which may occur as a dimer.

The recommended daily dietary allowance for adults is 1.4 to 2.0 mg.

Pyridoxine is used in the form of salts, such as hydrochloride (Hexavibex), ascorbate (Pyridoscorbine), aspartate (Aspardoxin), pyridofylline (Theodoxine).

Several derivatives and analogues of pyridoxine are available: (a) pyridoxine iodomethylate, used by oral or parenteral routes; (b) pyritinol (Bonifen, Encephabol), devoid of vitaminic activity but with neurotrophic activity; (c) 3-lauroyl-4,5-diacetylpyridoxine (Epixine), indicated for treatment of dermatoses.

Pyridoxine Hydrochloride

It occurs as a white, odorless, crystalline compound, which is relatively stable to air and light. Its aqueous solutions are stable at pH below 5 but unstable when irradiated at pH 6.8 or above. The usual dosage for adults is 20 to 200 mg daily.

E. Niacin

Niacin and niacinamide occur in meat of several species, yeast, some fruits, and vegetables. The dietary sources are liver, yeast, meat, and legumes.

The recommended daily dietary allowance for adults is 13 to 20 mg equivalents, which include dietary sources of the vitamin itself plus 1 mg equivalent for each 60 mg of dietary tryptophan.

Excessive dosage causes transient flushing of the face and neck, urticaria, rash, and gastrointestinal disturbances.

Magnesium, aluminum, betaine, papaverine, xanthinol, and other salts of niacin have been prepared and are marketed. A derivative of nicotinamide is nicastubine, which is nicotinamide L-ascorbate (Ascorbamide).

Nicotinic acid is used also as a hypocholesterolemic agent (see Chapter 22).

Niacin

It occurs as odorless, or almost so, white crystals or powder with a slight acidic taste. Under normal conditions it is stable, but it should be kept in well-closed containers away from light. The usual dosage, by oral route, is 50 mg 3 to 10 times daily; by injection, 25 mg 2 or more times daily.

Niacinamide

It is an odorless, or nearly so, white, crystalline powder with a bitter taste. The usual therapeutic dosage is 50 mg 3 to 10 times daily.

F. Pantothenic Acid

Pantothenic acid is widely distributed in nature. Liver, dairy products, eggs, yeast, cereal brans, leafy vegetables, and crude cane molasses are the richest sources. It occurs as an unstable viscous oil, which is extremely hygroscopic. For this reason, in pharmaceutical preparations its calcium salt is used, which is moderately hygroscopic and stable to air and light.

Besides calcium pantothenate and racemic calcium pantothenate, some other derivatives of pantothenic acid are marketed: panthenol, dexpanthenol (Cozyme), and menthyl pantothenate.

A recommended dietary allowance has not been established, but a daily intake of 5 to 10 mg is advisable.

Calcium Pantothenate

It is a white, odorless, slightly hygroscopic, bitter-tasting powder, that is stable in air. It is included in multiple-vitamin preparations, since its use alone is not recommended.

III. MECHANISM OF ACTION

Ascorbic acid owes its action to the important oxidation-reduction processes in which it is involved in the cell, such as (*a*) oxidation of phenylalanine and tyrosine via *p*-hydroxyphenylpyruvate; (*b*) hydroxylation of aromatic compounds; (*c*) conversion of folic acid to folinic acid; (*d*) regulation of the respiratory cycle in mitochondria and microsomes; (*e*) hydrolysis of alkyl monothioglycosides; (*f*) development of odontoblasts and other specialized cells, including collagen and cartilage; and (*g*) maintenance of the mechanical strength of blood vessels, particularly the venules.

Thiamine, in the form of pyrophosphate, functions as a coenzyme in carbohydrate metabolism for three types of enzymatic reactions: nonoxidative decarboxylation of α-keto acids, oxidative decarboxylation of α-keto acids, and formation of α-ketols (acyloins).

In the form of flavin mononucleotide (FMN) and flavin dinucleotide (FAD), riboflavin acts as a coenzyme or prosthetic group of the flavoproteins. Such enzymes are widely distributed in nature. They play an important role in biological oxidations, functioning as dehydrogenation catalysts. Substrates that are dehydrogenated by flavoproteins include saturated compounds, pyridine nucleotides, mono- and dithiols, α-amino acids, α-hydroxy acids, and aldehydes. In the oxidation process the isoalloxazine ring system of FMN or FAD is reduced:

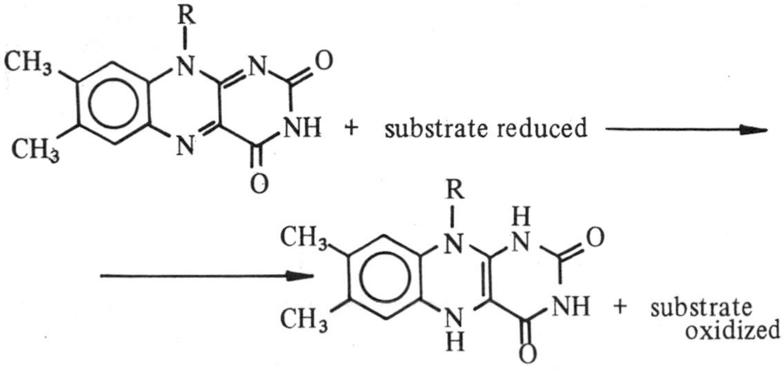

+ substrate reduced ⟶

⟶ + substrate oxidized

The oxidized form of the coenzyme is regenerated in the course of reactions in which the reduced flavoproteins function as substrates for other electron acceptors.

In the form of pyridoxal phosphate and pyridoxamine phosphate, vitamin B$_6$ acts as a coenzyme in a great number of different enzymatic reactions, mostly those involved in transformations of amino acids, such as racemization, transamination, dehydration, decarboxylation, and elimination. The mechanism of the amino acid transformation catalyzed by pyridoxal occurs via a Schiff base:

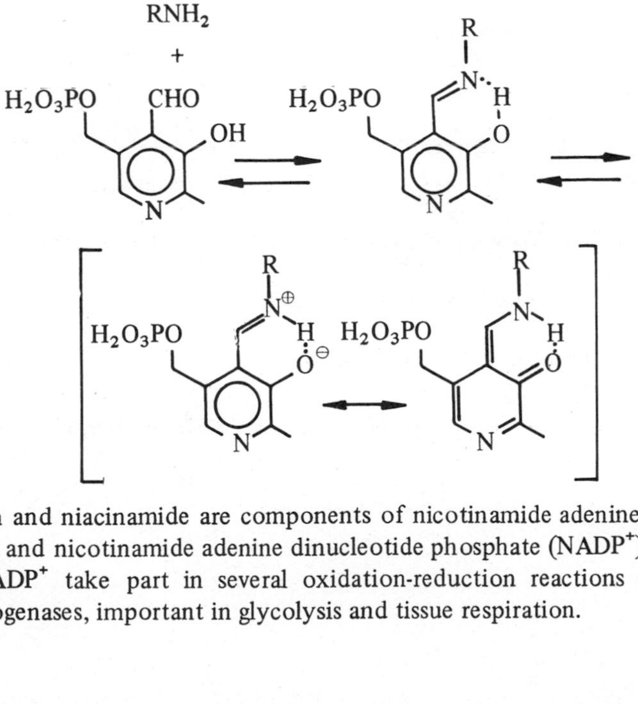

Niacin and niacinamide are components of nicotinamide adenine dinucleotide (NAD$^+$) and nicotinamide adenine dinucleotide phosphate (NADP$^+$). Both NAD$^+$ and NADP$^+$ take part in several oxidation-reduction reactions catalyzed by dehydrogenases, important in glycolysis and tissue respiration.

Pantothenic acid is the prosthetic group of coenzyme A, the most prominent acyl group transfer agent. Since sulfhydryl is the terminal and reactive group of this coenzyme, coenzyme A is abreviated to CoASH. It takes part in several reactions, which may involve (a) acyl group interchange; (b) nucleophilic attack at the acyl carbon atom; (c) condensation at the α-carbon of the acyl-SCoA; (d) additions to unsaturated acyl derivatives of CoASH.

REFERENCES

Classification

C. G. King and J. J. Burns, Eds., "Second conference on vitamin C," *Ann. N. Y. Acad. Sci.,* **258**, 1-552 (1975).

J. L. Napoli, *Annu. Rep. Med. Chem.,* **10**, 295 (1975).

H. Kamin, Ed., *Flavins and Flavoproteins,* University Park Press, Baltimore, 1971.

E. E. Harris *et al., J. Org. Chem.,* **34**, 1993 (1969).

M. A. Kelsall, Ed., "Vitamin B_6 in the Metabolism of the Nervous System," *Ann. N.Y. Acad. Sci.,* **166**, 1-364 (1969).

K. Yagi, Ed., *Flavins and Flavoproteins,* University Park Press, Baltimore, 1969.

E. V. Cox, *Vitamins Hormones,* **26**, 635 (1968).

D. L. Horrigan and J. W. Harris, *Vitamins Hormones,* **26**, 549 (1968).

P. György *et al.,* "International Symposium on Vitamin B_6," *Vitamins Hormones,* **22**, 359-855 (1964).

P. Holtz and D. Palm, *Pharmacol. Rev.,* **16**, 113-178 (1964).

Mechanism of Action

C. Woenckhaus, *Top. Curr. Chem.,* **52**, 209 (1974).

M. S. Ebadi and E. Costa, Ed., *Role of Vitamin B_6 in Neurobiology,* Raven, New York, 1972.

H. R. Mahler and E. H. Cordes, *Biological Chemistry,* 2nd ed., Harper & Row, New York, 1971.

L. O. Krampitz, *Thiamine Diphosphate and its Catalytic Functions,* Dekker, New York, 1970.

A. Kröger and M. Klingenberg, *Vitamins Hormones,* **28**, 533 (1970).

C. G. King, *Nutr. Rev.,* **26**, 33 (1968).

E. E. Snell *et al.,* Eds., *Pyridoxal Catalysis: Enzymes and Model Systems,* Wiley, New York, 1968.

PART 7

HORMONES

The term *hormone* derives from the Greek word ὁρμῶν, present participle of the verb which means urge on. It was first used in 1905 by Starling and Bayliss to denote secretin, because it stimulates the secretory activity of the pancreas. Many other substances were subsequently found with similar function.

By definition, hormones are substances secreted either in endocrine glands or in tissues and released into the bloodstream by which they are transported to other tissues where, by selectively binding to specific receptors, they exert their effects. Those secreted in endocrine glands, such as pituitary, thyroid, parathyroid, pancreas, adrenals, gonads, are called glandular hormones. Those produced in tissues receive the name of tissue hormones, for example, histamine, norepinephrine, and serotonin, already studied in preceding chapters.

Hormones belong to a wide range of chemical structures: peptides, steroids, and derivatives of aromatic amino acids and of fatty acids. Some are small organic molecules, whereas others are large polypeptides or glycoproteins.

Some glandular hormones extracted from natural sources are used in medicine, mainly in replacement therapy. Synthetic glandular hormones and their analogues are also being used clinically. These hormones, their synthetic derivatives and congeners, as well as their antagonists, are considered drugs. They are studied in the next three chapters.

The mechanism of action of hormones at the molecular level is not completely understood. However, there is evidence that they bind selectively to specific receptors located in target cells and that this interaction enables them to exert their effects.

There are two primary modes of action of hormones. Some hormones produce their action by activating the enzyme adenyl cyclase in the target cells. Activation of this enzyme by such a hormone, the first messenger, results in the formation of cyclic AMP, the second messenger (Figure 22.2). Examples of these hormones are adrenocorticotropic hormone, insulin, luteinizing hormone, melanocyte-stimulating hormone, oxytocin, parathyroid, thyroid-stimulating hormone. Other hormones induce enzyme synthesis by acting specifically on genetic material (Figure 40.1). Hormones acting by this mechanism are adrenocorticotropic hormone, androgens, estrogens, growth hormone, glucocorticoids, insulin, thyroid hormones, thyroid-stimulating hormone.

REFERENCES

H. Breuer, D. Hamel and H. L. Krüskemper, Eds., *Methods of Hormone Analysis,* Wiley Biomedical, New York, 1976.

A. Labhart, Ed., *Clinical Endocrinology,* Springer, Berlin, 1974.

S. Levine, Ed., *Hormones and Behavior,* Academic, New York, 1972.

R. Schauer, *Angew. Chem., Int. Ed. Engl.,* **11**, 7 (1972).

G. Litwack, Ed., *Biochemical Actions of Hormones,* 2 vols., Academic, New York, 1970, 1972.

H. Rasmussen, Ed., *Pharmacology of the Endocrine System and Related Drugs,* Pergamon, Oxford, 1970.

M. Tausk, *Pharmakologie der Hormone,* Thieme, Stuttgart, 1970.

G. M. Tomkins and D. W. Martin, Jr., *Annu. Rev. Genetics,* **4**, 91 (1970).

P. Karlson, *Humangenetik,* **6**, 99 (1968).

P. Karlson, Ed., *Mechanisms of Hormone Action,* Academic, New York, 1965.

G. Litwack and D. Kritchevsky, Eds., *Action of Hormones on Molecular Processes,* Wiley, New York, 1964.

G. Pincus and K. V. Thimann, Eds., *The Hormones,* 5 vols., Academic, New York, 1948-1964.

CHAPTER 38

HORMONES OF THE PITUITARY, THYROID, PARATHYROID, AND PANCREAS

I. INTRODUCTION

For convenience, these hormones are studied separately. Related compounds as well as antagonists of hormones are also included.

II. PITUITARY AND HYPOTHALAMIC HORMONES

Pituitary gland secretes several polypeptide hormones that play important physiological roles. Synthesis and secretion of these hormones are regulated by hypothalamic hormones. Both pituitary and hypothalamic hormones, as well as certain synthetic analogues, are being used in medicine, some for treatment of diseases and others as diagnostic agents only.

A. Classification

According to their origin, hormones studied in this section are classified into three groups: posterior pituitary hormones and related substances, anterior pituitary hormones, and hypothalamic hormones.

1. *Posterior Pituitary Hormones and Related Substances*

Generally known as neurohypophyseal hormones, posterior pituitary hormones are secreted by the hypothalamus but are stored in the posterior lobes of the mammalian pituitary gland, from which they are extracted. Those of clinical interest are oxytocin, arginine-vasopressin (from most mammals), and lysin-vasopressin (from pigs). Another analogue, arginine-vasotocin, is found in non-mammalian vertebrates. They are closely related cyclic octapeptides (Figure 38.1).

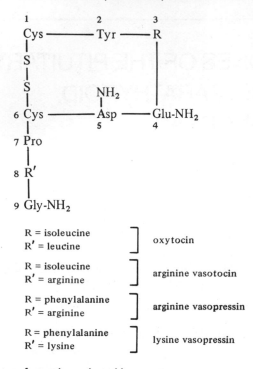

Figure 38.1   Structures of neurohypophyseal hormones.

Besides natural preparations, synthetic products are available. These are preferable, because they are pure, whereas natural products always contain contaminants. Several synthetic analogues of oxytocin and vasopressin are known and some have found limited application. Analogues of oxytocin are the following: 1-deamino-oxytocin, α-hydroxy-oxytocin, 4-leucine oxytocin, methyl-oxytocin, 4-threonine-mesotocin, 4-threonine oxytocin. Analogues of vasopressin include 1-deaminovasopressin, felypressin, and lypressin.

In therapy the following preparations are most commonly used: oxytocin (USP) (Pitocin, Syntocinon), deaminooxytocin, posterior pituitary, lypressin (Diapid), and vasopressin injection (USP) (Pitressin). The first is oxytocic, and the remaining three are antidiuretics used in neurohypophyseal diabetes insipidus, but they are not effective in nephrogenic diabetes insipidus.

2. Anterior Pituitary Hormones

Hormones secreted by the anterior pituitary stimulate various target organs or tissues. They are polypeptides with high molecular weight. Their synthesis and secretion are regulated by two groups of hormones from the thalamus: hypothalamic-releasing and hypothalamic-inhibiting hormones. The following hormones from the anterior lobe of the pituitary gland have been isolated and identified and some of them have been synthesized:

1. Adrenocorticotropin (ACTH)—it regulates secretion of adrenal cortico-sterois, mainly cortisol.

2. Follicle-stimulating hormone (FSH)—it promotes maturation of the graafian follicle and regulates secretion of estrogen in women.

3. Growth hormone (GH), also called somatotropin—it stimulates growth and regulates activities of many tissues.

4. Luteinizing hormone (LH)—it stimulates ovulation and regulates progester-one secretion in women and testosterone secretion in man.

5. Melanocyte-stimulating hormone (MSH)—it is responsible for pigmentary alterations.

6. Prolactin—it promotes secretion of milk.

7. Thyroid-stimulating hormone (TSH), also called thyrotropin—it regulates secretion of thyroid hormones.

Synthetic analogues of these hormones have been prepared, and some have found clinical application. Only a few anterior pituitary hormones or synthetic analogues are marketed: corticotropin (Acthar, Cortrophin), cosyntropin (Cortrosyn), thyrotropin (Thytropar), chorionic gonatropin (Pregnyl), menotropins (Pergonal).

Corticotropin

It is an open-chain polypeptide made up of 39 amino acids. Its biological activity depends on the first 24 amino acids, whose sequence is identical in ACTH which is extracted from man, pigs, sheep, and cattle. It is useful in the management of severe myasthenia gravis. Since it is destroyed by proteolytic enzymes, it is administered only by intramuscular, intravenous, or subcutaneous routes.

3. Hypothalamic Hormones

Release or inhibition of pituitary hormones is controlled by hypothalamic hormones. Some of these hormones or factors are already known. They are usually polypeptides. Synthesis of a number of them, as well as of their many analogues, has been achieved.

Nine substances from the hypothalamus are already known to control secretion of pituitary hormones: corticotropin-releasing hormone (CRH or CRF), follicle-stimulating-hormone-releasing hormone (FSH-RH or FSH-RF), growth-hormone-releasing-inhibiting hormone (GH-RIH or GIF), growth-hormone-releasing hormone (GH-RH or GH-RF), luteinizing-hormone-releasing hormone (LH-RH or LH-FR), melanocyte-stimulating-hormone-release-inhibiting hormone (MRIH or MIF), melanocyte-stimulating-hormone-releasing hormone (MRH or

MRF), prolactin-release-inhibiting hormone (PRIH or PIF), prolactin-releasing hormone (PRH or PRF), thyrotropin-releasing hormone (TRH or TRF).

No pharmaceutical preparations of these hormones or factors are presently available. However, they or their synthetic derivatives may find important therapeutic applications in the near future. Actually, lopremone, a TRH synthetic prohormone, is already marketed in England.

B. Mechanism of Action

Oxytocin causes strong contractions of the uterus by stimulation of adenyl cyclase.

Vasopressin and similar substances increase reabsorption of water in the distal tubules. Several hypotheses have been presented in explanation of the mechanism of this antidiuresis: (a) stimulation of the release of hyaluronidase, which would decrease the density and increase the permeability of intercellular substance and basement membrane; (b) activation of adenyl cyclase and consequently increase of cyclic 3',5'-AMP production, which exerts an antidiuretic action.

Corticotropin produces many different effects by various mechanisms. It stimulates steroidogenesis by interfering with the translation of stable mRNA. Stimulation of adenyl cyclase to synthesize cyclic AMP is achieved by an extracellular action. It also controls DNA synthesis by stimulating nuclear DNA polymerase and thymidine kinase.

III. THYROID HORMONES AND ANTITHYROID DRUGS

Thyroid hormones include active principles of the thyroid gland as well as their synthetic preparations. They are used mainly for treatment of hypothyroidism or myxedema and simple nonendemic goiter. However, they are also useful in the diagnosis of hyperthyroidism and in cases of thyrotropin-dependent carcinoma of the thyroid and in chronic fibrous or lymphocytic thyroiditis. Thyroid hormones are a lifelong replacement therapy for hypothyroid patients, but usually they do not cause adverse reactions.

Antithyroid drugs are substances that inhibit synthesis or release of thyroid hormones or that interfere with their metabolic actions at a cellular level. Some physical agents, such as external radiation and internal isotopic radiation, can also produce similar effects, and for this reason they are regarded by some authors as antithyroid agents. All these drugs or agents are used for treatment of hyperthyroidism. The alternative is surgery, for which patients are usually prepared by administration of antithyroid drugs.

Iodine, which is frequently used, causes several adverse gastrointestinal effects.

Propylthiouracil, the prototype of antithyroid drugs, may rarely cause agranulocytosis, rash, and minor discomforts.

A. Classification

1. Thyroid Hormones

Pharmaceutical preparations containing thyroid hormones most widely used are thyroid, thyroglobulin (Proloid), levothyroxine sodium, liothyronine sodium, and liotrix (Euthroid, Thyrolar) (Table 38.1). The first two are extracts of thyroid gland: thyroid is obtained from domesticated animals used for food: thyroglobulin is a purified extract of frozen hog thyroid. Levothyroxine and liothyronine were once isolated from thyroid gland but they now are obtained by synthesis. Liotrix is a mixture (4:1) of levothyroxine and liothyronine, as sodium salts.

Several analogues, derivatives, and isomers of thyroid hormones are available; for instance, acetiromate, ratironine, and DL-triiodothyronine.

2. Antithyroid Drugs

Antithyroid drugs belong to the following chemical groups:

1. Thioamides: carbimazole (Neomercazole), iothiouracil (Itrumil), methimazole, methylthiouracil, propylthiouracil, thiobarbital, thiouracil, thiourea
2. Iodine and its salts: iodine, sodium iodide, potassium iodide
3. Aromatic amines: p-aminobenzoic acid, p-aminosalicylic acid, amphenone B, carbutamide, sulfadiazine, sulfaguanidine
4. Phenols: 2,6-dihydroxybenzoic acid, phloroglucinol, resorcinol, salicyclic acid
5. Inorganic compounds: potassium perchlorate, potassium thiocyanate
6. Miscellaneous   compounds: aminothiazole,   aminotriazole,   calcium cyanamide, goitrin, tricyanoaminopropene

Antithyroid drugs most widely used are potassium iodide, sodium iodide, strong iodine solution, propylthiouracil, methimazole, methylthiouracil, potassium perchlorate, and sodium iodide I^{131} (Table 38.1).

Propylthiouracil

It is the prototype of antithyroid drugs. It occurs as a slightly water-soluble, white, crystalline powder with a bitter taste. Since its action is short-lived, it must be administered frequently. The usual dosage is 200 to 300 mg daily in divided doses every 8 hours.

Table 38.1 Thyroid Hormones and Antithyroid Drugs

Official Name	Proprietary Name	Chemical Name	Structure
thyroid*			
levothyroxine sodium*	Letter Synthroid	sodium L-3-[4-(4-hydroxy-3,5-diiodophenoxy)-3,5-diiodo-phenyl]alanine	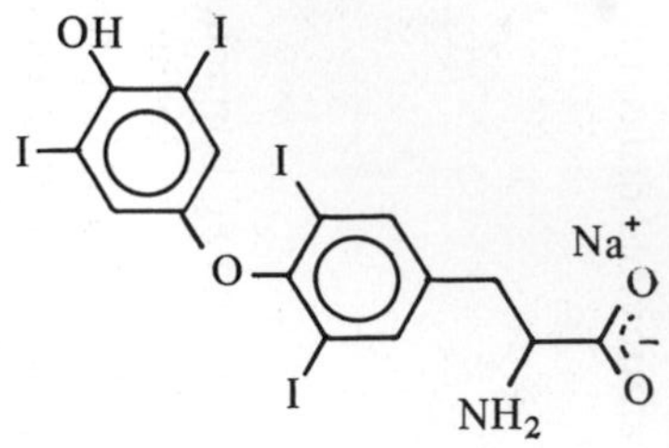
liothyronine sodium*	Cytomel	sodium L-3-[4-(4-hydroxy-3-iodophenoxy)-3,5-diiodophenyl]-alanine	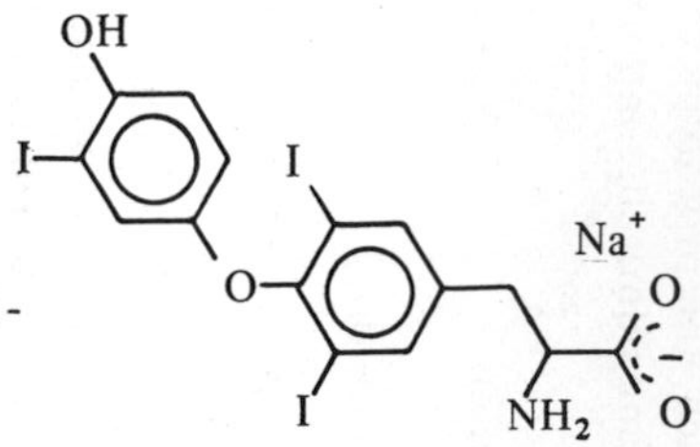
liotrix*	Euthroid Thyrolar	4:1 mixture of levothyroxine and liothyronine	

methylthiouracil[+]	Methiacil Thimecil	6-methyl-2-thiouracil	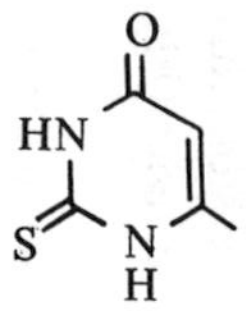
propylthiouracil*	Propacil	6-propyl-2-thiouracil	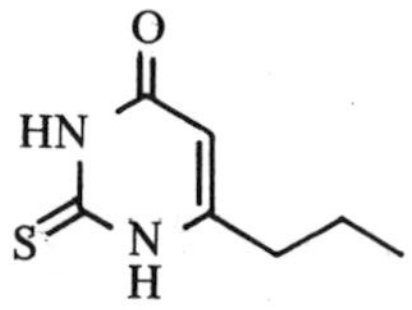
methimazole*	Tapazole	1-methylimidazole-2-thiol	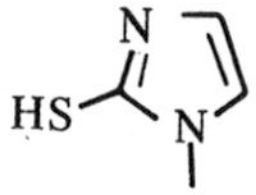

*In *USP XIX* (1975).
+ In *NF XIV* (1975).

B. Mechanism of Action

Thyroid hormones regulate several biological processes, such as respiration (by induction or a number of special enzymes, for example, succinate dehydrogenase and cytochrome oxidase), cell growth and differentiation, and changes in permeability of mitochondrial membranes. The diverse biological effects they produce result from a primary effect on the control of gene expression, that is, from induction of enzyme synthesis through activation of specific genes (Figure 40.1).

Antithyroid compounds act by three main mechanisms, as pointed out by Selenkov and Wool: (a) competitive antagonism with thyroid hormones or interference with associated enzymatic or energy mechanisms at peripheral tissue sites: 3,3'-diiodothyronine and 3,3',5'-triiodothyronine are competitive antagonists; reserpine and guanethidine reduce partially the peripheral sympathomimetic effects in hyperthyroidism; (b) alteration of plasma or membrane transport of thyroid hormones: salicylates, dinitrophenol, hydantoins, androgens, anabolic agents, estrogens, oral contraceptives; (c) inhibition of synthesis or release of thyroid hormones through intrathyroidal action. Intrathyroidal inhibitors, which are called goitrogens, can be divided into the following groups: (1) trapping inhibitors: fluoroborate, hypochlorite, nitrate, perchlorate, periodate, thiocyanate; (2) inorganic iodination inhibitors: iodides; (3) organic iodination inhibitors: aromatic amines, thioamides; (4) releasing inhibitors: iodides.

Radioactive iodine accumulates in the thyroid gland, as does nonradioactive iodine. The former's action is ascribed to ionizing beta radiation that in a short time destroys functional and regenerative capacities of the cells of the thyroid gland. How nonradioactive iodine acts is still a mystery.

IV. PARATHYROID HORMONE AND CALCITONIN

Parathyroid hormone is one of the hormones secreted by the parathyroid gland. Its physiological role is to increase blood concentration of calcium by stimulating adenyl cyclase. It is a polypeptide. Bovine parathyroid hormone contains 32 amino acids.

Parathyroid injection was once used for temporary control of tetany in acute hypoparathyroidism. Its use is no longer justified, because its biological activity is uncertain, since it is a preparation of animal origin.

Calcitonin is a hypocalcemic hormone; that is, it lowers calcium concentration. The source of this hormone is the thyroid gland. The name thyrocalcitonin was proposed to distinguish it from a possible parathyroid hormone. It was first described by Copp in 1961. Calcitonin is a polypeptide containing 32 amino acids. Its synthesis has been performed. Calcitonin's main physiological role is to regulate the concentration of calcium in bone. It acts by preventing loss of calcium from bone. Thyrocalcitonin has been used in the treatment of acute hypercalcemia, and it is being tested in Paget's disease and in osteoporosis.

V. HORMONE OF THE PANCREAS AND OTHER HYPOGLYCEMIC AGENTS

The hormone secreted by pancreas is insulin. This hormone, as well as certain insulin substitutes, is classified as a hypoglycemic agent.

Hypoglycemic agents are drugs used in the treatment of diabetes mellitus, a hereditary disorder characterized by a relative or absolute deficiency of insulin. It affects several million people in the United States. Insulin deficiency causes excessive mobilization of protein and lipid stores, a depressed ability for carbohydrate assimilation, and a depressed capacity to limit hepatic glucose production. One in every four persons has diabetogenic genes, and diabetes is increasing three times as rapidly as the population in general.

There are two clinical varieties of diabetes: (a) juvenile, or growth-onset, diabetes, which is ketoacidosis prone, and (b) maturity, or adult-onset, diabetes, which is ketoacidosis resistant. The first variety appears prior to age 20; keto-acidosis is prevented by insulin therapy. The second one appears after the age of 40; the absence of exogenous insulin does not result in ketoacidosis, since dietary regulation alone may suffice to eliminate hyperglycemia and glycosuria.

Diabetes is controlled by (a) diet alone in 22% of cases; (b) insulin therapy in 33% of cases; (c) oral hypoglycemic agents in 45% of cases. Maturity-onset diabetes should be controlled first with dietary regulation and, in case of failure, with insulin therapy; oral hypoglycemic agents are drugs of last resort.

Adverse effects of insulin include hypoglycemia (manifested by nervousness, hunger, warmth, sweating, headache, and other discomforts), allergic reactions, and visual disturbances. The first two are caused also by sulfonylureas; these agents are contraindicated in nondiabetic patients with renal glycosuria. Phenformin causes frequent gastrointestinal reactions.

A. Classification

Hypoglycemic agents can be divided into two classes: insulins and oral hypoglycemic agents.

1. Insulins

Insulin is a polypeptide hormone made up of two chains of amino acids linked by two intermolecular disulfide bridges. Its molecular weight is about 6000. It is synthesized and stored in the beta cells of the pancreatic islets of Langherans, from which it is released to carry out its many physiological roles, such as (a)

Table 38.2 Oral Hypoglycemic Agents

Official Name	Proprietary Name	Chemical Name	Structure
acetohexamide*	Dymelor	1-[(p-acetylphenyl)sulfonyl]-3-cyclohexylurea	
chlorpropamide*	Diabinese	1-[(p-chlorophenyl)sulfonyl]-3-propylurea	
tolbutamide*	Orinase	1-butyl-3-(p-tolylsulfonyl)urea	

tolazamide*	Tolinase	1-(hexahydro-1*H*-azepin-1-yl)-3-(*p*-tolylsulfonyl)urea 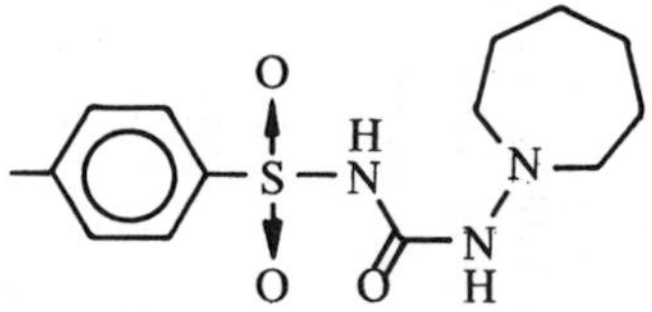
phenformin hydrochloride*	DBI Meltrol	1-phenethylbiguanide monohydro- chloride

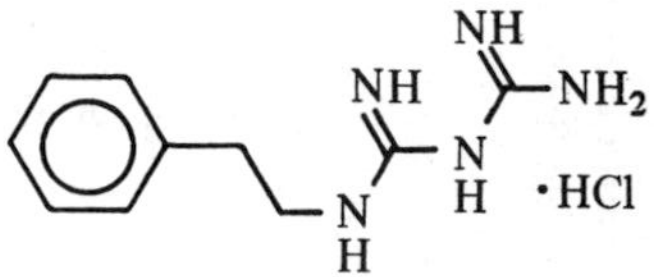

*In *USP XIX* (1975).

activation of a specific transport system to facilitate entry of glucose and certain other sugars into adipose or muscle tissue; (b) facilitation of the entry of specific amino acids into muscle; (c) increase in protein synthesis; (d) inhibition of the breakdown of neutral fat into free fatty acids and prevention of the release of these acids from adipose tissue; (e) activation of certain enzymes involved in increased glucose utilization, glycolysis, glycogenesis, and lipogenesis.

Inactivation of insulin *in vivo* occurs through the action of (a) insulinase, a proteolytic enzyme; (b) glutathione, which reduces the number of disulfide bridges; (c) an immunochemical system in the blood of insulin-treated patients.

Insulin release, which is modulated by cyclic AMP, can be induced by several agents: (a) certain analogues of amino acids: 2-aminobicycloheptane-2-carboxylic acid (BCH), guanidinoacetic acid, γ-guanidinobutyramide; (b) gut hormones: secretin, pancreozymin, and gastrin; (c) certain ions: K^+, Mg^{2+}, Ca^{2+}, Na^+, Ba^{2+}, and Li^+; (d) theophylline; (e) β-agonists; (f) cholinomimetics.

Hyperglycemia, on the other hand, is produced by the following substances, some of which act as antagonists of the effects of insulin: alloxan, corticotropin, cortisone, epinephrine, glucagon, glucocorticoids, oral contraceptives, pituitary growth hormone, thyroid hormone.

For treatment of diabetes mellitus, insulin is the drug of choice, because it controls hyperglycemia more effectively than oral hypoglycemic agents and it is safer with respect to mortality caused by cardiovascular complications. However, being a protein, it cannot be given by mouth, because in the gastrointestinal tract it is digested by proteases and other enzymes. For this reason, it is administered by injection.

Insulin is obtained from fresh beef pancreas. It occurs as an amorphous powder, although it can be crystallized. It is insoluble in water. All the official insulin preparations are stable up to about 30 or more months when stored at 5°C.

In the United States insulin is available in seven different forms. These forms are divided into the following groups, according to their rate of onset and duration of action: (a) rapid-acting—onset of action less than 1 hour and duration of action from 5 to 16 hours: insulin injection (Regular Iletin), prompt insulin zinc suspension (Semilente Iletin); (b) intermediate-acting—onset of action from 1 to 4 hours and duration of action from 12 to 24 hours: globin zinc insulin injection, isophane insulin suspension (NPH Iletin), insulin zinc suspension (Lente Iletin); (c) long-acting—onset of action from 4 to 6 hours and duration of action from 24 to 36 hours: protamine zinc insulin suspension (Protamine, Zinc & Iletin), extended insulin zinc suspension (Ultralente Iletin).

2. *Oral Hypoglycemic Agents*

Recently, doubt has been raised concerning the effectiveness of these agents. Those most widely used belong to different chemical classes: (a) sulfonylureas and related compounds and (*b*) biguanides. Besides those listed in Table 38.2, others are also available. Several other compounds are considered to have hypoglycemic activity, but most of them are not used as hypoglycemic agents because of undesired effects.

B. Mechanism of Action

Insulin causes its hypoglycemic effect by promoting carbohydrate utilization in peripheral tissues. It acts facilitating glucose entry as well as influencing its intracellular fate. At the molecular level, insulin acts by stimulating adenyl cyclase and inducing enzyme synthesis.

Sulfonylureas are believed to stimulate insulin release from the pancreatic beta-islet cells, reduce hepatic uptake of endogenously secreted insulin, and suppress directly glucagon release.

Biguanides act by blocking energy transfer at the cytochrome b site of the electron transport chain, and in this way they inhibit oxidative phosphorylation. This accounts for the metabolic effects of the drugs, such as inhibition of pyruvate and citrate oxidation, inhibition of hepatic gluconeogenesis, increased lactate and pyruvate levels, and decreased serum triglyceride and cholesterol levels.

REFERENCES

Introduction

H. S. Tager and D. F. Steiner, *Annu. Rev. Biochem.*, **43**, 509 (1974).

M. Margoulies and F. C. Greenwood, Eds., *Structure-Activity Relationships of Protein and Polypeptide Hormones*, Excerpta Medica, Amsterdam, 1971.

B. Berde, Ed., "Neurohypophyseal Hormones and Similar Polypeptides," *Handbuch der experimentellen Pharmakologie*, Vol. XXIII, Springer, Berlin, 1968.

Pituitary and Hypothalamic Hormones

J. Meienhofer, *Annu. Rep. Med. Chem.*, **10**, 202 (1974).

R. L. Holmes and J. N. Ball, *The Pituitary Gland*, Cambridge University Press, New York, 1974.

H. D. Niall *et al.*, *Recent Progr. Hormone Res.*, **29**, 387 (1973).

J. C. Porter *et al.*, *Recent Progr. Hormone Res.*, **29**, 161 (1973).

W. H. Sawyer and M. Manning, *Annu. Rev. Pharmacol.*, **13**, 5 (1973).

A. V. Schally *et al.*, *Science*, **179**, 341 (1973).

D. N. Ward *et al.*, *Recent Progr. Hormone Res.*, **29**, 533 (1973).

W. F. White, *Annu. Rep. Med. Chem.*, **8**, 204 (1973).

A. G. Frantz *et al.*, *Recent Progr. Hormone Res.*, **28**, 527 (1972).

C. Gual et al., Recent Progr. Hormone Res., 28, 173 (1972).

A. J. Kastin et al., Recent Progr. Hormone Res., 28, 201 (1972).

J. Rudinger et al., Recent Progr. Hormone Res., 28, 131 (1972).

B. B. Saxena, C. G. Beling, and H. M. Gandy, Eds., Gonadotropins, Wiley-Interscience, New York, 1972.

A. Kapoor, J. Pharm. Sci., 59, 1 (1970).

G. W. Harris and B. T. Donovan, Eds., The Pituitary Gland, 3 vols., University of California Press, Berkeley, 1966.

J. Rudinger, Ed., Oxytocin, Vasopressin and Their Structural Analogues, Pergamon, Oxford, 1964.

Thyroid Hormones and Antithyroid Drugs

V. Cody, J. Med. Chem., 18, 126 (1975).

L. J. DeGroot and J. B. Stanbury, The Thyroid and Its Diseases, 4th ed., Wiley Biomedical, New York, 1975.

N. Camerman and A. Camerman, Proc. Nat. Acad. Sci. U.S.A., 69, 2130 (1972).

P. A. Lehmann F., J. Med. Chem., 15, 404 (1972).

J. E. Dumont, Vitamins Hormones, 29, 287 (1971).

P. Liberti and J. B. Stanbury, Annu. Rev. Pharmacol., 11, 113 (1971).

K. Sterling, Recent Progr. Hormone Res., 26, 249 (1970).

R. D. Leeper, Advan. Clin. Chem., 12, 387 (1969).

H. A. Selenkov and M. S. Wool, Top. Med. Chem., 1, 241, 273 (1967).

J. T. Potts et al., Recent Progr. Hormone Res., 22, 101 (1966).

R. Pitt-Rivers and W. R. Trotter, Eds., The Thyroid Gland, Butterworth, London, 1964.

G. W. Anderson, Med. Chem., 1, 1 (1951).

Parathyroid Hormone and Calcitonin

G. D. Aurbach et al., Recent Progr. Hormone Res., 28, 353 (1972).

H. F. DeLuca, Recent Progr. Hormone Res., 27, 479 (1971).

J. T. Potts, Jr., et al., Vitamins Hormones, 29, 41 (1971).

P. L. Munson et al., Recent Progr. Hormone Res., 24, 589 (1968).

A. Tenenhouse et al., Annu. Rev. Pharmacol., 8, 319 (1968).

Hormone of the Pancreas and Other Hypoglycemic Agents

A. Y. Chang, Annu. Rep. Med. Chem., 9, 182 (1974).

P. Cuatrecasas, Annu. Rev. Biochem., 43, 169 (1974).

H. D. Breidahl et al., Drugs, 3, 79, 204 (1972).

H. Maske, Ed., "Oral wirksame Antidiabetika," Handbuch der experimentellen Pharmakologie, Vol. XXIX, Springer, Berlin, 1972.

A. M. Tomkins et al., Brit. Med. J., 1, 649 (1972).

T. L. Blundell et al., Recent Progr. Hormone Res., 27, 1 (1971).

E. Dörzbach, Ed., "Insulin I," *Handbuch der experimentellen Pharmakologie,* Vol, XXXII/1, Springer, Berlin, 1971.

A. R. Feinstein, *Clin. Pharmacol. Ther.,* **12**, 167 (1971).

P. G. Katsoyannis *et al., J. Amer. Chem. Soc.,* **93**, 5877 (1971).

E. Cerasi and R. Luft, Eds., *Pathogenesis of Diabetes Mellitus,* Wiley-Interscience, New York, 1970.

C. Villar-Palasi and J. Larner, *Annu. Rev. Biochem.,* **39**, 639 (1970).

M. J. Adams *et al., Nature (London),* **224**, 491 (1969).

G. D. Campbell, Ed., *Oral Hypoglycaemic Agents: Pharmacology and Therapeutics,* Academic, New York, 1969.

E. Samols *et al., Lancet,* **I**, 174 (1969).

R. Cherner, *Top. Med. Chem.,* **2**, 185 (1968).

T. S. Danowski, Ed., "Diabetes Mellitus and Obesity: Phenformin Hydrochloride as a Research Tool," *Ann. N. Y. Acad. Sci.,* **148**, 573-962 (1968).

F. Kurzer and E. D. Pitchfork, *Biguanides: The Chemistry of Biguanides,* Springer, Berlin, 1968.

A.-L. Loubatières, *Actual. Pharmacol.,* **21**, 174 (1968).

M. Rodbell *et al., Recent Progr. Hormone Res.,* **24**, 215 (1968).

I. G. Wool *et al., Recent Progr. Hormone Res.,* **24**, 139 (1968).

CHAPTER 39

ADRENAL CORTICAL HORMONES

I. INTRODUCTION

Adrenal cortical hormones are substances synthesized from cholesterol by the adrenal cortex, also called suprarenal cortex, whose activity is largely controlled by the adrenocorticotropic hormone (ACTH) released by the anterior pituitary. These hormones are also called corticoids, corticosteroids, adrenocorticoids, or adrenocorticosteroids.

In the adrenal cortex, three histological layers or zones can be distinguished: (a) glomerular (outer) layer—it secretes mineralocorticoids, which primarily act as regulators of electrolyte and water balance; the main representatives are aldosterone and desoxycorticosterone; (b) fascicular (middle) layer—it synthesizes glucocorticoids, which are primarily involved in carbohydrate metabolism but have also antiinflammatory, corticotropin-suppressing, and anabolic activities; the main representatives are cortisone and hydrocortisone; (c) reticular (inner) zone—it secretes sexual hormones, which are studied in the next chapter.

Drugs are known which inhibit the synthesis of adrenal steroids: aminoglutethimide, amphenone, metyrapone, mitotane, spironolactone.

Adrenal corticosteroids, as well as their synthetic derivatives, are used both in substitution therapy and in the treatment of many clinical conditions, mainly collagen diseases, inflammatory conditions, allergic states, and certain other disorders. For instance, glucocorticoids are effective in acute rheumatic fever; systemic lupus erythematosus; rhinitis; reactions to drugs, serum, and transfusions; pruritic dermatoses; pemphigus vulgaris; certain ocular diseases; hypercalcemia; thyroid ophthalmopathy; cerebral edema; neoplastic diseases; and several other disorders.

Mineralocorticoids cause enhancement of sodium retention and potassium excretion by the renal tubules of the kidneys. Glucocorticoids produce gluconeogenesis, antiinflammatory action, maintenance of muscle strength, inhibition of bone growth, and increase of calcium excretion, besides several other biological effects. However, certain glucocorticoids have mineralocorticoid activity as well.

Adrenocorticoids are administered by topical, oral, or parenteral routes. Whenever possible, topical application is preferred: it may give better results, and many adverse reactions are avoided.

Corticosteroids are dangerous if improperly used. They induce hypercorticism, causing many severe adverse reactions, such as hypokalemia, peptic ulcer, suppression of growth, osteoporosis, myopathy, suppression of corticotropin secretion, atrophy of the skin, aggravation of diabetes, impairment of host defenses against infections, behavioral and personality changes, glaucoma, acne, hirsutism, menstrual disorders, hypertension, headache, intestinal perforation, vertigo, asthenia, besides a great number of other untoward effects. For these reasons, "long term use of pharmacologic doses of systemic corticosteroids should be reserved for patients with life-threatening conditions or severe symptoms that fail to respond satisfactorily to more benign palliative measures," according to the American Medical Association.

II. CLASSIFICATION

Adrenocorticoids are steroids. Their molecules are approximately planar and inflexible. Their basic skeleton is relatively rigid. In it three stereochemical aspects are important:

1. Bonds: those that lie more or less in the plane of the ring are called *equatorial;* those that are perpendicular to this plane are called *axial.*

2. Position of substituents: Those above the plane of the page are said to be in β configuration, which is represented by a full line; those that lie under the same plane are in α configuration, which is pictured by a dotted line.

3. Conformation of cyclohexane: it can be *chair,* which is rigid and more stable, or *boat,* which is flexible and thermodinamically less stable.

All adrenocorticoids currently used are obtained by synthesis, because the demand is great and isolation from natural sources is not practical. Starting materials are mainly sterols (stigmasterol and ergosterol) and sapogenins (diosgenin, sarsasapogenin). They are derivatives of the cyclopentaneperhydrophenanthrene nucleus. The structural characteristics common to all these drugs and essential for all adrenocortical activity are 21 carbon atoms, a double bond between atoms C-4 and C-5, a ketone group at C-3, and an α-ketol group at C-20 and C-21. In mineralocorticoids an oxygen atom at C-11 is lacking. Glucocorticoids contain an oxygen atom at C-11 and an α-ketol (OH) at C-17:

corticosterone

Most adrenocorticoids are used as esters (acetate, valerate, pivalate, cypionate, tebutate, diacetate), acetonides (mono- or hexacetonide), or salts (sodium phosphate, sodium succinate).

Adrenocorticoids are white or creamy-white crystalline powders, odorless, stable in air. Most of them are insoluble in water, but some esters are water soluble. They can be divided into three classes: mineralocorticoids, glucocorticoids, and sexual hormones. In this chapter only the first two classes are studied; the third one is dealt with in Chapter 40.

A. Mineralocorticoids

The main mineralocorticoids are aldosterone, desoxycorticosterone (acetate and pivalate), and fludrocortisone acetate (Table 39.1). Aldosterone has no clinical use, but the other two are used as replacement therapy in the treatment of chronic adrenocortical insufficiency. Fludrocortisone is 20 times more potent than desoxycorticosterone.

B. Glucocorticoids

The following glucocorticoids are used in inflammatory and allergic conditions and other glucocorticoid-responsive diseases: betamethasone, cortisone acetate, dexamethasone, fluprednisolone, hydrocortisone, methylprednisolone, paramethasone, prednisolone, prednisone, triamcinolone. Others are effective in the treatment of steroid-responsive dermatoses and are "used under occlusive dressings for the management of resistant diseases, such as nummular dermatitis, chronic neurodermatitis, and psoriasis" (American Medical Association): desonide, flumethasone pivalate, fluocinolone acetonide, fluocinonide, fluorometholone, flurandrenolide.

Glucocorticoids, used primarily for substitution therapy in chronic adrenocortical insufficiency, including congenital adrenogenital syndromes, are the following: cortisone acetate, hydrocortisone, hydrocortamate hydrochloride (Ulcort).

Structures and names of certain glucocorticoids appear in Tables 8.3 and 34.5. Table 39.1 lists others. However, several other glucocorticoids are marketed: some are mentioned in Chapter 8, Section II.E; others follow: beclometasone (Becotide), clocortolone pivalate, chloroprednisone acetate (Topilan), cortivazol acetate, dichlorisone acetate (Dicloderm), fluocortholone (Topicorte), fluperolone acetate (Methral), formocortal, halcinonide (Halciderm), medrisone (Visudrisone), meprednisone (Bétalone), prednilidene (Dacortilen), tralonide. A new glucocorticoid, 25-hydroxydihydrochysterol$_3$, may become the drug of choice in the treatment of hypoparathyroidism and other similar bone diseases. A great number of glucocorticoids are being investigated: abeoprednisolone, dimesone, fluclorolone acetonide, flumethasone, hederagenin, papavallarinol, prednacinolone, sodium nimbinate.

Table 39.1 Adrenocorticoids

Official Name	Proprietary Name	Chemical Name	Structure
Mineralocorticoids			
desoxycorticosterone acetate (desoxycortone acetate)*	Percorten Cortate	21-(acetyloxy)pregn-4-ene-3,20-dione	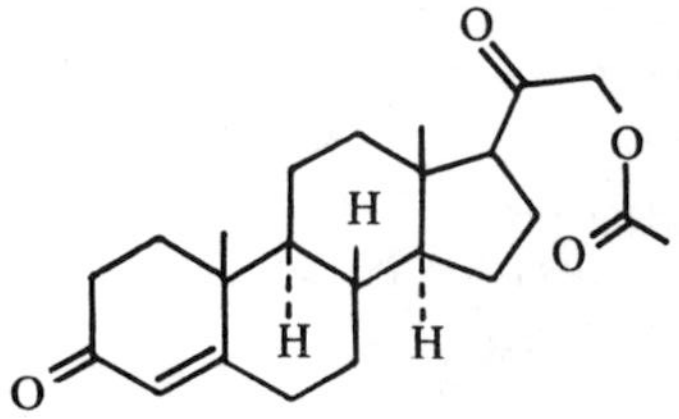
fludrocortisone acetate*	Florinef Acetate	9-fluoro-11β,17,21-trihydroxy-pregn-4-ene-3,20-dione 21-acetate	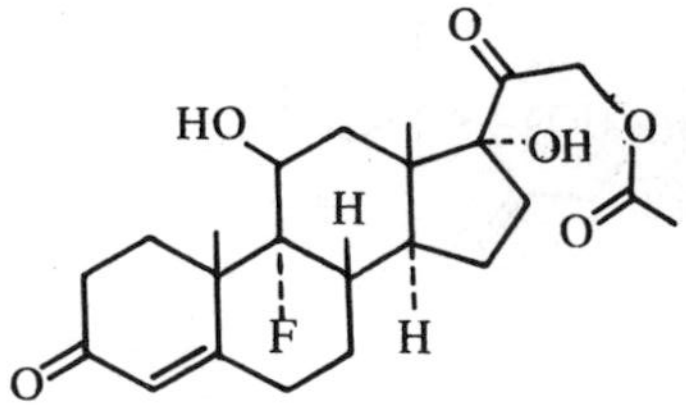

Table 39.1 Adrenocorticoids (continued)

Official Name	Proprietary Name	Chemical Name	Structure
Glucocorticoids			
betamethasone[+]	Celestone	9-fluoro-11β,17,21-trihydroxy-16β-methylpregna-1,4-diene-3,20-dione	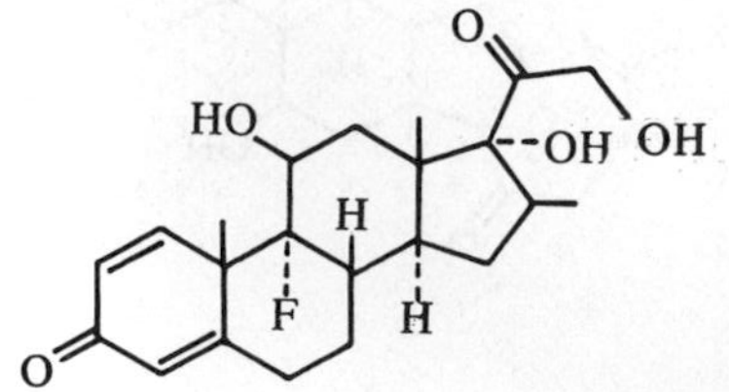
triamcinolone[+]	Aristocort Kenacort	9-fluoro-11β,16α,17,21-tetrahydroxypregna-1,4-diene-3,20-dione	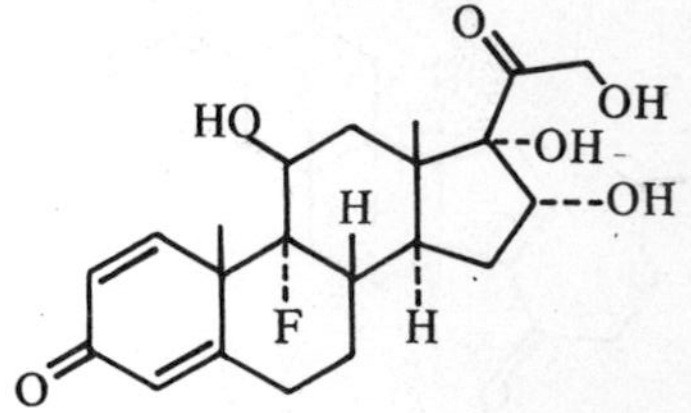
triamcinolone diacetate[+]		above as the 16,21-diacetate	
flumethasone pivalate[+]	Locorten	6α,9-difluoro-11β,17,21-trihydroxy-16α-methylpregna-1,4-diene-3,20-dione 21-pivalate	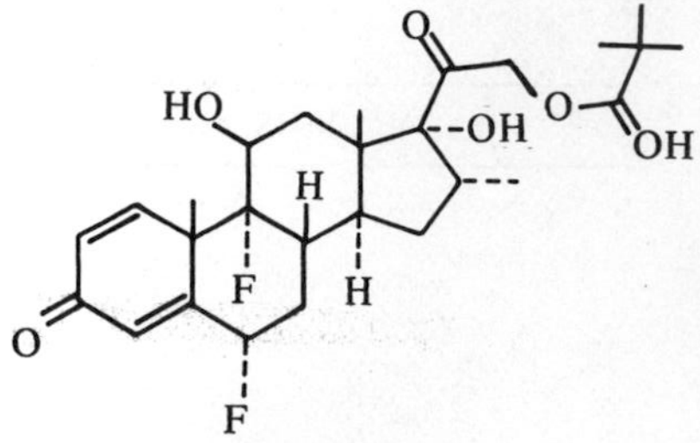

fluocinolone acetonide*

Fluonid
Synalar

6α,9-difluoro-11β,16α,17,21-
tetrahydroxypregna-1,4-diene-
3,20-dione, cyclic 16,17-acetal
with acetone

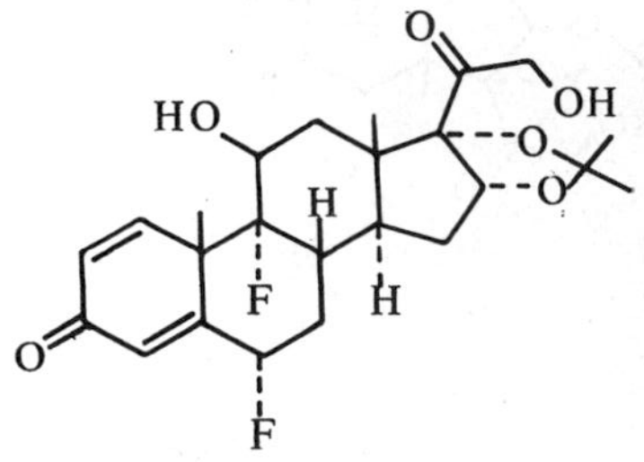

fluocinonide*

Lidex

6α,9-difluoro-11β,16α,17,21-tetra-
hydroxypregna-1,4-diene-3,20-
dione, cyclic 16,17-acetal with
acetone, 21-acetate

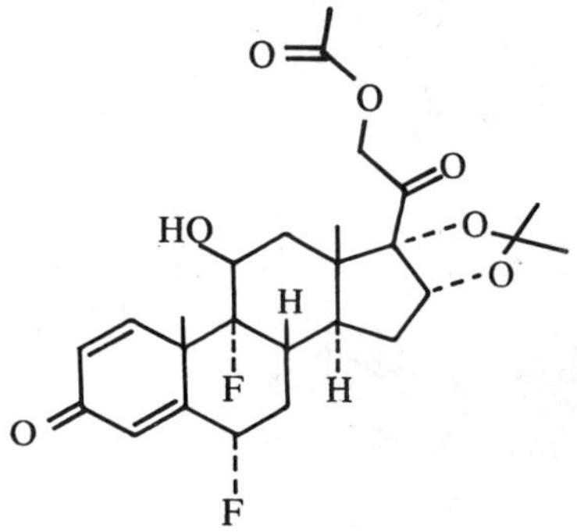

Table 39.1 Adrenocorticoids (continued)

Official Name	Proprietary Name	Chemical Name	Structure
fluorometholone[+]	Oxylone	9-fluoro-11β,17-dihydroxy-6α-methylpregna-1,4-diene-3,20-dione	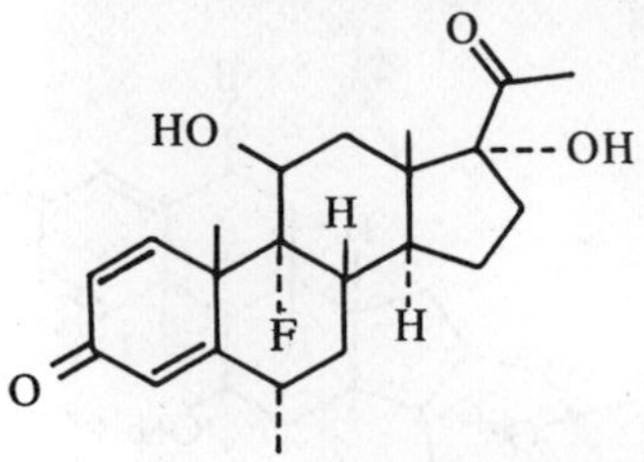
fluprednisolone	Alphadrol	6α-fluoro-11β-17,21-trihydroxy-pregna-1,4-diene-3,20-dione	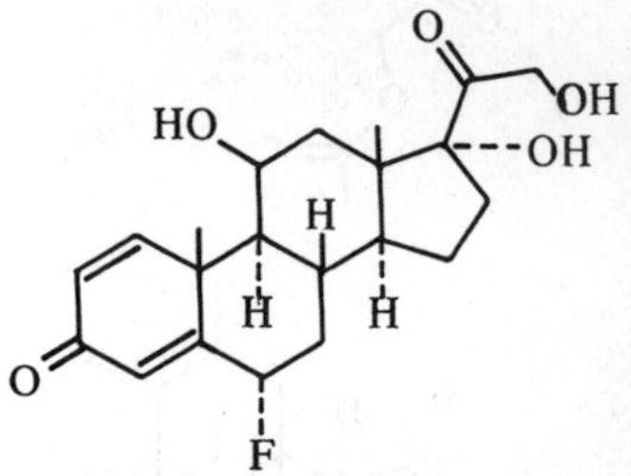
flurandrenolide* (fludroxycortide)	Cordran Drenison	6α-fluoro-11β,16α,17,21-tetra-hydroxypregn-4-ene-3,20-dione, cyclic 16,17-acetal with acetone	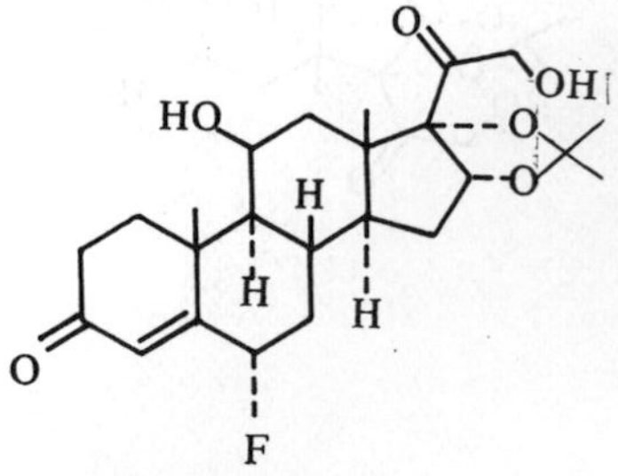

methylprednisolone*[+]	Medrol	11β,17,21-trihydroxy-6α-methyl-pregna-1,4-diene-3,20-dione 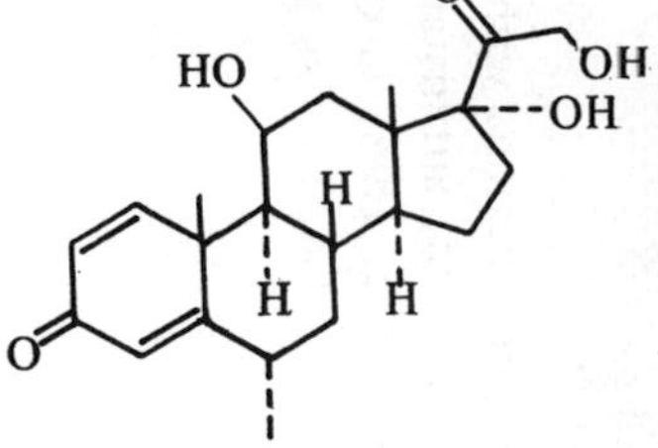
paramethasone acetate[+]	Haldrone Stemex	6α-fluoro-11β,17,21-trihydroxy-16α-methylpregna-1,4-diene-3,20 dione 21-acetate 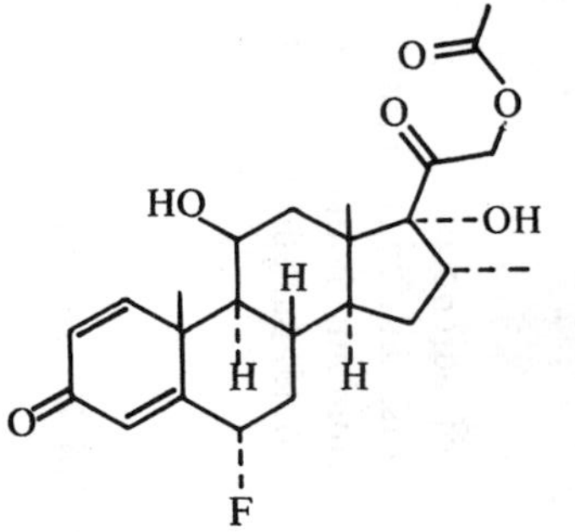

*In *USP XIX* (1975).
[+] In *NF XIV* (1975).

C. Mixtures

Several mixtures of adrenocorticoids in fixed combinations with other drugs (antibacterial agents, antifungal agents, ascorbic acid, antihistaminic agents, salicylates) are commonly marketed for systemic use. These mixtures are not recommended. Should an additional drug be needed, its dosage should be individualized.

III. MECHANISM OF ACTION

Several hypotheses have been advanced to explain the mechanism of action of adrenal cortical hormones, but only the most recent ones will be mentioned.

Aldosterone was once thought to act as permease to facilitate the entry of sodium into the epithelial cells. However, experimental evidence supports the hypothesis that this hormone is directly involved in the nuclear synthesis of RNA and ribosomal synthesis of proteins, via induction of a specific molecular receptor located in the nuclear protein fraction of the kidney. Formation of a reversible aldosterone-receptor complex triggers the sequency of events that results in mineralocorticoid effects.

As to glucocorticoids, there is evidence that they induce enzyme synthesis by the mechanism represented in Figure 40.1

REFERENCES

General

R. I. Dorfman, *Steroid Hormones,* Elsevier, New York, 1975.

M. H. Briggs and J. Brotherton, *Steroid Biochemistry and Pharmacology,* Academic, London, 1970.

N. Applezweig, *Steroid Drugs,* McGraw-Hill, New York, 1962.

I. E. Bush, *Pharmacol. Rev.,* 14, 317 (1962).

Introduction

T. Temple and G. Liddle, *Annu. Rev. Pharmacol.,* 10, 199 (1969).

J. S. Jenkins, *An Introduction to Biochemical Aspects of the Adrenal Cortex,* Arnold, London, 1968.

Classification

E. J. Ross, *Clin. Pharmacol. Ther.,* 6, 65 (1965).

J. R. Pasqualini and M. F. Jayle, Eds., *Structure and Metabolism of Corticosteroids,* Academic, London, 1964.

Mechanism of Action

R. M. S. Smellie, *The Biochemistry of Steroid Hormone Action,* Academic, London, 1971.

T. Uete and N. Shimano, *J. Biochem.,* **70**, 723 (1971).

D. M. Avioedo and R. D. Carrillo, *J. Clin. Pharmacol.,* **10**, 3 (1970).

J. H. Exton *et al., Recent Progr. Hormone Res.,* **26**, 411 (1970).

D. D. Fanestil, *Annu. Rev. Med.,* **20**, 223 (1969).

I. S. Edelman and G. M. Fimognari, *Recent Progr. Hormone Res.,* **24**, 1 (1968).

CHAPTER 40

SEX HORMONES

I. INTRODUCTION

Sex hormones and their derivatives, analogues, and antagonists may be conveniently studied if divided into three main classes: (*a*) androgens and anabolic steroids; (*b*) estrogens and progestagens; (*c*) oral contraceptives and ovulatory agents.

II. ANDROGENS AND ANABOLIC STEROIDS

Androgenic hormones are secreted mainly by the testes and, to a lesser degree, by the adrenal cortex and ovary. The most potent androgen is testosterone. It is secreted by the cells of Leydig of the testis. This secretion is controlled by the luteinizing hormone (LH) of the anterior pituitary gland. Adrenal cortex and ovary secrete very little testosterone. Normally they secrete less potent androgens, such as (+)-androstenodione and dehydroepiandrosterone, which are metabolized to testosterone.

The primary use of androgenic hormones is in development or maintenance of secondary sexual characteristics and other physiological functions. Another use of androgens is as anabolic agents, especially in the treatment of osteoporosis.

Adverse reactions which androgens can cause include virilism, hypercalcemia, premature epiphyseal closure, liver dysfunction, edema, and other disorders.

A. Classification

According to their therapeutic applications, masculine hormones may be classified as (*a*) androgens; (*b*) anabolic steroids; (*c*) antiandrogens. Most of these drugs are obtained by synthesis, although some are isolated from natural sources.

1. *Androgens*

Androgens are usually water-insoluble, white, odorless, tasteless, crystalline powders. Their esters are less soluble in water and have longer action, since they are latent forms of the parent compounds. Androgens most widely used are

Table 40.1 Adrogens

Official Name	Proprietary Name	Chemical Name	Structure
testosterone*+			see Table 34.5
fluoxymesterone*			see Table 34.5
methandrostenolone+	Metandren Neo-Hombreol (M) Oreton Methyl	17β-hydroxy-17-methyl- androsta-1,4-dien-3-one	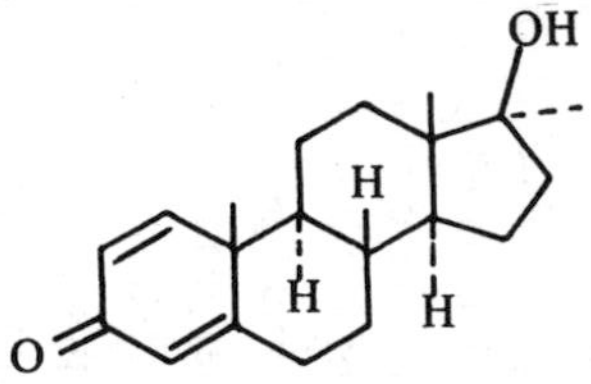

*In *USP XIX* (1975).
+In *NF XIV* (1975).

listed in Table 40.1. These drugs have also anabolic (myotropic) activity. Testosterone is inactive orally, short-acting when injected intramuscularly but very long-acting when administered by subcutaneous implantation. The other two of the table are short acting and are used only by the oral route. Testosterone is used in the free form and also as cypionate, phenpropionate, enanthate, and propionate. Other androgens of this list are used mainly in the free form.

Other steroids that have androgenic activity are mesterolone (Proviron), methenolone, norethandrolone, penmesterol.

Mixtures of androgens with estrogens are marketed and prescribed for certain conditions of menopausal and postmenopausal women. This therapy and mixtures of sexual hormones with other drugs is discouraged by the American Medical Association.

2. Anabolic Steroids

Anabolic steroids are weak androgens. Those mostly used are listed in Table 40.2. Nandrolone is used as the decanoate, laurate, phenpropionate, undecylate, and other esters. Several other steroids have anabolic (myotropic) activity, and some of them are marketed.

3. Antiandrogens

Antiandrogens are drugs that antagonize the effects of androgens. In this class estrogens and progestagens can be included, as partial pharmacological antagonists to androgens. However, substances with selective antiadrogenic effect are being developed for the treatment of benign prostatic hypertrophy, androgen-dependent acne, and prostatic cancer. The following are already being investigated: chlormadinone acetate, clanterone acetate, cyproterone acetate, dimethisterone, 17α-hydroxyprogesterone caproate, melengestrol acetate.

B. Mechanism of Action

Androgens and other sexual hormones cause male sexual differentiation and increase protein synthesis by acting on DNA of the cell nucleus and inducing the synthesis of fresh enzymes, as shown in Figure 40.1.

III. ESTROGENS AND PROGESTAGENS

Estrogens and progestagens are female sex hormones. Natural estrogens are secreted mainly by the ovary and placenta. The principal ones are estradiol, estrone, and estriol. Many synthetic estrogens are also available; some, but not all, have steroid structures.

Table 40.2 Anabolic Steroids

Official Name	Proprietary Name	Chemical Name	Structure
ethylestrenol	Maxibolin	19-nor-17-pregn-4-en-17-ol	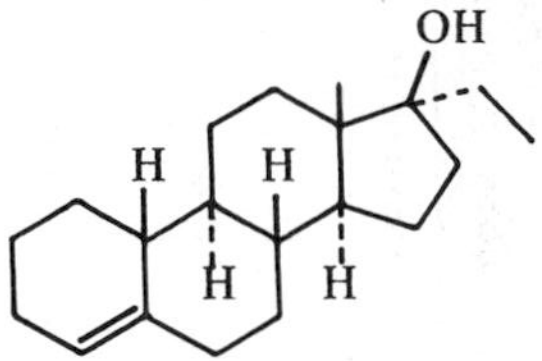
methandrostenolone[+] (metandienone)	Dianabol Primobolan	17β-hydroxy-17-methyl-androsta-1,4-dien-3-one	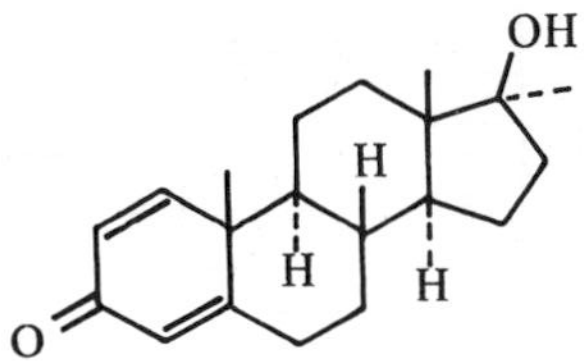
nandrolone[+]	Durabolin	17β-hydroxyestr-4-en-3-one	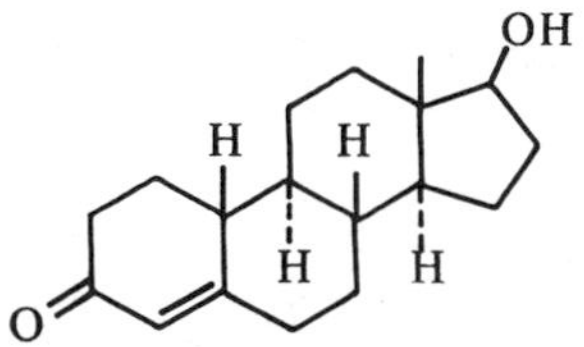

[+] In *NF XIV* (1975).

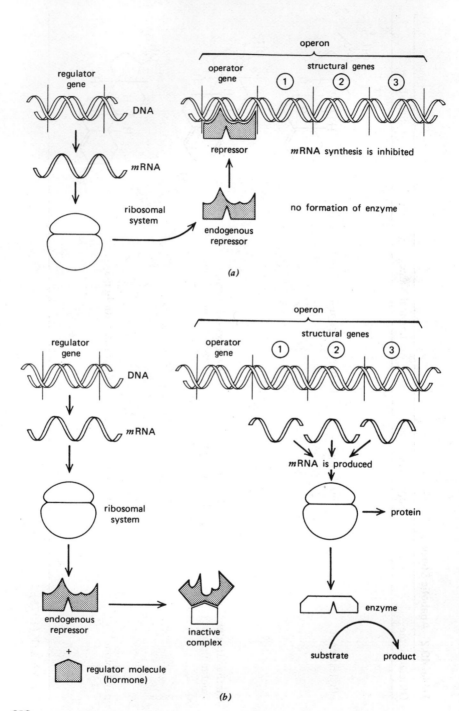

regulator gene
DNA
mRNA
ribosomal system

operon
operator gene
structural genes
1 2 3

repressor
mRNA synthesis is inhibited

endogenous repressor
no formation of enzyme

(a)

regulator gene
DNA
mRNA
ribosomal system

operon
operator gene
structural genes
1 2 3

mRNA is produced

protein

endogenous repressor

inactive complex

+

regulator molecule (hormone)

enzyme

substrate → product

(b)

Natural progestagens are secreted chiefly by the corpus luteum (during the menstrual cycle of the nonpregnant females and during early pregnancy) and placenta (after the first weeks of pregnancy). The most abundant natural progestagen is progesterone. A great number of synthetic analogues have been prepared and marketed.

Figure 40.2 shows the female sexual cycle.

The main clinical uses of natural and synthetic estrogens and progestagens are in hypogonadism and in menopausal and postmenopausal women as replacement therapy, in the treatment of certain carcinomas (see Chapter 34) and of various dysfunctions of the female reproductive system, and as oral contraceptive agents.

Adverse reactions caused by these drugs are malaise, nausea, depression, weight gain, irritability, gymnecomastia, thrombophlebitis, thromboembolism.

A. Classification

Female sex hormones and their synthetic analogues may be divided into three classes: (*a*) estrogens: (*b*) progestagens or progestins; and (*c*) antiestrogens.

1. *Estrogens*

According to their source and chemical structure, estrogens are classified as:

1. Natural estrogens: estradiol, estrone, estriol
2. Esterified estrogens: estradiol benzoate, estradiol dipropionate, estradiol cyclopentylpropionate, estradiol valerate, estradiol cypionate, estradiol enanthate, estradiol undecylate, polyestradiol phosphate
3. Conjugated estrogens: estrogenic substances, conjugated piperazine estrone sulfate
4. Semisynthetic derivatives: cloxestradiol acetate (Genovul), estrazinol hydrobromide, ethinyl estradiol, estrofurate, mestranol, quinestradol (Pentovis), quinestrol
5. Synthetic estrogens: benzestrol, broparestrol (Acnesol), chlorotrianisene, diethylstilbestrol, dienestrol, hexestrol, methallenestril (Vallestril), monomestrol, promethestrol dipropionate (Meprane Dipropionate).

Estrogens most widely used are listed in Table 40.3. Estradiol is used as such

Figure 40.1 Mechanism of action of certain hormones on the regulation of enzyme synthesis. (a) Repressor formed by regulator gene combines with operator gene and inhibits *m*RNA synthesis; the information is not expressed. (b) Repressor interacts with inductor; this complex is no longer able to block operator gene, *m*RNA is produced, which, in turn, directs the synthesis of enzymes. Inductor can be either a substrate or a hormone. [After F. Jacob and J. Monod, *J. Mol. Biol.*, 3, 318 (1961).]

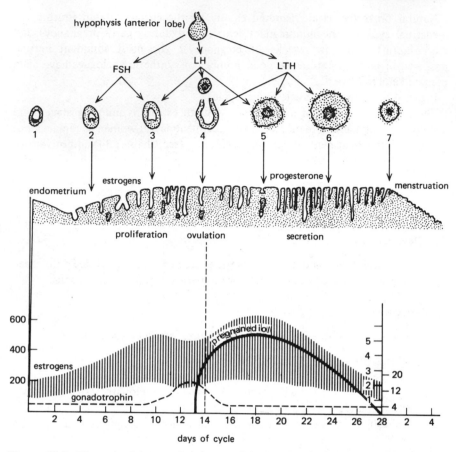

Figure 40.2 Diagram of human female sexual cycle. The follicle-stimulating hormone (FSH) of the adenohypophysis provokes ripening of the ovarian follicle. 1, primary follicle; 2, ripening folicle; 3, ripe follicle; 4, rupture of follicle (ovulation) as a result of luteinizing hormone (LH) action. LH stimulates corpus luteum formation (5), and luteotropic hormone (LTH) induces secretion by corpus luteum of progesterone (6). The mature corpus luteum (6) then degenerates (7) if there is no pregnancy. The endometrium proliferates under the action of estrogens in the first part of the cycle and secretes under the action of progester-one during the latter part of the cycle, which ends in menstruation. Below, urinary excretion of hormone; vertical lines, estrogens; broken line, gonadotropins; solid line, pregnanediol. Left, the units are international units of estrogens eliminated in 24 hr; right, milligrams of pregnanediol (1 to 5) and international units of gonadotropins (4 to 20) excreted in 24 hr. (From B. A. Houssay et al., Human Physiology, 2nd ed., McGraw-Hill-Blakiston, New York, 1955; used with permission of McGraw-Hill Book Co.) (See N. Applezweig. Steroid Drugs, McGraw-Hill, New York, 1962, p. 175.)

and also in the latent form of esters. Estrone is used in the free form and as the piperazine sulfate, potassium sulfate, sodium sulfate. Diethylstilbestrol is used as

Table 40.3 Estrogens

Official Name	Proprietary Name	Chemical Name	Structure
ethinyl estradiol*			see Table 34.5
diethylstilbestrol*+			see Table 34.5
estradiol*+	Aquadiol Progynon	estra-1,3,5(10)-triene-3,17β-diol	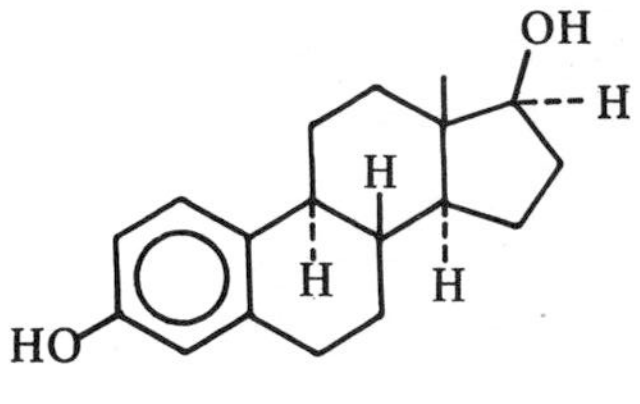
conjugated estrogens* esterified estrogens*	Premarin Amnestrogen Evix Menest SK-Estrogens		A mixture of sodium salts of sulfate esters of estrogenic substances
mestranol*+		3-methoxy-19-nor-17α-pregna-1,3,5(10)-trien-20-yne-17-ol	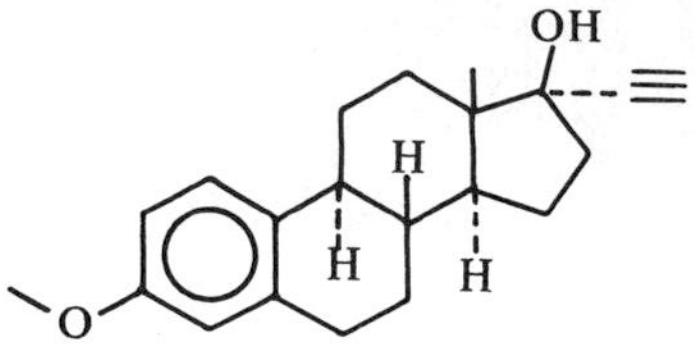

629

Table 40.3   Estrogens (continued)

Official Name	Proprietary Name	Chemical Name	Structure
benzestrol[+]	Chemestrogen	4,4'-(1,2-diethyl-3-methyl-trimethylene)diphenol	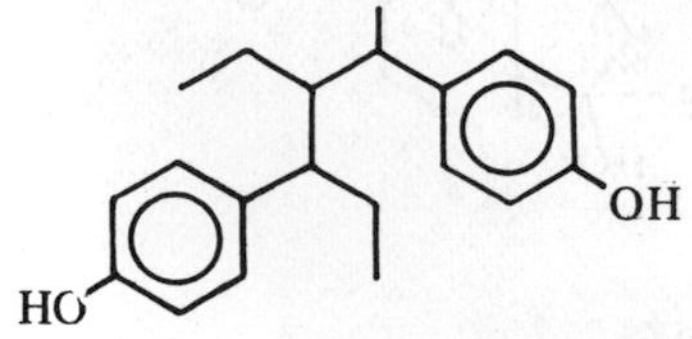
chlorotrianisene[+]	TACE	chlorotris(*p*-methoxyphenyl)-ethylene	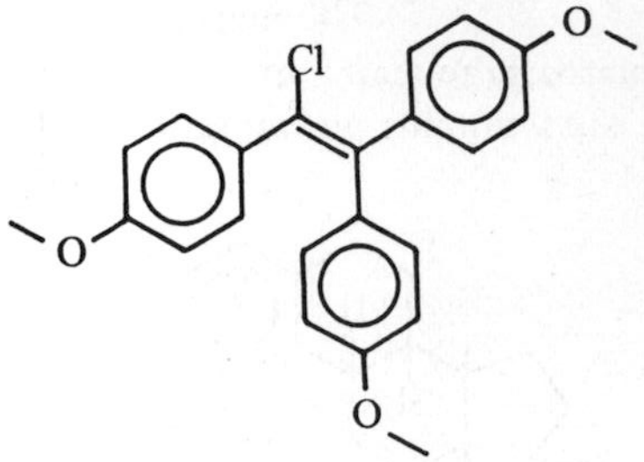
dienestrol[+]	Synestrol	4,4'-(diethylideneethylene)-diphenol	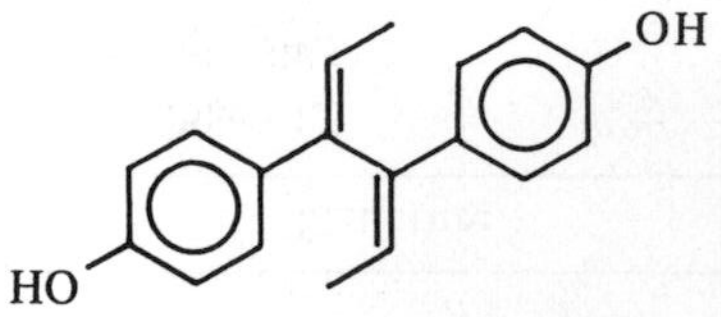

estrone[*] Estrusol 3-hydroxyestra-1,3,5(10)-
            Glyestrin   trien-17-one
            Wynestron

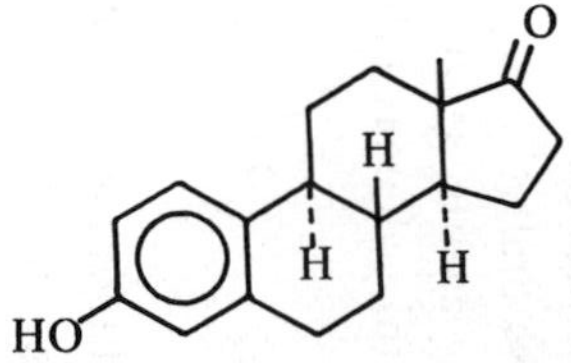

[*]In *USP XIX* (1975).
[+]In *NF XIV* (1975).

such and also as the dipropionate and diphosphate. Conjugated estrogens are stable in the gastrointestinal tract, being therefore effective orally.

Mixtures of estrogens with different drugs (antianxiety agents, sedatives, and miscellaneous substances) are marketed. Some are useful, whereas others are of doubtful therapeutic value.

2. Progestagens

Those most widely used are listed in Table 40.4. Norethindrone is used in the free form and also as the acetate. Recent progestagens are algestone aceto-phenide, allylestrenol (Gestanin), anagestone acetate (Anatropin), 8-azaquines-trol, dydrogesterone (Duvaron), ethynodiol diacetate, flugestone acetate (Cronolone), gestonorone caproate (Depostat), gestronal (Depostat), 17α-hydro-xyprogesterone, lynestrenol, medrogestone (Colpro), megestrol acetate (Megace), methylestrenolone (Orga-Steron), 7α-methylnorethindrone, moxe-strol, norethisterone enanthate (Micronor, Noriday), norgestrel, pentagestrone acetate (Gestovis), quingestanol acetate (Demovis), vinylestrenolone.

Mixtures of progestagens and estrogens are used as oral contraceptives and in various menstrual disorders, such as primary and secondary amenorrhea, dysmenorrhea, endometriosis, and dysfunctional uterine bleeding.

3. Antiestrogens

Antiestrogens are drugs that are able to inhibit or modify the effects produced by estrogens. Progestagens and androgens exert this action and, therefore, are regarded as antiestrogens by some authors. Other antiestrogens are the following synthetic compounds: clomiphene, ethamoxytriphetol, dimethylstilbestrol, chlorotrianisene, tamoxifen.

B. Mechanism of Action

Estrogens promote female sexual differentiation and stimulate lipid synthesis through induction of the synthesis of enzymes (Figure 40.1).

IV. ORAL CONTRACEPTIVES AND OVULATORY AGENTS

Oral contraceptives are drugs used in the control of fertility. Those commercially available are progestagens and estrogens.

Ovulatory agents are agents used to stimulate ovulation and, therefore, to treat infertility.

Adverse reactions produced by oral contraceptives involve almost all systems of the body. They increase the risk of thromboembolic disorders and cause hypertension, changes in carbohydrate, lipid, and protein metabolism, besides

Table 40.4 Progestagens

Official Name	Proprietary Name	Chemical Name	Structure
progesterone*	Gesterol Proluton	pregn-4-ene-3,20-dione	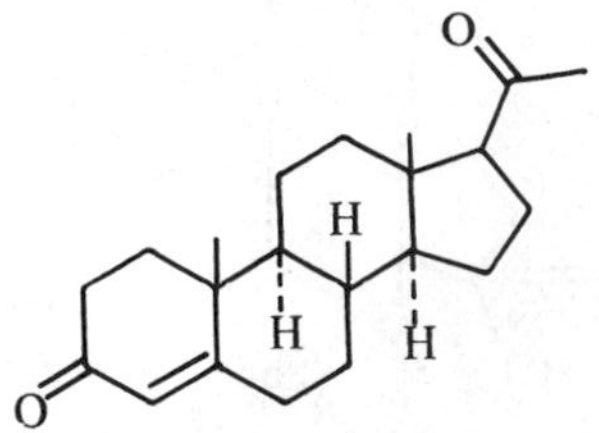
hydroxyprogesterone caproate* medroxyprogesterone acetate*			see Table 34.5 see Table 34.5
ethynodiol diacetate*		19-nor-17α-pregn-4-en-20-yne-3β-17-diol diacetate	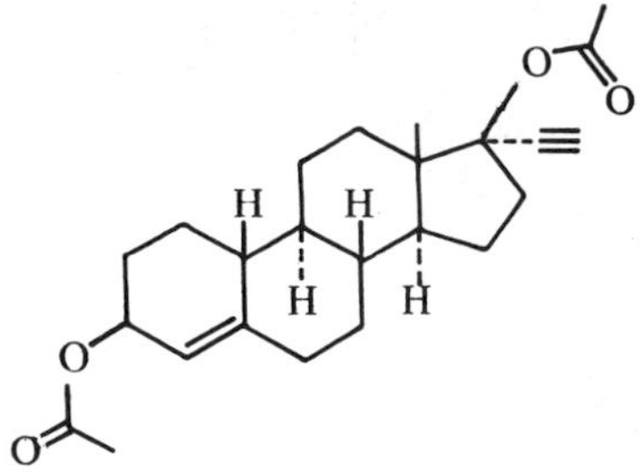
norethindrone*⁺	Norlutin	19-nor-17α-ethinyltestosterone	

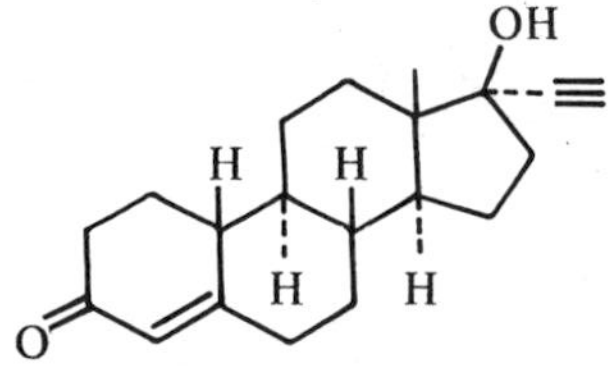

633

Table 40.4 Progestagens (continued)

Official Name	Proprietary Name	Chemical Name	Structure
norgestrel*		13β-ethyl-17α-ethenyl-17β-hydroxygen-4-en-3-one	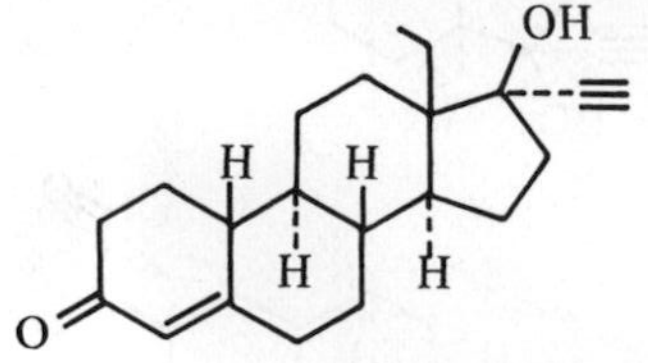
norethynodrel+		17-hydroxy-19-nor-17α-pregn-5(10)-en-20-yn-3-one	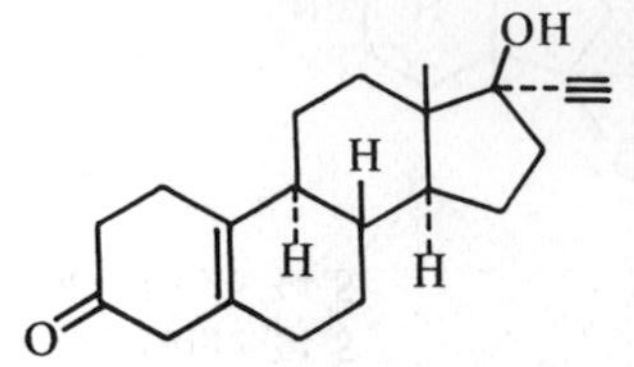
dimethisterone+	Secrosterone	17β-hydroxy-6α-methyl-17-(1-propynyl)androst-4-en-3-one	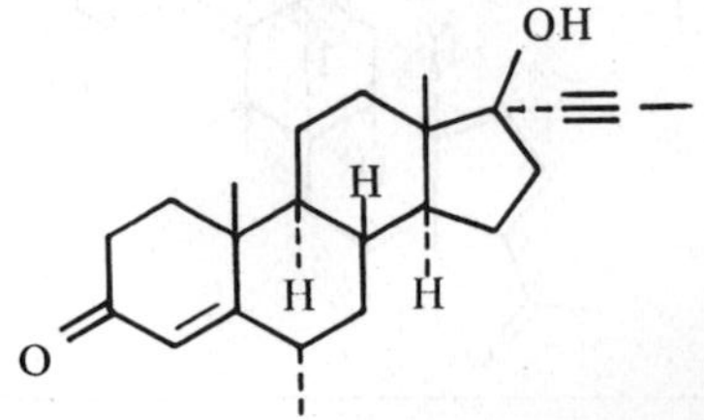

dydrogesterone[+]

Duphaston
Gynorest

9β,10α-pregna-4,6-diene-3,20-dione

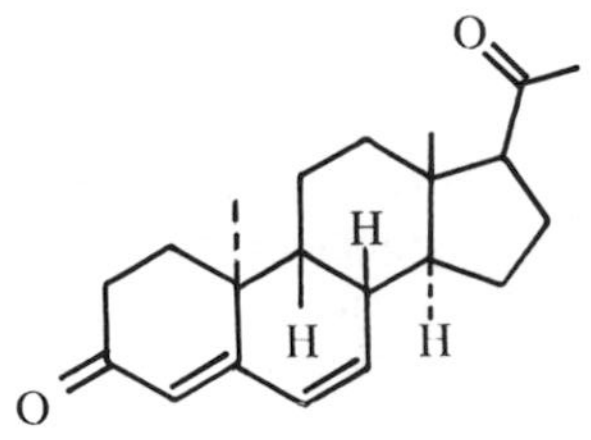

ethisterone

17α-ethynyltestosterone

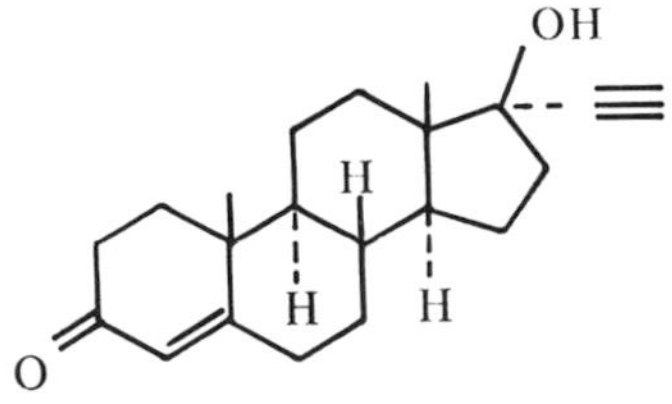

*In *USP XIX* (1975).
[+] In *NF XIV* (1975).

Table 40.5 Oral Contraceptives Marketed in the United States

Progestagen	Dose (mg)	Estrogen	Dose (μg)	Trade Names	Administration
Single-entity					
norethindrone*+	0.35			Micronor Nor-Q.D.	one tablet daily
Sequential					
dimethisterone+	25	ethinyl estradiol†	100	Oracon	estrogen for 16 days, then progestagen plus estrogen for 5 or 6 days
norethindrone*+	2	mestranol*+	80	Norquen Ortho-Novum SQ	estrogen for 14 days, then progestagen plus estrogen for 6 days
Combination					
ethynodiol diacetate†	1	ethinyl estradiol†	50	Demulen	from 5th to 24th or 25th day of cycle
	1	mestranol*+	100	Ovulen	or
norethindrone*+	1	mestranol*+	50	Norinyl 1+50 Ortho-Novum 1/50	21 days of treatment followed by 7 days during which no pills are taken or inert or iron-containing (75 mg ferrous fumarate) tablets are taken (the numbers 20, 21, or 28 following the trade name indicate number of tablets in package)
	1		80	Norinyl 1+80 Ortho-Novum 1+80	
	2		100	Norinyl 2 mg Ortho-Novum 2 mg	
	10		60	Norinyl 10 mg Ortho-Novum 10 mg	

norethindrone acetate[†]	1	ethinyl estradiol[†]	50	Norlestrin 1 mg
	2.5		50	Norlestrin 2.5 mg
norethynodrel[+]	2.5	mestranol[*][+]	100	Enovid-E
	5		75	Enovid 5 mg
norgestrel[†]	0.5	ethinyl estradiol[†]	50	Ovral

Source. Adapted from AMA Department of Drugs, *AMA Drug Evaluations*, 2nd ed., Publishing Sciences Group, Acton, Mass., 1973.

[*] In *USP XIX* (1975).
[†] In *USP XIX* (1975) as combinations.
[+] In *NFX XIV* (1975).

other minor side effects, such as nausea, vomiting, headache, altered libido, weight gain, amenorrhea, and breast engorgement.

In 1938 Parkes, Nobles and Dodds proposed the use of certain orally active estrogens, such as ethinyl estradiol and diethylstilbestrol, for interruption of pregnancy. But nothing was done immediately to test this hypothesis in women. In 1954 Djerassi and coworkers introduced the orally active gestagens of the 17α-ethinyl-19-nortestosterone type. This important discovery prompted Pincus to use 19-norsteroids associated with estrogens as contraceptives. The effectiveness of this combination, first tested in Puerto Rico (1956-1957), led eventually to further development and introduction of these agents as anticontraceptive agents.

Prostaglandins, first discovered and described independently in the 1930s by Goldblatt and Euler and revived by Bergström, are another group of substances that find application in the field of contraception. Presently more than a dozen naturally occurring human prostaglandins are known, all containing 20 carbon atoms and the same basic carbon skeleton—prostanoic acid. Several natural prostaglandins have already been totally synthesized, as well as many of their analogues, metabolites, and antagonists. Two prostaglandins are already being marketed for the interruption of pregnancy: dinoprost tromethamine (Prostin $F_2\alpha$) and dinoprostone (Prostin E_2). Cyclofenil, an ovulatory agent, is also inducer of abortion.

Research is now being directed toward finding a male contraceptive, which might either suppress spermatogenesis or act directly on the testis. Some drugs, such as cyproterone acetate, have shown promising results. Menphegol (Neo Sampon) is already being marketed as a spermicide.

A. Classification

1. *Oral Contraceptives*

Oral contraceptives are derivatives of one of the following steroids: estradiol, testosterone, nortestosterone, pregna-4,6-diene-3,20-dione, pregnadiene. Another class is comprised of synthetic compounds of miscellaneous structure.

The most widely contraceptive regimens currently used are the single-entity (or minipill), the sequential, and the combination methods (Table 40.5). Several other regimens are under study: (*a*) long-acting vaginal, parenteral, and oral preparations containing either only a progestagen or a progestagen-estrogen combination; (*b*) precoital pill, containing progestagen; (*c*) postcoital pill, containing estrogens; (*d*) weekend pill, containing an antiprogestational steroid.

Besides compounds listed in Table 40.5, several others are available or under investigation, either as contraceptives or as abortifacients: clogestone acetate, dehydroepiandrosterone, deladroxone, dydrogesterone combined with quin-

Table 40.6 Ovulatory Agents

Official Name	Proprietary Name	Chemical Name	Structure
clomiphene citrate*	Clomid	1[2-diethylaminoethoxy)-phenyl]-1,2-diphenyl-2-chloroethylene citrate	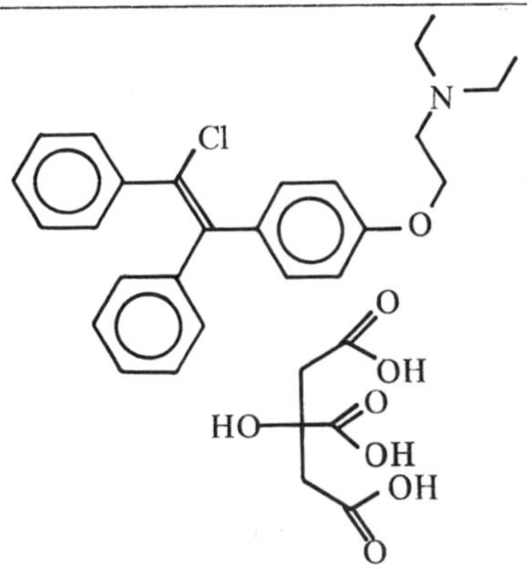
cyclofenil	Ondogyne Sexovid	4,4'-(cyclohexylidenemethyl-ene)diphenol diacetate ester	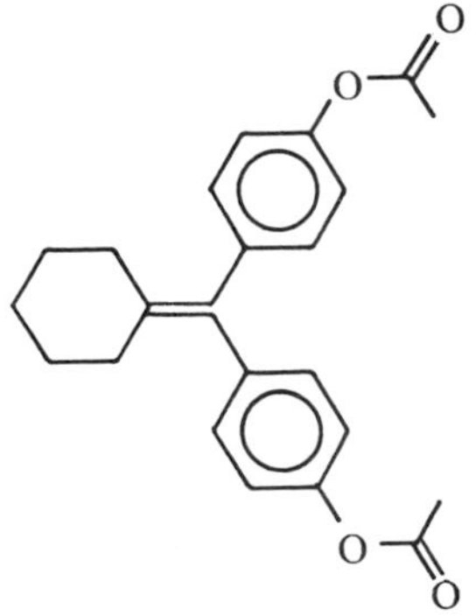
menotropins	Pergonal	pituitary hormones	

*In *USP XIX* (1975).

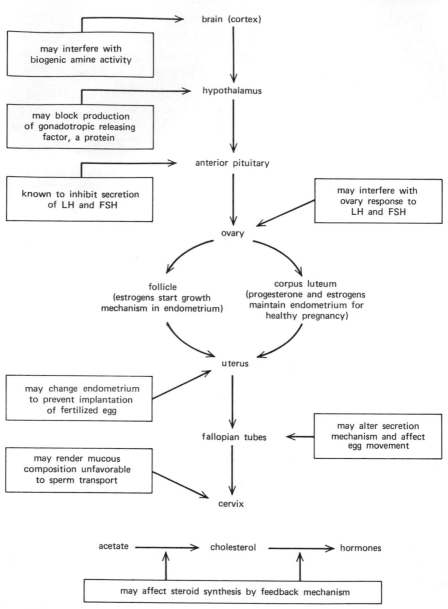

Figure 40.3 The various sites in which oral contraceptives that mimic pregnancy may act. (From *Chem. Eng. News,* **March 27, 1967, p. 46)**

estrol, ethinylestradiol, medrogestone, megestrol acetate, melengestrol, norgestrel, quinestrol combined with quingestanol, stilbestrol.

2. Ovulatory Agents

Only three ovulatory agents are presently used (Table 40.6).

B. Mechanism of Action

Theoretically, contraception can be achieved by administration of the proper preparation before, during, or after coitus. Contraception can be obtained in one of the following ways: inhibition of ovulation, prevention of ovule migration, inhibition of hormonal secretion that induces the preparation of endometrium, prevention of sperm transport, inhibition of ovule fecundation, inhibition of nidation, blockade of embryo development. Oral contraceptives may interfere, therefore, at various sites (Figure 40.3). However, it is believed that both the combined and sequential methods prevent ovulation by inhibiting release of gonatropin-releasing factors from the hypothalamus, or they may act on a higher brain center. As to the single-entity regimen, the mechanism of action of low-dose progestagens is complex and not well understood. It is known, however, that they usually act after ovulation. Recent experiments suggest that they act by releasing leucocytic lysozomal enzymes.

The mechanism of action of clomiphene is not known, but it is likely that it competes with estrogen at the hypothalamic level. This results in ovarian stimulation.

REFERENCES

Androgens and Anabolic Steroids

A. Korolkovas, *Rev. Paul. Med.,* 81, 169 (1973).

E.-E. Baulieu et al., *Recent Progr. Hormone Res.,* 27, 351 (1971).

K. B. Eik-Nes, *Recent Progr. Hormone Res.,* 27, 517 (1971).

K. W. McKerns, Ed., *The Sex Steroid: Molecular Mechanisms,* Appleton-Century-Crofts, New York, 1971.

J. A. Vida, *Androgens and Anabolic Agents,* Academic, New York, 1969.

Estrogens and Progestagens

B. W. O'Malley and A. R. Means, *Science,* 183, 610 (1974).

G. C. Mueller et al., *Recent Progr. Hormone Res.,* 28, 1 (1972).

A. E. Kellie, *Annu. Rev. Pharmacol.,* 11, 97 (1971).

J. Gorski et al., *Curr. Top. Develop. Biol.,* 4, 149 (1969).

E. V. Jensen et al., *Develop. Biol.,* Suppl., 3, 151 (1969).

K. Junkmann, Ed., "Die Gestagene," *Handbuch der experimentellen Pharmakologie,* Vol. XXII/I & II, Springer, Berlin, 1968-1969.

P. Morand and J. Lyall, *Chem. Rev.,* 68, 85 (1968).

Oral Contraceptives and Ovulatory Agents

N. F. Farnsworth *et al., J. Pharm. Sci.,* **64**, 535, 717 (1975)

J. P. Bennett, *Chemical Contraception,* Columbia University Press, New York, 1974.

M. H. Briggs and E. Diczfalusy, Eds., *Pharmacological Models in Contraceptive Development,* WHO Symposium Geneva 1973, WHO Research and Training Centre on Human Reproduction, Karolinska Institutet, Stockholm, 1974.

R. A. Mueller and L. E. Flanders, *Annu. Rep. Med. Chem.,* **9**, 162 (1974).

A. S. Bingel and P. S. Benoit, *J. Pharm. Sci.,* **62**, 179, 349 (1973).

H. Jackson, *Amer. Sci.,* **61**, 188 (1973).

J. B. Josimovich, Ed., *Uterine Contraction: Side Effects of Steroidal Contraceptives,* Wiley New York, 1973.

J. L. Marx, *Science,* **179**, 1222 (1973).

J. ApSimon, Ed., *Prostaglandins,* Wiley-Interscience, New York, 1972.

C. Bell, *Pharmacol. Rev.,* **24**, 657 (1972).

E. M. Southern, Ed., *The Prostaglandins: Clinical Applications in Human Reproduction,* Futura Publishing Company, Mount Kisco, N. Y., 1972.

M. P. L. Caton, *Progr. Med. Chem.,* **8**, 317 (1971).

F. Fuchs, Ed., *Endocrinology of Pregnancy,* Harper & Row, New York, 1971.

V. Petrow, *Progr. Med. Chem.,* **8**, 171 (1971).

P. W. Ramwell and J. E. Shaw, Eds., "Prostaglandins," *Ann. N.Y. Acad. Sci.,* **180**, 1-568 (1971).

E. Steinberger, *Physiol. Rev.,* **51**, 1 (1971).

C. Djerassi, *Science,* **169**, 941 (1970).

V. Petrow, *Chem. Rev.,* **70**, 713 (1970).

Pharmacology Society Symposium, "Antifertility Drugs, " *Federation Proc.,* **29**, 1209-1242 (1970).

R. L. Vande Wiele, *Recent Progr. Hormone Res.,* **26**, 63 (1970).

S. M. Kalman, *Annu. Rev. Pharmacol.,* **9**, 363 (1969).

S. Bergström *et al., Pharmacol. Rev.,* **20**, 1 (1968).

E. Diczfalusy, *Amer. J. Obstet. Gynecol.,* **100**, 136 (1968).

J. Gorski *et al., Recent Progr. Hormone Res.,* **24**, 45 (1968).

M. J. K. Harper, *Progr. Drug Res.,* **12**, 47 (1968).

P. W. Ramwell and J. E. Shaw, Eds., *Prostaglandins,* Wiley-Interscience, New York, 1968.

S. Bergström and B. Samuelson, *The Prostaglandins,* Interscience, New York, 1967.

C.-R. Garcia, *Amer. J. Med. Sci.,* **253**, 718 (1967).

G. Pincus, *Science,* **153**, 493 (1967).

H. W. Rudel and J. Martinez-Manautou, *Top. Med. Chem.,* **1**, 339 (1967).

H. Jackson, *Antifertility Compounds in the Male and Female,* Thomas, Springfield, Ill., 1966.

G. Pincus, *Control of Fertility,* Academic, New York, 1965.

PART 8

MISCELLANEOUS AGENTS

Drugs that were not studied in the preceding forty chapters are seen in the remaining two chapters. They can be divided into two groups: (*a*) diagnostic aids and (*b*) miscellaneous agents.

CHAPTER 41

DIAGNOSTIC AIDS

I. INTRODUCTION

Diagnostic aids are substances used to examine the body in order to detect impairment of its normal functions.

II. CLASSIFICATION

Diagnostic aids can be divided into (*a*) radiopaques and (*b*) miscellaneous diagnostic aids.

A. Radiopaques

Radiopaques, the most important class of diagnostic aids, are substances that absorb X-rays and, consequently, produce a shadow of positive contrast in soft-tissue structures, such as urinary bladder, gallbladder, and stomach, during roentgenographic examination. On the other hand, air produces a shadow of negative contrast.

These drugs are used in diagnostic roentgenography: angiography (blood vessels), aortography (aorta), bronchography (bronchial tree), cholecistography (gallbladder), hepatography (liver), hysterosalpingography (uterus and fallopian tubes), lymphagiography (lymphatic ducts), myelography (subarachnoid space of the spinal canal), sialography (salivary glands), urography (urinary tract). This last procedure, also called pyelography, can be either retrograde urography or excretion urography.

Iodine-containing radiopaques may alter the results of thyroid-function tests. Therefore, these tests should be performed prior to administration of these substances. Caution should be taken when these drugs are administered to patients with advanced renal disease, hepatic or biliary tract diseases, gastro-intestinal disorders, anuria, multiple myeloma and pheochromocytoma, because radiopaques can cause toxic reactions, especially to kidneys and heart, and also damage to spinal cord and small intestine, for instance.

Several dozen radiopaques are marketed. Those most widely used are listed in Table 41.1. They belong to five groups: inorganic compounds (barium sulfate),

Table 41.1 Radiopaques

Official Name	Proprietary Name	Chemical Name	Structure
barium sulfate*	Barobag Barosperse Rugar		$BaSO_4$
iodized oil+	Lipiodol		
ethiodized oil*	Ethiodol		
methiodal sodium+	Skiodan sodium	sodium iodomethane-sulfonate	ICH_2SO_3Na
acetrizoic acid	Cystokon	3-acetamido-2,4,6-tri-iodobenzoic acid	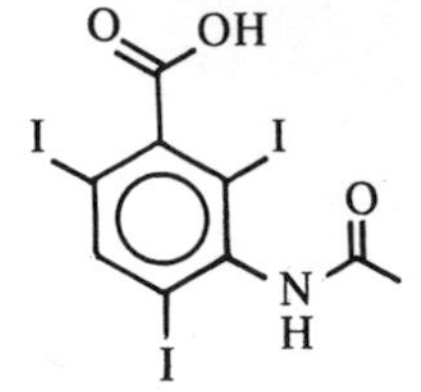
diatrizoic acid*	Gastrografin Hypaque	3,5-diacetamido-2,4,6-tri-iodobenzoic acid	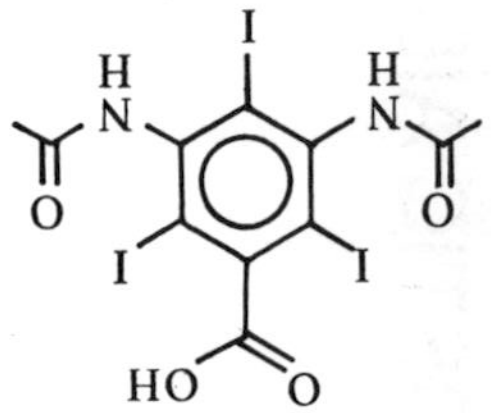

iothalamic acid* Conray 5-acetamido-2,4,6-triiodo-*N*-methylisophthalamic acid

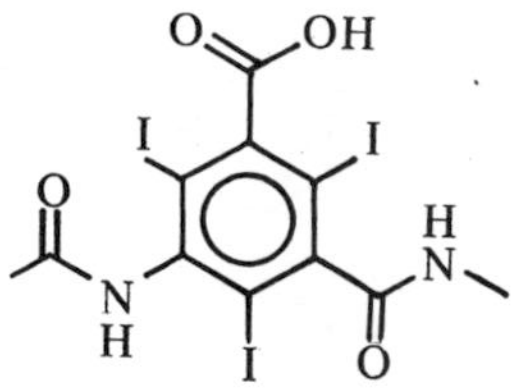

iodipamide* Cholografin 3,3'-(adipoyldiimino)-bis[2,4,6-triiodobenzoic acid]

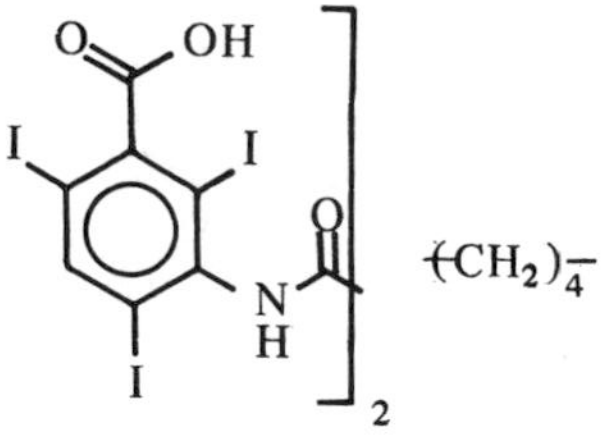

iopanoic acid* Telepaque 3-amino-α-ethyl-2,4,6-triiodo-hydrocinnamic acid

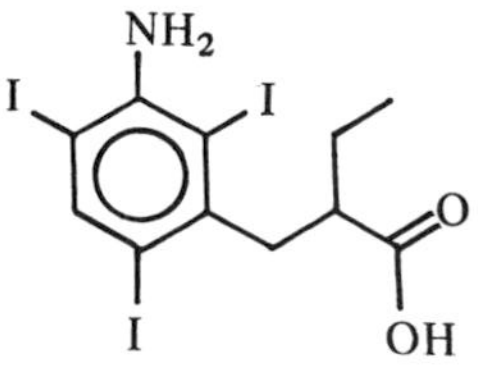

Table 41.1 Radiopaques (continued)

Official Name	Proprietary Name	Chemical Name	Structure
iophendylate*	Pantopaque	ethyl 10-(iodophenyl)unde-canoate	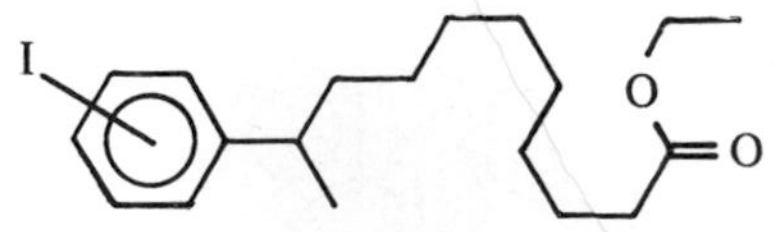
ipodaic acid*	Oragrafin	3[[(dimethylamino)methylene]-amino]-2,4,6-triiodohydrocin-namic acid	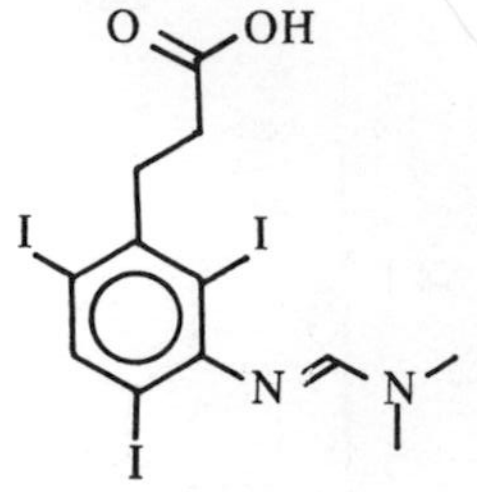
tyropanoic acid	Bilopaque	3-butyramido-α-ethyl-2,4,6-triiodo-hydrocinnamic acid	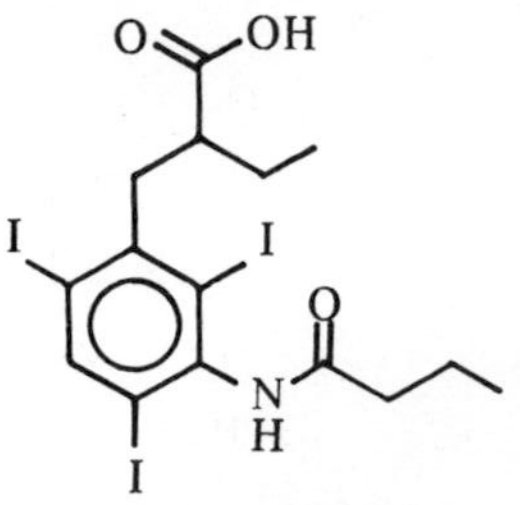

propyliodone* Dionosol propyl 3,5-diiodo-4-oxo-1(4H)-pyridineacetate

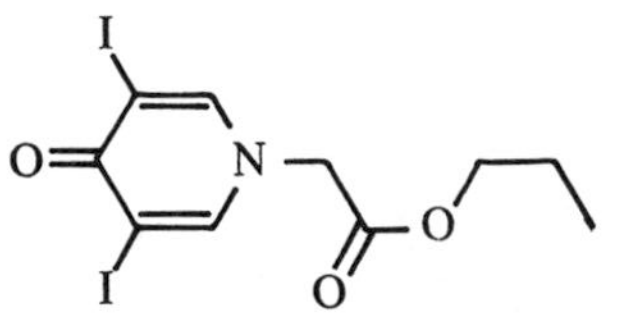

iodopyracet Diodrast 3,5-diiodo-4-oxo-1(4H)-pyridine-acetate diolamine

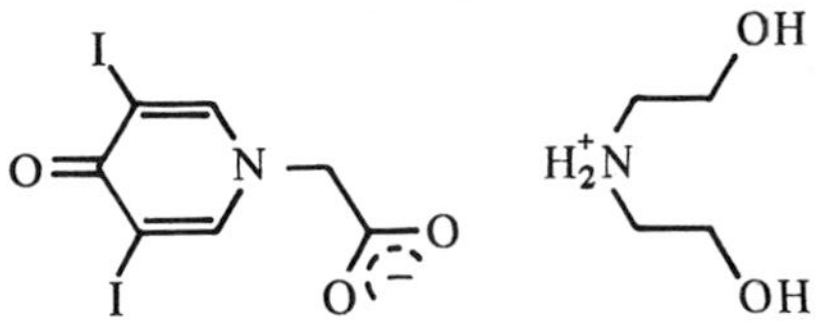

*In *USP XIX* (1975).
⁺In *NF XIV* (1975).

iodinated aliphatic derivatives (methiodal sodium), iodized oils (ethiodized oil, iodized oil), iodinated pyridone derivatives (propyliodone, iodopyracet), and iodinated benzene derivatives (the rest of those listed in Table 41.1). Thus, with the exception of barium sulfate, all radiopaques are organic iodine compounds.

Iodinated organic substances can be grouped into three classes: carboxylic acids and their salts; water-immiscible oils; suspensions.

The first class is the most important. To it belong sodium acetrizoate, sodium diatrizoate, meglumine diatrizoate, sodium iothalamate, sodium iodipamide, meglumine iodipamide, sodium ipodate, calcium ipodate, sodium methiodal, iodopyracet. New representatives of this class are sodium, monoethanolamine, or glumine salts of the following acids: iocarmic, iodoxanic, iomeglamic, ioxitalamic, tiropanoic. These radiopaques are used primarily for urography and cholecystography but also in many other roentgenographic examinations (angiocardiography, aortography, and nephrography) either as aqueous solutions of the salts or orally as acids or salts.

The second class is comprised of iodine-containing esters and acids (iopanoic acid) as well as iodized oils. These oils are vegetable oils or fatty acids containing about 40% iodine, which is incorporated into these oils by covalent addition to the double bonds through chemical treatment. Compounds of this class are used especially for myelography, lymphangiography, sialography, hysterosalpingography, and bronchography.

The third class consists of water-insoluble compounds suspended in aqueous or oily medium. Certain radiopaques of this class are esters of one of the acids of the first class, for instance, propyliodone. They are used for bronchography and other radiographic examinations.

Barium Sulfate

It is the agent of choice for gastrointestinal roentgenography. It is administered by oral or rectal route as an aqueous suspension. However, it is contraindicated in the presence of hemorrhagy and obstructions. The usual dosage is 200 to 300 g in a suitable suspension.

Diatrizoate

Marketed as sodium meglumine salt, it contains 65.8% iodine, being one of the best tolerated and least toxic of the radiopaques. It is used for most roentnographic examinations.

Iodipamide Meglumine

It is the radiopaque of choice in cholecystography, used by intravenous route.

Iopanoic Acid

It contains 66.6% iodine. It is administered orally for cholecystography.

Iothalamate

Available as sodium or meglumine salt, it is one of the best radiopaques for roentgenography of the circulatory system and urinary tract.

Tyropanoate

It is used orally for cholecystography and cholangiography.

Propyliodone

It occurs as water-insoluble, white, crystalline powder, being used as suspension in peanut oil for bronchography.

B. Miscellaneous Diagnostic Aids

Miscellaneous diagnostic aids are used to detect abnormalities or pathological disfunction of several organs of the body. Table 41.2 lists some of the most widely used.

According to their use, these aids are grouped by the American Medical Association into the following classes:

1. Agents for liver-function tests: indocyanine green, sulfobromophthalein sodium

2. Agents for kidney-function tests: aminohippuric acid and its sodium salt, indigotindisulfonate sodium, inulin, mannitol, phenolsulfonphthalein

3. Agents used to test gastric-acid secretion: azuresin, betazole hydrochloride, histamine phosphate

4. Agents for cardiovascular measurements: evans blue, fluorescein sodium, indocyanine green, sodium dihydrocholate

5. Agents for cutaneous immunity tests: diagnostic diphtheria toxin, mumps test antigen, purified derivative of tuberculin, old tuberculin, scarlet fever streptococcus toxin

6. Agents used as biological test allergens or antigens: allergens, blastomycin, coccidioidin, histoplasmin, trichinella extract

7. Agent for pituitary-function tests: metyrapone

8. Agents for detection of pheochromocytoma: phentolamine, histamine

9. Agent for detection of amyloid disease: congo red

10. Radiopharmaceuticals: sodium chromate Cr 51, cyanocobalamin Co 57, cyanocobalamin Co 60, sodium iodide I 125, sodium iodide I 131, iodohippurate sodium I 131, iodinated I 125 serum albumin, iodinated I 131 serum albumin, rose bengal sodium I 131, thyroxine I 125, thyroxine I 131, liothyronine I 131, ferrous citrate Fe 59, chlormerodrin Hg 197, chlormerodrin Hg 203, sodium phosphate P 32, sodium pertechnetate Tc 99m, gold Au 198

Table 41.2 Miscellaneous Diagnostic Aids

Official Name	Proprietary Name	Chemical Name	Structure
aminohippuric acid*		*p*-aminohippuric acid	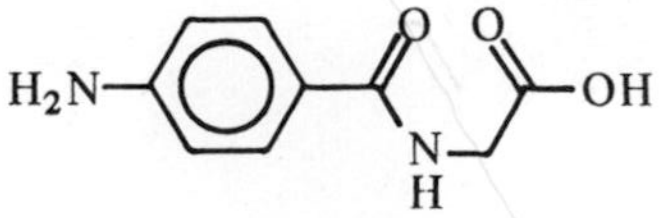
indigotindisulfonate sodium* (indigo carmine)		disodium 5,5'-indigotindisulfonate	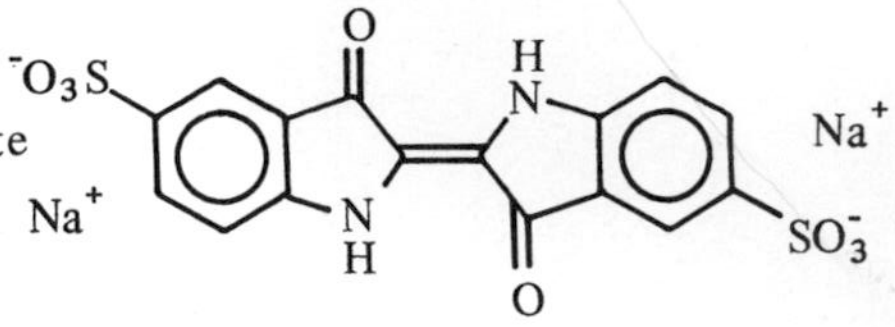
phenolsulfonphthalein+ (phenol red)		4,4'-(3H-2,1-benzoxathiol-3-ylidene)diphenol S,S-dioxide	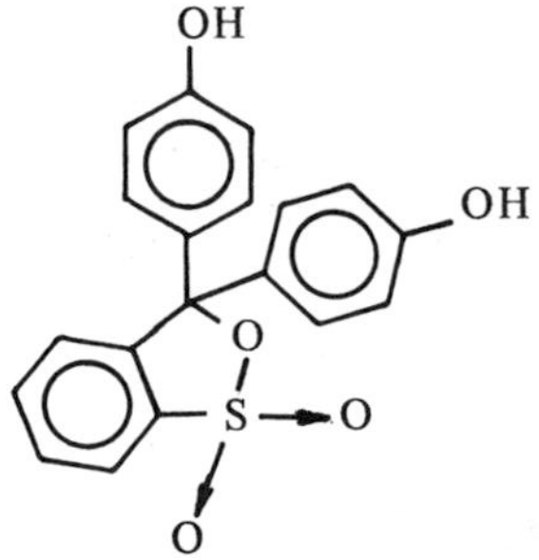

sulfobromophthalein sodium*

Bromosulphalein

4,5,6,7-tetrabromo-3',3''-disulfophenolphthalein disodium salt

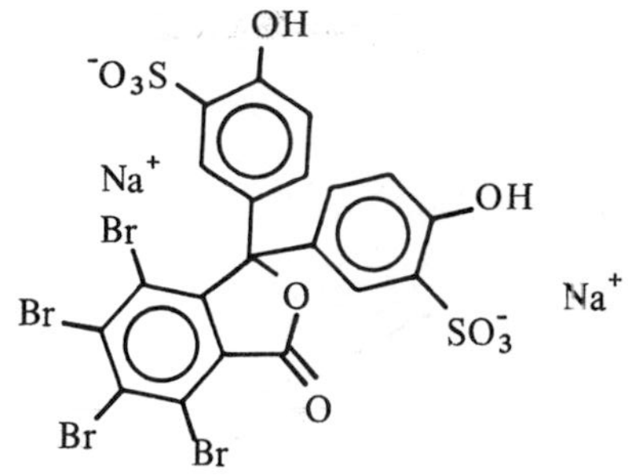

evans blue*

6,6'-[(3,3'-dimethyl[1,1'-biphenyl]-4,4'-diyl)bis(azo)]bis[4-amino-5-hydroxy]-1,3-naphthalene-disulfonic acid, tetrasodium salt.

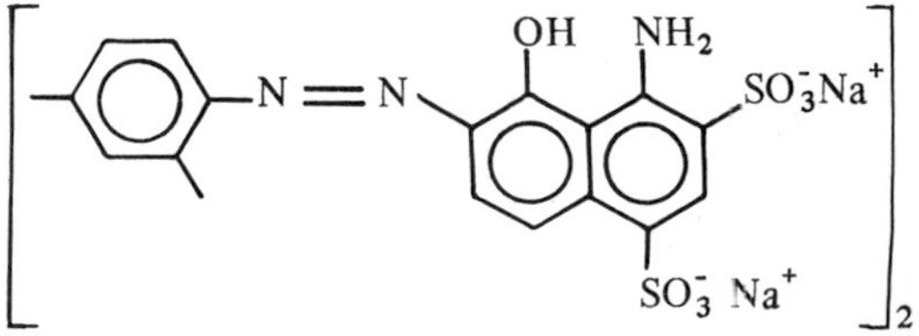

metyrapone*

Metopirone

2-methyl-1,2-di-3-pyridyl-1-propanone

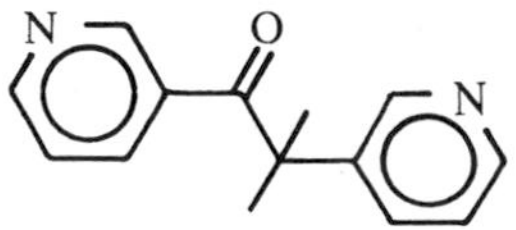

653

Table 41.2 Miscellaneous Diagnostic Aids (continued)

Official Name	Proprietary Name	Chemical Name	Structure
indocyanine green*	Cardio-Green	2-[7-[1,1-dimethyl-3-(4-sulfobutyl)benz-[e]indolin-2-ylidene]-1,3,5-heptatrienyl]-1,1-dimethyl-3-(4-sulfobutyl)-1H-benz[e]-indolium hydroxide, inner salt, sodium salt	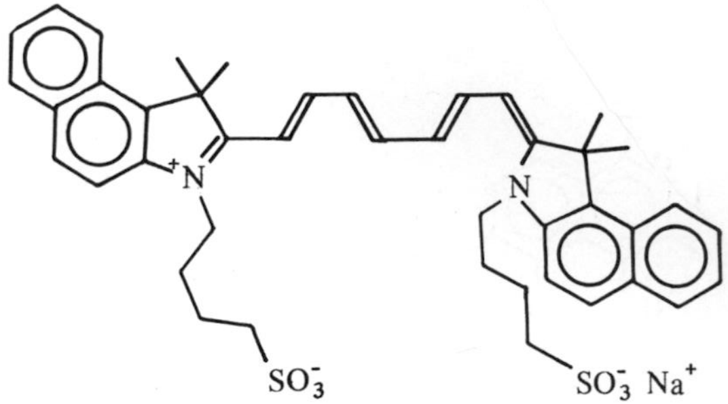
fluorescein sodium*	Fluorescite	fluorescein disodium salt	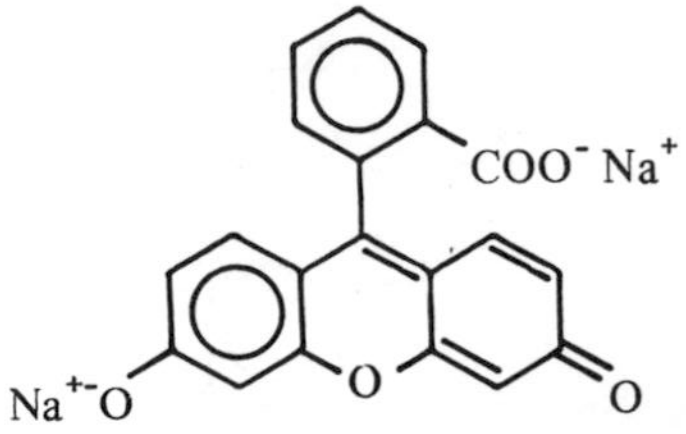

dehydrocholic acid[+]

Decholin

3,7,12-trioxo-5β-cholan-24-oic acid

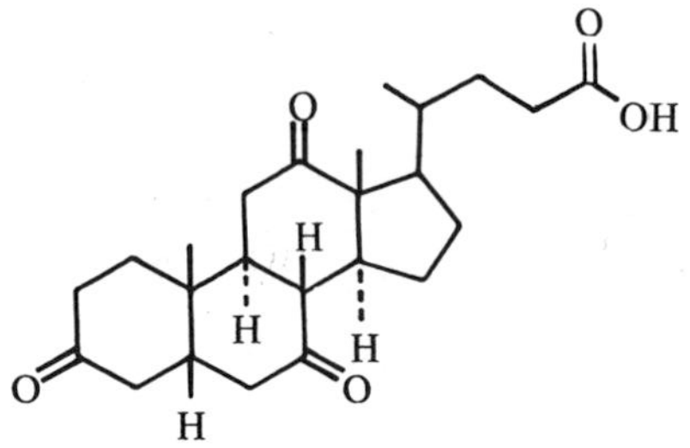

azuresin[+]

Diagnex Blue

azure A carbacrylic resin

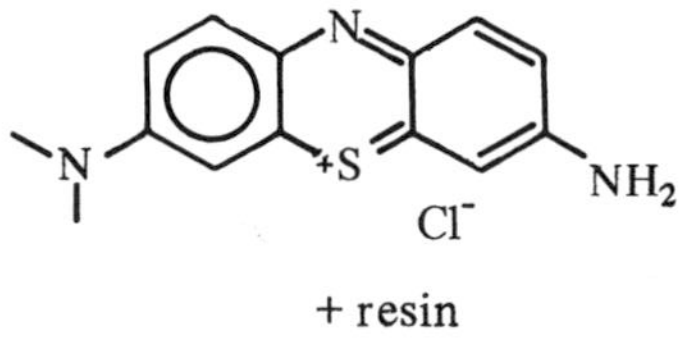

congo red

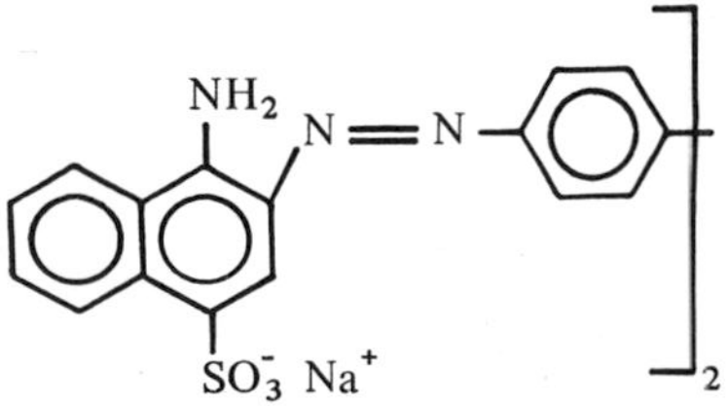

655

656

Table 41.2 Micellaneous Diagnostic Aids (continued)

Official Name	Proprietary Name	Chemical Name	Structure
inulin	Dahlin		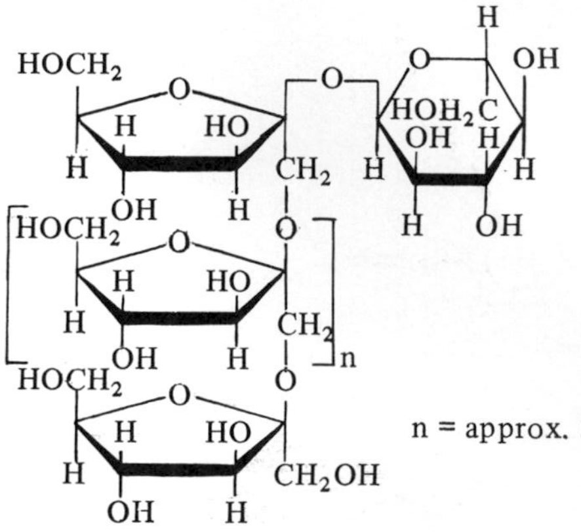 n = approx. 35
mannitol*	Osmitrol	hexahydroxy alcohol	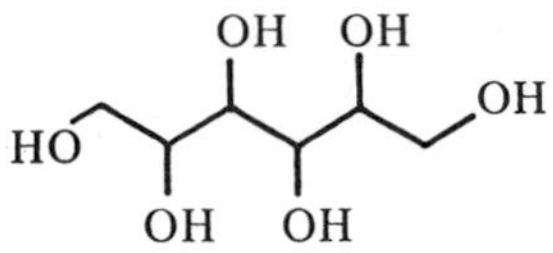

*In *USP XIX* (1975).
+ In *NF XIV* (1975).

REFERENCES

H.-J. Herms and V. Taenzer, "Design of X-ray Contrast Media," in E. J. Ariëns, Ed., *Drug Design,* Vol. VI, Academic, New York, 1975, pp. 261-295.

P. K. Knoefel, Ed., *Radiocontrast Agents,* 2 vols., Pergamon, Oxford, 1971.

S. A. Levinson and R. P. McFate, *Clinical Laboratory Diagnosis,* 7th ed., Lea and Febiger, Philadelphia, 1969.

J. L. Rabinowitz and G. A. Bruno, *Top. Med. Chem.,* 1, 357 (1967).

P. K. Knoefel, *Annu. Rev. Pharmacol.,* 5, 321 (1965).

J. O. Hoppe, *Med. Chem.,* 6, 290 (1963).

CHAPTER 42

MISCELLANEOUS AGENTS

I. DERMATOLOGICAL AGENTS

Dermatological agents are drugs used in the treatment of skin disorders and infestations. They may be applied both systemically and topically but usually they are not curative. Most drugs of this class used for other purposes were already seen in the preceding chapters. Here we discuss only those not studied elsewhere. Some dermatological agents are listed in Table 42.1.

A. Protectants

Protectants are drugs that protect skin and mucous membranes from contact with potential irritating agents. They are generally used as powders but occasionally as creams, lotions, ointments, and pastes. They act by absorbing irritating agents.

The most widely used protectants are pectin, peruvian balsam, zinc oxide, zinc salts, aluminum powder, starch, magnesium stearate, titanium dioxide, talcum, collodion, lycopodium, silicone, benzoin, bismuth subcarbonate, calamine, petrolatum.

B. Demulcents

Demulcents are drugs that coat and sooth mucous membranes and thus exert antiirritating action. They are helpful in cases of ingestion of corrosive or irritant poisons, peptic ulcer, gastritis, sore throat, and cough (see Chapter 9).

To this class belong glycerin, starch glycerolate, propylene glycol, poly-ethylene glycols (Carbowaxes), methylcellulose, and carboxymethylcellulose.

C. Emollients

Emollients are drugs used topically to soften and lubricate the skin and mucous membranes, thus exerting a protective action on these surfaces. Most of them are oily or fatty substances. Examples are olive oil, cotton oil, lanolin, paraffin, liquid paraffin, vaseline, white wax, cold cream.

D. Counterirritants

Counterirritants are drugs applied topically to cause reactive hyperemia. The result is pain relief in viscera or muscles, very likely by vasodilation at the site of the pain. Use of these agents is presently restricted to the relief of pain in arthritis, myalgias, and related discomforts.

Rubbing alcohol, isopropyl rubbing alcohol, peruvian balsam, pine tar, camphor, methyl salicylate, ether, chloroform, eucalyptol, menthol, anthralin, chrysarobin, nicotinic acid esters (e.g., nicotafuryl) are the main counterirritants.

E. Caustics and Styptics

Caustics are dermatological agents used to remove certain tissues, to destroy noxious agents, and to stop bleeding. Examples of these drugs are nitric acid, silver nitrate, carbon dioxide, and tetraquinone (Kelox).

Styptics are drugs that, besides the above-mentioned effects, cause precipitation of cellular protein and promote an inflammatory exudate that leads to the formation of a scab. This group is comprised of podophyllum, glacial acetic acid, trichloroacetic acid, phenol, and other substances.

F. Astringents

Astringents are drugs that decrease cell membrane permeability by producing mild coagulation of tissue proteins. They are used to dry, harden, and protect the skin. This group is made up mainly of two sorts of drugs:

1. Metallic salts, especially of zinc, bismuth, aluminum, iron and silver. The official preparations are alum, aluminum acetate, aluminum chloride, aluminum subacetate, calcium hydroxide, white lotion (contains sulfurated potash and zinc sulfate), zinc chloride, zinc oxide, zinc oxide paste with salicylic acid, zinc sulfate. Other astringents are silver nitrate, cupric citrate, selenium sulfide, cadmium sulfide.

2. Organic products, such as tannic acid.

G. Keratolytic Agents

Keratolytic agents are drugs used to treat various dermatoses, such as chronic lichenified and papulosquamous eruptions, atopic and seborrheic dermatitis, eczema, psoriasis, acne, condyloma acuminatum, multiple premalignant actinic keratoses, and superficial fungal infections. They are also useful in several other disorders; for instance, salicyclic acid is efficient in removing calluses.

Drugs of this class include coal tar (Banetar, Tarsum, Zetar), anthralin, benzoyl peroxide, salicyclic acid, benzoic acid, resorcinol, fluorouracil (Efudex, Fluoroplex), ichthammol, juniper tar, pine tar, podophyllum resin, trichloroacetic acid, chrysarobin, cantharidin, xenysalate, tetraquinone (Kelox), tretinoin (Aberel, Retin-A).

Table 42.1 Dermatological Agents

Official Name	Proprietary Name	Chemical Name	Structure
silicone			$-\overset{|}{\underset{|}{Si}}-\left[O-\overset{|}{\underset{|}{Si}}-\right]_n O-\overset{|}{\underset{|}{Si}}-$
carboxymethylcellulose*	Carmethose		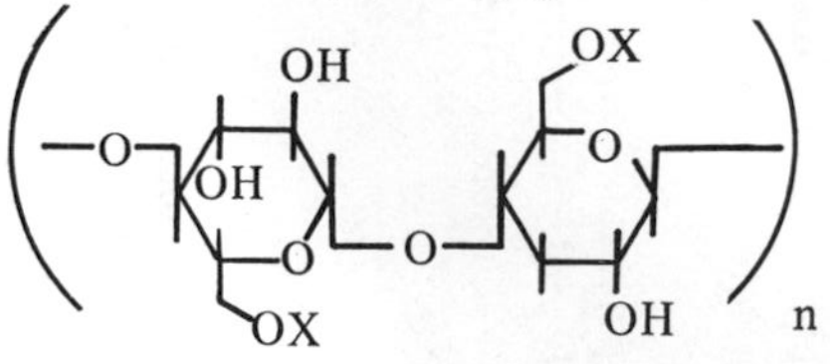 $X = H, CH_2COONa$
bismuth subcarbonate			$(BiO)_2CO_3$
titanium dioxide*			TiO_2
glycerin*		propanetriol	$HOCH_2-CH-CH_2OH$ $\overset{|}{OH}$
anthralin*	Anthra-Derm Lasan	1,8,9-anthracenetriol	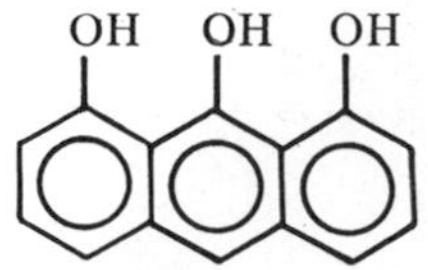
trichloroacetic acid*			CCl_3COOH
xenysalate		2-(diethylamino)ethyl-3-phenyl-salicylate	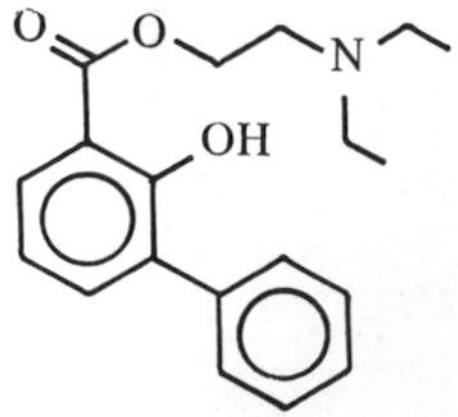

660

oxybenzone*	Spectra-Sorb UV 9	2-hydroxy-4-methoxybenzophenone 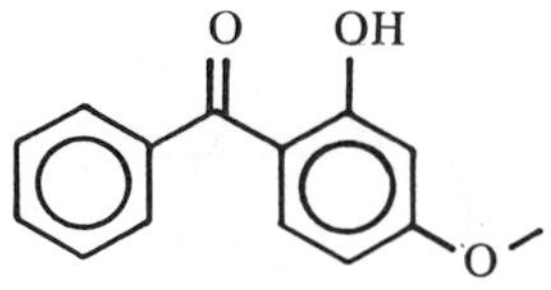
methoxsalen*	Oxsoralen	8-methoxypsoralen 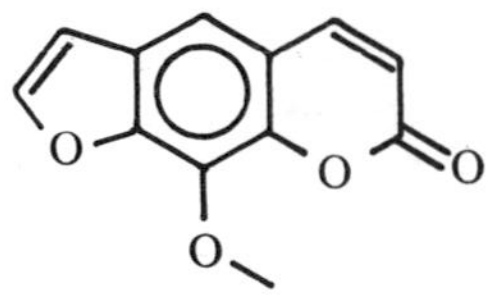
trioxsalen*	Trisoralen	4,5',8-trimethylpsoralen 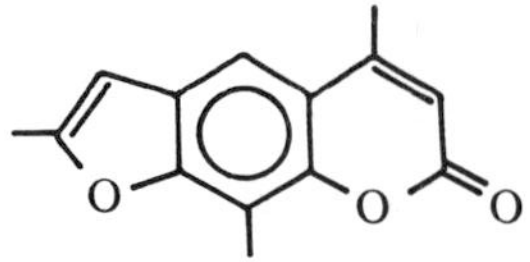
hydroquinone*	Eldoquin	1,4-benzenediol 

Table 42.1 Dermatological Agents (continued)

Official Name	Proprietary Name	Chemical Name	Structure
monobenzone[+]	Benoquin	p-(benzyloxy)phenol	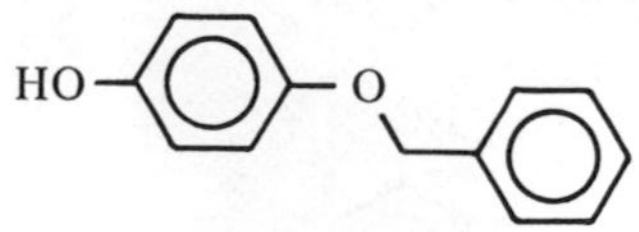
chlorophenothane[+]	DDT	1,1,1-trichloro-2,2-bis(p-chloro-phenyl)ethane	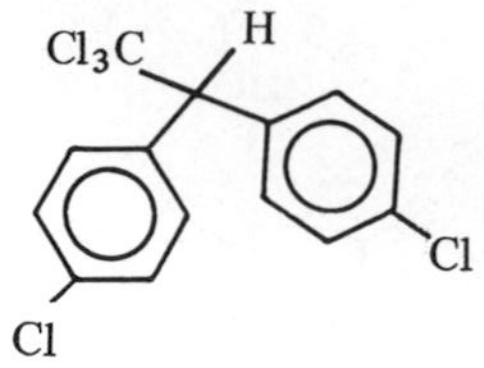
gamma benzene hexachloride*	Kwell Lindane	γ-1,2,3,4,5,6-hexachlorocyclo-hexane	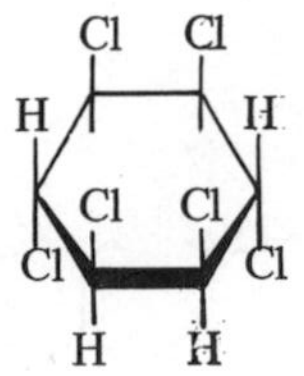
benzyl benzoate*[+]			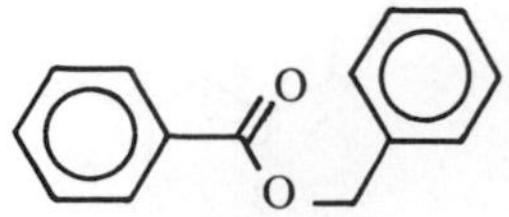

crotamiton[*]		N-ethyl-o-crotonotoluidide	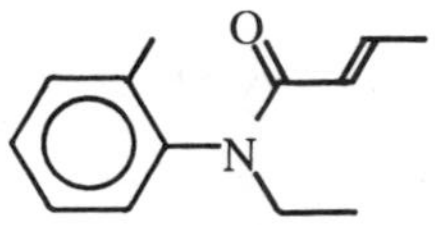 Eurax
diethyltoluamide[+]		N,N-diethyl-m-toluamide	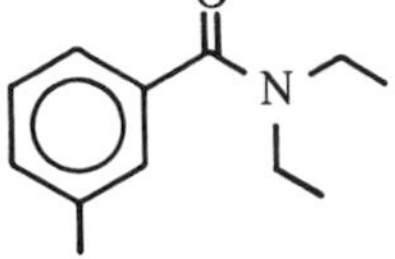 Autan
ethohexadiol		2-ethyl-1,3-hexanediol	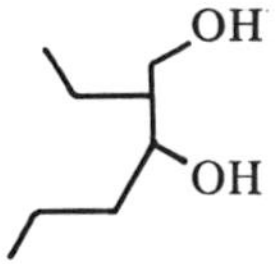 Rutgers 612
dimethyl phthalate			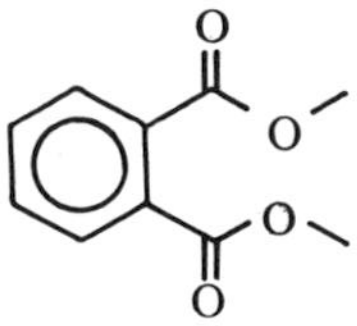

[*]In *USP XIX* (1975).
[+]In *NF XIV* (1975).

H. Sclerosing Agents

Sclerosing agents are drugs used to obliterate the lumen of varicose veins by destroying their endothelium locally. They are administered by intravenous injection directly into the varicose veins.

To this class belong sodium morrhuate, sodium tetradecyl sulfate (Sodium Sotradecol), invert sugar solutions, dextrose, ethanol, quinine and urea hydrochloride, and iron salts.

I. Deodorants and Antiperspirants

Deodorants are drugs that diminish or mask body odors. They are applied topically and exert their effect by one of the following mechanisms: (a) direct—inhibition of the bacterial growth; (b) indirect—reduction of the moisture which is ideal for bacterial proliferation.

Antibiotics and antiinfective agents have been used as deodorants, owing to their bacteriostatic action. However, use of neomycin is not indicated, because it produces sensitization and contact dermatitis.

Antiperspirants are drugs that diminish sweating, especially in axillae. They may also have deodorant properties. They are applied topically. Their effect results from occlusion of the openings of sweat glands.

Active ingredients most frequently used in antiperspirants are aluminum chlorhydroxide and aluminum chloride. Other drugs with similar action are aluminum sulfamate, aluminum sulfate, zinc phenolsulfonate, zirconium hydroxychloride.

Burning and irritation are the most common adverse effects caused by these preparations. These effects disappear spontaneously if application of the preparations is discontinued.

J. Sunscreens

Sunscreens are products used in the prevention of sunburns. They belong to two different classes:

1. Total screen products. They are opaque to all wavelengths of light and prevent sunburn by the so-called umbrella effect. Examples of these products are ointments containing zinc oxide, titanium oxide, barium sulfate, calcium carbonate, magnesium carbonate.

2. Drugs having more or less selective filtering properties. Examples of these drugs are p-aminobenzoic acid, padimate, glycerol p-aminobenzoate, benzophenones (dioxybenzone, octabenzone, oxybenzone, and sulisobenzone).

K. Pigmenting Agents

Pigmenting agents are substances used to increase pigmentation in hypopigmented skin, as in the case of vitiligo.

The most widely used are trioxsalen, methoxsalen, and dioxyacetone (Man-Tan). The pigmenting action of the first two results from a photosensitizing effect. As a consequence, the responsiveness of the skin to light increases, and this increases production of melanin, which is the skin pigment. Dioxyacetone reacts chemically with pigment to darken it.

L. Depigmenting Agents

Depigmenting agents are products that help to reduce hyperpigmentation in certain disorders, such as chronic skin inflammation, severe freckling, chloasma gravidarum, and photosensitization caused by certain perfumes.

Hydroquinone and monobenzone are the most frequently used. Other derivatives of monobenzone are also used: dioxybenzone, octabenzone, oxybenzone, sulisobenzone. All these benzones act by a similar mechanism. They inhibit tyrosinase and in this way prevent the formation of melanin.

M. Scabicides

Scabicides are drugs used in the treatment of scabies, a parasitic skin infestation caused by the itch mite, *Sarcoptes scabiei* var. *hominis*. The drug of choice is gamma benzene hexachloride. Other useful drugs are benzyl benzoate and crotamiton.

Pediculicides are agents used in the treatment of pediculosis, a parasitic skin infestation caused by *Pediculus capitis* (head louse), *Pediculus corporis* (body louse), or *Phthirus pubis* (crab louse). The most effective drugs are chlorophenothane and gamma benzene hexachloride. In pediculosis capitis and pubis, benzoyl benzoate is also useful.

N. Insect Repellents

Insect repellents are drugs applied on the skin to prevent the approach and biting by mosquitoes, flies, and other arthropodes.

Among several other drugs, mention should be made of diethyltoluamide, ethohexadiol, dimethylphthalate, and butopyronoxyl.

II. DETOXIFYING AGENTS

Detoxifying agents are drugs used in the treatment of intoxication caused by ingestion of drugs or other chemicals. Presently, most intoxications, mainly self-poisoning, result from drugs, especially barbiturates, aspirin, sedatives, tranquilizers, and antidepressants. Drugs of this class may be divided into two groups:

1. Specific pharmacological antidotes. These drugs are used in the treatment

of intoxication caused by specific substances. The most widely used are antihistamines, atropine, nalorphine, pralidoxime, and chelating agents. These drugs were studied in preceding chapters.

2. Nonspecific pharmacological antidotes. These drugs are used in the treatment of intoxication caused by nonspecific substances. Thus, analeptics are used in narcotic, and anticonvulsants in convulsant, poisoning.

III. GASTROINTESTINAL AGENTS

Useful gastrointestinal agents are listed in Table 42.2.

A. Antispasmodics

Antispasmodics are drugs that reduce the frequency and strength of gastrointestinal smooth-muscle contractions and thus relieve the resulting pain. Those most widely used are anticholinergic agents, which were extensively studied in Chapter 16.

B. Antacids

Antacids are drugs used to reduce gastric acidity and to relieve pain in several disorders of the stomach and duodenum, such as peptic ulcer and gastritis. They may be divided into three classes:

1. Alkaline salts. These drugs are inorganic salts with alkaline properties. They act by raising the pH of the gastric contents to 5. Those most commonly used are sodium bicarbonate (Glycate), calcium carbonate, calcium bicarbonate, tribasic calcium phosphate, magnesium hydroxide, magnesium oxide, magnesium carbonate, milk of magnesia.

2. Colloidal antacids. These drugs are insoluble salts with buffering properties. They act by buffering gastric acidity at about pH 5. Examples are aluminum hydroxide gel (Amphojel), aluminum phosphate gel (Phosphaljel), dihydroxy-aluminum aminoacetate (Robalate), magaldrate USP (Riopan), magnesium trisilicate, tribasic magnesium phosphate, activated attapulgite, aluminum histidine, ammonium and bismuth citrate, hydrotalcite, tripotassium dicitratebismuthate (De-Nol).

3. Other antacids. They are not so widely used. Examples are polyethadine, gastric mucin, and polyaminemethylene resin.

C. Antidiarrheals

Antidiarrheals are drugs used in the treatment of diarrhea. This disorder results from several causes: infection, drugs, poisoning, allergy, gastrointestinal lesions,

Table 42.2 Gastrointestinal Agents

Official Name	Proprietary Name	Chemical Name	Structure
diphenoxylate hydrochloride*	Lomotil	ethyl 1-(3-cyano-3,3-diphenyl-propyl)-4-phenylisonipecotate monohydrochloride	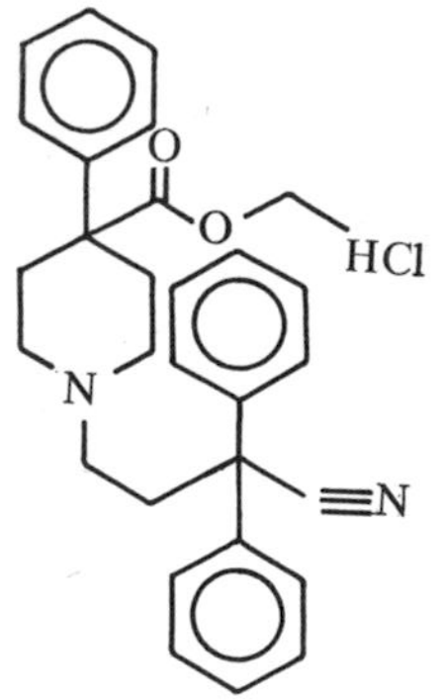
phenolphthalein[+]	Laxin	3,3-bis(p-hydroxyphenyl)phthalide	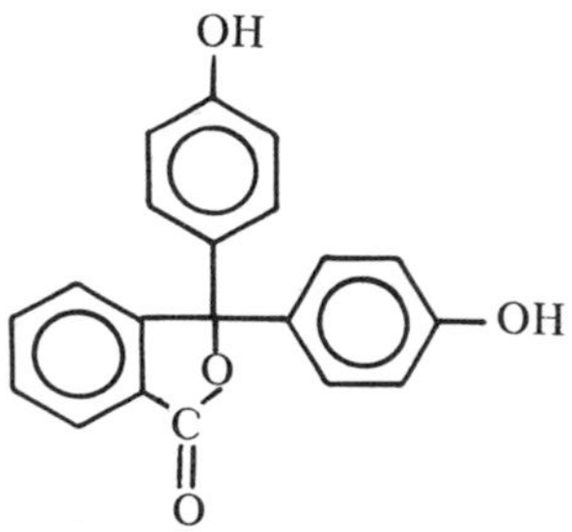

Table 42.2 Gastrointestinal Agents (continued)

Official Name	Proprietary Name	Chemical Name	Structure
bisacodyl*	Dulcolax	4,4'-(2-pyridylmethylene)diphenol diacetate	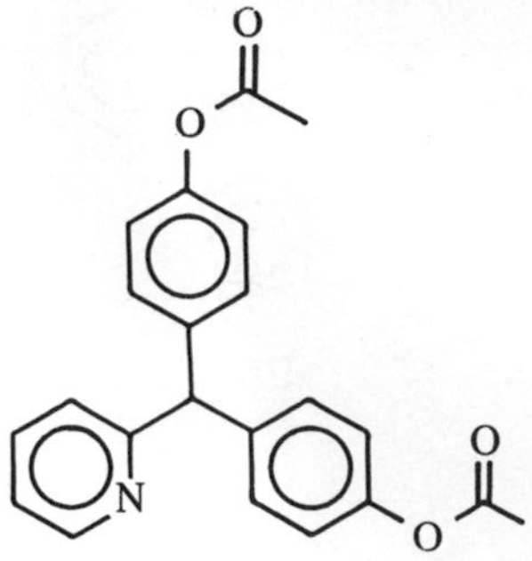
danthron[+]	Dorbane	1,8-dihydroxyanthraquinone	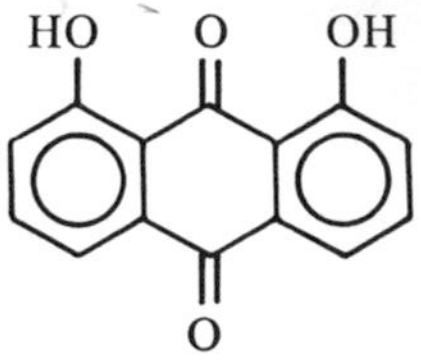
oxyphenisatin acetate	Isacen	3,3-bis(4-acetoxyphenyl)oxindole	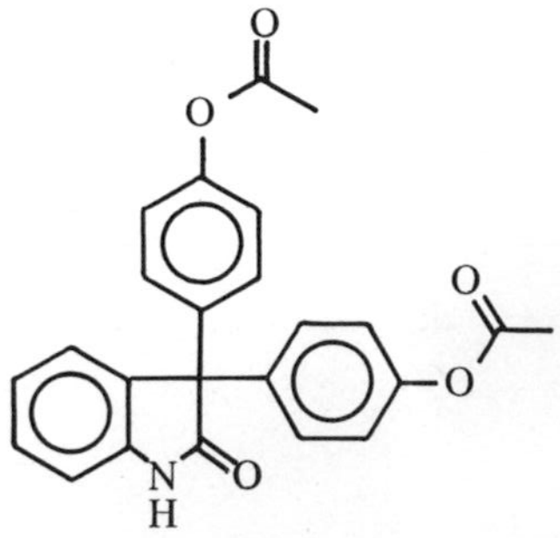

dioctyl sodium sulfosuccinate*	Doxinate	sodium 1,4-bis(2-ethylhexyl) sulfosuccinate 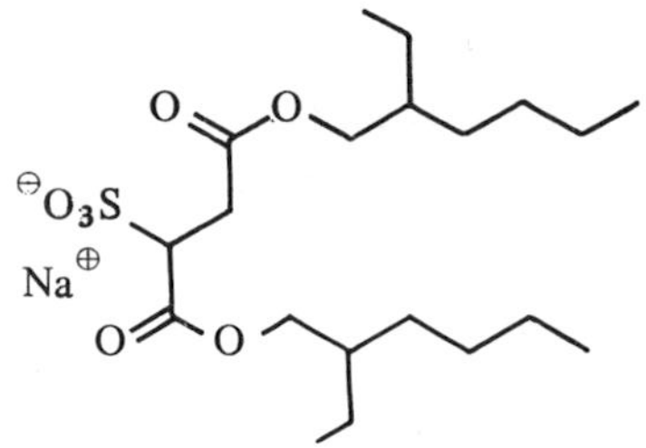
difenidol	Ventrol	α,α-diphenyl-1-piperidinebutanol 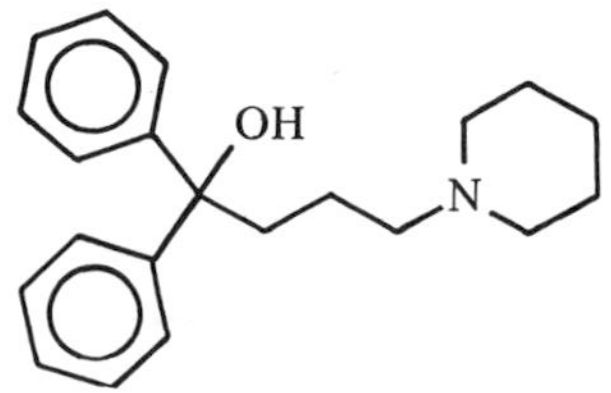
trimethobenzamide[+]	Tigan	N-(p-[2-(dimethylamino)ethoxy]-benzyl)-3,4,5-trimethoxybenz-amide 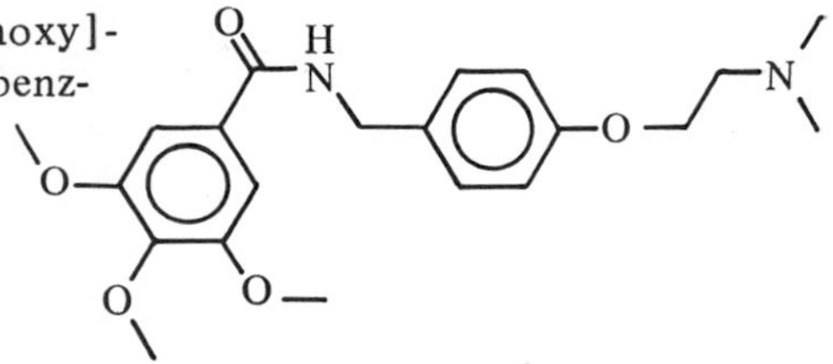

*In *USP XIX* (1975).
[+]In *NF XIV* (1975).

malabsorption, altered distribution of bile acids.

Drugs of this class are:

1. Nonspecific antidiarrheals. These are used only to relieve the symptoms of diarrhea. The most commonly prescribed are opiates (opium tincture and paregoric), diphenoxylate, bismuth salts (subcarbonate, subgallate, and subnitrate), activated charcoal, cholestyramine resin (USP), kaolin, silicate clays, polycarbophil, fluperamide, loperamide (Imodium). Several potential antidiarreals are being investigated; among them are the following two: difenoxin (a metabolite of diphenoxylate) and *N*-hydroxysuccinimide.

2. Specific antidiarrheals. They are active against pathogenic microorganisms that cause diarrhea. These drugs were studied in certain chapters of Part V, Chemotherapeutic Agents.

D. Laxatives

Laxatives are drugs that facilitate the elimination of stools. Based on their mechanism of action they are classified as:

1. Stimulants. They act by stimulating peristalsis. Examples are castor oil, casanthranol, cascara sagrada, senna, danthron, rhubarb, aloe, phenolphthalein, oxyphenisatin, bisoxatin (Wylaxine), bisacodyl, cofisatin, Laxilal, glycerin, sodium picosulfate (Laxoberal).

2. Bulk-forming agents. They act by absorbing water and expanding. The resulting expanded bulk causes a reflex peristalsis. Examples are bran, gum tragacanth, bassora gum, sodium carboxymethylcellulose, methylcellulose, agar, plantago seed, psyllum hydrophylic colloid.

3. Saline cathartics. Owing to their poor absorption, through osmotic pressure they draw a great amount of fluid into the intestine, thus increasing peristalsis. Examples are magnesium compounds (oxide, hydroxide, citrate, carbonate, sulfate), sodium salts (sulfate or phosphate), potassium salts (bitartrate, phosphate).

4. Lubricants. Their mechanism is unknown. Examples are mineral oil, olive oil, and cottonseed oil. Currently, only mineral oil is used.

5. Wetting agents. They lower the surface tension of feces and consequently facilitate the penetration of fecal mass by intestinal fluids. Examples are dioctyl sodium sulfosuccinate, dioctyl calcium sulfosuccinate (Surfak), and poloxalkol (Polykol).

E. Anorectal Preparations

Anorectal preparations are mixtures of drugs used to relieve pruritus and pain in cases of hemorrhoids, fissures, and related discomforts. Unfortunately, none is curative. Usually they contain a local anesthetic, emollient, and sometimes a corticosteroid.

The most commonly prescribed are Anusol, Anusol-HC, Rectal Medicone, Rectal Medicone-HC, Wyanoids, Wyanoids-HC. There are also several other mixtures. A newly introduced drug for treatment of hemorrhoids is dioxopromethazine (Prothanon).

F. Emetics

Emetics are drugs that induce vomiting. They are used especially in cases of poisoning. They act directly on the chemoreceptor trigger zone in the medulla. Their effects are enhanced if the patient drinks 200 to 300 ml of water concomitantly. Emetics are contraindicated in the following cases: unconscious, inebriated, or semicomatous patient; in patient who has ingested a caustic substance; after ingestion of petroleum distillates or volatile oils.

The most widely used emetics are apomorphine hydrochloride, which acts centrally, and ipecac syrup, which should not be confused with ipecac fluidextract, which is 14 times more concentrated and has caused several deaths.

G. Antiemetics

Antiemetics are drugs that prevent or relieve nausea and vomiting. They act by exerting their effects on one of the following targets: vomiting center, cerebral cortex, aural vestibular apparatus, or chemoreceptor trigger zone. Several types of drugs already studied have antiemetic effect:

1. Antihistamines. They affect neural pathways originating in the labyrinth. Examples are cyclizine, dimenhydrinate, diphenhydramine, hydroxyzine, meclizine.

2. Phenothiazines. They act on the vomiting center, the chemoreceptor trigger zone, or both. Examples are chlorpromazine, fluphenazine, perphenazine, prochlorperazine, promazine, promethazine, thiethylperazine, triflupromazine.

3. Anticholinergic agents. The most widely used is scopolamine. Apparently it reduces the excitability of the labyrinth receptors and depresses the conduction in the vestibular cerebellar pathways.

4. Sedatives and hypnotics. They act on the central nervous system by depressing the vomiting center or the cerebral cortex. An example is phenobarbital.

5. Miscellaneous agents. The most widely used are (a) difenidol, which acts on the aural vestibular apparatus, and (b) trimethobenzamide, which acts on the chemoreceptor trigger zone. Two new antimetics are metoclopramide (Plasil) and sulpiride (Dobren), which have also neuroleptic activity.

H. Digestants

Digestants are drugs that aid the process of digestion in the gastrointestinal tract. They are actually products used as replacement therapy in deficiency states. For instance, diluted hydrochloric acid, glutamic acid hydrochloride, pepsin, pancreatin (Panteric), pancrelipase (Cotazym), peppermint, fenipentol (Pancoral), synthetic secretin.

To this class belong also choleretics and cholagogues. These drugs, however, are more used in Europe than in North America. Choleretics increase bile production. Cholagogues facilitate bile excretion. Examples of these drugs are bile salts, bile acids, ox bile extract, dehydrocholic acid (Decholin), sodium dehydrocholate (Decholin Sodium), florantyrone (Zanchol), tocamphyl (Syncuma), cicrotoic acid (Accroibile), magnesium salt of dimecrotic acid (Hepadial), cinametic acid (Transoddi), Droctil, moquizone, piprozolin (Peristil), quenodesoxycholic acid.

I. Carminatives

Carminatives are drugs used for the relief of gaseous distension resulting from aerophagia or occurring after surgical intervention. Most of these drugs are mild irritants and volatile oils. The mixture of dimethylpolysiloxanes and silica gel, called simethicone (Mylicon), is the most widely used. Its efficacy, however, is questionable.

J. Lipotropic Drugs

Lipotropic drugs are substances used to prevent or decrease an abnormal accumulation of lipids in the liver. They act by various mechanisms. Those mostly used are N-acetylmethionine, arginine, arginine oxoglutarate, betaine salts, choline and derivatives, citiolone (Thioxidrene), cysteine, α,α-dithiodicapronic acid, inositol (Inosital) and its esters, lipocaic (Alipid), α-mercaptopropionylglycine (Thiola), methionine, silimarine (Legalon).

K. Antiulcer Agents

Antiulcer agents are drugs used in the treatment of gastric and duodenal ulcers. Antacids are also used for the same purpose, but antiulcer agents act by other mechanisms. Some of these agents are amylopectin sulfated (Depepsen), carbenoxolone (Biogastrone), chlorbenzoxamine (Libratar), sodium chondroitinsulfate, clocanfamide (Clamiren), deglycyrrhizinated liquorice (Caved-S), gefarnate (Gefarnyl), glycopeptide sulfate (Gliptide), sodium lignosulfate, polygeenan (Ebimar), proglumide (Milide), Solcoseryl, xilamide (Milid), zolimidine. Certain prostaglandins and antihistaminics acting on the H_2 receptor (burimamide, metiamide) are being investigated as potential antiulcer agents.

IV. SYNTHETIC SWEETENING AGENTS

Synthetic sweetening agents (Table 42.3) are substances used in low-caloric or

Table 42.3 Synthetic Sweetening Agents

Official Name	Proprietary Name	Chemical Name	Structure
saccharin*	Sweeta	1,2-benzisothiazolin-3-one 1,1-dioxide	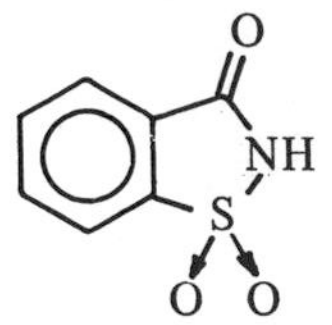
cyclamate	Sucary¸	cyclohexanesulfamate	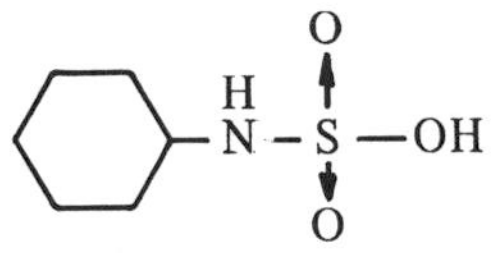
aspartame		3-amino-N-(β-carboxyphenethyl)-succinamic acid N-methyl ester	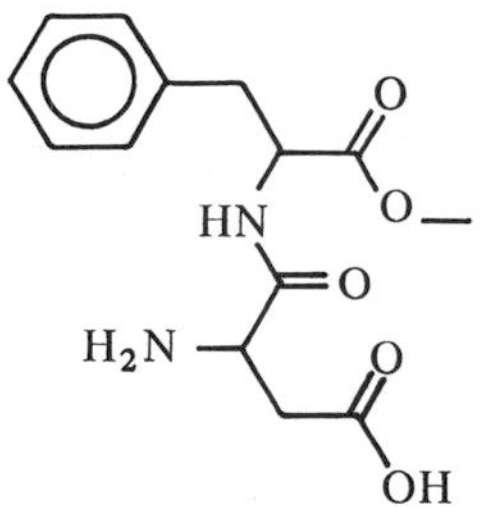

*In *USP XIX* (1975).

low-carbohydrate diets as substitutes for sugar. Their main usefulness is in the treatment of diabetes or obesity. Since cyclamate can cause cancer, in many countries it was withdrawn from the market and in others its use is restricted to special preparations. Aspartame has been recently approved by the FDA for use on the table and in ready-mix products but not for soft drinks or foods that have to be cooked.

REFERENCES

Dermatological Agents

G. W. van Ham and W. P. Herzog, "The Design of Sunscreen Preparations," in E. J. Ariëns, Ed., *Drug Design*, Vol. IV, Academic, New York, 1973, pp. 193-235.

M. Katz, "Design of Topical Drug Products: Pharmaceutics," in E. J. Ariëns, Ed., *Drug Design*, Vol. IV, Academic, New York, 1973, pp. 93-148.

W. B. Shelley and H. Goldschmidt, *Top. Med. Chem.*, 3, 285 (1970).

J. A. Moncreif, *Clin. Pharmacol. Ther.*, 10, 439 (1969).

Detoxifying Agents

A. J. Smith, *Brit. Med. J.*, 4, 157 (1972).

M. B. Chenoweth, *Clin. Pharmacol. Ther.*, 9, 365 (1968).

A. Soffer, *Chelation Therapy*, Thomas, Springfield, Ill., 1964.

T. N. Pullman, *Annu. Rev. Med.*, 14, 175 (1963).

Gastrointestinal Agents

A. C. Playle, *Med. Actual. (Drugs of Today)*, 10, 208 (1974).

F. A. Jones and E. W. Golding, Eds., *Management of Constipation*, Blackwell, Oxford, 1972.

P. P. Nair and D. Kritchevsky, Eds., *The Bile Acids*, 2 vols, Plenum, New York, 1971-1972.

J. F. Stokes, *Practitioner*, 206, 35 (1971).

S. Holtz, *Annu. Rev. Pharmacol.*, 8, 171 (1969).

Synthetic Sweetening Agents

M. G. J. Beets, "Structure-Response Relationships in Chemoreception," in C. J. Cavallito, Ed., *Structure-Activity Relationships*, Vol. I, Pergamon, Oxford, 1973, pp. 225-295.

L. B. Kier, *J. Pharm. Sci.*, 61, 1394 (1972).

W. Guild, Jr., *J. Chem. Educ.*, 49, 171 (1972).

B. S. Shallenberger and T. E. Acree, *Nature (London)*, 216, 480 (1967).

LIST OF DRUGS FROM
NATIONAL FORMULARY XIV

Italics indicate page number of chemical structure.
* Indicates drug is both USP and NF; reference is made to USP list.

Acenocoumarol, 335, *337*
Acetophenazine, 149, *150*
Acetylcysteine, 143
Acetyldigitoxin, 303, *304*
Acrisorcin, 451-452, *455*, 459
Alphaprodine, 17, *92, 120*, 449
Amantadine, *180*, 181-182, 370, 571, *572*, 574-575
Ambenonium chloride, *16*, 215
Aminocaproic acid, *331*, 332, 334, 340
Aminosalicylic acid, 7, 58, 106, 369, 372, 481, *482*, 493
Ammonium chloride, 97, 106, 143, *357*, 358, 445
Amobarbital*
Amyl nitrite, 312, *313*
Anileridine, *120*
Antazoline, 276, *280*
Apomorphine, *181*, 182
Aspirin*
Atropine, 29, 62, 92, 182, 206
Azuresin, 651, *655*

Bendroflumethiazide, 352, *353*
Benoxinate, *288*
Benzestrol, 627, *630*
Benzethonium chloride, *443*, 447
Benzocaine, 284, *288*
Benzonatate, *145*
Benzoylpas calcium, *482*
Benzthiazide, 352, *353*
Benzyl benzoate*
Betamethasone, 614, *616*
Bethanechol, 214, *217*

Biperiden, *177*
Bismuth, 381
Bromodiphenhydramine, *270*
Brompheniramine, *274*
Butabarbital, *97*, 101
Butacaine, 284, *287*
Butamben, *289*

Camphor spirit, 186
Candicidin, 451, 453, *456*, 459, 498, 579
Carbarsone, 377, *378*, 425, 427, 429-431, 436, 437
Carbinoxamine, *271*, 272
Carphenazine, 149, *151*
Cephaloglycin, 507, *508*, 510, 528-530
Cephaloridine, 495, 497, 507, *508*, 510, 528-530
Cetylpyridinium chloride, 48, *447*
Chloral betaine, *95*, 98
Chlorcyclizine, 277
Chlordiazepoxide*
Chlorophenothane, 406, *662*, 665
Chloroprocaine, 284, *287*
Chlorothiazide*
Chlorotrianisene, 13, 627, *630*, 632
Chlorphenoxamine, *178*
Chlorprothixene, 152, *153*
Chlortetracycline, 369, 425, 431, 437, 439, 488, 496, 514, *516*
Clindamycin*
Cocaine*
Codeine*
Cyclizine*
Cyclomethycaine, 284

LIST OF DRUGS FROM
U. S. PHARMACOPEIA XIX

Italics indicate page numbers of chemical structures.

106, *107,* 108-109, 459
Phenol, 4, 368
Phenoxybenzamine, *261*
Phentolamine, 92, 165, 253, *256,* 260, 309
Phenylbutazone, 7, 10, 13-14, *134,* 139-140, 340
Phenylephrine, 239, 241, *243,* 291
Phenytoin, 7, 106, *107,* 108, 303
Physostigmine, 16, 20, *215,* 217-218, 222-223
Phytonadione, *331,* 332-333, 337-338
Pigmenting agents, 664
Pilocarpine, 206, 213, *216,* 219, 223
Piperazine, 13, 369, 390-391, 393, *394*
Plasma albumin, 341
Polymyxin B, 77, 460, 495, 497-498, *517,* 518-519, 531
Polysorbate 80, 48
Potassium chloride, 303
Potassium hydroxide, 444
Potassium iodide, 143, *454*
Potassium permanganate, 446, 451
Povidone, 427, 446
Pralidoxime chloride, *41,* 219
Prednisolone, 6-7, *33, 136,* 137, 542, *559,* 614
Prednisone, 33, *136,* 547, *559,* 614
Primaquine, 404, 405, 407, 410, *412,* 415-416
Primidone, *110*
Probenecid, 29, *30, 138,* 139, 466, 504
Procainamide, *25,* 92, 303, *306*
Procaine, 21-22, *25,* 284-285, *287,* 291, 296-298, 305, 349
Procarbazine, *560*
Prochlorperazine, 149, *150*
Progesterone, 628, *633*
Promethazine, 29, 91, 92, 277, *278*
Propantheline bromide, *24, 229*
Proparacaine, 284, *289*
Propoxyphene, *122,* 123
Propranolol, *24,* 92, *254,* 258, 303, 307, 309-310, *313,* 316
Propylene glycol, 445
Propyliodone, *649,* 650, 651
Propylparaben, 444
Propylthiouracil, 30, 466, 601, *602*
Protamine sulfate, 332, 333, *337*
Protamine zinc insulin, 32
Protectants, 658

Pyrantel, 370, 390, *395,* 400
Pyrazinamide, *482,* 492
Pyridostigmine bromide, *215,* 217
Pyridoxine, 136, 586, *588,* 590, *591,* 593
Pyrimethamine, *40,* 59, 75, 404-405, 407, 410, 413, 415-416, 418, 420-421, 427, 435-438, 476, 544
Pyrvinium pamoate, 320, 393, *394,* 400

Quinacrine, 55, 71, 109, 368, 391, 396, *397,* 401, 404, 407, 410, *411,* 416, 426-427, 430, 435, 437, 560
Quinidine, 92, 303, 305, *306, 414*
Quinine, 55, 368, 404, 407, 410, *411,* 412, *414,* 415, 418, 419, 471, 476

Radiopharmaceuticals, 651
Reserpine, 106, 153, *154,* 165, 308-309
Riboflavin, *7,* 586, *588,* 590, 592
Rifampin, 59, *483,* 487-488, 492, 496-498, 523, 532, 575

Saccharin, *673*
Salicylic acid, 19, 37, 76, 444
Sclerosing agents, 664
Scopolamine hydrobromide, 29, 92, 182, *227*
Secobarbital, 7, 91, *97,* 101
Selenium sulfide, 452, *454*
Silver nitrate, 446, 449
Sodium benzoate, 444
Sodium borate, 444
Sodium chloride, 97, 106
Sodium hydroxide, 444
Sodium iodide I, 131, 561
Sodium lauryl sulfate, 436
Sodium phosphate P32, 561
Sodium salicylate, 131
Sodium stearate, 48
Sodium sulfobromophthalein, 653
Sorbitol, 356
Spectinomycin, 460, 497, *524,* 532
Spironolactone, 347, 356, *357,* 612
Stibophen, 368
Streptomycin, 59, 72-74, 359, 372, 453, 460, 481, *483,* 487-488, 492-493, 495-498, *524,* 532-533
Styptics, 659
Succinylcholine chloride, 92, *233,* 234-235, 291

GENERAL INDEX